DC/AC
Foundations of
Electronics

R. Jesse Phagan

Woodstock Academy
Woodstock, Connecticut

Publisher
THE GOODHEART-WILLCOX COMPANY, INC.
Tinley Park, Illinois

ABOUT THE AUTHOR

R. Jesse Phagan has an extensive teaching and writing background in areas of math and electronics. He is the author of two electronics math books and an electronics book. He holds degrees in Technology Education, Vocational Education, and Electronics Technology. In addition to his teaching experience, he has worked in manufacturing and for the federal government.

Cover: Kitagawa II/SuperStock

Copyright 1997

by

THE GOODHEART-WILLCOX COMPANY, INC.

Library of Congress Catalog Card Number 96-22362
International Standard Book Number 1-56637-341-7

2 3 4 5 6 7 8 9 10   97   01 00 99 98

Library of Congress Cataloging-in-Publication Data

Phagan, R. Jesse
DC/AC foundations of electronics / by R. Jesse Phagan.
    p.   cm.
Includes index.
ISBN 1-56637-341-7
1. Electronics.   I. Title.
TK7816.P428   1996
621.381—dc20   96-22362
                                                              CIP

Introduction

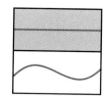

We are in a technological world where amazing changes are made so rapidly it is nearly impossible to keep up. An individual wishing to get into a career in electronics can be overwhelmed. Yet, it is exciting to be a part of a field that is so challenging and rewarding.

DC/AC Foundations of Electronics provides you with more than just a place to begin. This textbook establishes a background so solid that advanced electronics will be much easier to understand. Even the most advanced circuitry follows the rules established at this level of study.

The textbook is broken into four sections.

- Introductory Topics.
- Direct Current.
- Alternating Current.
- Advanced Topics.

Introductory Topics explores such items as careers and safety. The safety chapter includes simplified house wiring, a must for anyone working with electricity. In addition, an entire chapter explains how to read multimeter and oscilloscope scales clearly and in detail. You will find this very useful when beginning laboratory activities.

Direct Current and *Alternating Current* are the major sections forming the heart of this textbook. These chapters contain the laws of electrical theory. You will prove these laws using mathematics. You will also learn a language of technical words that describes the concepts used.

A section on *Advanced Topics* is provided for two reasons. First, this section introduces semiconductors and their applications. Second, this section will help you to fully understand the need for a strong foundation in the basics.

DC/AC Foundations of Electronics was developed with the student in mind. Basic electronics can be a difficult subject to study. It has a new technical vocabulary, there are new electrical principles to learn, and math is used extensively. In **DC/AC Foundations of Electronics,** the technical terms are clearly defined and used repeatedly. This gives the words meaning as they are applied to an electrical principle. Many of the technical terms learned in early chapters are used in subsequent chapters. This will build your technical vocabulary. This technical vocabulary will help your understanding in many subject areas, not just electronics. Developing a good technical vocabulary will support advancement in many fields

Mathematics is required to understand electrical laws and solve electrical circuits. However, sometimes math can bog down the learning process. Math errors made in solving problems can make you wonder if the theories were misunderstood. **DC/AC Foundations of Electronics** is designed to simplify the use of the required mathematics. This is done through the extensive use of sample problems. These problems are worked out step-by-step. They include the formula used, the substitution of numbers, and the answer. Many sample problems show the formula rearranged to solve for each different variable. All of the sample problems show possible applications of the formula and help to further explain the electrical principles.

At the beginning of each chapter are objectives. These are the goals you are expected to meet. At the end of the chapter is a summary, a glossary of the new terms learned, and set of review questions. All of the objectives are challenged in the review questions. This is to give clear feedback on the important material contained in each chapter.

Welcome to the exciting world of electronics. Remember, the understanding of the most advanced topics depends on the development of a solid foundation.

Contents

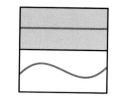

Section 3   Alternating Current

Section 4   Advanced Topics

Chapter 1
The Electronics Industry

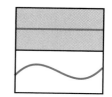

Upon completion of this chapter, you will be able to:
- Write job descriptions for workers in several areas of the electronics industry.
- State specific skills and training requirements for technicians and engineers.

When you begin a new program of study, it is not always clear what type of jobs will be available. Another important question is how much training and education does a particular career require?

Nobody can be sure of just what job openings will be available, especially in different areas of the country. What this chapter will do, however, is help you learn how to recognize the pros and cons of different career opportunities.

1.1 THE ELECTRONICS INDUSTRY

In the electronics industry there are three broad categories in which workers are employed: electronic equipment manufacturing, scientific research and development, and the service and repair of equipment.

These categories offer a wide range of opportunity, with a variety of skill levels and job duties. The electronics industry has a positive employment outlook, with some ups and downs depending on the economy. It is projected that the demand for future workers will be best in equipment servicing and repair. Highly skilled technicians will be in more demand as electronic equipment continues to increase in complexity.

Electronic Equipment Manufacturing

Manufacturing requires workers in many skill levels. In areas related to electronics training, career progression is through education and on-the-job training. The career path is generally as follows:

- Electronic parts assembly worker.
- Assembly line testing and inspection worker.
- General technician.
- Specialized technician (trained in specific areas).
- Engineering technician.
- Engineer.

There are many other positions available, depending on the specific type of manufacturing.

Assembly workers are the people who place components on circuit boards, attach wires, and hand-solder components. Some of these jobs are being taken over by machines. Computer-controlled machines are being *trained* to perform more and more of these tasks. However, people are still needed to perform the tasks not possible or not repetitive enough to be done with machines. Advancement for an assembly worker requires on-the-job training. Inspection and assembly-line testing are the next steps. Further advancement requires classroom training.

Technicians are required at many different stages of the manufacturing process. Entry-level jobs are available for an individual with classroom training, but no work experience. As a technician gains experience, testing is performed more accurately and repairs are completed more quickly. Employers usually offer a variety of on-the-job training. Individuals with a greater number of skills are considered more valuable.

Most companies also have an engineering department. Engineering technicians and engineers work together to perform two basic tasks, monitoring the manufacturing process and developing new products and applications for a customer's needs.

Research and Development

Research and development (R & D) involves the design, building, and testing of new products, figure 1-1. Often, it all starts with an idea. A customer service representative talks with a prospective customer who has a specific need. The customer describes a situation in which there are no products available to meet the application.

The R & D department investigates the idea and gets back to the customer with an estimated cost. If the cost is acceptable, an engineer designs a new product that will meet the customer's needs. A technician will build a **prototype,** which is a working model. The prototype is tested under the conditions in which it will be used. This allows the engineers to work out any problems. The prototype is then presented to the customer for approval. After approval, R & D personnel prepare the manufacturing plant to begin assembly.

With the high cost of engineering, it is usually best to do a market survey. A **market survey** is an investigation to find if there is any significant interest in the product. Selling more of the same item reduces the cost per item needed to cover the R & D expenses.

Equipment Service and Repair

As electronic equipment becomes more sophisticated it also becomes more reliable. Workers are needed to make initial adjustments on calibration controls to ensure it is

Figure 1-1. Computers are valuable tools for R & D departments. (Tektronix)

operating as intended. In certain situations, a technician may travel to the site where the equipment is to be used. The technician sets up the equipment, connects cables, and instructs the user in basic operations.

Even though there will be few equipment failures, repairs will still be needed. Some repairs are made at the factory, some at a repair center, and other repairs are made at the equipment's location.

1.2   ELECTRONICS CAREER OPPORTUNITIES

Electronics are involved in most every industry, providing good employment opportunities. As shown in figure 1-2, an electronics technology program trains workers for many major industries. It is exciting to realize how many different options there are for a person working in electronics. New doors of opportunity continue to open as new developments occur. Currently, there are opportunities in such diverse fields as computers, consumer products, manufacturing, medical, communications, space exploration, defense, and transportation.

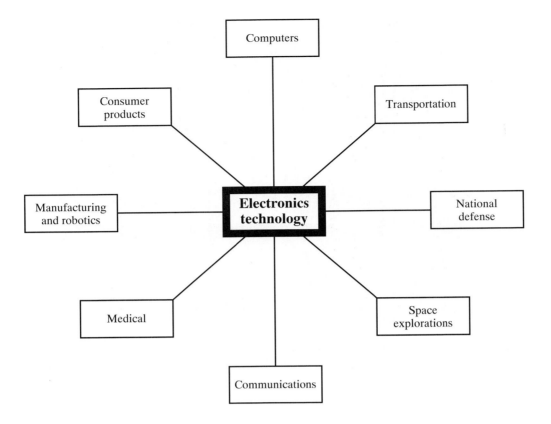

Figure 1-2. An individual trained in electronics technology will have career opportunities in many other fields.

Computers

Desktop computers have become quite common, figure 1-3. They have made significant changes in many aspects of business and have many home applications as well. One important invention that made the computer affordable for the average consumer was the microprocessor. A **microprocessor** is an integrated circuit that serves as the computer's brain.

Microprocessors have been applied to many products making them computer-operated. As a challenge, examine the many different things with which you come into contact. How many of them use the microprocessor as their source of intelligence.

When a computer fails to operate normally, an electronics technician is called in to do the repairs.

Figure 1-3. Many businesses have become dependent on computers for operation.

Consumer Products

Consumers are people who buy products for their personal use. Typical examples of electronic consumer products are: televisions, video cassette machines, stereos, home computers, computer games, and burglar alarms.

Many vocational schools offer some form of consumer product training as part of their curriculum. Reasons for offering this training vary. The most common reason is that all of the students in the class own one or more of these products. These classes are a chance to connect theory and lab exercises with the real world.

Manufacturing

Manufacturing products for sale continues to be one of the largest industries throughout the world. The manufacturing environment needs to be efficient to be competitive. Electronic controls for machinery, computer inspections and inventory control, and many other aspects of the electronics industry have improved the competitive edge.

Robotics are a major influence in manufacturing. A **robot** is a machine that can be programmed to perform work to make life simpler for humans. The manufacturing industry has improved production considerably with the use of robots on assembly lines.

A computer programmer teaches the robot its tasks, figure 1-4. Then, the computer operates the robotic functions. Robots are especially useful for repetitive tasks. Robots do not get bored and make mistakes. Robots are also useful in performing tasks that are dangerous, such as handling materials in extremely hot places.

Figure 1-4. Robotics is a hot topic in the electronics industry. (DeVry Institutes)

Medical

Advances are made in the medical field at an astounding rate. During a visit to the hospital, a person cannot help but see many different types of electronic equipment, figure 1-5. All of this equipment must be kept in peak condition. It is the job of an electronics technician to maintain, calibrate, and repair medical equipment.

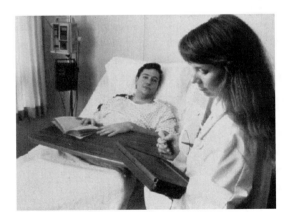

Figure 1-5. Patient monitoring systems help save lives in hospitals. (Dauphin)

Communications

Telephones, radios, televisions, facsimile machines, and satellites are examples of the electronics communications industry. There is virtually no house or business that does not have some form of electronic communications. All of this equipment needs to be manufactured, inspected, and in time repaired.

Space Exploration

With the federal government providing the funding for space exploration, research and development takes place with less regard to producing a profit, figure 1-6. As a result of space research, many other industries have received new products. Space travel has improved communications, navigation, medical technology, power and energy sources, comfort control, and many other areas.

Figure 1-6. Space exploration has created new technologies in many fields. (NASA)

National Defense

Like space exploration, the federal government spares little expense in creating new developments in defense. In the past few years the government has began cutting defense contract. However, a great deal of money is still being invested in the industry. With such a huge amount of money being spent, many businesses and individuals are supported. The new developments are shared with private industry and eventually are used to improve other consumer products.

Transportation

Airplanes, boats, trains, trucks and automobiles all benefit from the advances made in electronics. Airplanes and boats need navigation and communication equipment. Trains, buses, and other forms of mass transportation use computers to improve their efficiency in scheduling.

The automobile is an excellent example of electronics in transportation. All new cars are controlled by some form of computer or electronic ignition. A top-paid automotive technician needs a working knowledge of electronic controls and test equipment. A course in electronics technology will certainly help the technician to move up the career ladder.

1.3   CAREERS IN ELECTRONICS

Study in the field of electronics can lead to numerous job titles in many fields. Some of the most common positions fall into the divisions of electronics technician, field service technician, engineer, and engineering technician. These opportunities are well worth exploring.

Electronics Technician

The job title, *Electronics Technician,* is a general title relating to many positions in the electronics industry. Electronics technicians have a wide range of duties, and they often have more specific job titles. Generally speaking, a **technician** is a skilled worker trained in basic electronics, with advanced training in a specialized area, figure 1-7.

Figure 1-7. This technician's specialty is motor repair.

Technical Skills and Training

The skills needed by a technician can be taught through formal education. Vocational/technical schools and community colleges offer formal education. While work experience and on-the-job training are extremely valuable, formal education provides the fastest, and sometimes only, way to be hired as a technician.

Specialized courses are offered in areas such as: electronic communications, computer repair, robotics, and instrumentation. Specialty classes are added as needed by local industry.

Paperwork is an essential part of a technician's duties. A technician is required to keep accurate records of work performed on equipment, order replacement parts, and handle customer orders. Written and spoken communication skills are necessary for a technician to succeed. Electronics is a field where the worker comes in contact with both machinery and people.

Opportunity and Advancement

An electronics technician can find employment in so many different types of industry that advancement can often be a matter of taking a new position. Manufacturing electronic equipment offers many opportunities as discussed earlier. Another field where electronics is playing a major part is the biomedical field. Electronic equipment helps doctors and hospitals diagnose and monitor patients. Technicians are needed to keep the biomedical equipment at its peak efficiency.

Small business also employ or are run by technicians. Small business is said to be the backbone of the American economy. Technicians commonly deal with consumer repairs on products such as: televisions, VCRs, video games, and audio equipment. Being a small business owner is very demanding work, but the rewards can be worth the effort. Formal education is not needed to own a business, but it is the fastest way to obtain many of the skills not easily learned on the job.

Advancement usually requires additional education. The next step up for a technician is to enter the field of engineering. An engineer needs a bachelor's degree.

Field Service Technicians

A field service technician is usually a highly experienced individual who has the ability to work well without supervision. A field service position requires extensive travel, which means additional travel pay. With electronic equipment running fairly reliably, one person can service a large number of machines. A technician may be required to travel to several different states. Some jobs require travel overseas. This travel can be a problem. Field service technicians may not be able to spend a lot of time with their families.

When first hired, the technician is trained on one type of machine. With experience and further training, the technician becomes more valuable and works on a greater variety of equipment. Field service results in higher pay than working in one location. For this reason, it is considered an important career step. It is a step that does not require going back to college.

Office equipment technicians service copy machines, electronic typewriters, personal computers, and other equipment found in a business office. If the equipment in an office is purchased from different companies, several different technicians are needed to service the office.

Field service technicians are needed for almost every type of equipment sold. Biomedical equipment, network computers, weather service equipment, flight service equipment, and two-way radios all need field service technicians.

Engineers and Engineering Technicians

Engineering is a broad title that reflects an education through a bachelor's degree or higher. An engineering technician may hold an associate's degree, but the B.S. degree is preferred. The actual job title reflects an individual's specialty area. Engineers are usually more involved with the paperwork part of engineering. The engineering technician tends to be more involved with the hands-on aspects.

Research engineers study basic sciences from which they derive theories and develop ideas. Development engineers, working with the research engineers, use the ideas to produce prototypes. The prototypes are used to test and further develop the products. Design engineers work closely with research and development engineers, helping to make a product one that can be manufactured and sold.

Applications engineers, also called consultant engineers, work with the marketing department to find ways in which a product can be sold. The engineer also helps a customer set up the equipment, train the operators, and develop further ideas on how the equipment can be made more useful.

The engineer is considered an expert and is frequently called upon to assist technicians when field repairs are needed. The engineer may direct the technician over the phone.

The four-year or six-year degree needed for an engineer also qualifies the individual to apply for teaching positions, on the job and in the classroom. On-the-job training has opened many positions for someone wishing to move into training. This move means working more with people and less with equipment. Colleges also need engineers to step into the classroom and teach others to be engineers or technicians.

1.4   OTHER CAREERS IN ELECTRONICS

The job descriptions that follow relate to the career ladder. This helps when reviewing help-wanted ads listed in newspapers and posted on bulletin boards. The jobs shown here start from the bottom of the career ladder and work up the steps. In some situations, jobs can be filled by individuals without the minimum education requirements. Companies list qualifications that they would like the applicants to have. In addition, it may be difficult for an individual with advanced training to be hired in a lower position. Companies worry that over-qualified personnel may not stay for very long or their salary requirements may be too high.

Assembly Worker

An assembly worker builds products in a manufacturing environment. This work can be tedious, with long periods of time spent doing repetitive activities. Pay for the assembly worker is usually the lowest of all electronics workers. The primary skill needed for this position is the ability to work with extreme accuracy. An assembly worker should have the ability to:

• Use drawings, written instructions, and visual models.
• Place parts on circuit boards.
• Assemble mechanical components.
• Use hand tools.

Tester

A tester is the person who tests products after they have been assembled. Sometimes, sections of a product are tested before they are installed in a larger, more complete unit. Testing is often a promotion for an assembly worker. Some on-the-job training is provided in the use of test equipment. Working conditions are slightly better than for the assembly department. A tester should have the ability to:

• Follow written instructions for test procedures.
• Use test equipment to ensure proper operation.
• Identify defective products.

Repair Technician

A repair technician corrects defects in equipment rejected by the testing department. The repair technician must understand the complete operation of the equipment and follow systematic repair procedures. A thorough understanding of the circuitry is usually not necessary. Extremely difficult repairs are performed by more experienced technicians.

Figure 1-8. Electronic scopes provide great insight to the inner workings of electronic equipment.

A repair technician should have the ability to:
• Use test equipment to locate the problem, figure 1-8.
• Change defective parts.
• Retest to ensure proper operation.

Electrician's Helper

Manufacturing companies use their own maintenance personnel to maintain the equipment and machinery used in production. The electrician's helper works with a licensed electrician to maintain the wiring and electrical equipment.

Electrician's helpers are also employed in jobs other than manufacturing. For example, wiring in a house is usually done by the helper, while the licensed electrician supervises. An electrician's helper should have the ability to:
• Work with machinery that may have live voltages.
• Work long hours to repair nonfunctional equipment.
• Work under the supervision of a licensed electrician.

Quality Control Technician

The quality control department is responsible for ensuring that products meet the customer's needs and specifications. The technician performs many advanced tests on products that have already been tested and certified as working properly.

The quality control technician devises tests for the production department to perform and continually monitors routine testing. These technicians must have advanced training and knowledge of the complete operation. A quality control technician should have the ability to:
• Use test equipment for extensive testing.
• Use accurate measurement tools.
• Develop new testing procedures.

Inspector

The inspector is usually a member of the quality control department. The inspector's main function is to make routine checks of testing by all persons. The inspector position is frequently a step up within the quality control department. An inspector should have the ability to:
• Perform all tests being inspected.
• Train others in proper testing techniques.
• Coordinate testing by other personnel.

Repair Troubleshooter

The troubleshooter is usually a top-rated technician with years of experience performing repairs. This individual may work as a supervisor or may work alone. Primary duties include difficult repairs and finding the causes of problems a technician has been unable to solve. A repair troubleshooter should have the ability to:

- Read and understand all types of schematics and written instructions.
- Pay attention to details.
- Perform specific tests based on symptoms.

Repair and Service Technician

Electronic products break down and require trained technicians to make repairs and adjustments. Products include such items as: office machines, televisions, VCRs, and desktop computers. A repair and service technician should have the ability to:
- Use a variety of test equipment to troubleshoot circuits.
- Estimate time and materials needed to correct problem.
- Complete repairs with a minimum of supervision.

Electrician

An electrician must be licensed by the state in which the work is performed. Programs of study for electricians are somewhat different from those for technicians. The electrician is more involved with the wiring and operation of machinery. They deal with live voltages. Technicians are usually more involved with electronic circuitry. An ideal combination for a manufacturing electrician is to also be a technician. Electricians are also involved with the wiring of buildings. An electrician should have the ability to:
- Install wiring.
- Repair defective machinery.
- Supervise an electrician's helper.

1.5   FURTHER INFORMATION

For further information about careers in the electronics industry contact your local vocational school, community college, or university. College catalogs usually have descriptions of jobs available upon graduation.

For information from agencies write to the following:

Electronics Technicians Association
602 North Jackson Street
Greencastle, Indiana 46135

Junior Engineering Technical Society
1420 King Street, Suite 405
Alexandria, Virginia 22314-2715

Institute of Industrial Engineers
25 Technology Park/Atlanta
Norcross, Georgia 30092

SUMMARY

- There are three broad categories in the electronics industry: equipment manufacturing, scientific research, and service and repair.
- Research and development involves the design, building, and testing of new products.
- Some positions in the electronics field are electronics technician, field service technician, engineer, and engineering technician.
- Electricians are involved with the wiring of buildings. Electricians must be licensed by the state.

KEY WORDS AND TERMS GLOSSARY

market survey: An investigation to find out if there is any significant interest in a product.

microprocessor: An integrated circuit that is the central processing unit in a computer.

prototype: A working model of a product build for a customer's approval to ensure its design is satisfactory. The prototype is tested under the conditions in which it will be used, allowing the engineers to work out any problems.

research and development (R & D): Involves the design, building, and testing of new products.

robot: A machine that can be programmed to do work to make life simpler for humans.

technician: A skilled worker trained in basic electronics with advanced training in a specialized area.

TEST YOUR KNOWLEDGE

Do not write in this text. Please use a separate sheet of paper.
1. List the three categories in which electronics workers are employed.
2. List six industries that employ workers with electronics training.
3. Who teaches a robot to perform its tasks?
4. What three skills does a tester need to possess?
5. What three things should a quality control technician be able to do?

Chapter 2
Electrical Safety

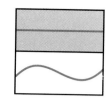

Upon completion of this chapter, you will be able to:
* Describe the effects of electrical shock.
* Explain the factors that influence the severity of electrical shock.
* Recognize ways in which to prevent electrical shock.
* Make simple repairs to a plug, outlet, and light socket.
* Demonstrate safe soldering techniques by making solder connections.
* Define technical words used in conjunction with electrical safety.

Safety is a life skill that a person uses to protect himself and others from possible harm. In a classroom and laboratory setting, it is the responsibility of the instructor to demonstrate safe working techniques. It is the responsibility of the student to learn and use safe operating procedures at all times.

Electricity has become a part of modern society. Yet electricity brings danger along with convenience. Electricity cannot be seen nor can it be safely tested with any of our human senses. To detect if electricity is present requires the use of electrically operated equipment.

2.1 FACTORS AFFECTING ELECTRICAL SHOCK

Electrical shock is the physical sensation of the nerves and muscles reacting to electricity passing through the body. In mild cases, there is a harmless jerking of the muscles. This contraction may leave the muscles sore for a short time. A severe shock can result in paralysis of the nerves and muscles, including respiratory and heart muscles. **Electrocution** is the term used when the exposure to electric shock results in death.

The severity of electrical shock is determined by the amount of electricity flowing through the body, the path it takes through the body, and the length of time of exposure. Greater amounts of electricity, paths that cross the heart, and longer periods of time all increase the likelihood of electrocution.

Amount of Electricity

To best understand how much electricity can flow, it is necessary to understand three basic principles: voltage, current, and resistance.

Voltage

Voltage is potential energy of the electric source. An automotive battery produces 12 volts. Household electricity is 120 volts. *Normally,* voltages less that 30 volts are not a shock hazard. As a general rule, higher voltages are more dangerous than lower voltages.

Household voltage, which is not considered high voltage, kills more people than any other voltage. Part of the problem is a lack of respect for electricity in the home. Most people have received a mild shock at one time or another without a problem. This can lead to a careless attitude that causes severe shocks.

Current

It is the flow of current that causes electrical shock. It takes a very small amount of current to cause severe electric shock. **Current** is the flow of electricity. Current requires a voltage source and a path to follow.

Coming in contact with a voltage is not enough to cause a shock. The body must act like a conductor, providing a path for the current to flow.

Resistance

Resistance is the opposition a current path offers to the flow of electricity. A high resistance allows less current flow than a lower resistance.

There are two extremes of resistance, zero and infinity. **Conductors**, such as a piece of wire, offer almost zero resistance. This allows very large amounts of current flow. **Insulators**, offer an infinite resistance, allowing no current to flow. An example of an insulator is the plastic insulation on a wire.

The resistance of the human body, which results in more or less current, varies greatly depending on the part of the body affected and its condition, figure 2-1. A large value of resistance decreases the chance of severe shock. Low resistance, such as wet skin, allows more current to flow through the body. This increases the chance for severe shock. Low resistance also means a lower voltage can cause a severe or fatal shock.

Figure 2-1. The human body offers varies greatly in the amount of resistance it offers. These are approximate resistances that the body offers to electric current.

Area of Body	Resistance (Ohms)
Skin (Dry)	100,000 to 600,000
Skin (Wet)	1,000
Internal body—hand to foot	400 to 600
Ear to ear	100

Current Path through the Body

The path taken by the electricity determines which body parts receive the greatest damage. When the body acts like a conductor, current only travels through those parts that offer a complete circuit. Any path that crosses the heart is extremely dangerous.

An example of extreme danger: a person is standing barefoot in water attempting to use an electric tool. If the tool is defective, current will flow through the hands, arms, chest, and legs. This situation is likely to cause a very serious shock.

An example of slight danger: a person wishing to use an electric tool near water finds a way to stand on an insulated object, such as a plastic or dry wooden box. The person starts operating the tool using only one hand on the handle. If the tool is defective, a shock will occur only in the arm. This situation may cause some muscle pain, but is less likely to cause any serious damage.

Note: normally when a tool has defective wiring, it will be found the moment it is turned on. It is for this reason that operation with one hand is suggested at startup. Once a tool is found to be operating normally, use both hands for safety.

Duration of Exposure

The length of time the victim is exposed usually determines the damage caused by the shock. It is necessary, therefore, to remove the victim from the voltage source as quickly as possible.

Use extreme caution. Touching the victim causes the rescuer to also get a shock. To free the victim, use a material that is nonconductive. You can wrap a length of cloth around the person and pull, or a length of dry wood can be used to push the victim free.

2.2 PREVENTION OF ELECTRICAL SHOCK

When dealing with electricity, the ideal is to prevent the shock from ever happening. This can be accomplished in two ways. Either do not provide a path for electricity to flow, or turn off the electricity at the source.

Even though electricity has the potential to cause severe shock, it is a necessary part of modern life. When working with electricity, it is everyone's responsibility to be sure there are no accidents. Being aware of the dangers and methods of prevention is best, as shown in figure 2-2.

Figure 2-2. Safety with electricity is no accident.

Factors Affecting Severity of Electrical Shock		
Amount of Electricity	Current Path through Body	Duration of Exposure

Prevention of Electrical Shock		
Use insulated tools to increase resistance.	Insulate body from becoming and electrical path	Work with a buddy for safety.

No-Current Path Protection

Rubber gloves are frequently worn by persons working with live high-voltage circuits. A **live circuit** is a circuit with voltage present. A circuit can be live even if it is not connected to a load. It is safest to work only on **dead circuits,** circuits containing no voltage. However, this is not always possible.

Many technicians who work with wires use tools with plastic-coated handles. Other safety tricks of the trade are working with only one hand and working while standing on an insulated surface.

Portable electric tools have been a frequent source of electric shock in the past. As a result, many manufacturers now sell double insulated tools. **Double insulated tools** can be identified by a plastic coating on their outside surface. Also, they only require a two-prong plug. Tools that are not double insulated must have a three-prong plug.

Circuit Interruption

Household electrical circuits are protected through the use of **overcurrent protection devices,** referred to as circuit breakers or fuses. The primary purpose of these devices is to protect a house from fire. If too much current flows in a wire, it gets hot. A circuit breaker or fuse will interrupt (disconnect) the current flow before a fire can result.

Circuit breakers also protect people from electrical shock. If a wire comes loose and touches the metal handle of a tool, it results in a shock hazard. However, the third wire of a three-prong plug is a ground wire. The ground wire will short circuit a loose wire and trip the circuit breaker in such a situation. This prevents current flow and the tool from operating.

Ground fault interrupters (GFI) function as a circuit interrupter in the event of electric shock. Ground fault interrupters operate by sensing a difference in current between the two current carrying wires. Under normal operation, there is no difference in current between the two wires. When a person is receiving a shock, it results in a difference in the two currents. Ground fault interrupters interrupt a circuit with a very small amount of current difference.

2.3 SAFELY MAKING REPAIRS TO HOUSE WIRING

With household electricity being such an important part of modern living, everyone comes in contact with its potential dangers. When handled and maintained properly, the use of electrical tools and appliances is quite safe.

It is the responsibility of the user to recognize potential dangers. However, the repair of electrical equipment should be performed by a qualified person. As an electronics technician, you are expected to be the qualified person who performs the repairs.

Standard Color Code

The house wiring running to switches, outlets, and light sockets has a standard color code and uniform assignments to the function of the wires. It is important to observe the standard color code to ensure protection from shock and fires. The wiring used in houses is in the form of a cable, with very stiff solid wire, see figure 2-3. The wire

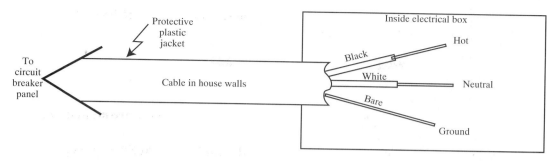

Figure 2-3. The cable used in the walls of a home usually contain color-coded solid copper wire.

connected from an electrical appliance to its plug is also in the form of a cable, except the wire inside is stranded and is flexible, figure 2-4.

A **hot wire** is a current carrying wire. This means that it is one of the wires providing a current path for the circuit to operate. The hot wire carries the voltage. It is the hot wire that causes all electrical shocks. The standard color code for hot is black. The hot wire will have black insulation. It is connected to the brass-colored screw on the electrical device. Note: it is critical that the wires are connected to the proper screw.

The **neutral wire** is also a current carrying wire. However, it has zero voltage, the same as ground. When wired properly it does not cause shocks. *Caution: never trust your life to guessing if the circuit is wired properly.* The standard color code for neutral is white. The neutral wire will have white insulation and should be connected to the silver-colored screw on the electrical device.

The **ground wire** inside the house walls is a bare wire. Its main function inside the walls is to prevent fire. The ground wire is also used to protect against electrical shock. If the hot wire were to come loose, it would contact the bare wire, tripping the circuit breaker stopping all current flow. The ground wire in flexible cords has green insulation.

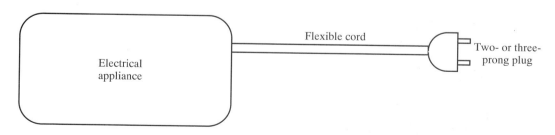

Figure 2-4. Portable electrical equipment use flexible cords with two- or three-prong plugs.

Electrical Plugs

The electrical plug used for tools and other portable equipment is wired for 120 volt operation. This plug comes in two different forms, two-prong and three-prong.

The two-prong plug is used with double insulated tools. Some appliances also use the two-prong plug, such as kitchen products, table lamps, and clocks. Electrical devices using a two-prong plug must be manufactured in such a way as to provide additional insulation between the user and the wiring.

On a polarized two-prong plug, one of the blades is larger than the other, figure 2-5. The larger blade is the neutral wire. The smaller blade is the hot wire. These plugs provide an added protection. They ensure that the internal wiring of the device has maximum protection on the hot wire.

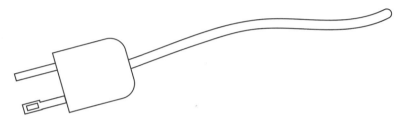

Figure 2-5. On a polarized plug, one blade is larger than the other.

A three-prong plug has a round ground prong in addition to the other two prongs. The ground wire is connected to the part of the tool where the user's skin will come in contact. The function of the ground prong is to trip the circuit breaker if a current-carrying wire becomes loose and comes in contact with the casing of the device. This protects the operator from serious electrical shock.

There are situations where a tool has a three-prong plug, but the outlet is only the two-prong type. When this problem arises, an adaptor, figure 2-6, is available. The **adaptor** has two prongs and a ground wire. The ground wire should be attached to the screw on the outlet's cover plate.

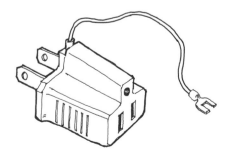

Figure 2-6. Remember to attach the ground wire to the screw on the outlet cover plate. If the wire is not attached, the appliance in use is not grounded.

Faulty Wiring

After appliances have been in operation for a period of time, defects may occur. The plugs may become cracked or the wiring may be pulled loose. In either of these situations, it is necessary to make a repair to avoid contact with exposed wires.

Wiring Two-Prong Electrical Plugs

Flexible two conductor cord has the same color insulation covering both wires. Sometimes, as a means of color coding, the copper wires under the insulation are marked. One wire will be silver and the other wire will be copper colored. If a two-prong *polarized* plug is to be wired, it is important to use the two-colored wire. If the two-prong plug is *non-polarized,* it makes no difference with the wires.

To wire a polarized two-prong plug, connect the copper colored wire to the smaller blade. It should have a brass screw. Connect the silver colored wire to the larger blade. It should have a silver screw. When wiring a non-polarized plug, the wires can be connected to either screw.

Wiring Three-Prong Electrical Plugs

Electrical cords intended for a three-prong plug have three wires: hot, neutral, and ground. Inside the protective outer jacket, the three wires are individually color coded with insulation. Black is hot, white is neutral, and green is used for ground. It is critical that these wires are connected to the proper screws.

In a manner similar to two-prong plugs, the black wire is connected to the brass screw. The white wire is connected to the silver screw. The green ground wire is connected to the round prong, which usually has a green screw.

Wiring a Light Socket

Floor and desk lamps are subject to extreme abuse and quite often the light socket or the wires inside become damaged. Loose wires present a shock hazard. The user often touches the base of the socket in order to operate the switch.

The light socket follows the color code for wires as observed with two-prong plugs. The brass screw (hot) is connected to the center of the light bulb base. The silver screw (neutral) is connected to the bulb's threads, see figure 2-7. The light will work if the wires are connected either way. However, if the threads are incorrectly connected to the

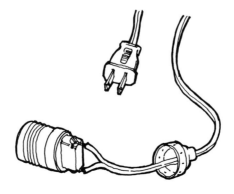

Figure 2-7. Most electrical light sockets are wired like the one shown.

hot wire, a shock hazard is present when the bulb is twisted in and out of the socket, partially exposing the threads.

Wiring an Electrical Duplex Outlet

Electrical duplex outlets used for household electricity are made with plastic. Cords are constantly plugged in and out of them, and there is a tendency for the plastic to break. If the plastic breaks, live wires are exposed.

When an outlet needs to be changed, the solid wires inside of the electrical box are reused. The standard color code (black to brass, white to silver, and bare to the green screw) is used, figure 2-8.

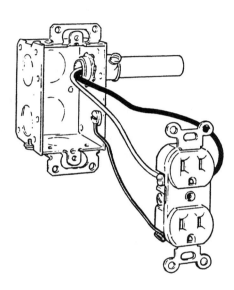

Figure 2-8. It is important that wall outlets are wired correctly. Incorrectly wired outlets may leave live wires in appliances that have been switched off.

2.4 PROPER SOLDERING TECHNIQUES

During the study of electricity and electronics, you are expected to learn to solder wires and electrical components used in the construction of circuits. The primary danger with soldering is burns. Burns can be avoided with proper use of the tools. *Warning: Hot solder can cause severe burns. Hot solder and hot rosin can splatter into the eyes. Always wear safety glasses when soldering.*

Positioning the Soldering Iron

The soldering iron should be held so that the tip makes contact with both surfaces being soldered. For example, if a transistor is to be soldered to a printed circuit board, the soldering iron must heat the copper on the circuit board and, at the same time, the lead of the transistor to the same temperature.

When the surfaces are hot enough, the solder should be touched to the surface being soldered, not the iron's tip. The solder will make a puff of smoke and melt. The melted solder should flow to cover both surfaces. Once again, be sure to wear safety glasses when soldering to prevent an eye injury.

Creating a Good Mechanical Connection

The components being soldered should first have a good **mechanical connection.** This means attaching them together in such a way that they will not come apart.

The surfaces to be soldered must be clean. Bare copper will oxidize. This is a process similar to the rusting of iron or steel. The copper can be cleaned with light steel wool. However, this is usually not necessary. The solder used for electronic applications has a rosin core. The rosin turns to a liquid when heated and flows ahead of the solder, cleaning the surface. This is the reason you should not reuse a piece of solder and or melt the solder on the iron and then touch it to the surface. Having a drop of solder on the tip of the iron aids in the soldering process. But, do not expect that drop to do the work of fresh solder.

When soldering two or more wires together, the wires should be twisted together first. Another method used is to make a small bend on the end of each wire and hook them together. Be sure to allow all surfaces being soldered to reach the same temperature. When one wire is very large in comparison another wire, the larger wire requires much more heat. Therefore, the larger wire should be heated longer than the smaller wire.

When soldering to a printed circuit board, follow this procedure for the safest method to ensure a good mechanical connection:

1. Insert the component through the circuit board. The component should be pressed flat against the board on the top, and the leads should stick through on the bottom.
2. Bend the component's leads tightly against the bottom of the circuit board. Make certain to bend the lead so that it lays in the same direction as the circuit path. Note: do not allow the lead to cross over and touch another path.
3. Using diagonal cutting pliers, snip the lead to a length of approximately 1/8 in. to 1/4 in. long.
4. When applying heat from the soldering iron, heat both the component's lead and the circuit path at the same time. Solder should flow to cover the entire hole.

Recognizing a Good/Bad Solder Joint

If the temperature of the surface is correct, the solder melts and flows onto the joint smoothly and evenly. Remove the solder, then the iron and the joint will cool slowly and evenly. Do not blow on it or move it in any way.

The good solder joint is shiny and smooth. It looks as though it is a part of the objects soldered together.

Bad solder joints are formed in several ways:
- Surface temperatures were uneven. Often, the soldering iron did not even touch one of the surfaces. This is very common with circuit boards.
- Solder was touched to the soldering iron tip to start it melting. The solder melts and appears to flow. However, it does not attach to the surfaces intended to be soldered.
- The soldered joint was moved before the joint cooled. If one of the objects is moved, the solder fractures, resulting in a cracked joint. The fracture may not appear on the surface.
- The solder was cooled too quickly. Solder should be allowed to cool evenly. Uneven cooling can also fracture the joint.

A **cold solder joint** is the name used to describe a bad solder connection. It can be recognized by its dull appearance and it may appear to have tiny cracks. On circuit boards, a cold solder joint appears as a ball that does not look like it is a part of the surface it is connected to. A cold solder joint can usually be made good by reheating and applying another drop of solder.

2.5 SAFETY RULES

Safety rules in the laboratory and classroom help to provide a safety-conscience atmosphere. In the work place, employers do not want the employees to get injured. Everyone is much happier when safety procedures are learned and followed.
- Tell your instructor or supervisor immediately when there is an accident or injury, regardless of how slight.
- Always unplug or disconnect the electrical power before handling any wires.
- When it is necessary to make tests on the inside of energized equipment, work with only one hand. Make certain no part of the body touches a grounded surface.
- The *only* safe way to determine if electricity is present is with proper test equipment.
- Be extra cautious when working near wet areas.
- Keep all equipment in proper repair. Frequently inspect all cords and plugs.
- Wear safety glasses when soldering.
- Do not place cords underneath carpeting. The wire can get hot and start a fire.
- Turn off the power or use an insulated push/pull stick before touching an electrical shock victim.
- Use common sense. This is always the best form of accident prevention.

SUMMARY

- Safety is a life skill that everyone must develop as a part of everyday work habits.
- Electrical shock results from electricity passing through the body. It affects the muscles, nervous system, and other organs.
- Voltage is the driving force behind electricity. The higher the voltage, the more danger of a serious electrical shock.

- Current is the flow of electricity. It is the current flowing through the body that causes electrical shock.
- The path the current follows through the body is a major factor in the severity of the shock.
- Electrical shock can be prevented by not providing a path for the electricity and by installing automatic circuit interrupters, such as a GFI.
- In house wiring, the standard color code is: black for hot, white for neutral, and bare (green) for ground.
- When soldering, always wear safety glasses. Allow the soldering iron to touch both surfaces and heat them enough to melt the solder.

KEY WORDS AND TERMS GLOSSARY

cold solder joint: The name used to describe a bad solder connection. It can be recognized by its dull appearance and it may have tiny cracks.

conductor: A material which allows the easy flow of electricity.

current: The flow of electricity.

dead circuit: A circuit containing no voltage.

double insulated tool: Tools on which the outside surfaces are plastic coated. They require only a two-prong plug.

electrical shock: The physical sensation of the nerves and muscles reacting to electricity passing through the body.

electrocution: The term used when the exposure to electric shock results in death.

ground fault interrupter (GFI): A circuit breaker that protects against electric shock by sensing an unbalanced condition between the two current carrying wires. The circuit is interrupted with very small amounts of current difference.

ground wire: The bare wire in house wiring. Its main functions are to prevent fire and to protect against electrical shock. If the hot wire comes loose, it would contact the bare wire, which would trip the circuit breaker stopping all current.

insulator: Materials which offer an infinite resistance, allowing no current to flow. An example is the plastic insulation on a wire.

live circuit: A circuit with voltage present, even if it is not connected to a load.

neutral wire: In house wiring, it is a current carrying wire with a voltage potential equal to ground. The standard color code for the neutral wire is white. It is wired to the silver-colored screw on the electrical device.

overcurrent protection device: Device that prevents too much current from flowing in a circuit. Fuses and circuit breakers are examples of overcurrent protection devices.

resistance: The opposition a current path offers to the flow of electricity.

safety: A life skill in which people protect themselves and others from possible harm. In a classroom and laboratory setting, it is the responsibility of the teacher to demonstrate safe techniques. It is the responsibility of the student to make every effort to learn and use safe operating procedures at all times.

voltage: The potential energy of an electric source.

TEST YOUR KNOWLEDGE

Do not write in this text. Please use a separate sheet of paper.

1. Describe the effects of electrical shock.
2. List the three electrical terms that affect electric shock.
3. List three factors that influence the severity of electrical shock.
4. State two methods of preventing electrical shock.
5. Identify the three colors in the standard color code for house wiring.
6. State the function of the three standard wires used in house wiring.
7. What safety device must be worn when soldering?
8. Describe the best method of soldering an electronic component to a circuit board.
9. What can be done to correct a cold solder joint?
10. Make a list of safety rules that apply to your home or apartment.

Chapter 3
Reading Meter and Oscilloscope Scales

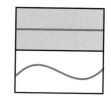

Upon completion of this chapter, you will be able to:
- State the value of numbers written with engineering notation.
- Convert numbers containing engineering notation.
- Identify digital and analog multimeters.
- Interpret the value of each line on an analog scale.
- Analyze the readings on an ohmmeter scale.
- Distinguish between the ac and dc arcs on a multimeter.
- Adjust the decimal place on a multimeter scale for each range.
- Read an oscilloscope screen for dc voltages.

In every technical field, it is necessary to read scales on all types of meters properly. Understanding how to read the scales of the two most popular electronic testing instruments, multimeters and oscilloscopes, provides the skills to read any meter scale.

A **multimeter** is used for electrical measurements. The meter is usually capable of four different types of measurements: ohms, milliamps, dc volts, and ac volts. Some meters also offer other combinations of measurements. Meters that can check capacitance are quite common.

A multimeter has a number of resistors built into its internal circuitry. This allows the meter to check many scales and ranges. For example, turning the range switch can change a voltmeter from a maximum reading of 10 volts to a maximum of 1000 volts.

An **oscilloscope** draws a picture of the voltage in a circuit. Oscilloscopes are used to do many things including measuring a voltage, checking a voltage frequency, and comparing two voltage signals.

3.1 ENGINEERING NOTATION

In electronics and other scientific subject areas, the numbers used are frequently very large or very small. This places many zeros on one side of the decimal point or the other. This situation can make arithmetic operations difficult. Errors result from misplacement of the decimal point. Engineering notation simplifies the use of very large and very small numbers. It is used on both multimeters and oscilloscope scales. **Engineering notation** moves the decimal point using powers of 10 and multiplier names.

Scientific electronic calculators have special keys to handle engineering notation. A calculator, however, is not used when reading meters and oscilloscopes. Therefore, it is necessary to understand the relationships of engineering notation.

Writing Numbers Times a Power of 10

Scientific notation is a method of moving the decimal point of some number and multiplying that number by 10 raised to a power (10^n) to maintain an equivalent value. In scientific notation, the decimal point is moved to follow the first significant digit. The number of places that the decimal point is moved to the right becomes the power of 10. If the decimal point is moved to the left, the number is simply the negative of the number of positions traveled. Thus, the number 103.5 in scientific notation would be 1.035×10^2. The number 0.001035 would become 1.035×10^{-3}.

Engineering notation uses multiplier names in place of the powers of 10 when moving the decimal point. These multiplier names are used for both positive and negative powers of 10 in multiples of three (one thousandth, one thousand, one million, etc.). Figure 3-1 is a table of the engineering notation multiplier names and symbols with the decimal value and power of 10.

Figure 3-1. These are the most commonly used multipliers. Also shown are their symbols and appearance in engineering notation.

Multiplier Name	Symbol	Multiply by	Power of 10	E Notation
tera	T	1,000,000,000,000 (one trillion)	10^{12}	E12
giga	G	1,000,000,000 (one billion)	10^9	E09
mega	M	1,000,000 (one million)	10^6	E06
kilo	k	1,000 (one thousand)	10^3	E03
basic units	1	1 (decimal remains without change)	10^0	E00
milli	m	0.001 (one one-thousandth)	10^{-3}	E–03
micro	μ (Greek letter mu)	0.000 001 (one one-millionth)	10^{-6}	E–06
nano	n	0.000 000 001 (one one-billionth)	10^{-9}	E–09
pico	p	0.000 000 000 001 (one one-trillionth)	10^{-12}	E–12

Exponential notation or **E notation** writes the power of 10 as an E followed by two digits. This format is used by computers and calculators. In Chapter 4, the calculator is used to solve electrical problems and to understand the basic concepts. E notation will be discussed in detail there. It is an important tool when using the calculator.

The rules of engineering notation are summarized as follows:

Numbers Larger than 1000

- For numbers larger than 1000, move the decimal to the *left* in groups of three.
- Write the number with a *positive* power of 10 equal to the number of places moved.
- A number may be expressed with several different powers. The *best power* has the number between 1 and 1000.

Examples:

$$2500 = 2.5 \times 10^3 = 2.5 \text{ E03} = 2.5 \text{ kilo } \textit{(best power)}$$

$$2500 = 0.0025 \times 10^6 = 0.0025 \text{ E06} = 0.0025 \text{ mega}$$

$$603{,}000{,}000 = 603 \times 10^6 = 603 \text{ E06} = 603 \text{ mega } \textit{(best power)}$$

$$603{,}000{,}000 = 603{,}000 \times 10^3 = 603{,}000 \text{ E03} = 603{,}000 \text{ kilo}$$

$$603{,}000{,}000 = 0.603 \times 10^9 = 0.603 \text{ E09} = 0.603 \text{ giga}$$

Numbers Smaller than 1

- For numbers smaller than 1, move the decimal to the *right* in groups of three.
- Write the number with a *negative* power of 10 equal to the number of places moved.
- A number may be expressed with several different powers. The *best power* writes the number between 1 and 1000.

Examples:

$$0.035 = 35 \times 10^{-3} = 35 \text{ E-03} = 35 \text{ milli } \textit{(best power)}$$

$$0.035 = 35{,}000 \times 10^{-6} = 35{,}000 \text{ E-06} = 35{,}000 \text{ micro}$$

$$0.00067 = 670 \times 10^{-6} = 670 \text{ E-06} = 670 \text{ micro } \textit{(best power)}$$

$$0.00067 = 0.67 \times 10^{-3} = 0.67 \text{ E-03} = 0.67 \text{ milli}$$

Engineering Notation Conversion Bar

The **engineering notation conversion bar** is a visual aid showing the decimal point being moved and replaced by a multiplier name. To use the conversion bar, place the decimal point under its proper multiplier symbol. Then, move the decimal point to the multiplier you wish to use.

Between each multiplier name are three places for digits. Each place must be filled. Place zeros as needed for place holders. The spot in the center of the conversion bar, with no multiplier name, represents an exponent of zero, the basic unit without a multiplier.

Rules for the Engineering Notation Conversion Bar

Positive ←   → Negative

```
T   G   M   k       m   μ   n   p
***^***^***^***^***^***^***^***^***^***
```

- Always move the decimal point in groups of three.
- When the decimal point is moved to the *left*, it is moved in the *positive* direction.
- When the decimal point is moved to the *right*, it is moved in the *negative* direction.
- A number can have its decimal point moved to any location, provided the power is raised or lowered accordingly.
- To use the conversion bar, place the decimal point under the appropriate multiplier name. Move to other locations as required.
- Significant figures remain in position under the bar. The decimal point moves and zeros are used as place holders.

Sample problem 1. _____

Use the conversion bar to change 23 kilo to: mega, basic units, and milli.

```
T   G   M   k       m   μ   n   p
***^***^***^***^***^***^***^***^***^***
```

23000×10^6

23.	23. kilo (E03)
0.023	0.023 Mega (E06)
23,000.	23,000. (E00)
23,000,000.	23,000,000. milli (E-03)

23000000 milli

Answer: 23 kilo = 0.023 Mega = 23,000 = 23,000,000 milli

Sample problem 2. _____

Use the conversion bar to change 4.5 in basic units to: kilo and milli.

```
T   G   M   k       m   μ   n   p
***^***^***^***^***^***^***^***^***^***
```

.0045

4,5	4.5 (E00)
0.004 5	0.0045 kilo (E03)
4 500.	4500 milli (E-03)

Answer: 4.5 = 0.0045 kilo = 4500 milli

Sample problem 3. _____

Use the conversion bar to change 60 milli to: mega, basic units, and milli.

```
T   G   M   k       m   μ   n   p
***^***^***^***^***^***^***^***^***^***
```

60×10^{-3}

60000×10^{-6}

60.	60 milli (E-03)
60,000.	60,000 micro (E-06)
0.06	0.06 (E00)

Answer: 60 milli = 60,000 micro = 0.06 E00

Sample problem 4.

Convert the numbers on the left to the underlined units shown. Each of the units given will be studied at some time in this text. Note: a unit with no multiplier name is the basic unit (E00).

Given	*To Find*		*Answers*
a. 9,800,000 ohms	_____kΩ	_____MΩ	9,800 kΩ and 9.8 MΩ
b. 4300 watts	_____mW	_____kW	4,300,000 mW and 4.3 kW
c. 0.0062 volts	_____mV	_____μV	6.2 mV and 6200 μV
d. 0.000 000 32 amps	_____μA	_____nA	0.32 μA and 320 nA
e. 250 millihenrys	_____H	_____μH	0.25 H and 250,000 μH
f. 7800 milliwatts	_____kW	_____W	0.0078 kW and 7.8 W
g. 1250 kilohertz	_____Hz	_____MHz	1,250,000 Hz and 1.25 MHz
h. 0.0035 kilovolts	_____mV	_____V	3,500 mV and 3.5 V
i. 25 microfarads	_____mF	_____nF	0.025 mF and 25,000 nF
j. 64 megohms	_____kΩ	_____GΩ	64,000 kΩ and 0.064 GΩ

3.2 DIGITAL VS ANALOG MULTIMETERS

A **digital multimeter** displays numbers, as shown in figures 3-2 and 3-3. **Analog multimeters** use a needle, as shown in figures 3-4 and 3-5.

Digital meters have certain advantages. They are usually lighter and more compact than analog meters. Digital meters are also easier to read and their cost is fairly low. They have internal adjustments to calibrate the meter. This can be a benefit—or a problem if not adjusted correctly. Generally speaking, the digital meter has a good selection of ranges and their accuracy is fairly good.

Figure 3-2. This is a bench model digital multimeter. (BK Precision)

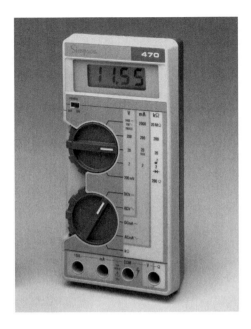

Figure 3-3. Hand-held digital multimeters assist in checking circuits that cannot be brought to a work area. (Simpson Electric Co.)

Figure 3-4. A standard analog multimeter. (Simpson Electric Co.)

Figure 3-5. This is a smaller hand-held analog multimeter. (BK Precision)

The digital meter shown in figure 3-2 is a bench model. It is a fairly large meter with large digits and a wide selection of ranges to choose from. The advantage of having several ranges is increased accuracy.

The digital meter shown in figure 3-3 is a hand-held model. This particular meter has many of the same features as the bench model, especially important is the range selection. The disadvantage of the smaller meter is the smaller numbers.

Many digital meters offer automatic ranging. This means the operator only chooses the type of measurement that is taken: voltage, current, or resistance. The electronics inside the meter select the best range based on the value being measured. When extreme accuracy is not necessary, this autoranging may be preferred.

The biggest disadvantage of the digital meter is its need for batteries. An analog meter uses batteries only when measuring resistance. If the batteries go dead in an analog meter, it will still measure voltage and current. The digital meter will not operate at all without batteries.

Both types of meters are used in most applications. There are certain applications where one meter is better than the other, but generally speaking, it is a matter of personal preference.

The analog meter shown in figure 3-4 is a fairly large meter. It is easily carried but best as a bench model. The meter face is large for easy reading, and it has a good selection of ranges to provide accuracy.

The analog meter shown in figure 3-5 is a smaller hand-held model. As with most analog meters, the range selection provides accuracy. However, the smaller face is more difficult to read.

3.3 VALUE OF EACH LINE ON AN ANALOG SCALE

On a digital multimeter, you simply look and the LED or LCD display and a readout is given. Reading an analog multimeter is not as easy.

A **scale** on an analog meter is a set of numbers belonging to a selected range. The markings on a meter scale are numbered at intervals to allow for easy reading. The unmarked scale lines have a different value for each meter range. It is necessary to determine the value of the unmarked lines to properly read the meter.

Figures 3-6 through 3-22 include an expanded scale not found on the meter face. The expanded scale is a ruler-drawn number line showing more details. These will assist you in determining the value of the unmarked lines.

Steps in Finding the Value of Each Line

By following a standard series of actions, you can calculate the value of the unmarked lines on any of the scales. These actions will work on all analog multimeters. Though, there may be different divisions on each multimeter.

a. Select the range that you need. The **range** is the maximum value allowed on the selected scale on an analog meter.

b. The meter scale you will be using has a number on the right side corresponding to the selected range. Find the correct scale.

c. Count the number of lines from zero to the first marked number.

d. Divide the first marked number by the number of lines.

Note: be sure to count the line corresponding to the marked number.

Range: 12 AC/DC

Refer to the meter face in figure 3-6. The range has been selected for a maximum of 12. This could be 12 volts ac, 12 volts dc, 12 amps, or 12 milliamps. For the moment, the substance being measured is unimportant. Find the scale with 12 located at maximum on the meter, the far right. The numbers corresponding to the 12 scale are the only ones to be examined at this time.

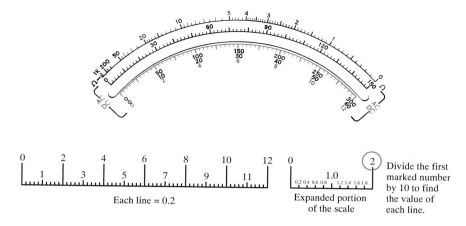

Each line = 0.2

Expanded portion of the scale

Divide the first marked number by 10 to find the value of each line.

Figure 3-6. AC/DC 12 scale has a value of 0.2 for each line.

    a.   Scale = 12 AC/DC.
    b.   Numbers marked on the 12 scale: 0, 2, 4, 6, 8, 10, and 12.
    c.   Count of lines between 0 and first number: 10.
    d.   First marked number: 2.

$$\textbf{Value of lines:} \quad \frac{\text{First marked number}}{\text{Count}} = \frac{2}{10} = 0.2$$

Sample problem 5.

    Determine the readings of needles a and b of the sample meter in figure 3-7. Make use of the expanded scales.

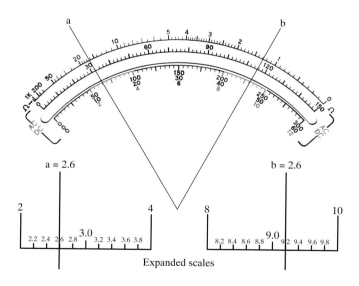

a = 2.6

b = 2.6

Expanded scales

Figure 3-7. Sample meter set on AC/DC 12 scale. (Sample problem 5)

Scale: 12 AC/DC
Value per line: 0.2
Needle a: 2.6
Needle b: 9.2

Range: 60 AC/DC

Refer to the meter face in figure 3-8. The range has been selected for a maximum of 60. Find the scale with 60 located at maximum on the meter. The numbers corresponding to the 60 scale are the only ones to be examined at this time.

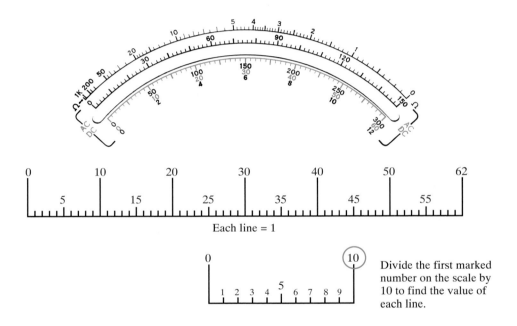

Figure 3-8. AC/DC 60 scale has a value of one for each line.

a. Scale = 60 AC/DC.

b. Numbers marked on the 60 scale: 0, 10, 20, 30, 40, 50, and 60.

c. Count of lines between 0 and first number: 10.

d. First marked number: 10.

$$\textbf{Value of lines: } \frac{\text{First marked number}}{\text{count}} = \frac{10}{10} = 1 \text{ per line}$$

Sample problem 6. _____

Determine the readings of needles a and b of the sample meter in figure 3-9.

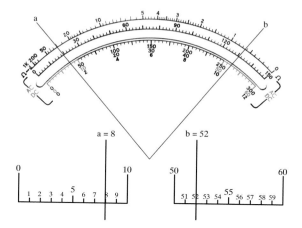

Figure 3-9. Sample meter set on AC/DC 60 scale. (Sample problem 6)

Scale: 60 AC/DC
Value per line: 1
Needle a: 8
Needle b: 52

Range: 300 AC/DC

Refer to the meter face in figure 3-10. The range has been selected for a maximum of 300. Find the scale with 300 located at maximum on the meter. The numbers corresponding to the 300 scale are the only ones to be examined at this time.

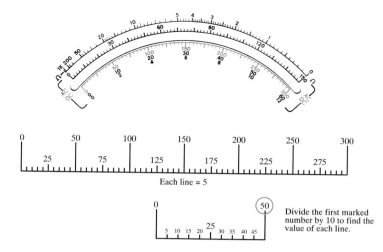

Figure 3-10. AC/DC 300 scale has a value of five for each line.

a. Scale = 300 AC/DC.

b. Numbers marked on the 300 scale: 0, 50, 100, 150, 200, 250, and 300.

c. Count of lines between 0 and first number: 10.

d. First marked number: 50.

Value of lines: $\dfrac{\text{First marked number}}{\text{count}} = \dfrac{50}{10} = 5$ per line

---

Sample problem 7. _____

Determine the readings of needles a and b of the sample meter in figure 3-11.

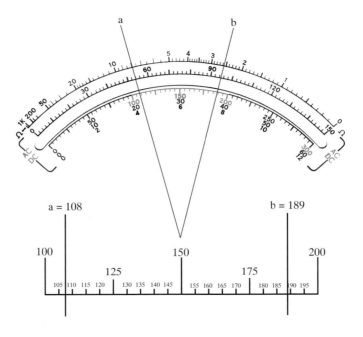

Figure 3-11. Sample meter set on AC/DC 300 scale. (Sample problem 7)

Scale: 300 AC/DC
Value per line: 5
Needle a: 108
Needle b: 189

Range: 150 AC/DC

Refer to the meter face in figure 3-12. The range has been selected for a maximum of 150. Find the scale with 150 located at maximum on the meter. The numbers corresponding to the 150 scale are the only ones to be examined at this time.

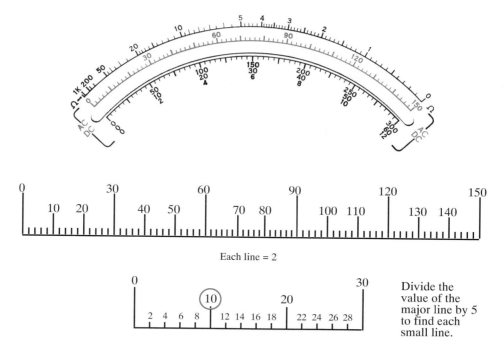

Each line = 2

Divide the
value of the
major line by 5
to find each
small line.

Figure 3-12. AC/DC 150 scale has a value of two for each line.

This scale is different from the other AC/DC scales. The numbers on the other scales are marked on every other heavy line. This 150 scale has the numbers marked on every third heavy line. Also, rather than having 10 lines between each marked number, the 150 scale has five between each heavy line. Notice that each heavy line goes in steps of 10. The five lines between have a value of two.

a.   Scale = 150 AC/DC.

b.   Numbers marked on the 150 scale: 0, 30, 60, 90, 120, and 150. Medium-sized unmarked lines: steps of 10.

c.   Count of lines between 0 and first step: 5.

d.   Value of lines: $\dfrac{\text{First step}}{\text{Count}} = \dfrac{10}{5} = 2$ per line

Sample problem 8. _____

Determine the readings of needles a and b of the sample meter in figure 3-13.

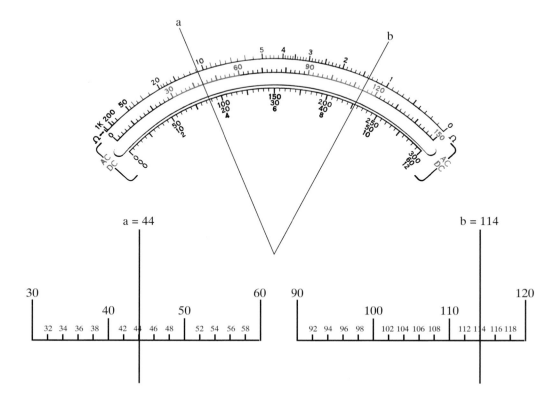

Figure 3-13. Sample meter set on AC/DC 150 scale. (Sample problem 8)

Scale: 150 AC/DC
Value per line: 2
Needle a: 44
Needle b: 114

_____

Ohms Scale (Ω)

The ohms scale is a nonlinear scale. In a **nonlinear** scale the spacing between each line is not the same across the entire scale. Another unique feature about the ohms scale is the placement of zero. The zero is on the right side. The scale is numbered up, heading to the left. The largest number is the infinity symbol, ∞, on the far left.

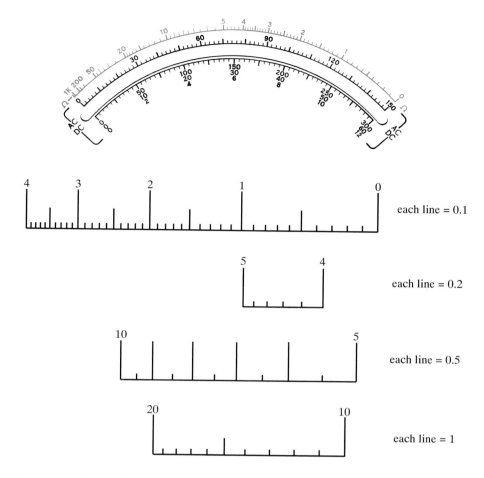

Figure 3-14. The value of each line in the ohms section varies.

Figure 3-14 shows the ohms scale is subdivided into four different sections, each having a different value for the markings. Remember, when counting lines between two numbers, be sure to count *one* of the two numbers. Looking at the scale:

0 to 4: 10 lines between each number... Value per line = 0.1.

4 to 5: 5 lines... Value per line = 0.2.

5 to 10: 2 lines... Value per line = 0.5.

10 to 20: 10 lines... Value per line = 1.

Sample problem 9. _____

Determine the readings of needles a and b in of the Ohms scale in figure 3-15.

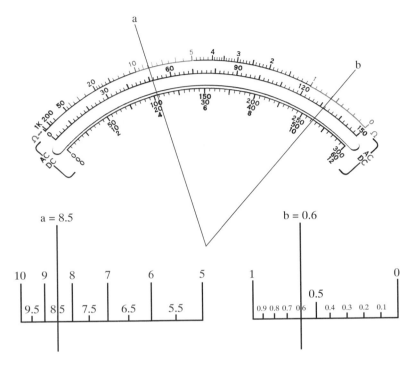

Figure 3-15. Sample meter set on the ohms scale. (Sample problem 9)

a = 8.5

b = 0.6

_____

3.4 ADJUSTING THE DECIMAL PLACE OF A METER SCALE

Resistors built into a meter allow one set of numbers on a meter face to represent more than one range. Turning the switch on the range selector changes these resistors. The set of numbers is modified by adjusting the decimal place to the right or to the left.

As a suggestion, it is helpful to write down the numbers with their decimal place adjusted. Also, it is always a good idea to write down the value of each line. After determining the value of the lines, count them out to verify that everything is worked out correctly.

Adjusting the Decimal for Milliamps, AC Volts, and DC Volts scale

The meter used in the next set of examples has scales of 10, 50, and 250.

Scale: 10 ... Also Used for 1.0, 100, 1000

Figure 3-16 shows how the scale marked with a maximum of 10 can be used on four different ranges. The procedure for finding the value of each line is to move the decimal place as follows:

a.   One place to the left, the scale changes to 1 from 10.
b.   One place to the right, the scale changes to 100 from 10.
c.   Two places to the right, the scale changes to 1000 from 10.

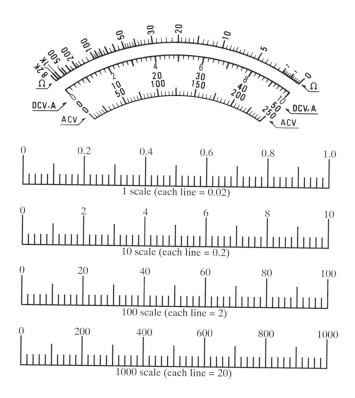

Figure 3-16. Adjusting the decimal of the 10 scale.

Scale: 50 ... Also Used for 0.5, 5, 500

The 50 scale is the middle set of numbers on the meter shown in figure 3-17. This scale can be used on four ranges by moving the decimal point as follows:

a.   Two places to the left, the scale changes to 0.5 from 50.
b.   One place to the left, the scale changes to 5 from 50.
c.   One place to the right, the scale changes to 500 from 50.

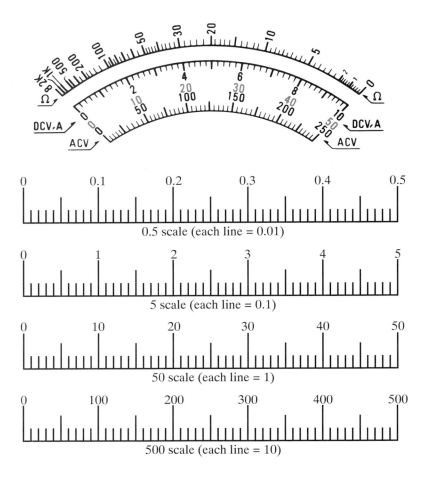

Figure 3-17. Adjusting the decimal of the 50 scale.

Scale: 250 ... Also Used for 0.25, 2.5, 25

The 250 scale is the bottom set of numbers in figure 3-18. The scale has four ranges to select from by moving the decimal, as follows:

    a.   One place to the left, the scale changes to 25 from 250.

    b.   Two places to the left, the scale changes to 2.5 from 250.

    c.   Three places to the left, the scale changes to 0.25 from 250.

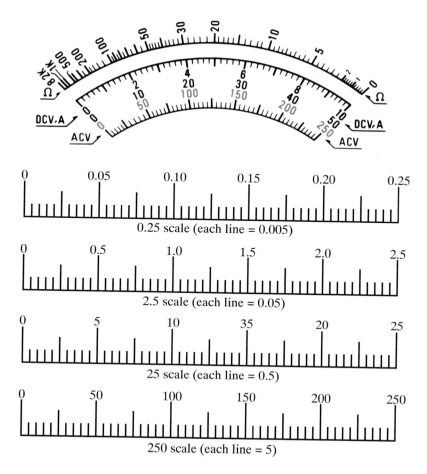

Figure 3-18. Adjusting the decimal of the 250 scale.

Adjusting the Decimal for the Ohms Scale

Multimeters have more than one range for making measurements on the ohms scale. Range selection is different from the method used for volts or amps. With volts and amps, the range is the maximum that can be read on a particular scale. With ohms, the range is *multiplied* by the reading on the needle.

Sample problem 10. _____

Determine the readings on the meter in figure 3-19 for needles a and b.

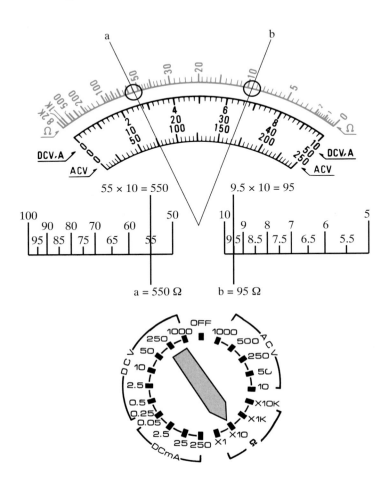

Figure 3-19. Sample meter on ohms scale. (Sample problem 10)

Range:	$\Omega \times 10$
Needle a:	$55 \times 10 = 550 \ \Omega$
Needle b:	$9.5 \times 10 = 95 \ \Omega$

3.5 AC AND DC SCALE ARCS

It is common for meters to have two arcs of lines, such as those shown in the previous figures. One arc is for readings taken of ac voltages. Another arc is used for dc voltages and amps. There is a slight difference between the readings on the ac and dc arcs. This can be a problem when a high degree of accuracy is required.

When using a meter, it is necessary to determine the set of numbers making up the scale being used. All other scales are ignored. It is also necessary to select the proper arc and ignore the other one. The meter used for figures 3-20 to 3-22 has an arrow on the side pointing to the arc to use for either ACV or DCV and DCA. Figures 3-20 to 3-22 also include the range switch to show which range has been selected.

Sample problem 11.

Determine the readings on the meter in figure 3-20 for needles a and b.

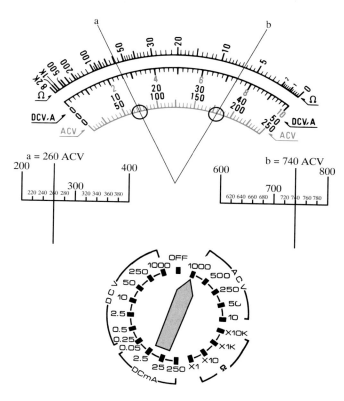

Figure 3-20. Sample meter on 1000 ACV scale. (Sample problem 11)

Range:	1000 ACV
Arc used:	Bottom (ACV)

Numbers after adjusting the decimal: 0, 200, 400, 600, and 1000

Value of each line: 20

Needle a = 260 ACV

Needle b = 740 ACV

Sample problem 12. _____

Determine the readings on the meter in figure 3-21 for needles a and b.

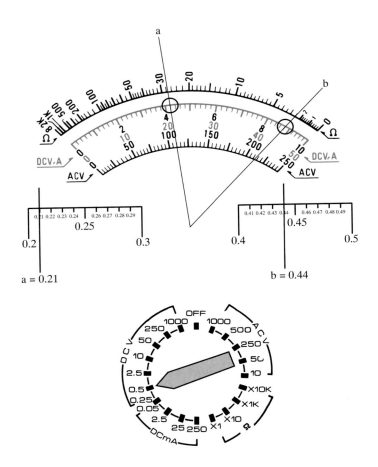

Figure 3-21. Sample meter on 0.5 DCV scale. (Sample problem 12)

Range:	0.5 DCV
Arc used:	Top (DCV,A)

Numbers after adjusting the decimal: 0, 0.1, 0.2, 0.3, 0.4, and 0.5

Value of each line:	0.01
	Needle a = 0.21 DCV
	Needle b = 0.44 DCV

Sample problem 13.

Determine the readings on the meter in figure 3-22 for needles a and b.

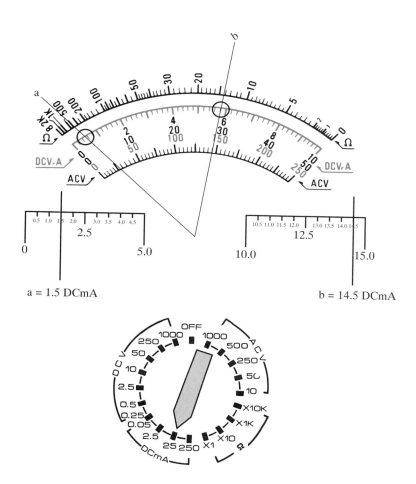

Figure 3-22. Sample meter on 25 DCmA scale (Sample problem 13).

Range:	25 DCmA
Arc used:	Top (DCV,A)
Numbers after adjusting the decimal:	0. 5, 10, 15, 20, and 25
Value of each line:	0.5
	Needle a = 1.5 DCmA
	Needle b = 14.5 DCmA

3.6 READING AN OSCILLOSCOPE SCREEN

The oscilloscope is a voltmeter that draws a picture of the voltage being measured. Two different models of oscilloscopes are shown in figures 3-23 and 3-24. Although there are many different models to choose from, they all plot the waveform of a voltage with respect to time.

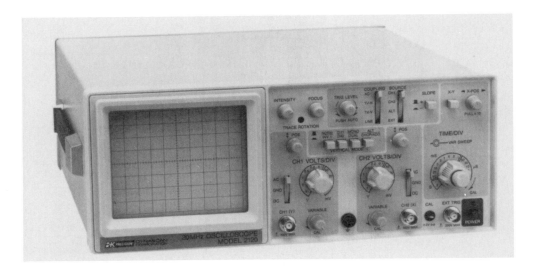

Figure 3-23. Typical oscilloscope. (BK Precision)

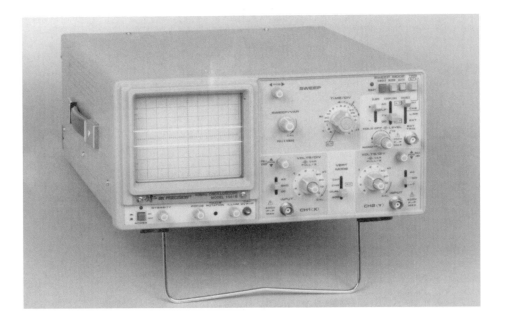

Figure 3-24. This oscilloscope has two straight lines on the screen. This is typical of dc voltage measurements. (BK Precision)

Voltage and time measurements can be broken into small segments, allowing an extremely accurate measurement for all or part of a voltage waveform. This section discusses how to read the oscilloscope screen and how to take measurements of dc voltages.

This section discusses measurements of dc voltages only. A dc voltage is displayed on the oscilloscope screen as a straight, horizontal line. Other voltage waveforms are discussed in Chapter 15.

Oscilloscope Screen

The screen is divided into squares called divisions. Each division is subdivided into smaller segments of 0.2 division. In figure 3-25, notice the screen has two center axis with subdivision marks. One axis is vertical, up-and-down, the other axis is horizontal, or across. Voltage is measured in the vertical direction. Time is measured on the horizontal axis.

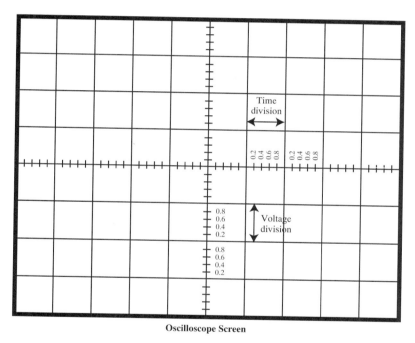

Oscilloscope Screen

Figure 3-25. Lines forming blocks make up one whole division. Small markings indicate 0.2 division.

Volts per Division and Time per Division

Like a voltmeter, the oscilloscope has a variety of ranges to assist in making the most accurate measurements. With an oscilloscope, each division is assigned a specific amount of voltage or time depending on the setting of the controls.

- Typical voltage/division settings are: 20 V, 10 V, 5 V, 2 V, 1 V, 500 mV, 200 mV, 100 mV, 50 mV, 20 mV, and 10 mV.
- Typical time/division settings are: 1 s, 500 ms, 100 ms, 50 ms, 10 ms, 5 ms, 1 ms, 500 μs, 100 μs, 50 μs, 10 μs, 5 μs, and 1 μs.

To determine the measurement for voltage, measure from one point to another in the vertical direction. To measure time, measure from one point to another in the horizontal direction. Multiply the units per division times the number of divisions.

Measuring DC Voltages

In figures 3-26 to 3-29, the screen has a ground reference line indicated. This point is equal to 0 V and is adjusted according to the operator's preference. The center line on the scope does not have to be the ground reference. All voltages are measured from the ground reference to the trace. The **trace** is the line that is produced by the incoming voltage signal.

The figures also show settings for channel 1 and channel 2 volts/division. You select one or both channels, as needed. Both voltage channels use a common time/division control. These sample oscilloscope drawings do not show a setting for time. A time setting is not necessary when making dc voltage measurements.

Following are steps to take when measuring dc voltages.

   a.   Count the number of divisions from the ground reference line to the trace. Be accurate to one decimal place.

   b.   Determine the setting of the volts/division for the channel in use.

   c.   Multiply the number of divisions times the volts/division.

Sample problem 14. _____

Determine the dc voltage being measured in the oscilloscope drawing in figure 3-26.

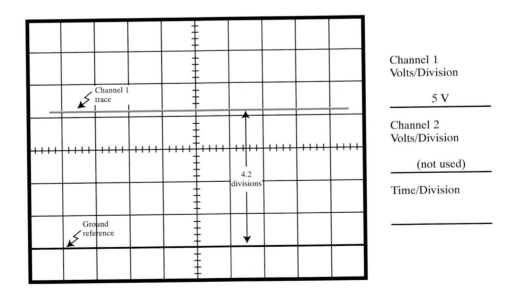

Figure 3-26. DC voltage = 4.2 divisions x 5 volts per division = 21 volts. (Sample problem 14)

a. Number of divisions to one decimal place: 4.2 divisions, from ground to the trace

b. Volts/division setting: Channel 1 = 5 V/div

c. Voltage measurement: 4.2 div × 5 V/div = 21 V

Sample problem 15.

Determine the dc voltage being measured in the oscilloscope drawing in figure 3-27.

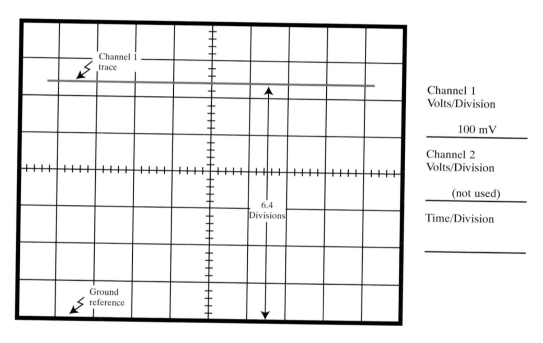

Figure 3-27. DC voltage = 6.4 divisions x 100 millivolts per division = 0.64 volts. (Sample problem 15)

a. Number of divisions to one decimal place: 6.4 divisions, from ground to the trace

b. Volts/division setting: Channel 1 = 100 mV/div

c. Voltage measurement: 6.4 div × 100 mV/div = 640 mV = 0.64 V

Sample problem 16. _____

Determine the dc voltage being measured in the oscilloscope drawing in figure 3-28.

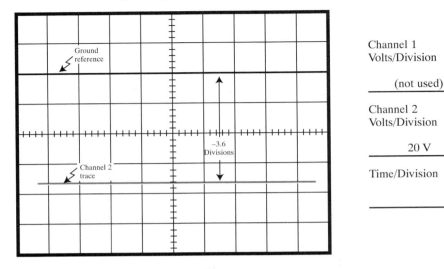

Channel 1
Volts/Division

_____(not used)_____

Channel 2
Volts/Division

_____20 V_____

Time/Division

_____

Figure 3-28. DC voltage = -3.6 divisions x 20 volts per division = –72 volts. (Sample problem 16)

a. Number of divisions to one decimal place: –3.6 divisions, from ground to the trace. (Negative because it is below ground.)

b. Volts/division setting: Channel 2 = 20 V/div

c. Voltage measurement: –3.6 div × 20 V/div = –72 V

Sample problem 17. _____

Determine the dc voltage being measured in the oscilloscope drawing in figure 3-29.

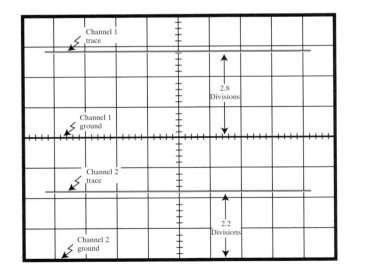

Channel 1
Volts/Division

_____2 V_____

Channel 2
Volts/Division

_____10 V_____

Time/Division

_____

Figure 3-29. Channel 1 DC voltage = 2.8 divisions x 2 volts per division = 5.6 volts. Channel 2 DC voltage = 22 millivolts. (Sample problem 17)

Channel 1

a. Number of divisions to one decimal place: 2.8 divisions, from ground to the trace

b. Volts/division setting: Channel 1 = 2 V/div

c. Channel 1 voltage measurement: 2.8 div × 2 V/div = 5.6 V

Channel 2

a. Number of divisions to one decimal place: 2.2 divisions, from ground to the trace

b. Volts/division setting: Channel 2 = 10 mV/div

c. Channel 2 voltage measurement: 2.2 div × 10 mV/div = 22 mV

SUMMARY

- Engineering notation is a system of writing numbers with multiplier names for very large and very small numbers.
- Multimeters are available in two styles, digital and analog.
- Analog multimeters have different values assigned to each line on the different scales.
- A scale on an analog multimeter can be used for more than one range by adjusting the decimal place.
- The ohms scale on an analog multimeter is non-linear, with the spacing between the lines changing across the scale. The ohms scale has zero on the right, infinity on the left.
- The oscilloscope is a voltmeter that draws a picture of the voltage waveform.
- Voltage is measured on the oscilloscope vertically.
- Time is measured on the oscilloscope horizontally.

KEY WORDS AND TERMS GLOSSARY

analog multimeter: Analog-type meters have a needle that moves along a scale.

digital multimeter: Meters with a digital display have numbers that change with the reading.

engineering notation: A system in mathematics which uses multiplier names and powers of 10 to move the decimal point and label quantities.

exponential notation: A system in mathematics which writes the power of 10 as E followed by two digits. This format is used by computers and calculators.

multimeter: A meter for electrical measurements, usually capable of four different types of measurements: ohms, milliamps, dc volts, ac volts.

nonlinear scale: The spacing between each line is not the same across the scale. An example is the ohms scale on a multimeter.

oscilloscope: Device that draws a picture of the voltage in a circuit.

range: The maximum value that can be read on an analog meter scale.

scale: In reference to a meter, it is the set of numbers belonging to the selected range.

scientific notation: A mathematic process of moving the decimal point and
multiplying by 10 raised to a power to maintain an equivalent value.
trace: The line on the oscilloscope produced by the incoming voltage signal.

TEST YOUR KNOWLEDGE

Do not write in this text. Please use a separate sheet of paper.
1. Convert the following numbers into their proper scientific notation form.
    a.   4900
    b.   6,333,333
    c.   0.0001023
    d.   –0.0505
    e.   10
2. Convert the numbers on the left into the engineering notation units that are shown
   on the right.

	Given		*Find*	
a.	68,000 ohms	____ kΩ	____ MΩ.
b.	2100 watts	____ mW	____ kW.
c.	0.0039 volts	____ mV	____ μV.
d.	0.000 042 amps	____ μA	____ nA.
e.	490 millihenrys	____ H	____ μH.
f.	6100 milliwatts	____ kW	____ W.
g.	23,000 kilohertz	____ Hz	____ MHz.
h.	0.0012 kilovolts	____ mV	____ V.
i.	360 microfarads	____ mF	____ nF.
j.	6400 megohms	____ kΩ	____ GΩ.

For problems 3 through 9, answer the following:
    a.   range.
    b.   arc (ACV, DCV, Ω).
    c.   numbers of scale after adjusting the decimal.
    d.   value of each line.
    e.   reading of needle #1.
    f.   reading of needle #2.

3. Use the meter in figure 3-30.

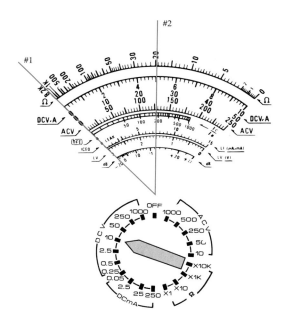

Figure 3-30. Multimeter for Test Your Knowledge problem 3.

4. Use the meter in figure 3-31.

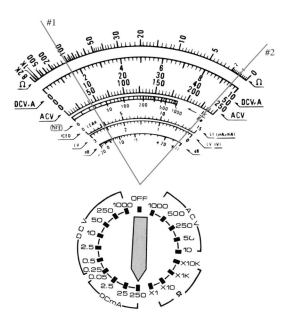

Figure 3-31. Multimeter for Test Your Knowledge problem 4.

5. Use the meter in figure 3-32.

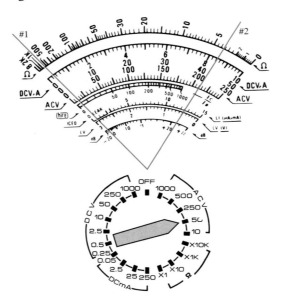

Figure 3-32. Multimeter for Test Your Knowledge problem 5.

6. Use the meter in figure 3-33.

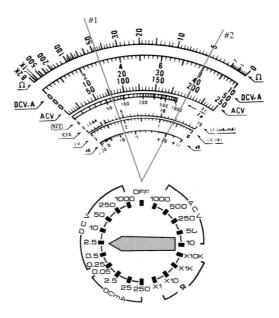

Figure 3-33. Multimeter for Test Your Knowledge problem 6.

7. Use the meter in figure 3-34.

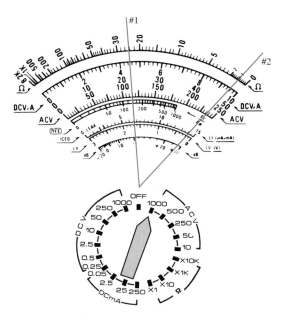

Figure 3-34. Multimeter for Test Your Knowledge problem 7.

8. Use the meter in figure 3-35.

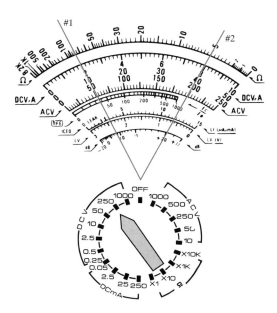

Figure 3-35. Multimeter for Test Your Knowledge problem 8.

9. Use the meter in figure 3-36.

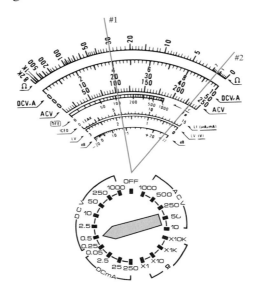

Figure 3-36. Multimeter for Test Your Knowledge problem 9.

For problems 10 and 11, answer the following:
    a.   range.
    b.   arc (ACV, DCV, Ω).
    c.   reading of needle #1.
    d.   reading of needle #2.

10. Use the meter in figure 3-37.

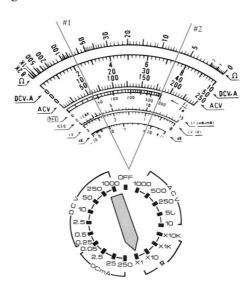

Figure 3-37. Multimeter for Test Your Knowledge problem 10.

11. Use the meter in figure 3-38.

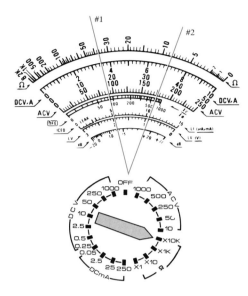

Figure 3-38. Multimeter for Test Your Knowledge problem 11.

For problems 12 through 15, answer the following:
   a.   Volts per division for the selected channel.
   b.   Number of divisions from ground to the trace. Be accurate to one decimal place.
   c.   Value of dc voltage being measured.

12. Use the oscilloscope in figure 3-39.

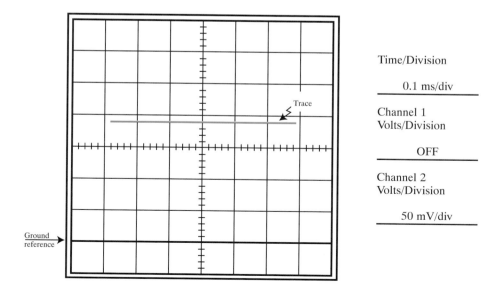

Time/Division

0.1 ms/div

Channel 1
Volts/Division

OFF

Channel 2
Volts/Division

50 mV/div

Figure 3-39. Oscilloscope for Test Your Knowledge problem 12.

13. Use the oscilloscope in figure 3-40.

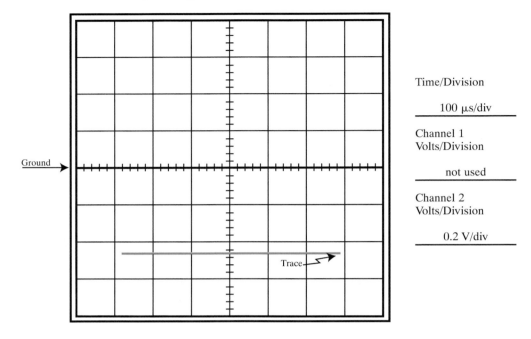

Figure 3-40. Oscilloscope for Test Your Knowledge problem 13.

14. Use the oscilloscope in figure 3-41.

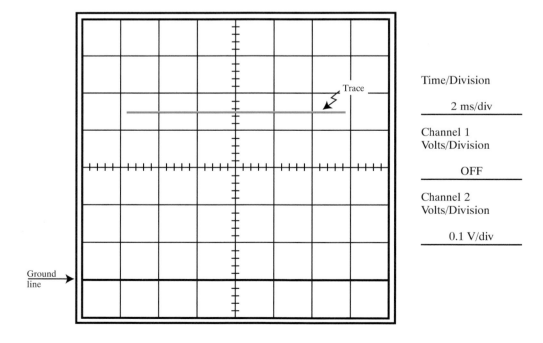

Figure 3-41. Oscilloscope for Test Your Knowledge problem 14.

15. Use the oscilloscope in figure 3-42.

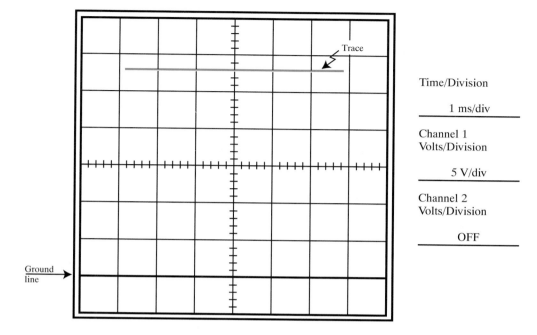

Time/Division

1 ms/div

Channel 1
Volts/Division

5 V/div

Channel 2
Volts/Division

OFF

Figure 3-42. Oscilloscope for Test Your Knowledge problem 15.

Laboratory activities are an important part of learning electronics theory.

Chapter 4
Voltage, Current, Resistance, and Power

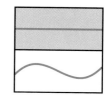

Upon completion of this chapter, you will be able to:
* Describe the role of the electron in electricity.
* Define the terms: voltage, current, resistance, and power.
* Identify the units of measure and their symbols for the four basic electrical terms.
* Use engineering notation as a mathematical tool.
* Analyze the Ohm's law and power formulas in terms of direct and inverse relationships.
* Solve problems using the Ohm's law and power formulas.

Voltage, current, resistance, and power are four important terms used to describe electrical circuits. People use these items from an early age, and often without understanding their meaning. This chapter establishes clear definitions that can be built upon in later training. Understanding basic electrical terminology involves definitions, units of measure, and mathematics.

4.1 STRUCTURE OF AN ATOM

To understand electricity, you must start at the atomic level. An **atom** is the smallest portion of an element that still maintains the properties of that element. The atom is made up of electrons, protons, and neutrons, as shown in figure 4-1.

Protons have a positive charge. **Neutrons** have a neutral charge. Their mass is approximately equal to that of the proton. The combination of the protons and neutrons form the nucleus of the atom.

Electrons have a negative charge. They orbit the nucleus in a manner similar to the planets of the solar system orbiting the sun. There is an equal number of protons and electrons for each atom. In terms of physical size, the proton has a mass 1845 times greater than an electron.

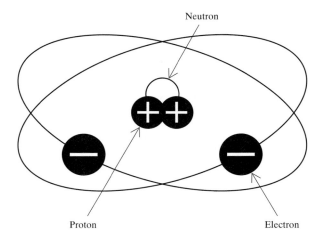

Figure 4-1. Negatively charged electrons orbit around the positively charged nucleus in this simple atom.

The large mass of the nucleus makes it very difficult to move. In comparison, the lightweight electrons are very mobile. Electron movement from one atom to another is the basic concept of electricity.

Like and Unlike Charges

The charged particles, protons and electrons, follow a law: *like charges repel and unlike charges attract.* Refer to figure 4-2.

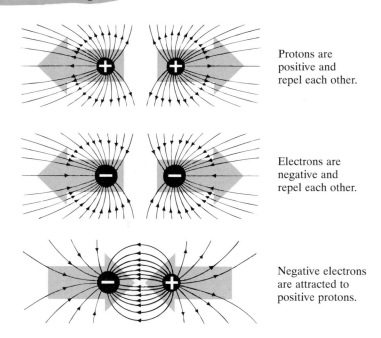

Protons are positive and repel each other.

Electrons are negative and repel each other.

Negative electrons are attracted to positive protons.

Figure 4-2. Protons and electrons follow a law. Like charges repel and unlike charges attract.

The protons, although they all have like charges, are bound together by the structure of the nucleus. As a result, the nucleus has the characteristic of a strong positive charge.

Electrons are held in orbit around the nucleus through two forces: the law of charges and centrifugal force. The electrons are held by the attraction to the unlike, positively charged protons. This force is canceled by centrifugal force, a mechanical outward force that tries to keep an object moving in a straight line. The combined forces allow the electrons to spin in stable concentric orbits as shown in figure 4-3.

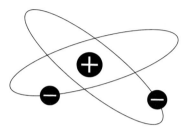

Figure 4-3. Electrons are held in concentric orbits.

Electron Shells

An **element** is the simplest form of a substance. In an element, all atoms have the same atomic number. The **atomic number** represents how many protons are in an atom. A stable atom has an equal number of electrons. Each element has a different atomic number.

An **electron shell** is the area in which electrons are likely to be found in orbit. Each shell has a specific number of electrons to make it full. Electrons fill the shells closest to the nucleus first, then move toward outer shells. Figure 4-4 shows the shells of three

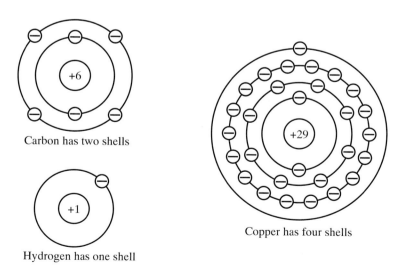

Carbon has two shells

Hydrogen has one shell

Copper has four shells

Figure 4-4. Different atoms have different numbers of electrons and different numbers of electron shells.

different atoms. Some elements have one, two, or three partially filled outer shells. These elements combine easily with other elements to form new compounds. A **compound** is created when two or more elements combine to form a new substance.

Electrons in the same shell of an atom follow different orbital paths. Figure 4-5 shows how this is possible.

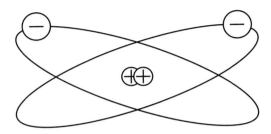

Figure 4-5. Electrons in the same shell follow different orbital paths.

4.2 CONDUCTORS AND INSULATORS

The atomic structure of a material determines how it responds to electricity. Energy is used to cause the electrons to flow from one atom to another. If very small amounts of energy are needed to cause movement, the substance is a conductor. If current does not flow, even with large amounts of energy, the material is an insulator.

Some materials fall between conductors and insulators. They allow current to flow, but with some resistance. These materials may be used as resistors or used to generate heat or light in an electrical circuit. Some materials, called semiconductors, have a wide range of resistances. Their characteristics vary with the current that flows through them.

A list of selected elements, showing atomic numbers and the number of electrons in each band, is shown in figure 4-6. A list of electrical applications is also provided for reference.

Figure 4-6. Selected elements commonly used in electrical applications.

| Atomic Number | Element | Electrons per Shell | | | | | | Application |
		1	2	3	4	5	6	
1	Hydrogen, H	1						Purifying metals
5	Boron, B	2	3					Doping of semiconductors
6	Carbon, C	2	4					Resistor
8	Oxygen, O	2	6					Burning
13	Aluminum, Al	2	8	3				Conductor
14	Silicon, Si	2	8	4				Semiconductor
17	Chlorine, Cl	2	8	7				Insulator
29	Copper, Cu	2	8	18	1			Conductor
32	Germanium, Ge	2	8	18	4			Semiconductor
33	Arsenic, As	2	8	18	5			Doping of semiconductors
47	Silver, Ag	2	8	18	18	1		Conductor
79	Gold, Au	2	8	18	32	18	1	Conductor

Conductors

The **valence shell** is the outermost shell of electrons on an atom. The valence shell contains the electrons available to move to another atom or to be shared with another atom. The valence shell contains the electrons used for forming new compounds and for electrical applications.

The copper atom, shown in figure 4-7, has only one electron in its valence shell. The single electron is not tightly bound to the atom. It takes very little energy to be moved out of its orbit to be used for current flow. When there is only one electron in the valence shell, it is very easy to move out. The electron becomes a free electron. A **free electron** is not attached to an atom. The electron can move to an nearby atom and knock its valence electron free. A chain reaction results.

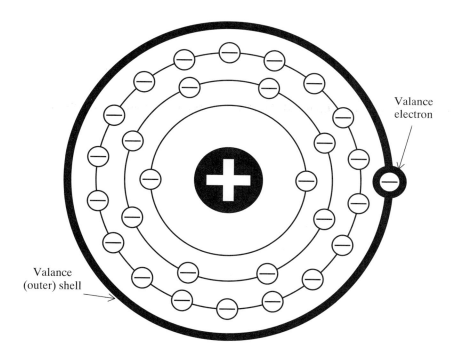

Figure 4-7. Copper has only one electron in its valence shell. This makes copper a good electrical conductor.

Copper, silver, and gold, three of the most commonly used conductors, are shown in figure 4-8. These metals allow easy flow of electricity. Silver is the best conductor of electricity based on the material's resistivity (to be discussed in detail in Ch. 5). The disadvantage of using silver is its high cost. Copper is less expensive, but it requires slightly more energy to release its electron. More energy has the effect of a higher resistance, the opposition to the flow of electricity. A higher energy requirement results in more heat buildup in the conductor.

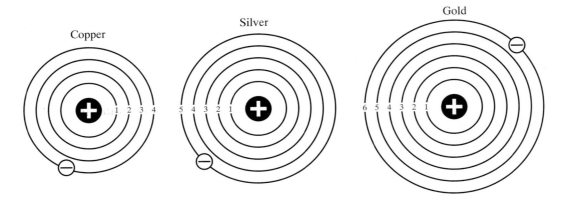

Figure 4-8. Atoms with one valence electron make good conductors. Silver has the highest conductance.

Insulators

An insulator has such high resistance that it has the effect of stopping electricity. Insulators have a full or nearly full valence shell, as shown in figure 4-9. The electrons are held tightly. A strong force is necessary to break any electrons free.

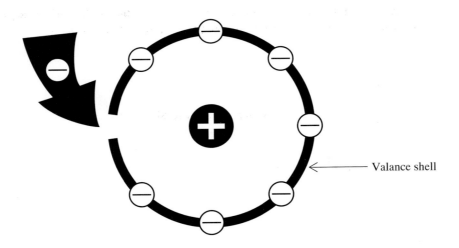

Figure 4-9. Atoms with full, or nearly full, valence shells make good insulators.

With a full valence shell, the free electrons from a nearby conductor cannot move into the area of the insulator. If the valence shell is partially vacant, free electrons from a conductor will become part of the orbit.

4.3 A BRIEF HISTORY

This brief history introduces five important individuals. Their research has earned them the honor of having electrical quantities named after them.

Charles A. Coulomb (1736–1806)—French physicist who worked with static charges. Coulomb made many discoveries and determined that a very large amount of electrons are needed for a useful measure of electric charge. The **coulomb (C)** is the unit of measure for electric charge. The symbol for total charge is Q (standing for quantity), is equated to a number of electrons.

$$1 \text{ coulomb (C)} = 6.25 \times 10^{18} \text{ electrons } (6{,}250{,}000{,}000{,}000{,}000{,}000)$$

Alessandro Volta (1745–1827)—Italian physicist who found that it takes one joule of work (0.7376 foot-pound) to move one coulomb of electrons between two points. The unit of measure for electrical potential is **voltage,** shortened to **volt.**

$$1 \text{ volt (V)} = 1 \text{ joule of work per coulomb of charge}$$

André M. Ampère (1775–1836)—French scientist who defined a measure for the rate of electrical current. A current flow of 6.25×10^{18} electrons past a given point in one second equals one **ampere (A).** The ampere is the basic unit of measure for current.

$$1 \text{ ampere (A)} = 1 \text{ coulomb of electrons per second}$$

Georg S. Ohm (1787–1854)—German physicist who found that with one ohm of opposition, one volt is needed for one ampere of current to flow. This became known as Ohm's law, discussed in detail later in this chapter. The **ohm** is the unit of measure for electrical resistance. The symbol for ohm is Ω, the Greek letter omega.

James Watt (1736–1819)—Scottish inventor whose work with steam engines found that electricity could be used to power machinery to perform work successfully. The name for electrical work performed is **power (P).** The unit of measure for power is **watt (W).** When one volt moves one ampere, the work performed is one watt. Compared to mechanical work, one horsepower is equal to 746 watts.

$$1 \text{ watt (W)} = 1 \text{ volt (V)} \times 1 \text{ amp (A)}$$

$$746 \text{ watts (W)} = 1 \text{ horsepower (hp)}$$

4.4 DEFINITIONS OF BASIC ELECTRICAL TERMS

Each of the basic electrical terms are described with four items. The four items are: the term, the symbol for the term, the unit of measure, and the symbol for the unit of measure. In addition, the definitions are summarized in the table in figure 4-10.

Voltage

Voltage (V or E) is the driving force that causes electricity to flow through a conductor. The negative side of a voltage source has a supply of electrons. The positive side of the voltage source attracts electrons.

Figure 4-10. Summary of electrical terminology.

Electrical Term	Unit of Measure	Definition
voltage (V or E)	volt (V)	Electrical driving force.
current (I)	ampere (A)	Flow of electrons.
resistance (R)	ohm (Ω)	Opposition to current.
power (P)	watt (W)	Electrical work performed.

Without voltage, there can be no work performed by electricity. The unit of measure is of voltage is the volt (V). Another name for voltage is **electromotive force (emf).** This name arises since voltage is the electrical pressure applied to a circuit.

The **potential difference** is the voltage measured between two points in a circuit. When current flows through a resistor, or any component that uses power, there will be a difference in potential between the two sides of the component. Potential difference is measured in volts.

Voltage has the *potential* to do work and is referred to as potential energy. Voltage can be present in a circuit even when there is no current. An example of this is seen in the wall sockets of a house. Voltage is present, with or without an electrical appliance plugged in.

Current

Current (I) is the flow of electrons through a conductor. The unit of measure is the ampere (A), abbreviated amp. Current travels at the speed of light. There are several other features that describe current.

- A voltage must be present in the circuit.
- A complete path must be available so that the current can flow from the negative terminal to the positive terminal.
- The amount of current is restricted by the resistance in the circuit.
- Current must be present for work (power) to be performed.

Electrons enter a conductor, which provides a path for the electrons to flow. The entering electrons are free to enter an atom, which bumps out electrons that are already in orbit. The displaced electrons bump into still more electrons. This cascading effect travels down the length of wire.

Figure 4-11 shows a pipe filled with balls to represent the effects of and electrical current. A push on the first ball immediately forces the last ball out of the pipe. However, it takes a period of time for the first ball to reach the end.

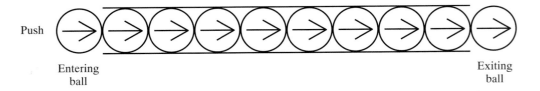

Push

Entering
ball

Exiting
ball

Figure 4-11. The balls in a pipe represent electrons flowing through a conductor.

Resistance

Resistance (R) is the opposition to the flow of electrical current. Its unit of measure is **ohm (Ω)**. *Resistance does not slow down current. Resistance restricts the volume of current traveling in a circuit.* Larger resistance results in less current, and smaller resistance results in more current. Resistance is a general term used to describe any electrical load. The word **resistor** is used for an electronic component.

Work performed by electricity takes place in the load. When current flows through a load, it causes a difference in potential (voltage), also called voltage drop, across the resistance. The voltage drop multiplied by the current through the load is power.

Power

Power (P) is work performed. The unit of measure for electrical power is the **watt (W)**. Power is the result of converting energy from one form to another. Electrical power is the conversion of electron flow to: heat, magnetism, electrical fields, semiconductor operation, and other applications.

Many electrical appliances are rated in watts to indicate their ability to produce heat or light from a hot filament. Examples of these appliances are toasters, hair dryers, and light bulbs. It should be noticed, however, that in order to perform at the rated power, the device must also operate at a specified voltage.

Direct Current (DC) vs Alternating Current (AC)

Sources of electricity fall into two very broad categories, dc and ac. This is the primary focus of this book, DC/AC FOUNDATIONS OF ELECTRONICS.

In a **direct current (dc)** circuit, the current is always in the same direction. The voltage and current can fluctuate up and down, but the polarity of the voltage remains the same and current flows in the same direction. **Polarity** is the positive and negative relationships of the voltage source. Examples of dc power sources are: batteries, solar panels, and automobile alternators.

In an **alternating current (ac)** circuit, the current changes direction due to the voltage changing polarity. Usually the voltage changes polarity at a fixed rate. The voltage is positive for a period of time, and then it switches to negative for a period of time, then back to positive to repeat the cycle. Examples of ac voltage sources are: household electricity, most forms of radio signals, and standard magnetic tape recordings.

4.5 SIGNIFICANT FIGURES

Performing arithmetic operations frequently produces a number with many digits. **Significant figures** include:
- All non-zero digits.
- Zeros used as place holders between two non-zero numbers.
- Zeros added to the right of a number that are not needed as place holders.
  (Example: 1.200 has four significant figures.)

Significant figures determine the accuracy of the number.
Examples:

356.8903 has seven significant figures

200.5 has four significant figures

70 has one significant figure

0.0213 has three significant figures

0.0003 has one significant figure

0.00030 has two significant figures

The question arises, how many significant figures are necessary? An electronic calculator can produce eight or more digits. Meters used to take measurements are usually accurate to three digits. For most practical purposes, three significant figures is considered enough.

4.6 ROUNDING NUMBERS

When a number has more significant figures than necessary, the number needs to be rounded. **Rounding** removes the unnecessary figures. Zeros are used as place holders to maintain the value of the significant figures.

To round a number, count (from the left) the number of significant places to be kept. Look at the next digit to the right:

- If the next digit is 5 or more, drop that digit and all others to the right. The last place kept is rounded up one.

Examples, round to 3 significant figures:

586.93 = 587

2,048,902 = 2,050,000

0.09038 = 0.0904

- If the next digit to the right is less than 5, drop that digit and all others to the right. The last place kept remains unchanged.

Examples, round to 3 significant figures:

10,240 = 10,200

2.091 = 2.09

0.003502 = 0.00350

4.7 OHM'S LAW

Ohm's law is a mathematical formula giving the relationships between voltage, current, and resistance. This section presents Ohm's law and variations created by rearranging the formula. Given any two of the variables (*V*, *I*, or *R*), Ohm's low solves for the third. Ohm's law is typically stated as:

Voltage = Current × Resistance

Find Voltage When Given Current and Resistance

Calculate voltage when given current and resistance. The basic formula states that if there is an increase in either current or resistance, there must be a proportionate increase in voltage. This is a **direct relationship,** when one side of an equation increases, the other side increases.

Formula 4.A

$$V = I \times R$$

Find: voltage

Given: current and resistance

Sample problem 1.

Find the voltage if there is current of 2 amps and resistance of 100 ohms. Refer to figure 4-12.

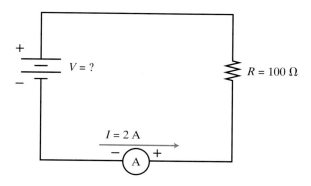

Figure 4-12. Given resistance and current, find the voltage. (Sample problem 1)

Formula: $V = I \times R$

Substitution: $V = 2\,A \times 100\,\Omega$

Answer: $V = 200$ volts

Sample problem 2. _____

How much voltage is needed to produce 0.5 amps of current through a 40 ohm load?

Formula: $V = I \times R$

Substitution: $V = 0.5 \text{ A} \times 40 \text{ }\Omega$

Answer: $V = 20$ volts

Find Current When Given Voltage and Resistance

It is easy to calculate current when given voltage and resistance. The formula comes from rearranging formula 4.A. Current is shown to be directly related to voltage. Current will increase with an increase in voltage.

An **inverse relationship** means that one side of the equation decreases when the other side of the equation increases. Current is inversely related to resistance. When there is an increase in resistance, there will be a decrease in current. A decrease in resistance results in an increase in current.

Formula 4.B

$$I = \frac{V}{R}$$

Find: current

Given: voltage and resistance

Sample problem 3. _____

With a voltage of 20 volts and a load of 5 ohms, calculate current flow. Refer to figure 4-13.

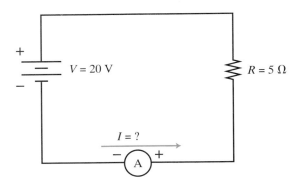

Figure 4-13. Given voltage and resistance, find the current. (Sample problem 3)

Formula: $I = \frac{V}{R}$

Substitution: $I = \dfrac{20 \text{ V}}{5 \text{ }\Omega}$

Answer: $I = 4$ A

Sample problem 4. _____

Determine current in a circuit with a 20 volt battery and 10 ohms of resistance.

Formula: $I = \dfrac{V}{R}$

Substitution: $I = \dfrac{20 \text{ V}}{10 \text{ }\Omega}$

Answer: $I = 2$ A

Find Resistance When Given Voltage and Current

Resistance can be calculated when given voltage and current. This formula is also derived from rearranging formula 4.A. The formula states that resistance is directly related to voltage and inversely related to current. With an increase in voltage, an increase in resistance is required to maintain a constant current. With an increase in current, there must be a decrease in resistance.

Formula 4.C

$$R = \frac{V}{I}$$

Find: resistance

Given: voltage and current

Sample problem 5. _____

In a circuit with 50 volts, how much resistance must be connected to allow a current flow of 5 amps? Refer to figure 4-14.

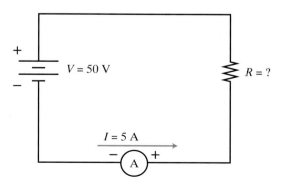

Figure 4-14. Given voltage and current, find the resistance. (Sample problem 5)

Formula: $R = \dfrac{V}{I}$

Substitution: $R = \dfrac{50 \text{ V}}{5 \text{ A}}$

Answer: $R = 10 \ \Omega$

Sample problem 6.

Current of 0.1 amps is measured in a circuit containing a power supply of 50 volts. How much resistance is in this circuit?

Formula: $R = \dfrac{V}{I}$

Substitution: $R = \dfrac{50 \text{ V}}{0.1 \text{ A}}$

Answer: $R = 500 \ \Omega$

4.8 POWER FORMULAS

As discussed earlier in the chapter, power is the name for electrical work performed. Power is measured in watts. The power formula expresses the relationship of power with voltage and current. The formula is commonly written $P = I \times V$.

Find Power When Given Current and Voltage

Power is directly related to current and voltage. If either current or voltage is increased, it results in an increase in power.

Formula 4.D

$$P = I \times V$$

Find: power

Given: current and voltage

Sample problem 7. _____

Find the power consumption in a circuit with 120 volts applied and a measured current of 2 amps. Refer to figure 4-15.

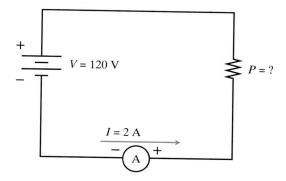

Figure 4-15. Given voltage and current, find the power. (Sample problem 7)

Formula: $P = I \times V$

Substitution: $P = 2\,A \times 120\,V$

Answer: $P = 240\,W$

Sample problem 8. _____

What power can be expected to be dissipated in an electric motor with a voltage rating of 24 volts and a current draw of 0.2 amps?

Formula: $P = I \times V$

Substitution: $P = 0.2\,A \times 24\,V$

Answer: $P = 4.8\,W$

Find Current When Given Power and Voltage

The power formula can be rearranged to find the current in a circuit. It is shown here that current is directly related to power and inversely related to voltage.

Formula 4.E

$$I = \frac{P}{V}$$

Find: current

Given: power and resistance

Sample problem 9. _____

How much current is drawn through a 40 watt light bulb when connected to a 100 volt source? Refer to figure 4-16.

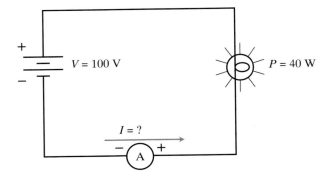

Figure 4-16. Given voltage and power, find the current. (Sample problem 9)

Formula: $I = \dfrac{P}{V}$

Substitution: $I = \dfrac{40 \text{ W}}{100 \text{ V}}$

Answer: $I = 0.4 \text{ A}$

Sample problem 10. _____

A light bulb used in an automobile is rated for 48 watts at 12 volts. How much current must be supplied to this bulb?

Formula: $I = \dfrac{P}{V}$

Substitution: $I = \dfrac{48 \text{ W}}{12 \text{ V}}$

Answer: $I = 4 \text{ A}$

Find Voltage When Given Power and Current

The power formula can also be rearranged to find the voltage. The formula states that voltage is directly related to power and inversely related to current. If there is an increase in power with no change in current, it is from an increase in voltage. If there is an increase in current, with no change in power, it is from a decrease in voltage.

Formula 4.F

$$V = \frac{P}{I}$$

Find: voltage

Given: power and current

Sample problem 11.

Determine how much voltage is needed to provide a current flow of 3 amps when the power dissipated is 12 watts. Refer to figure 4-17.

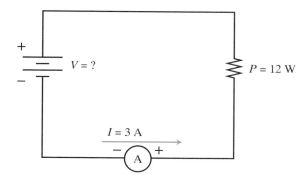

Figure 4-17. Given power and current, find the voltage. (Sample problem 11)

Formula: $V = \dfrac{P}{I}$

Substitution: $V = \dfrac{12 \text{ W}}{3 \text{ A}}$

Answer: $V = 4 \text{ V}$

Sample problem 12.

A circuit is measured to have 0.6 amps of current and 12 watts of power. How much voltage is applied to this circuit?

Formula: $V = \dfrac{P}{I}$

Substitution: $V = \dfrac{12 \text{ W}}{0.6 \text{ A}}$

Answer: $V = 20 \text{ V}$

4.9 COMBINING OHM'S LAW WITH THE POWER FORMULAS

The three forms of Ohm's law can be combined with the three basic power formulas. The result is twelve formulas that give the mathematical relationships between voltage, current, resistance, and power. The formula wheel, shown in figure 4-18, is used by selecting the quantity to be found in the center of the wheel. The formulas around the outside are selected depending on what values are known.

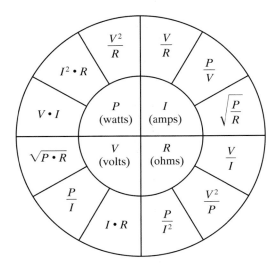

Figure 4-18. Ohm's law and power formula wheel provides a useful tool for calculating unknowns.

Find Power When Given Current and Resistance

This formula is a combination of:

$$V = I \times R \text{ and } P = I \times V$$

Substitute for V in the second equation.

$$P = I \times (I \times R) = I^2 \times R$$

This shows that power is directly related to both current and resistance. If the resistance is raised while the current stays the same, or if the current increases while the resistance stays the same, there is an increase in power. With both resistance and current directly related to voltage, there would also have to be an increase in voltage to allow an increase in either R or I.

Formula 4.G

$$P = I^2 \times R$$

Find: power

Given: current and resistance

Sample problem 13. _____

With 3 amps of current flowing through a 10 ohm resistor, how much power will be consumed? Refer to figure 4-19.

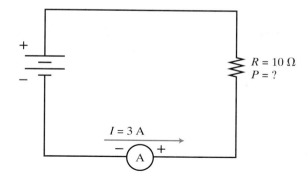

Figure 4-19. Given resistance and current, find the power. (Sample problem 13)

Formula: $P = I^2 \times R$

Substitution: $P = (3 \text{ A})^2 \times 10 \ \Omega$

Answer: $P = 90 \text{ W}$

Sample problem 14. _____

The resistance of a light bulb is measured to be 20 ohms. With this bulb in a certain circuit, current is measured to be 0.2 amps. How much power is in this circuit?

Formula: $P = I^2 \times R$

Substitution: $P = (0.2 \text{ A})^2 \times 20 \ \Omega$

Answer: $P = 0.8 \text{ W}$

Find Resistance When Given Power and Current

The formula for finding resistance results from rearranging formula 4.G. This formula shows a direct relationship between power and resistance. There is an inverse relationship between current and resistance.

Formula 4.H

$$R = \frac{P}{I^2}$$

Find: resistance

Given: power and current

Sample problem 15. _____

What is the resistance of a 1400 watt toaster if it draws 10 amps? Refer to figure 4-20.

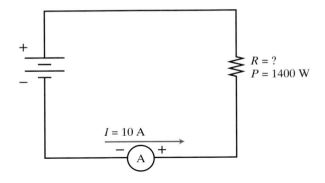

Figure 4-20. Given power and current, find the resistance.
(Sample problem 15)

Formula: $R = \dfrac{P}{I^2}$

Substitution: $R = \dfrac{1400\ W}{(10\ A)^2}$

Answer: $R = 14\ \Omega$

Sample problem 16. _____

An electric motor dissipates 800 watts while drawing 2 amps. What is the motor's resistance?

Formula: $R = \dfrac{P}{I^2}$

Substitution: $R = \dfrac{800\ W}{(2\ A)^2}$

Answer: $R = 200\ \Omega$

Find Current When Given Power and Resistance

This formula is also created by rearranging formula 4.G. The square root on the right side is necessary as I was squared in the starting equation. Current is inversely related to resistance and directly related to power.

Formula 4.I

$$I = \sqrt{\frac{P}{R}}$$

Find: current

Given: power and resistance

Sample problem 17.

A 100 ohm resistor is dissipating 1600 watts. What is the current flow? Refer to figure 4-21.

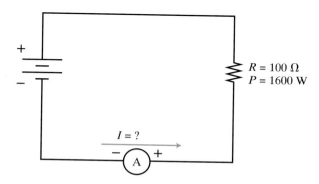

Figure 4-21. Given resistance and power, find the current. (Sample problem 17)

Formula: $I = \sqrt{\dfrac{P}{R}}$

Substitution: $I = \sqrt{\dfrac{1600 \text{ W}}{100 \text{ }\Omega}}$

Answer: $I = 4 \text{ A}$

Sample problem 18.

How much current is being drawn if a light bulb with 10 ohms is consuming 40 watts?

Formula: $I = \sqrt{\dfrac{P}{R}}$

Substitution: $I = \sqrt{\dfrac{40 \text{ W}}{10 \text{ }\Omega}}$

Answer: $I = 2 \text{ A}$

Find Power When Given Voltage and Resistance

The formula to find power uses a combination of these two formulas:

$$P = I \times V \text{ and } I = \frac{V}{R}$$

$\dfrac{V}{R}$ is substituted in place of I.

$$P = \frac{V}{R} \times V = \frac{V^2}{R}$$

Power is directly related to voltage and inversely related to resistance. If voltage stays constant with an increase in resistance, there is a decrease in power. With the resistance staying constant, an increase in voltage produces an increase in power.

Formula 4.J

$$P = \frac{V^2}{R}$$

Find: power

Given: voltage and resistance

Sample problem 19.

What is the power dissipated when 20 volts is applied to a 100 ohm resistor? Refer to figure 4-22.

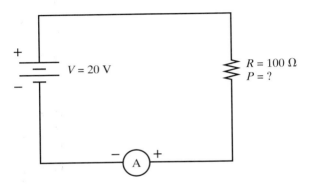

Figure 4-22. Given voltage and resistance, find the power.
(Sample problem 19)

Formula: $P = \dfrac{V^2}{R}$

Substitution: $P = \dfrac{(20 \text{ V})^2}{100 \text{ } \Omega}$

Answer: $P = 4 \text{ W}$

Sample problem 20.

When a 50 ohm resistor has 40 volts applied, what is the power consumed?

Formula: $P = \dfrac{V^2}{R}$

Substitution: $P = \dfrac{(40 \text{ V})^2}{50 \text{ } \Omega}$

Answer: $P = 32 \text{ W}$

Find Resistance When Given Voltage and Power

The formula to calculate the resistance is created by rearranging formula 4.J. Resistance is inversely related to power and directly related to voltage. If voltage is held constant, a decrease in resistance must result if there is to be an increase in power. If the power remains constant, there needs to be an increase in resistance when there is an increase in voltage.

Formula 4.K

$$R = \dfrac{V^2}{P}$$

Find: resistance

Given: voltage and power

Sample problem 21. _____

Calculate the value of resistance needed in a circuit with 10 volts applied and a dissipated power of 50 watts. Refer to figure 4-23.

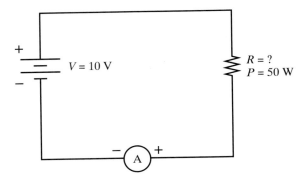

**Figure 4-23. Given voltage and power, find the resistance.
(Sample problem 21)**

Formula: $R = \dfrac{V^2}{P}$

Substitution: $R = \dfrac{(10\ V)^2}{50\ W}$

Answer: $R = 2\ \Omega$

_____

Sample problem 22. _____

What value of resistor should be used if a circuit is to consume 50 watts when 20 volts is applied?

Formula: $R = \dfrac{V^2}{P}$

Substitution: $R = \dfrac{(20\ V)^2}{50\ W}$

Answer: $R = 8\ \Omega$

_____

Find Voltage When Given Power and Resistance

The formula to find the voltage results from rearranging formula 4.J. The square root on the right side is required as V was squared in the starting formula. Voltage is directly related to both power and resistance. In order to have an increase in power or resistance while the other quantity remains constant, there must be an increase in voltage.

Formula 4.L

$$V = \sqrt{P \times R}$$

Find: voltage

Given: power and resistance

Sample problem 23.

How much voltage is required to dissipate 0.3 watts in a 30 ohm load? Refer to figure 4-24.

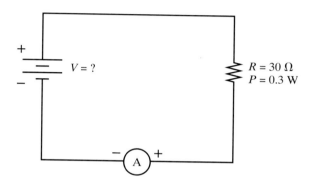

Figure 4-24. Given power and resistance, find the voltage. (Sample problem 23)

Formula:	$V = \sqrt{P \times R}$
Substitution:	$V = \sqrt{0.3\ \text{W} \times 30\ \Omega}$
Answer:	$V = 3\ \text{V}$

Sample problem 24.

What is the applied voltage in a circuit with 32 ohms dissipating 2 watts?

Formula:	$V = \sqrt{P \times R}$
Substitution:	$V = \sqrt{2\ \text{W} \times 32\ \Omega}$
Answer:	$V = 8\ \text{V}$

4.10 COST OF ELECTRICITY: KILOWATT-HOUR

The local electrical power company charges their customers for the use of electricity by the kilowatt-hour. One **kilowatt-hour (kWh)** is equivalent to using 1000 watts in a one hour time period. The electric company charges a few cents per kWh plus service charges. To estimate the cost of using an electric appliance, multiply the power consumed, in kilowatts, by the time it is operated. This gives the number of kWh used. Then, multiply the number of kWh by the cost per kWh and add service charges.

Sample problem 25. _____

Calculate the cost to operate a 1400 watt hair dryer for 30 minutes if the electric company charges 10 cents per kWh.

First convert given values to proper units.

$$1400 \text{ watts} = 1.4 \text{ kW}$$

$$30 \text{ minutes} = 0.5 \text{ hour}$$

Multiply the kW by hours and then by cents per kWh.

$$1.4 \text{ kW} \times 0.5 \text{ H} \times \$0.10 = \$0.07$$

Answer: 7 cents

4.11 ENGINEERING NOTATION AS AN ARITHMETIC TOOL

Engineering notation was introduced in Chapter 3 as a tool for handling very large and very small numbers. It also makes arithmetic easier by moving the decimal point and eliminating zeros as place holders. Scientific calculators have special keys to enter powers of 10. With the exponent entered, the calculator will perform the arithmetic and make adjustments to the decimal and power of 10 as needed.

Addition and Subtraction with Engineering Notation

Electronic circuits, such as those presented in Chapter 6, *Series Circuits*, use formulas with addition and subtraction. The rules for addition and subtraction are:
- Align the decimals in a column prior to performing addition or subtraction.
- Add/subtract in columns, each column having the same place value.
- Powers of 10 must be the same.
- Answer should be in best form of engineering notation.

The sample problems that follow, with engineering notation, show different changes for the decimal point and multiplier. All of these are acceptable.

Sample problem 26. _____

Add these numbers without powers of 10.

$$50.1 + 2.25 + 1000.403 =$$

Align decimals:

$$
\begin{array}{r}
50.1 \\
2.25 \\
+\ 1000.403 \\
\hline
1052.753
\end{array}
$$

Sample problem 27. _____

Add: 1.5 kilohms + 24.3 megohms

Option a. Change to kilohms:

$$1.5 \text{ k}\Omega \text{ (no change)}$$

$$24.3 \text{ M}\Omega = 24300. \text{ k}\Omega$$

Align the decimals:

$$
\begin{array}{r}
1.5 \text{ k}\Omega \\
+\ 24300. \text{ k}\Omega \\
\hline
24301.5 \text{ k}\Omega = 24.3015 \text{ M}\Omega
\end{array}
$$

Option b. Change to megohms:

$$1.5 \text{ k}\Omega = 0.0015 \text{ M}\Omega$$

$$24.3 \text{ M}\Omega \text{ (no change)}$$

Align the decimals:

$$
\begin{array}{r}
0.0015 \text{ M}\Omega \\
+\ 24.3 \text{ M}\Omega \\
\hline
24.3015 \text{ M}\Omega
\end{array}
$$

Option c. Change to basic units:

$$1.5 \text{ k}\Omega = 1,500. \ \Omega$$

$$24.3 \text{ M}\Omega = 24,300,000. \ \Omega$$

Align the decimals:

$$
\begin{array}{r}
1,500. \ \Omega \\
+\ 24,300,000. \ \Omega \\
\hline
24,301,500. \ \Omega = 24.3015 \text{ M}\Omega
\end{array}
$$

Sample problem 28. _____

Subtract: 45.9 millivolts – 601 microvolts

Option a. Change to millivolts:

$$45.9 \text{ mV (no change)}$$

$$601 \ \mu\text{V} = 0.601 \text{ mV}$$

Align the decimals:

$$
\begin{array}{r}
45.900 \text{ mV} \\
-\ 0.601 \text{ mV} \\
\hline
45.299 \text{ mV}
\end{array}
$$

Option b. Change to microvolts:

$$45.9 \text{ mV} = 45900. \ \mu\text{V}$$

$$601. \ \mu\text{V (no change)}$$

Align the decimals:

$$
\begin{array}{r}
45900. \ \mu\text{V} \\
-\ 601. \ \mu\text{V} \\
\hline
45299 \ \mu\text{V} = 45.299 \text{ mV}
\end{array}
$$

Option c. Change to basic units:

$$45.9 \text{ mV} = 0.0459 \text{ V}$$

$$601 \ \mu\text{V} = 0.000601 \text{ V}$$

Align the decimals:

$$
\begin{array}{r}
0.045900 \text{ V} \\
-\ 0.000601 \text{ V} \\
\hline
0.045299 \text{ V} = 45.299 \text{ mV}
\end{array}
$$

Multiplication and Division with Engineering Notation

The Ohm's law and power formulas are good examples of the use of multiplication and division in electronics. To make multiplication with engineering notation easier, replace the multiplier symbol with its E form to show the power of 10. The rules for multiplication are:

• It is not necessary to align the decimal points.
• Multiply the numbers, ignoring decimal points.
• To find the decimal place for the answer, add the decimal places of the numbers multiplied.
• If engineering notation is used, *add* the exponents.
• Answer should be in the best form of engineering notation.

Sample problem 29. _____

Multiply these numbers without powers of 10.

2.51 × 0.003 =

$$\begin{array}{r} 2.51 \text{ (2 decimal places)} \\ \times\ 0.003 \text{ (3 decimal places)} \\ \hline 0.00753 \text{ (5 decimal places)} \end{array}$$

Now, multiply the number with powers of 10.

2.51 E00 x 3.0 E–03 =

$$\begin{array}{r} 2.51 \text{ E00} \\ \times\ 3.0 \text{ E–03} \\ \hline 7.53 \text{ E–03} \end{array}$$

These sample problems use formulas from earlier in this chapter.

Sample problem 30. _____

Use formula: $V = I \times R$

I = 2.01 amps

R = 1.5 kilohms

$$\begin{array}{r} 2.01 \text{ E00 amps} \\ \times\ 1.5 \text{ E03 ohms} \\ \hline 3.015 \text{ E03 volts} \end{array}$$

V = 3.015 kV

Sample problem 31. _____

Use formula: $P = I \times V$

I = 3.6 milliamps

V = 150 volts

$$\begin{array}{r} 3.6 \text{ E–03 amps} \\ \times\ 150 \text{ E00 volts} \\ \hline 540 \text{ E–03 watts} \end{array}$$

P = 540 mW

The rules for division are:
- It is not necessary to align the decimal points.
- Divide the numbers, ignoring decimal points.
- Each of the following forms are read "A divided by B."

$$A \div B \qquad B\overline{)A} \qquad \frac{A}{B}$$

- If engineering notation is used, *subtract* the exponents.
- Subtract the engineering exponents, $A - B$.
- Answer should be in the best form of engineering notation.

Sample problem 32. _____

Formula: $R = \dfrac{V}{I}$

$V = 10$ volts

$I = 0.5$ milliamps

$$10 \text{ volts} \div 0.5 \text{ milliamps}$$

$$10 \text{ E00 V} \div 0.5 \text{ E–03 A} = 20 \text{ E03 } \Omega$$

$$R = 20 \text{ k}\Omega$$

Sample problem 33. _____

Formula: $I = \dfrac{V}{R}$

$V = 30$ volts

$R = 1.5$ kilohms

$$1.5 \text{ k}\Omega \overline{)30 \text{ volts}}$$

$$\begin{array}{r} 0.02 \text{ A} = 20 \text{ E–03 A} \\ 1500 \text{ } \Omega \overline{)30.00 \text{ V}} \end{array}$$

$$I = 20 \text{ mA}$$

Sample problem 34. _____

Formula: $P = \dfrac{V^2}{R}$

$V = 2$ kilovolts

$R = 8$ kilohms

$$\frac{2\text{E03} \times 2\text{E03}}{8 \text{ E03}} = \frac{4 \text{ E06}}{8 \text{ E03}} = 0.5 \text{ E03 watts}$$

$$P = 0.5 \text{ kW} = 500 \text{ W}$$

SUMMARY

- An atom contains protons, electrons, and neutrons.
- Like charges repel and unlike charges attract.
- Electrons orbit the nucleus in shells.
- Single electrons in the valence band are used as current carriers in a conductor.
- Electrons in insulators are tightly bound. Insulators have full valence bands.
- Voltage is the driving force of electricity. Other names for voltage include: electromotive force, potential difference, and voltage drop.
- Current is the flow of electrons.
- Resistance is the opposition to current flow.
- Power is electrical work performed.
- Generally speaking, numbers should be rounded to three significant figures.
- Ohm's law and the power formulas can solve for V, I, R, or P when any two of the values are given.
- Engineering notation is used to simplify arithmetic operations.

KEY WORDS AND TERMS GLOSSARY

alternating current (ac): Current in an electric circuit that periodically changes direction due to the voltage changing polarity. Examples of ac voltage sources include: household electricity, all forms of radio signals, magnetic tape recordings, sound converted to an electronic signal.

ampere (A): Unit measure of current.

atom: The smallest portion of an element that still maintains the properties of the element. An atom is made up of electrons, protons, and neutrons.

coulomb (C): Unit measure of electrical charge.

current: The flow of electricity through a conductor. The symbol for current is *I*. The unit of measure is the ampere (A).

direct current (dc): Current that is always in the same direction. The voltage and current can fluctuate up and down, but the polarity of the voltage remains the same.

direct relationship: When one quantity changes, it causes a change in another quantity in the same direction. An increase causes an increase and a decrease causes a decrease.

electromotive force (emf): Electrical pressure applied to a circuit. It is another name for voltage and has the same unit of measure, the volt.

electron: A particle of an atom having a negative charge. Electrons orbit the nucleus in a manner similar to that of planets orbiting the sun.

electron shell: The positions the electrons hold in orbit around the nucleus. Electrons fill the shells closest to the nucleus first, then they fill the outer shells.

element: The smallest particle of a substance containing atoms with the same atomic number.

free electron: An electron that moves easily between atoms to provide current.

inverse relationship: When one quantity increases, the related quantity decreases. Also, when one quantity decreases, the related quantity increases.

kilowatt-hour (kWh): The equivalent of using 1000 watts in a one hour time period.

neutron: A particle in an atom with neutral charge and a mass approximately equal to that of the proton.

ohm (Ω): Unit measure of resistance.

polarity: The positive and negative relationships of a voltage.

potential difference: The voltage measured between two points in a circuit. When current flows through a resistor, or any component that uses power, there is a difference in potential (voltage) from one side to the other. Unit of measure of potential difference is the volt.

power (*P*): Electrical work performed. Unit of measure is the watt.

proton: The part of an atom that has a positive charge. There are an equal number of protons as electrons for each atom. In terms of physical size, it is 1845 times larger than an electron.

resistance (*R*): The opposition to the flow of electrical current. Unit of measure is the ohm. It does not slow down the current, which flows at the speed of light. Instead, resistance restricts the *volume* of current flowing in a circuit. Larger resistance results in less current and smaller resistance results in more current.

resistor: Electrical component used to oppose the flow of electricity.

rounding: Changing a number to its approximate equivalent. Zeros are used as place holders to maintain the value of the significant figures.

significant figures: All non-zero digits and zeros used as place holders between two non-zero numbers. Significant figures determine the accuracy of the number.

valence shell: The outer-most orbit of an atom. Contains the electrons used in electrical current.

volt (V): Unit measure of voltage.

voltage (*V*): 1. The driving force that causes electricity to flow through a conductor. Without voltage, there can be no work performed by the electricity. Unit of measure is the volt. 2. Potential energy of the electric source. Also called electromotive force (emf).

watt (*W*): Unit measure of power.

KEY FORMULAS

Formula 4.A:

$$V = I \times R$$

Formula 4.B

$$I = \frac{V}{R}$$

Formula 4.C

$$R = \frac{V}{I}$$

Formula 4.D

$$P = I \times V$$

Formula 4.E

$$I = \frac{P}{V}$$

Formula 4.F

$$V = \frac{P}{I}$$

Formula 4.G

$$P = I^2 \times R$$

Formula 4.H

$$R = \frac{P}{I^2}$$

Formula 4.I

$$I = \sqrt{\frac{P}{R}}$$

Formula 4.J

$$P = \frac{V^2}{R}$$

Formula 4.K

$$R = \frac{V^2}{P}$$

Formula 4.L

$$V = \sqrt{P \times R}$$

TEST YOUR KNOWLEDGE

Do not write in this text. Please use a separate sheet of paper.
1. What two things are required to achieve a current in a circuit?
2. Round the following numbers to three significant figures.
    a. 27,030
    b. 5.715
    c. 0.03419802
    d. 0.0036226
    e. 100,219.3
3. How much voltage is needed to produce 5 amps of current through a resistive load of 20 ohms?
4. Calculate the current flow in a circuit with a 9 volt battery when connected to a lamp having a resistance of 300 ohms.
5. What is the resistance in a circuit when 1 amp flows from a 120 volt source?
6. An ammeter connected in a circuit powered by a 12 volt battery measures 0.2 amps. How much power is produced from the battery?
7. How much current flows through a 20 watt light bulb when it is connected to a 40 volt supply?
8. What is the voltage applied to a circuit with 0.3 amps of current in a 60 watt load?
9. Calculate the power consumed in a 10 ohm load with a current flow of 3 amps.
10. Find the resistance of a 10 watt load if it is drawing 5 amps.
11. How much current will flow through an 8 watt load with a resistance of 2 ohms?
12. If a 100 ohm resistor is connected to a 20 volt source, what is the power dissipated?
13. Determine the resistance of a 60 watt light bulb with 120 volts.
14. How much voltage must be applied to a 40 ohm load if it is to produce 10 watts?
15. How much will it cost to operate five 100 watt lights for 6 hours with a cost of 10 cents per kWh?
16. Add these three resistors connected in series to find their combined resistance: 4.7 kilohm, 23.1 kilohm, 0.5 megohm.
17. If a 90.8 millivolt microphone signal is reduced by 250 microvolts, how much voltage remains?
18. What is the voltage applied to a circuit if the load is 6 kilohms with a 3 milliamp current?
19. Calculate the current flow in a 100 volt circuit with a load of 10 kilohms.
20. If a 90 megohm resistor is connected across a 3 kilovolt supply, how much power will be dissipated?

Chapter 5
Circuit Components

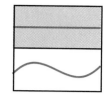

Upon completion of this chapter, you will be able to:
- Identify the characteristics of conductors.
- Describe how to select wire for an electrical conductor.
- Explain the ratings of fuses and circuit breakers.
- Identify the style and ratings of switches.
- Determine the value of resistors using the color code.
- Describe the ratings of variable resistors.

This chapter describes most of the dc circuits that you will calculate and build during the laboratory exercises of the next few chapters. The components that are used in these electrical circuits have ratings that describe their performance or the point at which they break down. When it is necessary to design a circuit or choose a component, it is necessary to know how the components are rated.

5.1 CONDUCTORS

As previously discussed, a conductor provides a path for electricity to flow. A perfect conductor would have zero resistance. There is no such thing as a perfect conductor. All materials offer some resistance. The resistance in a conductor results in the production of heat and the loss of some electrical energy.

The characteristics that rate the performance of a conductor fall into four groups.
- Material.
- Cross-sectional area.
- Resistivity/conductivity.
- Other variables.

Conductor Materials

The material from which a conductor is made determines how much resistance it offers. The four best conductors, (conductors that offer the least resistance) are silver, copper, gold, and aluminum.

Silver

As was stated in Chapter 4, silver is the best conductor. However, there are a couple of disadvantages to using silver for wiring, oxidation and cost. Silver oxidizes, combines with the oxygen in the air, very quickly. The oxidation of silver is very similar to the rusting of iron. Oxidation causes a high resistance surface on the silver. This leads to circuit failure. In addition, this surface cannot be soldered.

Silver can be mixed with tin to prevent oxidation. This compound is often used as a coating on copper wire to prevent the copper from oxidizing.

Copper

Copper is the second best conductor. Copper is also the most commonly used material for making wire. The price of copper is considerably less than either silver or gold, its two closest rivals in conductivity.

Gold

Gold is the third best conductor. The obvious disadvantage to using gold is its high cost. However, gold has one very useful quality. Gold is not subject to oxidation. As a result, it is an ideal material for plating sensitive electrical switch contacts. Computers, spacecraft, and switches for automobile air bags rely on gold-plated contacts for reliability.

Aluminum

Aluminum is the fourth best conductor. The advantage of using aluminum is that it is the cheapest of these four highly conductive materials. However, it takes a larger aluminum wire to allow the same amount of current as copper. Also, there are greater energy losses when using aluminum. Finally, aluminum is more affected by changes in temperature.

Cross-Sectional Area

The size of a wire is determined by its **cross-sectional area**, the surface area looking from the end. In figure 5-1, the diameters of two circles are compared to their cross-sectional area. The unit of measure in this figure is the mil. One **mil** is equal to 0.001 inch. The area of the circle is measured in **cmil**, which stands for circular mils. Notice, when the diameter of a circle is doubled, the cross-sectional area is four times larger. The formula used is the formula to calculate the area of a circle.

Diameter = 5 mil
Area = 20 cmil

Diameter = 10 mil
Area = 80 cmil

Figure 5-1. Doubling the area of a wire results in four times the cross-sectional area.

Formula 5.A

$$A = \frac{\pi d^2}{4}$$

A is the cross-sectional area.

d is the diameter of the wire.

The larger a wire is, the more current it can carry. Larger wires, however, are stiffer, heavier, and more costly than smaller wires. Wire is given a current rating based on the material it is made from and its size. Wire sizes are given as American Wire Gauge (AWG).

Resistivity/Conductivity

Resistivity is a characteristic of a material. The **resistivity** of a material gives the ohms of resistance for a one-foot length that has a cross-sectional area of one mil. Figure 5-2 compares the resistivity of the top four materials used for conductors. The resistance for a length of wire is calculated, based on its length, cross-sectional area, and resistivity. Formula 5.B calculates the resistance of a wire.

Figure 5-2. Resistivity chart for selected conductors.

Material	Resistivity
Silver	9.8
Copper	10.37
Gold	14.7
Aluminum	17.02

Formula 5.B

$$R = \rho \frac{L}{A}$$

R is the resistance, measured in ohms.

ρ (Greek letter rho) is the resistivity.

L is the length, measured in feet.

A is the area, measured in circular mils.

Sample problem 1.

Determine the resistance of 1000 feet of #10 copper wire. A section of #10 wire has a cross-sectional area of 10,400 cmil.

$$\rho = 10.37 \text{ (see figure 5-2)}$$

Formula: $R = \rho \dfrac{L}{A}$

Substitution: $R = 10.37 \, \rho \, \dfrac{1000 \text{ ft.}}{10,400 \text{ cmil}}$

Answer: $R = 0.997$ ohms

#10 copper wire has a resistance of approximately 1 Ω per 1000 feet.

Conductivity (G) is the ease with which a conductor allows electron flow. Conductivity is the reciprocal of resistance. The unit of measure used is the **siemen (S).** Another unit often used for conductance is the **mho (℧)** (ohm spelled backwards). Formula 5.C calculates conductivity.

Formula 5.C

$$G = \frac{1}{R}$$

G is the conductance, measured in siemens.

R is resistance, measured in ohms.

Sample problem 2.

Determine the conductance of wire with resistance of two ohms.

Formula: $G = \dfrac{1}{R}$

Substitution: $G = \dfrac{1}{2 \, \Omega}$

Answer: $G = 0.5$ S

Other Variables

There are other conditions that affect conductors and their ability to carry current. Two important factors are temperature and the length of the conductor.

Materials respond differently to changes in temperature. The **temperature coefficient** for a material is a factor that predicts if the resistance increases or decreases as the temperature changes. Copper and other materials commonly used as conductors have *positive* temperature coefficients. This means that their resistance increases with an increase in temperature.

Current flowing through a very long length of wire causes heat. Heat produced in a wire results in loss of energy, lowering the current and decreasing the heat. If the temperature decreases, there is a decrease in resistance, producing an increase in current. A positive temperature coefficient is a stable condition.

If the temperature of a conductor is made low enough, there will be zero resistance. For copper, this temperature is approximately –234.5°C. When a conductor has zero resistance, it is classified as a **superconductor**. Recent experiments in technology have lead to promising developments with superconductive materials. Materials have been developed that become superconductive at temperatures much higher than copper's –234.5°C.

Semiconductor materials, such as silicon and germanium have a *negative* temperature coefficient. The resistance decreases with an increase in temperature. As current flows, heat is produced, which in turn causes an increase in current due to the lower resistance. If this cycle continues, a condition known as **thermal runaway** can occur. This is destructive to the semiconductor device. External circuit parameters must be used as a prevention.

The length of a conductor also has a factor in the resistance of the conductor. For short lengths of a conductor wire, under a thousand feet or so, length has no noticeable effects. However, with longer wires, such as those used for power lines, telephone cable, and cable TV, it is necessary to compensate for losses in the line.

5.2 SELECTION OF WIRE

There are four general characteristics to examine when selecting wire for any application. These characteristics are:

- Current capacity.
- Solid or stranded conductor.
- Insulation.
- Type of wire/cable.

Cost considerations are also important in many applications.

Current Capacity

Excessive heat is the most frequent cause of failure in conductors. A buildup of heat is usually a result of too much current flowing through too small of a wire for too long. Problems also result from the wire being used in areas where heat cannot be dissipated. An extension cord run underneath carpeting is an all too common example.

The **American Wire Gauge (AWG)** is the standard for wire sizes. The AWG number is stamped on the wire or on the reel on which the wire is supplied. A tool used to measure the size of a wire is shown in figure 5-3. Common wire sizes range from approximately one-half inch diameter (AWG 0000) to three one-thousandths inch diameter (AWG 40), see figure 5-4. A wire's current rating is based on its size. When a current rating is given, it is the maximum current a conductor can carry when it is exposed to air. Other applications will lower the current rating.

Listed here are some examples of applications for wire of different gauges.

AWG 00: Used for 200 amp electrical service in homes. Two current-carrying conductors are used. Each one has a maximum of 100 amps.

AWG 10: Used for 30 amps of current to power electric dryers and stoves.

Figure 5-3. This tool gives the AWG number for most common wire sizes. (L. Starrett Co.)

Figure 5-4. Working table for selected wire sizes.

AWG No.	Diameter (mils)	Circular mils	Ohms (per 100 feet)	Current Rating (amps)
0000	460	212,000	0.00500	195
000	410	168,000	0.00630	165
00	365	133,000	0.00795	145
0	325	106,000	0.0100	125
1	289	83,700	0.0126	110
2	258	66,400	0.0159	95
3	229	52,600	0.0201	80
4	204	41,700	0.0253	70
6	162	26,300	0.0403	55
8	128	16,500	0.0641	40
10	102	10,400	0.102	30
12	81	6,530	0.162	20
14	64	4,110	0.258	15
16	51	2,580	0.409	7
18	40	1,620	0.651	5
20	32	1,020	1.04	3
22	25.3	642	1.65	
24	20.1	404	2.62	
26	15.9	254	4.16	
28	12.6	160	6.62	

Figure 5-4. Continued.

AWG No.	Diameter (mils)	Circular mils	Ohms (per 100 feet)	Current Rating (amps)
30	10.0	101	10.5	
32	8.0	63.2	16.7	
34	6.3	89.8	26.6	
36	5.0	25.0	42.3	
38	4.0	15.7	67.3	
40	3.1	9.9	107.0	

AWG 12: Used for 20 amp outlets for kitchen appliances.

AWG 14: Used for 15 amp outlets for most of a residence.

AWG 18: Used for a 5 amp current capacity for power supplies within electronic equipment and most automotive circuits.

AWG 20: Used for 3 amps of current within electronic equipment.

AWG 22: Used as "hookup" wire for low current applications such as hooking up electronic circuits.

AWG 36: Used in small electrical motor windings.

Solid and Stranded Conductors

Wire is available in solid and stranded form. Large wire sizes are only available in stranded form. Wire that is made of many strands is much more flexible than solid wire of the same thickness. However, stranded wire is more costly.

The size of the individual strands determines how many strands are needed to create a particular wire size. Smaller strands are more flexible but also more delicate.

For some applications solid wire is preferred, for other stranded is more useful. Some examples:

House wiring:	Solid wire
Coaxial cable:	Solid center conductor with a stranded shield
Hookup wire:	Solid or stranded
Lamp cord:	Stranded
Extension cords:	Stranded
Automotive wire:	Stranded

Insulation

Insulation is a protection placed on the wire. Depending on the material, it can serve many purposes. Insulation protects people from contact with electricity as well as protecting the wire from environmental conditions (water, sun, corrosive underground materials).

Insulation is also given a voltage breakdown rating. This rating, stamped on the wire, indicates the maximum voltage the insulation can safely contain. Exceeding this voltage can result in shock hazards, sparking, fire, and other problems.

Types of Wire and Cable

There are many different types of wire and cable available for as wide a range of applications. The examples shown in the following figures reflect what are found catalogs. As a matter of interest, you should observe the different types of wire and cable in actual situations.

Single conductors, figure 5-5, are used for point-to-point wiring within electrical/electronic equipment. This figure shows both solid and stranded wire.

Stranded

Solid

Figure 5-5. Single conductors (solid and stranded).

Multiconductor cables, figure 5-6, are used when it is necessary to bring a few, or many, conductors from one point to another. Examples include telephone wire within a house, the cable connecting a computer to a printer, and the cable used for electricity in the walls of a house.

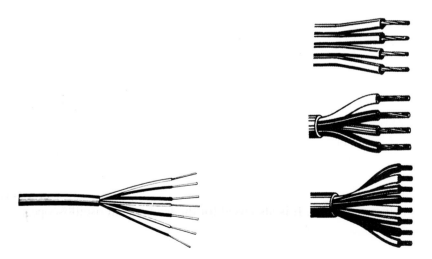

Figure 5-6. Multiconductor cable.

Ribbon cable, figure 5-7, is a flat form of multiconductor cable. It is frequently used *inside electronic* equipment because of its space-saving characteristics.

Figure 5-7. Ribbon cable.

Twin lead cable, figure 5-8, is used for antenna wire for televisions or FM antennas.

Figure 5-8. Twin lead cable.

Coaxial cable or coax, figure 5-9, is a round cable. Coaxial cable is made with a center conductor that is covered with a thick insulator. This insulator is then covered with a braided cable and finally an outside protective jacket. The braided cable is usually connected to ground. Its functions are to act as a conductor and to reduce electrical noise (static) for the signal carried on the center conductor. Coax is used for antenna wire in radio communications. It is also used for the leads on an oscilloscope.

Figure 5-9. Coaxial cable.

5.3 OVERCURRENT PROTECTION

When voltage is applied to a circuit, the load determines how much current is drawn. This is calculated with Ohm's law in Chapter 4. When the resistance of the load is too small, there is an excessive amount of current passing through the circuit. The wires must be protected from this current.

Most people are familiar with some situations where a circuit is overloaded. In your home, when too many appliances are plugged into the same circuit and used at the same time the circuit will overload. The combined current of the appliances trips a circuit breaker, removing the voltage and stopping the current.

Overcurrent protection is a means of preventing too much current from flowing in a circuit. **Fuses** and **circuit breakers** are examples of overcurrent protection devices. The schematic symbols for fuses and circuit breakers are shown in figure 5-10.

Circuit breaker   Fuse

Figure 5-10. Shown are the schematic symbols used for overcurrent protection devices.

Electronic equipment frequently comes equipped with its own current protection to safeguard the internal circuits from a malfunction. Current protection (usually a fuse) is placed in the power supply of the equipment to protect the circuitry in the event a component were to fail and draw too much current. Each component in a circuit has a current rating. The equipment is designed to never exceed these ratings. For this reason, never replace a fuse with one that has a larger current rating.

Style of Fuses and Circuit Breakers

Fuses are designed to melt when a specified amount of current flows through. When a fuse is blown, it cannot be reused. Fuses and circuit breakers are available in several different styles and types. **Cartridge fuses,** such as those shown in figure 5-11, are the

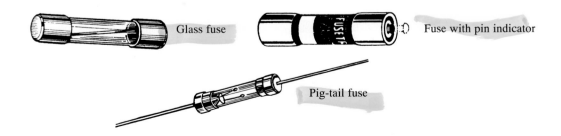

Glass fuse   Fuse with pin indicator

Pig-tail fuse

Figure 5-11. Drawn are three types of cartridge fuse: a glass fuse, a fuse with a pin indicator, and a pig-tail fuse.

most common type found in electronic equipment. The glass case reveals if the fuse has blown with a burned spot. Fuses with a pin-type indicator give a more reliable indication, a small button pops out.

Fuses need some way of being installed into the circuit. The pig-tail fuse has wires on the ends that allow it to be soldered easily into place.

Fuse holders, shown in figure 5-12, provide another way of installing a fuse in a circuit. Fuse holders allow for ease of changing fuses when necessary. Snap-in type fuse holders are found mounted on a circuit board. These fuse holders have the advantage of providing a visual inspection of the fuse without removing it from the circuit.

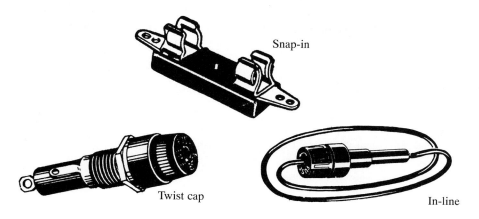

Figure 5-12. Drawn are three types of fuse holder: a snap-in holder, a twist cap holder, and an in-line holder.

Twist cap fuse holders are mounted through the side of the equipment cabinet. Electrical connections are made safely inside, while there is access to the fuse from the outside. The disadvantage of twist cap fuse holders is that the fuse must be removed from the circuit to be inspected.

An in-line fuse holder has wires attached. The fuse holder becomes a part of the voltage supply line.

A circuit breaker is similar to a switch, see figure 5-13. It turns electricity off by opening the current path. Most circuit breakers use magnetism. When current flows, a

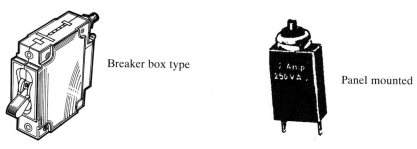

Figure 5-13. These are two commonly encountered styles of circuit breaker.

magnetic field is created. The stronger the current, the stronger the magnetic field it creates. If the field reaches a fixed level, it activates the breaker. A circuit breaker is reset by switching it off, then back on.

Ratings of Fuses and Circuit Breakers

Overcurrent protection devices have three basic ratings:

- Maximum current—Available sizes range from less than one amp to many amps.
- Reaction time—Options are either fast or time-delayed protection. Fast devices respond immediately to overcurrent. Time-delayed devices do not act on a momentary overload.
- Maximum voltage—Most applications use a 250 volt rating. As long as the applied voltage stays under this level, it is acceptable.

5.4 SWITCHES

Switches serve many purposes. They do more than turn a circuit on and off. A switch might turn off one part of a circuit, while switching on another. The same switch can be used to operate two or more different circuits. There are many different types of switches. Examples are: lever action, push button, and rotary.

Switch Ratings

Each style of switch has different characteristics that make it useful for a specific application. Many of the switches used in electronic circuits are miniaturized for the small circuits. They are easily damaged if used in the wrong application. Switch contacts have two basic electrical ratings, current and voltage. Usually, a miniature switch has a very low current rating.

A **current rating** states the maximum current allowable through the switch contacts. In any electrical device, current produces heat. This heat must be dissipated, or it can cause damage.

Switches also have **voltage ratings,** stating the maximum allowable voltage. Switches have voltage limits for two reasons. First, when the switch is open (off position), it acts like an insulator. Too high a voltage could break down the switch and allow current to pass. Second, when a switch is being switched from a closed (on position) to open, some circuits have a tendency to arc. A large spark jumps across the gap. If a spark does jump, it can damage the switch contacts leaving tiny burn marks.

Switch Styles and Characteristics

A toggle switch, figure 5-14, is often thought of as an on/off switch. It operates with a lever. As seen in the left side of the figure, a contact moves back and forth, sometimes with a stop in the middle. This type of switch can have more than one set of contacts.

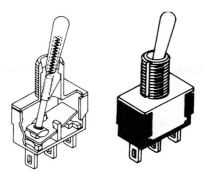

Figure 5-14. Toggle switch (on/off switch) with cut-away.

A rocker switch, figure 5-15, is often used on the front panel of electronic equipment. The contacts perform in the same manner as the contacts in a toggle switch.

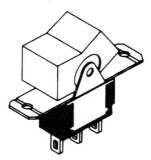

Figure 5-15. Rocker switch.

In a slide switch, figure 5-16, the contacts move in a sliding motion. There can be several different positions for the slide to activate.

Figure 5-16. Slide switch.

In a key switch, figure 5-17, the contacts are activated by turning a key. A key can be used with several different types of switches.

Figure 5-17. Key switch.

With a snap-action switch, figure 5-18, when the lever is depressed, the switch is activated. When the lever is released, it is deactivated. Some snap-action switches remain activated when the lever is released. These snap-action switches require the lever to be pressed a second time to be deactivated. A push-button switch, figure 5-19, is a type of snap-action switch available in many styles. The lighted push-button switch is popular for panel-mounted on/off applications.

Figure 5-18. Snap-action switch.

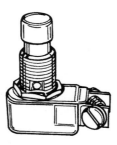  

Figure 5-19. Selection of push button switches.

The knife switch, figure 5-20, is used as a high current on/off switch. Current travels up through one contact, through the movable bar, and then down the other contact on the same side. Plastic insulation protects the operator. The model shown in this figure is used to switch two different wires, one on the left and one on the right.

Figure 5-20. Knife switch.

Dip switches, figure 5-21, are a set of on/off switches frequently found on a computer's printer. Many different circuits can be controlled from this set of switches that occupy a small space.

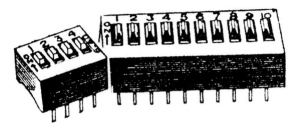

Figure 5-21. Dip switch set.

A rotary switch, figure 5-22, is made up of a set of wafer switches. Each wafer is electrically independent from the other wafers. As the shaft is turned, contact can be made in several different locations.

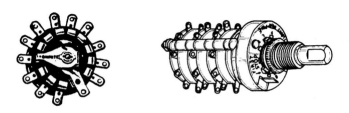

Figure 5-22. Two views of a rotary switch.

Switch Schematic Symbols

The symbol for a single-pole, single-throw switch (SPST) is shown in figure 5-23. This is the symbol for a simple on/off switch. The switch is either open or closed. **Single-pole** means the switch has only one common connection. **Single-throw** means the switch has only one direction in which to make contact.

Figure 5-23. Schematic symbol for single-pole, single-throw switch (SPST).

Figure 5-24 shows symbols for a single-pole double-throw switch (SPDT). The single-pole, again, indicates one common connection. The center terminal, connected to the lever, is common to both sides of the switch. **Double-throw** indicates the switch makes contact in either of two positions. An application of this switch is when either of two switches can be used to turn on the same light. This schematic symbol is drawn in two ways, as shown.

Figure 5-24. Schematic symbols for single-pole, double-throw switch (SPDT). This switch can be drawn two ways as shown.

Snap-action switches, figure 5-25, can either open the circuit or close the circuit when they are activated. A **normally open** switch requires operation to close the circuit. A **normally closed** switch requires operation to open the circuit.

Figure 5-25. Schematic symbols for normally open (left) and normally closed (right) snap-action switches.

Double-pole switches, figure 5-26, have two common terminals. These common terminals are connected to the activator. Double-pole, single-throw switches (DPST) are activated in only one direction. Double-pole, double-throw switches (DPDT) activate in two directions.

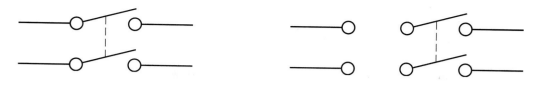

Figure 5-26. Schematic symbols for double-pole switches. The symbol on the left shows a double-pole, single-throw switch. The schematic on the right shows a double-pole, double-throw switch.

The symbols for rotary switches, figure 5-27, show the number of contacts that are available on the wafer. Only one terminal is common in this drawing. It is possible for a rotary switch to have more than one common terminal. If the switch is made of more than one wafer, they could be drawn anywhere on the schematic diagram.

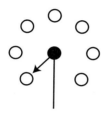

Figure 5-27. Symbol for rotary switch with one common terminal.

5.5 FIXED RESISTORS

Fixed resistors have an ohmic value determined during their manufacture. The values available range from less than one ohm to several megohms. The ohmic value of a resistor has nothing to do with its physical size. The size of the resistor is determined by its wattage rating.

Fixed Resistor Ratings

Resistors have a specified ohmic value and a percent tolerance. Resistors are manufactured to have a specific resistance, but to mass produce resistors with every one having an exact value of resistance is very expensive. Instead, the resistors are given values with a percent tolerance. The **tolerance** states the acceptable accuracy of the assigned value. Common tolerances are 5 percent, 10 percent, and 20 percent. If a resistor is manufactured to have a 1000 ohm resistance and a tolerance of 10 percent, the value of that resistor will be somewhere in the range of 900 to 1100 ohms. Precision resistors can have 2 percent, 1 percent, and even lower tolerances.

The wattage rating is also an important consideration with resistors. When current flows through a resistor, it produces a voltage drop. This voltage drop and the current through the resistor result in power. This power produces heat that the resistor must dissipate. The **wattage rating** of a resistor gives the maximum power that the resistor can safely manage continuously.

Construction of Fixed Resistors

The construction of a resistor determines what applications it is used for. Figures 5-28 to 5-30 show a few styles of resistors. Most fixed resistors fall into four construction groups.

- Wire wound.
- Carbon composition.
- Metal film.
- Molded resistor networks.

Other construction designs are available for specific applications.

Wire wound resistors, figure 5-28, are designed for high wattage applications. Some wire wound resistors are bolted to a metal **heat sink** to help dissipate the heat produced. Resistors with wattages of five watts and higher fall into this group. The ohmic value of a wire wound resistor is stamped on the side.

Figure 5-28. Wire wound resistors such as these, are used in high wattage applications.

Carbon composition resistors, figure 5-29, and metal film resistors, are the most commonly used resistors. These types of resistors have a series of color code bands that give their resistance. Wattage ratings are 1/4 W, 1/2 W, 1 W, and 2 W. The physical size of the resistor increases with the wattage rating.

Figure 5-29. Typical carbon composition resistor. The bands indicate the resistors ohmic value and tolerance.

Molded resistor networks, figure 5-30, have very low wattage ratings. They are frequently used in circuits having a very low current. Circuits used with integrated circuit chips in computers are one example.

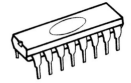

Figure 5-30. Molded resistor networks.

5.6 RESISTOR COLOR CODE

Color bands are placed on resistors to identify their ohmic value and percent tolerance. The standard four-color code is used for resistors with tolerances of 5 percent, 10 percent, and 20 percent. Refer to figure 5-31. Other codes are used for precision resistors. The resistor color code is used to find the nominal value of the resistor. The **nominal value** is the "name" of the resistor. It is the resistor's ideal value and tolerance.

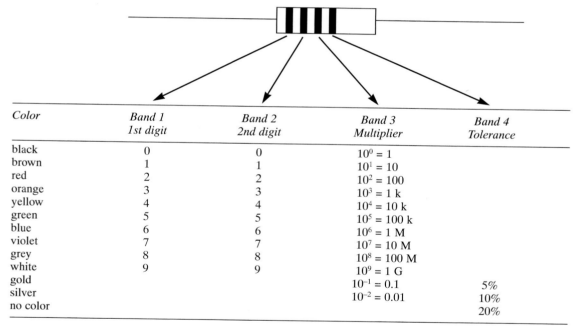

Color	Band 1 1st digit	Band 2 2nd digit	Band 3 Multiplier	Band 4 Tolerance
black	0	0	$10^0 = 1$	
brown	1	1	$10^1 = 10$	
red	2	2	$10^2 = 100$	
orange	3	3	$10^3 = 1\,k$	
yellow	4	4	$10^4 = 10\,k$	
green	5	5	$10^5 = 100\,k$	
blue	6	6	$10^6 = 1\,M$	
violet	7	7	$10^7 = 10\,M$	
grey	8	8	$10^8 = 100\,M$	
white	9	9	$10^9 = 1\,G$	
gold			$10^{-1} = 0.1$	5%
silver			$10^{-2} = 0.01$	10%
no color				20%

Figure 5-31. Standard four-color resistor color code. The first three bands give the resistor's ohmic value. The last band is the resistor's tolerance.

Finding the Nominal Value

The first three color bands on a resistor give the values of two significant figures and the number of zeros. The fourth color is the percent tolerance. You must be able to translate the color code into the nominal value and be able to translate a resistor's value back into its color code. There are four points to calculating a resistor's nominal value from the resistor's color code.

1. The first two color bands give two significant figures.
2. The third band gives the power of 10 of the multiplier.
3. The fourth band indicates the percent tolerance.
4. If the value is greater than 1000, it should be written in engineering notation.

Sample problem 3. _____

Determine the nominal value of the resistor shown in figure 5-32.

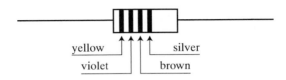

yellow silver

violet brown

Figure 5-32. Resistor for sample problem 3.

Band 1: Yellow = 4 (1st significant figure)

Band 2: Violet = 7 (2nd significant figure)

Band 3: Brown = 1 (multiply by 10^1 = 10)

Band 4: Silver = 10% (tolerance)

Nominal value = 47 × 10 = 470 Ω ±10%

Sample problem 4. _____

Determine the nominal value of the resistor shown in figure 5-33.

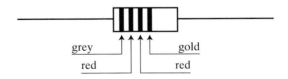

grey gold

red red

Figure 5-33. Resistor for sample problem 4.

Band 1: Gray = 8 (1st significant figure)

Band 2: Red = 2 (2nd significant figure)

Band 3: Red = 2 (multiply by 10^2 = 100)

Band 4:   Gold = 5% (tolerance)

Nominal value = 82 × 100 = 8200 Ω = 8.2 kΩ ±5%

Sample problem 5.

Determine the nominal value of the resistor shown in figure 5-34.

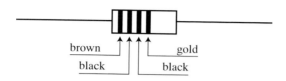

Figure 5-34. Resistor for sample problem 5.

Band 1:   Brown = 1 (1st significant figure)

Band 2:   Black = 0 (2nd significant figure)

Band 3:   Black = 0 (multiply by 10^0 = 1)

Band 4:   Gold = 5% (tolerance)

Nominal value = 10 × 1 = 10 Ω ±5%

Sample problem 6.

Determine the nominal value of the resistor shown in figure 5-35.

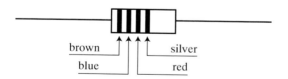

Figure 5-35. Resistor for sample problem 6.

Band 1:   Brown = 1 (1st significant figure)

Band 2:   Blue = 6 (2nd significant figure)

Band 3:   Red = 2 (multiply by 10^2 = 100)

Band 4:   Silver = 10% (tolerance)

Nominal value = 16 × 100 = 1600 Ω = 1.6 kΩ ±10%

Sample problem 7. _____

Determine the nominal value of the resistor shown in figure 5-36.

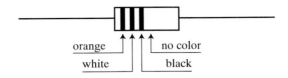

Figure 5-36. Resistor for sample problem 7.

Band 1:   Orange = 3 (1st significant figure)

Band 2:   White = 9 (2nd significant figure)

Band 3:   Black = 0 (multiply by $10^0 = 1$)

Band 4:   No color = 20% (tolerance)

Nominal value = $39 \times 1 = 39\ \Omega \pm 20\%$

Finding the Color Code from the Resistor Value

On schematic diagrams, resistor values are given in numerical form. Technicians must be able to change these numbers into the proper color code. To find the resistor color code from the resistor numerical value:

1. Change the engineering notation into basic units with a multiplier of 100.
2. The two significant figures give the first two color bands.
3. The number of zeros gives the color for the third band.
4. The percent tolerance is translated into the fourth band.

Sample problem 8. _____

Give the first three color bands (ignore tolerance) for a 270 ohm resistor.

1. Change the engineering notation to basic units. This resistance, $270\ \Omega$, is already in basic units.

2. Determine the colors for the two significant figures.

$$2 = red$$
$$7 = violet$$

3. Determine the third color band by the number of zeros.

$$1\ zero = brown$$

Combine the colors: red-violet-brown

Sample problem 9. _____

> Give the first three color bands (ignore tolerance) for a 3.9 kilohm resistor.
>
> 1. Change the engineering notation to basic units. 3.9 kΩ = 3900 Ω
>
> 2. Determine the colors for the two significant figures.
>
>
>
> 3 = orange
>
> 9 = white
>
>
>
> 3. Determine the third color band by the number of zeros.
>
>
>
> 2 zeros = red
>
>
>
> Combine the colors: orange-white-red

Sample problem 10. _____

> Give the first three color bands (ignore tolerance) for a 560 kilohm resistor.
>
> 1. Change the engineering notation to basic units. 560 kΩ = 560,000 Ω
>
> 2. Determine the colors for the two significant figures.
>
>
>
> 5 = green
>
> 6 = blue
>
>
>
> 3. Determine the third color band by the number of zeros.
>
>
>
> 4 zeros = yellow
>
>
>
> Combine the colors: green-blue-yellow

5.7 RESISTOR TOLERANCE

The tolerance of a resistor is its range of acceptable values. The range of values has a minimum and a maximum. The tolerance is given by a color band as a percentage.

Finding the Maximum and Minimum Values

Resistance values are important for the proper functioning and safety of most electronic equipment. Consequently, it is important to make sure that all the possible values of a resistor fall into an acceptable level. There are four steps in calculating the maximum and minimum values of a resistor.

1. Determine the nominal value.
2. Find the range of tolerance by multiplying the nominal value by the decimal equivalent of the percent tolerance.
3. Find maximum value by adding the range to the nominal value.
4. Find minimum value by subtracting the range from the nominal value.

For sample problems 11 through 15, find all of the following:

1. Nominal value.
2. Range of tolerance.
3. Maximum value.
4. Minimum value.

Sample problem 11. _____

Use the resistor shown in figure 5-37.

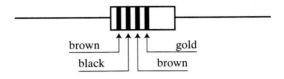

brown   gold

black   brown

Figure 5-37. Resistor for sample problem 11.

1. Nominal value:

   Band 1:   Brown = 1 (1st significant figure)

   Band 2:   Black = 0 (2nd significant figure)

   Band 3:   Brown = 1 (multiply by $10^1 = 10$)

   Band 4:   Gold = 5% (tolerance)

   Nominal value = $10 \times 10 = 100 \ \Omega \ \pm 5\%$

2. Range of tolerance = nominal × percent

   $100 \times 5\% = 100 \times 0.05 = 5$

   Range = 5 Ω

3. Maximum value = nominal + range

   $100 + 5 = 105$

   Maximum = 105 Ω

4. Minimum value = nominal − range

   $100 - 5 = 95$

   Minimum = 95 Ω

Sample problem 12.

Use the resistor shown in figure 5-38.

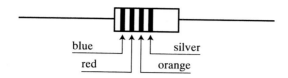

Figure 5-38. Resistor for sample problem 12.

1. Nominal value:

    Band 1:   Blue = 6 (1st significant figure)

    Band 2:   Red = 2 (2nd significant figure)

    Band 3:   Orange = 3 (multiply by 10^3 = 1000)

    Band 4:   Silver = 10% (tolerance)

    Nominal value = 62 × 1000 = 62,000 Ω = 62 kΩ ±10%

2. Range of tolerance = nominal × percent

    62,000 × 10% = 62,000 × 0.10 = 6200

    Range = 6200 Ω

3. Maximum value = nominal + range

    62,000 + 6200 = 68,200

    Maximum = 68.2 kΩ

4. Minimum value = nominal − range

    62,000 − 6200 = 55,800

    Minimum = 55.8 kΩ

Sample problem 13.

Use the resistor shown in figure 5-39.

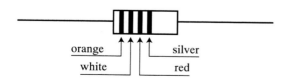

Figure 5-39. Resistor for sample problem 13.

1. Nominal value:

    Band 1:   Orange = 3 (1st significant figure)

    Band 2:   White = 9 (2nd significant figure)

    Band 3:   Red = 2 (multiply by 10^2 = 100)

    Band 4:   Silver = 10% (tolerance)

$$\text{Nominal value} = 39 \times 100 = 3900 \ \Omega = 3.9 \ k\Omega \ \pm10\%$$

2. Range of tolerance = nominal × percent

$$3900 \times 10\% = 3900 \times 0.10 = 390$$

$$\text{Range} = 390 \ \Omega$$

3. Maximum value = nominal + range

$$3900 + 390 = 4290$$

$$\text{Maximum} = 4.29 \ k\Omega$$

4. Minimum value = nominal - range

$$3900 - 390 = 3510$$

$$\text{Minimum} = 3.51 \ k\Omega$$

Sample problem 14. _____

Use the resistor shown in figure 5-40.

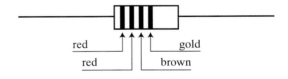

Figure 5-40. Resistor for sample problem 14.

1. Nominal value:

    Band 1:   Red = 2 (1st significant figure)

    Band 2:   Red = 2 (2nd significant figure)

    Band 3:   Brown = 1 (multiply by 10^1 = 10)

    Band 4:   Gold = 5% (tolerance)

$$\text{Nominal value} = 22 \times 10 = 220 \ \Omega \ \pm5\%$$

2. Range of tolerance = nominal × percent

$$220 \times 5\% = 220 \times 0.05 = 11$$

Range = 11 Ω

3. Maximum value = nominal + range 220 + 11 = 231

Maximum = 231 Ω

4. Minimum value = nominal – range

220 – 11 = 209

Minimum = 209 Ω

Sample problem 15.

Use the resistor shown in figure 5-41.

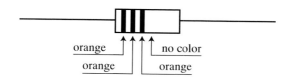

Figure 5-41. Resistor for sample problem 15.

1. Nominal value:

   Band 1:   Orange = 3 (1st significant figure)

   Band 2:   Orange = 3 (2nd significant figure)

   Band 3:   Orange = 3 (multiply by 10^3 = 1000)

   Band 4:   No color = 20% (tolerance)

Nominal value = 33 × 1000 = 33,000 Ω = 33 kΩ ±20%

2. Range of tolerance = nominal × percent

33,000 × 20% = 33,000 × 0.20 = 6600

Range = 6600 Ω

3. Maximum value = nominal + range

33,000 + 6600 = 39,600

Maximum = 39.6 kΩ

4. Minimum value = nominal – range

33,000 – 6600 = 26,400

Minimum = 26.4 kΩ

5.8 VARIABLE RESISTORS

A **variable resistor** has a wiper that is moved to change its resistance. There are two types of variable resistors: potentiometers and rheostats. A **potentiometer** is a variable resistor with three terminals, one for each end of the resistor and one for the wiper. A **rheostat** is a variable resistor with only two terminals. One terminal is connected to the wiper, and the other terminal is connected to one end of the resistor.

Rheostats are used to vary the current in a circuit. Potentiometers are usually used to control the voltage. A rheostat cannot be used as a potentiometer because there is no third lead. A potentiometer, however, can be used as a rheostat by not connecting the third lead into the circuit application. Several resistor schematic symbols are shown in figure 5-42.

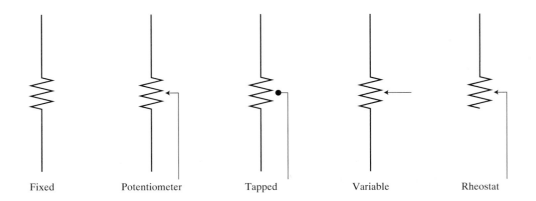

| Fixed | Potentiometer | Tapped | Variable | Rheostat |

Figure 5-42. Schematic symbols for a resistor and its many variations.

Rating a Variable Resistor

There are four ratings to be considered when examining variable resistors.
• Resistance value.
• Wattage rating.
• Number of turns.
• Percent tolerance.

Variable resistors can be purchased that have a tremendous range of resistance values. What must be considered when picking out a resistor is the expected resistance needed in the circuit. For example, if the resistance needed is expected to be around 200 ohms, it would not be wise to use a variable resistor with a range of zero ohms to two megohms. It may be possible to adjust the resistor to around 200 ohms, but it would be unnecessarily difficult. In addition, a great deal of accuracy is lost. The potentiometer must also encompass the expected resistance. If the expected resistance value needed is approximately 25 kilohms, a one kilohm potentiometer would obviously not be a good choice.

The entire range of resistances that might be needed has to be included in the resistor choice. Variable resistors are given a range of operation, minimum to maximum. A 50 kilohm variable resistor may have a range from 10 kilohm to 50 kilohm. If it is necessary to vary the resistance from zero to a maximum, a different 50 kilohm resistor would be necessary.

Wattage ratings are necessary with resistors of all types. When current flows through a resistor, a voltage is dropped across it. Current and voltage result in power. Power results in heat, and heat must be dissipated or it is destructive. Variable resistors mounted to circuit boards have ratings of 1/4 watt to one watt. Wire wound resistors have a much higher wattage rating.

The number of turns the adjustment on variable resistor is capable of making determines the ease at which an accurate setting can be achieved. Variable resistors are available in single turn and multi-turn forms. The multi-turn resistors usually have ten turns from the minimum to the maximum resistance.

The percent tolerance on a variable resistor is the same as the percent tolerance on a fixed resistor. It determines the accuracy of the specified ohmic rating. Multi-turn potentiometers are usually given this rating. Three percent and five percent are fairly common tolerances.

Variable Resistor Types

The three most common types of variable resistors are shown in figures 5-43, 5-44, and 5-45. Each of the different types of variable resistor are shown in both single turn and multi-turn models. Notice that all models shown have three terminals. Technically, these are all classified as potentiometers.

Circuit board mounted variable resistors, figure 5-43, have their legs pushed through holes in the board. Solder connections are made directly between the copper board and the resistor leads. Wattage ratings from 1/4 watt to one watt are common.

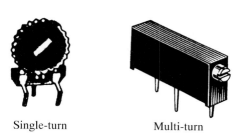

Single-turn   Multi-turn

Figure 5-43. Circuit board mount variable resistors.

Panel mounted variable resistors, figure 5-44, have wires attached behind a panel and a knob attached in the front. These resistors have a wattage rating of typically one or two watts. Although, higher ratings are possible.

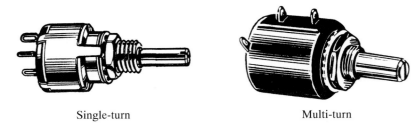

Single-turn   Multi-turn

Figure 5-44. Panel mount variable resistors.

Wire wound variable resistors, figure 5-45, can handle a great deal of power. Wire wound resistors with wattage ratings of 50 to 100 watts are common. High wattage also means a great deal of heat to dissipate. These potentiometers must not be located too close to other circuit components.

Sliding   Rotary

Figure 5-45. Wire wound variable resistors.

SUMMARY

- Conductors are low resistance paths for electrical current.
- The material, cross-sectional area, and length of a wire determine its resistivity.
- Wire is selected based on features such as: its current capacity, whether it is solid or stranded, the insulation used, and its type.
- Protection against too much current is provided by fuses and circuit breakers.
- Overcurrent protection devices are rated in: maximum current, reaction time, and maximum voltage.
- Switches are available in many different styles.
- The schematic symbols of switches reflect their operation.
- Fixed resistors are available in many values and types. Their actual ohmic value is determined during their manufacture.
- Certain types of resistors have a color code to mark the resistance value.
- Variable resistors are called potentiometers and rheostats.

KEY WORDS AND TERMS GLOSSARY

American Wire Gauge (AWG): The standard for wire sizes. Wire is commonly available from approximately one-half inch diameter (AWG 0000) to three one-thousandths inch diameter (AWG 40).

circuit breaker: An electromagnetic overcurrent protection device.

cmil: The abbreviation for circular mil, a measurement for the area of very small circles, such as the cross section of a wire.

conductivity (G): The ease with which a conductor allows a current to flow. It is the reciprocal of resistance. The unit of measure is the siemen (S) or mho.

cross-sectional area: The surface of an object that would be exposed by slicing the object.

current rating: States maximum current allowed through a device.

double-pole: Switches with two common positions.

double-throw: Switch makes contact in either of two positions.

fixed resistor: Resistor with an ohmic value determined during manufacture. The values available range from close to one ohm to several megohms. The ohmic value of a resistor has nothing to do with its physical size. The size of the resistor is determined by its wattage rating.

fuse: Overcurrent protection device that melts open to protect a circuit.

heat sink: Protects sensitive components by dissipating excess heat.

insulation: Protection placed on wiring.

mho (℧): Unit measure of conductivity.

mil: Unit of measure equal to 0.001 inches.

nominal value: The ideal value.

normally closed: A switch that requires operation to open.

normally open: A switch that requires operation to close.

potentiometer: A variable resistor with three terminals.

resistivity: The resistance for a one-foot length of a material, with a cross-sectional area of one mil.

rheostat: A variable resistor with only two terminals.

siemen (S): Unit measure of conductance.

single-pole: A switch with only one common connection.

single-throw: A switch with only one direction to make contact.

superconductor: A conductor with zero resistance. Recent experiments in technology have lead to promising developments with superconductive materials.

temperature coefficient: A mathematical factor used to predict how a material will respond to changes in temperature.

thermal runaway: A condition found in semiconductors. As current flows, heat is produced. This causes an increase in current due to the lower resistance. The increase in current produces more heat, less resistance, and more current.

tolerance: The acceptable accuracy of the assigned value.

variable resistor: A resistor that can have its resistance adjusted. It has a contact that can be moved to change its resistance in relation to the total resistance of the device.

voltage rating: The maximum voltage a device can safely handle.
wattage rating: The maximum power a device can safely handle.

KEY FORMULAS

Formula 5.A

$$A = \frac{\pi d^2}{4}$$

Formula 5.B

$$R = \rho \frac{L}{A}$$

Formula 5.C

$$G = \frac{1}{R}$$

TEST YOUR KNOWLEDGE

Do not write in this text. Please use a separate sheet of paper.
1. What is the difference between a conductor and an insulator?
2. State the four characteristics that rate the performance of a conductor.
3. List in order the four best conductor materials.
4. Describe how the cross-sectional area of a wire affects its ability to conduct electricity.
5. Calculate the resistance of 5000 feet of #14 copper wire.
6. If a length of wire has a resistance of five ohms, what is its conductance?
7. How is the resistance of a length of wire affected by the temperature?
8. Which wire can carry more current: AWG #12 or AWG #18?
9. Is the internal house wiring stranded or solid?
10. List the three ratings for a fuse.
11. What type of switch is most commonly used for light switches in homes?
12. What type switch is used in the keys on a computer keyboard?
13. Draw the schematic symbols for these switches: SPST, SPDT, PB, DPST, DPDT.
14. What are two ratings of a fixed resistor?
15. What are two ratings of a variable resistor?
16. Find the nominal values of the resistors with the following color codes:
    a. Orange - orange - red - silver.
    b. Brown - black - brown - gold.
    c. Red - violet - yellow - silver.

   d.   Green - white - orange - gold.

   e.   Yellow - blue - black - no color.

17. Find the first three color bands (ignore tolerance) of these resistors:

   a.   390 ohms.

   b.   2.7 kilohms.

   c.   10 kilohms.

   d.   39 ohms.

   e.   1 megohm.

18. Using the following color codes, determine: nominal value, range of tolerance, maximum value, minimum value.

   a.   Brown - gray - red - gold.

   b.   Blue - green - orange - silver.

   c.   White - black - black - gold.

   d.   Red - red - yellow - silver.

   e.   Orange - white - brown - no color.

Integrated circuits can be found in devices from televisions to electric toothbrushes. These *chips* can contain thousands of circuit components. The integrated circuits shown here help in the transmission of computer data. (Texas Instruments)

Chapter 6
Series Circuits

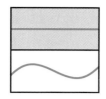

- Identify a series circuit by its schematic diagram and the circuit current.
- Calculate total resistance in a series circuit when given individual resistance values or voltage and current.
- Calculate the voltage drops in a series circuit.
- Predict the effects of large and small resistors on other parts of the circuit.
- Use the voltage divider formula to determine circuit voltages.
- Calculate the power in a series circuit.
- Recognize the symptoms created by opens and shorts in a series circuit.

A **schematic diagram** is a picture that shows how components are *electrically* connected. The physical connections of the components shown in the schematic can look quite different. This becomes obvious when looking at the length of wires on the schematic diagram and comparing them to an actual circuit.

Figure 6-1 shows some of the symbols used in drawing electrical schematics. Most dc voltage sources use the symbol of a battery. The long line is always positive and the short line is the negative end. This applies even if the +/− polarity symbol is not shown. The box shown can be used to represent any load. However, a resistor symbol is more commonly used and will be used for most of the sample problems that follow.

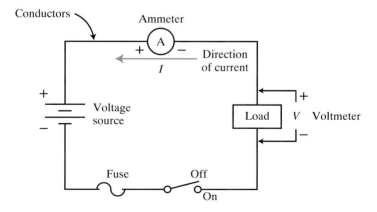

Figure 6-1. Familiarize yourself with the components drawn in this circuit. These symbols are found on many schematics.

6.1 IDENTIFICATION OF A SERIES CIRCUIT

The identification of a circuit type is important because there are different procedures for solving different types of circuits. Circuit configurations can be series, parallel, or a combination of the two. Figure 6-2 compares the schematic diagrams of series and parallel circuits. In the series circuit, current has only one path. All of the resistors in the circuit have the same current.

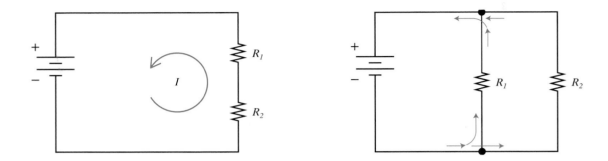

Figure 6-2. Circuit type is identified by the number of paths the current can take. A series circuit (left) has only one current path. A parallel circuit (right) has more than one current path.

      A **series circuit** can be identified by the following characteristics:
- There is only one path for the current.
- The voltage will drop across each resistor.
- The total resistance is the sum of the individual resistances.
- The total power is the sum of the individual powers.

      A parallel circuit has more than one path for the current to follow. The current splits at the junction, with the amount of flow determined by the resistance in the path. Parallel circuits are discussed in detail in Chapter 7. Series-parallel combination circuits are discussed in Chapter 8.

6.2 TOTAL RESISTANCE OF A SERIES CIRCUIT

The amount of current flowing from a dc voltage source is determined by the total resistance and the applied voltage. To the voltage source, it makes no difference if there is one large resistor or several smaller ones. The source only recognizes the total resistance for which it must supply current. As shown in figure 6-3, the total resistance can be represented by a single equivalent resistor.

      Since the current travels through all of the resistors in a series circuit, the total is found by *adding* all of the series resistors. The equation for the total resistance in a series circuit is written:

Formula 6.A

$$R_T = R_1 + R_2 + R_3 + ...R_N$$

R_T is the total resistance, measured in ohms.

R_1 through R_N are the individual resistances, measured in ohms.

N is the number of resistors in the circuit.

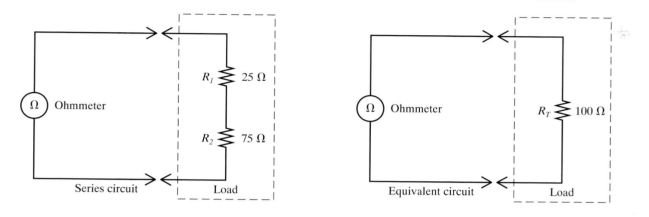

Figure 6-3. Resistors in series can be combined into an equivalent load.

Finding the Total Resistance Given Individual Resistances

When the resistance of each of the individual components is given, the total resistance is found using the basic addition formula.

Sample problem 1. _____

Find the total resistance of the circuit in figure 6-4.

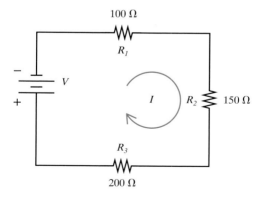

Figure 6-4. Find the total resistance. (Sample problem 1)

Formula: $R_T = R_1 + R_2 + R_3 + ...R_N$

Substitution: $R_T = 100\ \Omega + 150\ \Omega + 200\ \Omega$

Answer: $R_T = 450\ \Omega$

Sample problem 2. _____

Find the total resistance of the circuit in figure 6-5.

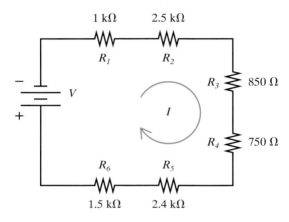

Figure 6-5. Find the total resistance. (Sample problem 2)

Note, this problem mixes different forms of engineering notation. Although this is no problem for a calculator, an extra step is included here to simplify the numbers.

Formula: $R_T = R_1 + R_2 + R_3 + ...R_N$

Substitution: $R_T = 1\ k\Omega + 2.5\ k\Omega + 850\ \Omega + 750\ \Omega + 2.4\ k\Omega + 1.5\ k\Omega$

Intermediate step: $R_T = 1000\ \Omega + 2500\ \Omega + 850\ \Omega + 750\ \Omega + 2400\ \Omega + 1500\ \Omega$

Answer: $R_T = 9000\ \Omega = 9\ k\Omega$

Finding an Individual Resistance Using Total Resistance

The basic formula for total resistance can be rearranged to find the value of any one resistance when the total resistance and the values of the other resistors are given.

Sample problem 3. _____

Find the value of R_3 in the circuit shown in figure 6-6. R_T is given with all resistor values except one.

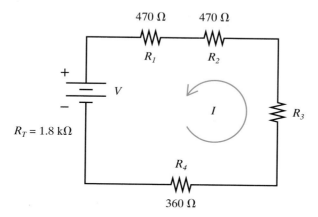

Figure 6-6. Find the unknown resistance value. (Sample problem 3)

Formula: $R_T = R_1 + R_2 + R_3 + ...R_N$

Substitution: 1.8 kΩ = 470 Ω + 470 Ω + R_3 + 360 Ω

Add the numbers on the right, leaving the unknown.

$$1.8 \text{ k}\Omega = 1.3 \text{ k}\Omega + R_3$$

Subtract the combined resistances from the total.

Answer: $R_3 = 500 \ \Omega$

Resistor Color Code in a Schematic Diagram

When working on an actual circuit, the resistor color code may be the only information known. To calculate the total resistance on these circuits, first change the color code to resistance values. If necessary, draw a schematic diagram before solving the circuit.

Sample problem 4. _____

Find the total resistance of the circuit in figure 6-7.

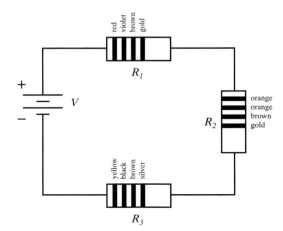

Figure 6-7. Find the total resistance. (Sample problem 4)

1. Find the resistor values using the color codes. Tolerances are not used because the calculations are made with exact values.
   R_1 = red - violet - brown = 270 Ω
   R_2 = orange - orange - brown = 330 Ω
   R_3 = yellow - black - brown = 400 Ω

2. Use series resistance formula.

   Formula: $R_T = R_1 + R_2 + R_3 + ...R_N$

   Substitution: $R_T = 270\ \Omega + 330\ \Omega + 400\ \Omega$

   Answer: $R_T = 1000\ \Omega = 1\ k\Omega$

6.3 CURRENT IN A SERIES CIRCUIT

The current in a circuit is dependent on the total resistance of the circuit. The same number of electrons that leave the negative terminal return to the positive terminal. If there is only one path for current, the amount of current must be the *same at all points in a series circuit.*

Figure 6-8 is a series circuit with three different value resistors. The total resistance is 200 ohms. With a 100 volt supply, Ohm's law is used to calculate the current at 0.5 amps. An arrow is used to indicate direction of current, from negative to positive. The arrow indicates 0.5 amps flowing from the power supply. Ammeters are shown in a circuit as a circle in series with the current path. In this circuit, ammeters measure current at three different points. Each of the ammeters read 0.5 amps.

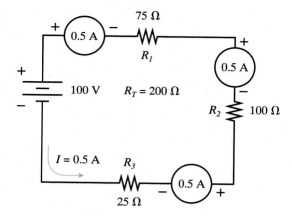

Figure 6-8. As the ammeters show, current is the same at all points in a series circuit.

Finding Current When Given Voltage and Resistance

Ohm's law and the power formulas are used to solve circuits with several resistors in the same manner as solving a single-load circuit. If two values are given, the third can be found. In circuits with multiple loads, it is usually necessary to find the equivalent of a single load by finding total resistance.

Sample problem 5. _____

Calculate the total current in the circuit shown in figure 6-9.

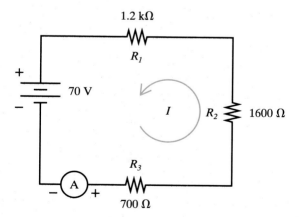

Figure 6-9. Find the total current. (Sample problem 5)

1. Total resistance must be found first.

    Formula: $R_T = R_1 + R_2 + R_3 + ...R_N$

Substitution: $R_T = 1.2\ \text{k}\Omega + 1600\ \Omega + 700\ \Omega$

Answer: $R_T = 3500\ \Omega = 3.5\ \text{k}\Omega$

2. Use Ohm's law to calculate current.

Formula: $I = \dfrac{V}{R}$

Substitution: $I = \dfrac{70\ \text{V}}{3.5\ \text{k}\Omega}$

Answer: $I = 0.02\ \text{A} = 20\ \text{mA}$

Finding Voltage Given Current and Resistance

The applied voltage in a series circuit is calculated at the power supply. The current in the circuit is dependent on the applied voltage and total resistance. With resistance and current known, the applied voltage can be found using Ohm's law.

Sample problem 6. _____

With the circuit shown in figure 6-10 and the current given, find the applied voltage.

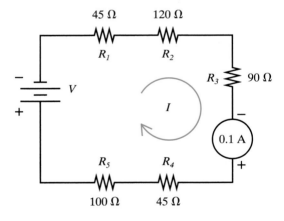

**Figure 6-10. With the current given, find the applied voltage.
(Sample problem 6)**

1. Total resistance must be found first.

Formula: $R_T = R_1 + R_2 + R_3 + ...R_N$

Substitution: $R_T = 45\ \Omega + 120\ \Omega + 90\ \Omega + 45\ \Omega + 100\ \Omega$

Answer: $R_T = 400\ \Omega$

2. Use Ohm's law to calculate applied voltage.

   Formula: $V = I \times R$

   Substitution: $V = 0.1 \text{ A} \times 400 \text{ } \Omega$

   Answer: $V = 40 \text{ V}$

Finding Resistance Given Current and Voltage

The load connected to a voltage source can be made of many different resistors. The current value will be determined by the total resistance of the circuit. Therefore, if voltage and current are given, total resistance is calculated with Ohm's law.

Sample problem 7. _____

When given current and applied voltage of the series circuit in figure 6-11, find total resistance.

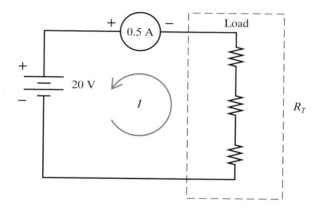

Figure 6-11. With the voltage and current given, find the total resistance. (Sample problem 7)

   Formula: $R_T = \dfrac{V}{I}$

   Substitution: $R_T = \dfrac{20 \text{ V}}{0.5 \text{ A}}$

   Answer: $R_T = 40 \text{ } \Omega$

Finding an Individual Resistance Given Voltage and Current

An unknown resistance in a series circuit can be found if voltage and current are given. The first step is finding the total resistance. Ohm's law is used to find the total resistance, then the addition formula is used to find the missing resistance.

Sample problem 8. _____

Given current and applied voltage, find the missing value of resistance in figure 6-12.

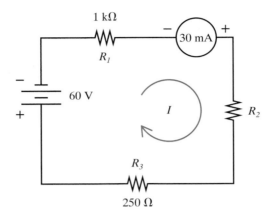

Figure 6-12. Find the missing value for R_2. (Sample problem 8)

1. Use Ohm's law to find R_T.

    Formula: $R_T = \dfrac{V}{I}$

    Substitution: $R_T = \dfrac{60 \text{ V}}{30 \text{ mA}}$

    Answer: $R_T = 2000 \ \Omega = 2 \text{ k}\Omega$

2. Substitute known values into R_T formula.

    Formula: $R_T = R_1 + R_2 + R_3 + ...R_N$

    Substitution: $2 \text{ k}\Omega = 1 \text{ k}\Omega + R_2 + 250 \ \Omega$

    Answer: $R_2 = 750 \ \Omega$

6.4 VOLTAGE DROPS IN SERIES CIRCUITS

When current flows through a resistor, a voltage drop occurs across the resistor. A **voltage drop** is the difference in potential from one side of the resistor to the other. The amount of the voltage drop is directly related to the amount of current and to the size of the resistor. A voltage drop is characterized by the following:

- There is a difference in potential from one side of the resistor to the other.
- The voltage polarity (+/–) is equivalent to the direction of the current.
- The sum of the applied voltage drops equals the applied voltage.

Refer to figure 6-13. The direction of the current is from negative to positive, as indicated by the arrow. Voltage drops across each resistor have a polarity in the same direction as the current. The negative side of the resistor is closest to the negative side of the power supply. The positive side of the resistor is closest to the positive side of the power supply.

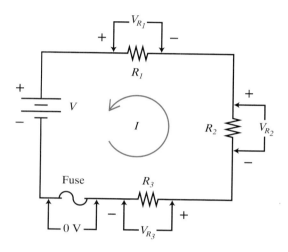

Figure 6-13. Voltage drops can be measured across each resistance in a series circuit. A fuse, in theory, has no resistance. Consequently, it has no voltage drop.

The fuse shown in figure 6-13 has a voltage drop of zero volts. If the fuse is good, it acts like a piece of wire, which has near zero resistance. Current flowing through zero resistance produces zero voltage drop. This is true with all conductors.

Finding Voltage Drops When Given Individual Resistances

To find the voltage drops, it is necessary to know the circuit current. Current is found using Ohm's law with the total resistance and applied voltage. The first step, therefore, is to find the total resistance. Ohm's law is then used to find the voltage across each resistance.

Sample problem 9. _____

Find the voltage drops in the circuit shown in figure 6-14.

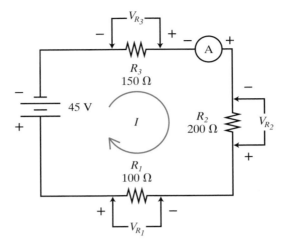

Figure 6-14. Find the individual voltage drops. (Sample problem 9)

1. Total resistance is found using the R_T formula.

   Formula: $R_T = R_1 + R_2 + R_3 + ...R_N$

   Substitution: $R_T = 150\ \Omega + 200\ \Omega + 100\ \Omega$

   Answer: $R_T = 450\ \Omega$

2. Use Ohm's law to find current.

   Formula: $I = \dfrac{V}{R}$

   Substitution: $I = \dfrac{45\ V}{450\Omega}$

   Answer: $I = 0.1\ A$

3. Use Ohm's law to find individual voltage drops.
   a. Voltage drop across R_1:

      Formula: $V_{R_1} = I \times R_1$

      Substitution: $V_{R_1} = 0.1\ A \times 100\ \Omega$

      Answer: $V_{R_1} = 10\ V$

   b. Voltage drop across R_2:

Formula: $V_{R_2} = I \times R_2$

Substitution: $V_{R_2} = 0.1 \text{ A} \times 200 \text{ }\Omega$

Answer: $V_{R_2} = 20 \text{ V}$

c. Voltage drop across R_3:

Formula: $V_{R_3} = I \times R_3$

Substitution: $V_{R_3} = 0.1 \text{ A} \times 150 \text{ }\Omega$

Answer: $V_{R_3} = 15 \text{ V}$

d. Compare the sum of the voltage drops to the total voltage by adding.

Formula: $V_A = V_{R_1} + V_{R_2} + V_{R_3}$

Substitution: $V_A = 10 \text{ V} + 20 \text{ V} + 15 \text{ V}$

Answer: $V_A = 45 \text{ V}$ (Equals applied circuit voltage.)

Finding Selected Other Values Using One Voltage Drop

A voltage drop across one resistor can be used to find the current through that resistor with Ohm's law. Since the current is the same at all points in a series circuit, the current throughout the circuit is known. The total resistance can be found using the current and applied voltage. All other unknown values can be found using similar techniques.

Sample problem 10.

With one voltage drop given, find the unknown values in figure 6-15.

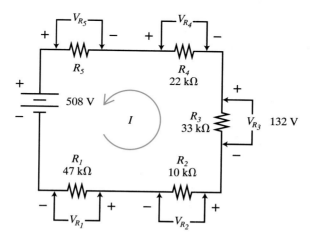

Figure 6-15. Given a voltage drop, find the unknown values.
(Sample problem 10)

1. Use the given voltage drop and its resistance to find current.

   Formula: $I = \dfrac{V_{R_3}}{R_3}$

   Substitution: $I = \dfrac{132 \text{ V}}{33 \text{ k}\Omega}$

   Answer: $I = 4$ mA

2. Use current with the applied voltage to find total resistance.

   Formula: $R_T = \dfrac{V_A}{I}$

   Substitution: $R_T = \dfrac{508 \text{ V}}{4 \text{ mA}}$

   Answer: $R_T = 127$ kΩ

3. Use the R_T formula to find R_5.

   Formula: $R_T = R_1 + R_2 + R_3 + \ldots R_N$

   Substitution: 127 kΩ = 47 kΩ + 10 kΩ + 33 kΩ + 22 kΩ + R_5

   Answer: $R_5 = 15$ kΩ

4. Find remaining voltage drops.
   a. Voltage drop across R_1:

      Formula: $V_{R_1} = I \times R_1$

      Substitution: $V_{R_1} = 4$ mA $\times$ 47 kΩ

      Answer: $V_{R_1} = 188$ V

   b. Voltage drop across R_2:

      Formula: $V_{R_2} = I \times R_2$

      Substitution: $V_{R_2} = 4$ mA $\times$ 10 kΩ

      Answer: $V_{R_2} = 40$ V

   c. Voltage drop across R_4:

      Formula: $V_{R_4} = I \times R_4$

      Substitution: $V_{R_4} = 4$ mA $\times$ 22 kΩ

      Answer: $V_{R_4} = 88$ V

d. Voltage drop across R_5:

Formula: $V_{R_5} = I \times R_5$

Substitution: V_{R_5} = 4 mA × 15 kΩ

Answer: V_{R_5} = 60 V

e. Compare the sum of the voltage drops to the total voltage by adding.

Formula: $V_A = V_{R_1} + V_{R_2} + V_{R_3} + V_{R_4} + V_{R_5}$

Substitution: V_A = 188 V + 40 V + 132 V + 88 V + 60 V

Answer: V_A = 508 V (Equals applied circuit voltage.)

6.5 EFFECTS OF LARGE AND SMALL VALUE RESISTORS

In a series circuit, when a resistor is ten times larger than the other resistors, it dominates the circuit. The largest resistor in the circuit has the largest voltage drop. When one resistor is ten times larger than the others, so much voltage is dropped across that resistor there is little left for the others. The current is also determined by the too-large resistor.

Sample problem 11.

Find the circuit values of figure 6-16, which contains a resistor more than 10 times larger than the other resistors.

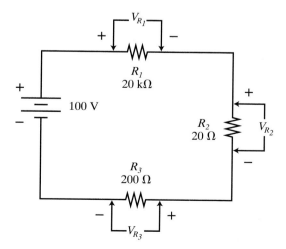

Figure 6-16. Examine the effects of a too-large resistor. (Sample problem 11)

1. Find total resistance:

    Formula: $R_T = R_1 + R_2 + R_3 + \ldots R_N$

    Substitution: $R_T = 20 \text{ k}\Omega + 20 \ \Omega + 200 \ \Omega$

    Answer: $R_T = 20{,}220 \ \Omega$

2. Find current:

    Formula: $I = \dfrac{V}{R}$

    Substitution: $I = \dfrac{100 \text{ V}}{20{,}220 \ \Omega}$

    Answer: $I = 4.95 \text{ mA}$

3. Find individual voltage drops:
    a. Voltage drop across R_1:

        Formula: $V_{R_1} = I \times R_1$

        Substitution: $V_{R_1} = 4.95 \text{ mA} \times 20 \text{ k}\Omega$

        Answer: $V_{R_1} = 99 \text{ V}$

    b. Voltage drop across R_2:

        Formula: $V_{R_2} = I \times R_2$

        Substitution: $V_{R_2} = 4.95 \text{ mA} \times 20 \ \Omega$

        Answer: $V_{R_2} = 0.099 \text{ V}$

    c. Voltage drop across R_3:

        Formula: $V_{R_3} = I \times R_3$

        Substitution: $V_{R_3} = 4.95 \text{ mA} \times 200 \ \Omega$

        Answer: $V_{R_3} = 0.99 \text{ V}$

    d. Check voltage drops by adding:

        Formula: $V_A = V_{R_1} + V_{R_2} + V_{R_3}$

        Substitution: $V_A = 99 \text{ V} + 0.099 \text{ V} + 0.99 \text{ V}$

        Answer: $V_A = 100.089$ (Difference from applied voltage is due to rounding.)

6.6 VOLTAGE DIVIDER CIRCUITS

A series circuit provides a convenient means of changing a high dc voltage into a lower voltage. For example, if a light bulb is rated for 6 V and the power source is 12 V, a series circuit can be used as a **voltage divider** to drop the voltage to the desired amount.

Voltage divider circuits have a **zero volt reference point,** with voltage measurements made with one lead on the reference. With two resistors in a voltage divider, the midpoint is compared to the zero volt reference point. Notice in figure 6-17 that this voltage divider circuit is a series circuit.

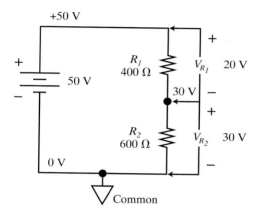

Figure 6-17. Calculate the voltage drop across each resistor. (Sample problem 12)

Calculating Voltages in a Simple Voltage Divider

Voltages in a voltage divider are calculated in the same manner as those in any series circuit. First, the total resistance is calculated. Second, the current is calculated. Third, individual voltage drops are calculated. However, a formula for the voltage divider circuit simplifies this process. In a voltage divider with two resistors, two formulas are used. One formula is used to find the voltage across R_1. The other formula is used to find the voltage across R_2.

To find the voltage across a resistor, multiply the applied voltage times the ratio of the desired resistance divided by the total resistance.

Formula 6.B

$$V_{R_1} = V_A \times \frac{R_1}{R_1 + R_2}$$

V_A is the applied voltage, measured in volts.

V_{R_1} is the voltage across resistor R_1, measured in volts.

R_1 and R_2 are individual resistances, measured in ohms.

Formula 6.C

$$V_{R_2} = V_A \times \frac{R_2}{R_1 + R_2}$$

V_A is the applied voltage, measured in volts.

V_{R_2} is the voltage across resistor R_2, measured in volts.

R_1 and R_2 are individual resistances, measured in ohms.

Sample problem 12.

Calculate the voltage across each resistor in figure 6-17, using the voltage divider formulas.

1. Voltage across R_1:

   Formula: $V_{R_1} = V_A \times \dfrac{R_1}{R_1 + R_2}$

   Substitution: $V_{R_1} = 50 \text{ V} \times \dfrac{400 \text{ } \Omega}{400 \text{ } \Omega + 600 \text{ } \Omega}$

   Answer: $V_{R_1} = 20 \text{ V}$

2. Voltage across R_2:

   Formula: $V_{R_2} = V_A \times \dfrac{R_2}{R_1 + R_2}$

   Substitution: $V_{R_2} = 50 \text{ V} \times \dfrac{600 \text{ } \Omega}{400 \text{ } \Omega + 600 \text{ } \Omega}$

   Answer: $V_{R_2} = 30 \text{ V}$

Calculating Voltages in Multi-Tap Voltage Dividers

A multi-tap voltage divider has several voltages in reference to common. To calculate the voltage at any point in the voltage divider circuit, there is a *general voltage divider formula*. In this formula, the applied voltage is multiplied times the ratio of the point's resistance to ground and divided by the total resistance. The general voltage divider formula is written:

Formula 6.D

$$V_X = V_A \times \frac{R_X}{R_T}$$

V_X is the voltage from a point to ground, measured in volts.

V_A is applied voltage, measured in volts.

R_X is the resistance from a point to ground, measured in ohms.

R_T is total resistance, measured in ohms.

Sample problem 13. _____

Use the general voltage divider formula to find the voltages at each of the tap points to ground in figure 6-18.

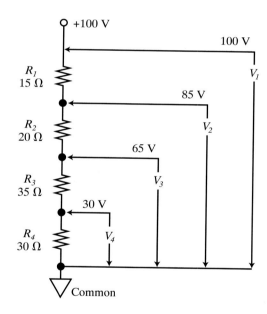

Figure 6-18. Solve for each of the tap points to ground. (Sample problem 13)

1. Voltage V_1 is the applied voltage. No calculations are necessary.

2. Voltage V_2:

    Formula: $V_2 = V_A \times \dfrac{R_2 + R_3 + R_4}{R_1 + R_2 + R_3 + R_4}$

    Substitution: $V_2 = 100 \text{ V} \times \dfrac{20 \ \Omega + 35 \ \Omega + 30 \ \Omega}{15 \ \Omega + 20 \ \Omega + 35 \ \Omega + 30 \ \Omega}$

Intermediate step: $V_2 = 100 \text{ V} \times \dfrac{85 \ \Omega}{100 \ \Omega}$

Answer: $V_2 = 85 \text{ V}$

3.   Voltage V_3:

Formula: $V_3 = V_A \times \dfrac{R_3 + R_4}{R_1 + R_2 + R_3 + R_4}$

Substitution: $V_3 = 100 \text{ V} \times \dfrac{35 \ \Omega + 30 \ \Omega}{15 \ \Omega + 20\Omega + 35 \ \Omega + 30 \ \Omega}$

Intermediate step: $V_3 = 100 \text{ V} \times \dfrac{65 \ \Omega}{100 \ \Omega}$

Answer: $V_3 = 65 \text{ V}$

4.   Voltage V_4:

Formula: $V_4 = V_A \times \dfrac{R_4}{R_1 + R_2 + R_3 + R_4}$

Substitution: $V_4 = 100 \text{ V} \times \dfrac{30 \ \Omega}{15 \ \Omega + 20 \ \Omega + 35 \ \Omega + 30 \ \Omega}$

Intermediate step: $V_4 = 100 \text{ V} \times \dfrac{30 \ \Omega}{100 \ \Omega}$

Answer: $V_4 = 30 \text{ V}$

Voltage Divider with Zero-Volt Reference

As shown in figure 6-19, a zero-volt reference can be placed in a circuit at a point other than at one end. This can be done with one or two power supplies. The four lights

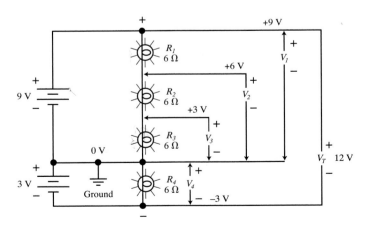

Figure 6-19. Zero-volt ground references can be placed in the middle of a voltage divider. They are not required to be on one end.

in figure 6-19 are exactly the same type, with the same voltage rating. This is an ideal way to make a voltage divider to reduce the voltage for light bulbs. If each of their resistances is the same, the voltage drops are equal. The voltages in reference to ground add, as they would in any voltage divider.

Light bulb R_4 is connected to a battery in such a way that the negative terminal of the battery is away from ground. Therefore, the voltage in reference to ground is a negative. Calculations for this type of voltage divider follow the general voltage divider formula. An alternate method is to use the rules for series circuits. All calculations are made in reference to ground, not from positive to negative.

6.7 POWER IN A SERIES CIRCUIT

Power is directly related to current and voltage. In order for a circuit to have power, both current and voltage must be present. Power can be calculated for the individual resistors and for the complete circuit. Total power, calculated for a complete circuit, is equal to the sum of the individual powers. Total power can also be calculated using Ohm's law with applied voltage and current.

Sample problem 14.
_____

Calculate all of the parameters of the circuit shown in figure 6-20. Circuit parameters are:
1. Total resistance.
2. Current.
3. Voltage drops.
4. Individual powers.
5. Total power.

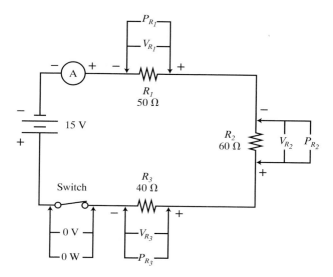

Figure 6-20. Find the unknown parameters. (Sample problem 14)

1. Find total resistance:

   Formula: $R_T = R_1 + R_2 + R_3 + \ldots R_N$

   Substitution: $R_T = 50\ \Omega + 60\ \Omega + 40\ \Omega$

   Answer: $R_T = 150\ \Omega$

2. Find current:

   Formula: $I = \dfrac{V}{R}$

   Substitution: $I = \dfrac{15\ V}{150\ \Omega}$

   Answer: $I = 0.1\ A = 100\ mA$

3. Find individual voltage drops:
   a. Voltage drop across R_1:

      Formula: $V_{R_1} = I \times R_1$

      Substitution: $V_{R_1} = 0.1\ A \times 50\ \Omega$

      Answer: $V_{R_1} = 5\ V$

   b. Voltage drop across R_2:

      Formula: $V_{R_2} = I \times R_2$

      Substitution: $V_{R_2} = 0.1\ A \times 60\ \Omega$

      Answer: $V_{R_2} = 6\ V$

   c. Voltage drop across R_3:

      Formula: $V_{R_3} = I \times R_3$

      Substitution: $V_{R_3} = 0.1\ A \times 40\ \Omega$

      Answer: $V_{R_3} = 4\ V$

   d. Check voltage drops by adding:

      Formula: $V_A = V_{R_1} + V_{R_2} + V_{R_3}$

      Substitution: $V_A = 5\ V + 6\ V + 4\ V$

      Answer: $V_A = 15\ V$

4. Find individual powers:
   Note, any of the power formulas can be used. Select the one that is easiest.
   a. Power dissipated in R_1:

   Formula: $P_{R_1} = I \times V_{R_1}$

   Substitution: $P_{R_1} = 0.1\ \text{A} \times 5\ \text{V}$

   Answer: $P_{R_1} = 0.5\ \text{W}$

   b. Power dissipated in R_2:

   Formula: $P_{R_2} = \text{I} \times V_{R_2}$

   Substitution: $P_{R_2} = 0.1\ \text{A} \times 6\ \text{V}$

   Answer: $P_{R_2} = 0.6\ \text{W}$

   c. Power dissipated in R_3:

   Formula: $P_{R_3} = I \times V_{R_3}$

   Substitution: $P_{R_3} = 0.1\ \text{A} \times 4\ \text{V}$

   Answer: $P_{R_3} = 0.4\ \text{W}$

5. Total power (using two methods):

   a. Sum of the individual powers:

   Formula: $P_T = P_{R_1} + P_{R_2} + P_{R_3}$

   Substitution: $P_T = 0.5\ \text{W} + 0.6\ \text{W} + 0.4\ \text{W}$

   Answer: $P_T = 1.5\ \text{W}$

   b. Current multiplied by applied voltage:

   Formula: $P_T = V_A \times I$

   Substitution: $P_T = 15\ \text{V} \times 0.1\ \text{A}$

   Answer: $P_T = 1.5\ \text{W}$

Sample problem 15. _____

Using the values given in figure 6-21, find the unknown circuit parameters.

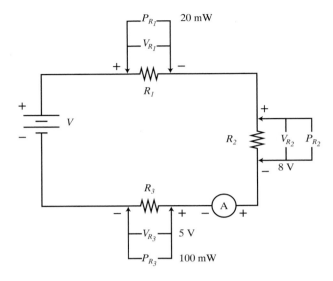

Figure 6-21. Find the unknown parameters. (Sample problem 15)

1. Find current for R_3:

   Formula: $I = \dfrac{P_{R_3}}{V_{R_3}}$

   Substitution: $I = \dfrac{100 \text{ mW}}{5 \text{ V}}$

   Answer: $I = 20 \text{ mA}$

2. Find the value of R_3:

   Formula: $R_3 = \dfrac{V_{R_3}}{I}$

   Substitution: $R_3 = \dfrac{5 \text{ V}}{20 \text{ mA}}$

   Answer: $R_3 = 250 \text{ }\Omega$

3. Find the value of R_2:

   Formula: $R_2 = \dfrac{V_{R_2}}{I}$

   Substitution: $R_2 = \dfrac{8}{20 \text{ mA}}$

Answer: $R_2 = 400 \ \Omega$

4. Find the power dissipated across R_2:

Formula: $P_{R_2} = V_{R_2} \times I$

Substitution: $P_{R_2} = 8 \text{ V} \times 20 \text{ mA}$

Answer: $P_{R_2} = 160 \text{ mW}$

5. Find the value of R_1:

Formula: $R_1 = \dfrac{P_{R_1}}{I^2}$

Substitution: $R_1 = \dfrac{20 \text{ mW}}{(20 \text{ mA})^2}$

Answer: $R_1 = 50 \ \Omega$

6. Find the voltage dropped across R_1:

Formula: $V_{R_1} = I \times R_1$

Substitution: $V_{R_1} = 20 \text{ mA} \times 50 \ \Omega$

Answer: $V_{R_1} = 1 \text{ V}$

7. Find applied voltage by adding all voltage drops:

Formula: $V_A = V_{R_1} + V_{R_2} + V_{R_3}$

Substitution: $V_A = 1 \text{ V} + 8 \text{ V} + 5 \text{ V}$

Answer: $V_A = 14 \text{ V}$

8. Find total power by adding the individual powers:

Formula: $P_T = P_{R_1} + P_{R_2} + P_{R_3}$

Substitution: $P_T = 20 \text{ mW} + 160 \text{ mW} + 100 \text{ mW}$

Answer: $P_T = 280 \text{ mW}$

9. Find total resistance by adding individual resistors:

Formula: $R_T = R_1 + R_2 + R_3$

Substitution: $R_T = 50 \ \Omega + 400 \ \Omega + 250 \ \Omega$

Answer: $R_T = 700 \ \Omega$

6.8 TROUBLESHOOTING SERIES CIRCUITS

Troubleshooting a circuit requires knowledge of how a circuit should work and knowledge of the symptoms that can be expected when there is a failure or defect. Problems can occur in circuit components, connecting wires, switches, and even the voltage source. There are two common problems, open circuits and short circuits. These occurrences can cause a wide variety of symptoms.

Open Circuits

An **open circuit** exists when some point on the current path has an infinite amount of resistance. Current cannot flow in an open circuit.

In an open circuit, the voltage at the point of infinite resistance is equal to the applied voltage. All other points on the circuit have zero volts. Remember, a voltage drop across a resistor is a result of the current through it. If there is no current, voltage is not dropped across the circuit components. Instead, the entire voltage drop appears at the point of the open.

As an example of the full voltage appearing at the point of an open, think of the wall sockets in a house when there is nothing plugged in. The full voltage drop appears across that socket.

As shown in figure 6-22, a switch in the off position is an open circuit. Current does not flow in this circuit. The ammeter reads 0 A. Current, or the lack or it, can be detected by measuring the voltages across the resistors. If there is no voltage drop, there is no current.

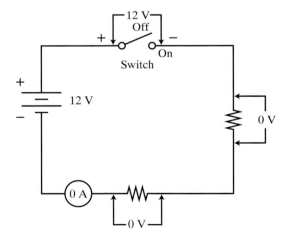

Figure 6-22. A circuit with a switch in the off position is an open circuit. The voltage drop across each resistor in this circuit is zero.

Figure 6-23 demonstrates that if a light bulb connected in series is burned out, it stops current through the entire circuit. The applied voltage is measured across the burned-out light. If there is no current, there is also no power developed.

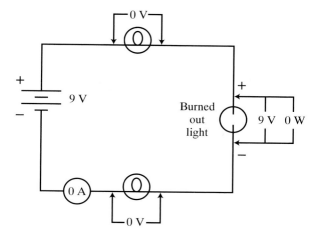

Figure 6-23. If a light bulb burns out in a series circuit, it acts an open.

Fuses are intended to stop the current when they blow. See figure 6-24. The fuse melts and opens when too much current is present, thus protecting the circuit. When troubleshooting a circuit, if there is no voltage available at the load, the fuse is a good place to check first. A normal fuse has zero resistance and would have zero voltage drop. If the fuse is found to have a voltage across it, it is defective.

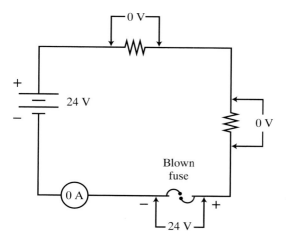

Figure 6-24. Blown fuses are open for a purpose. The fuses open to protect the rest of the circuit from too high a current.

Short Circuits

A **short circuit** is a defect where the load is bypassed. It can occur in a complete circuit, or it can develop in just a section of a circuit. A short in a circuit acts like a conductor, having very low resistance. If the entire load is shorted, there is no limit to the amount of current present. This is a danger to any circuit and an extreme fire hazard. A fuse is intended to protect a circuit from the dangers of a short circuit, as shown in figure 6-25.

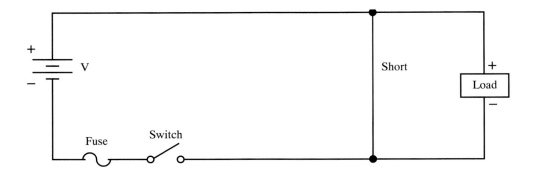

Figure 6-25. With the load shorted, the fuse will blow the instant the switch is closed.

In figure 6-26, the effect of part of a load being shorted is examined. Part A of the figure has all of the normal calculations. There is an equal resistance value for the light bulbs for simplicity. The voltage is distributed equally to the three loads.

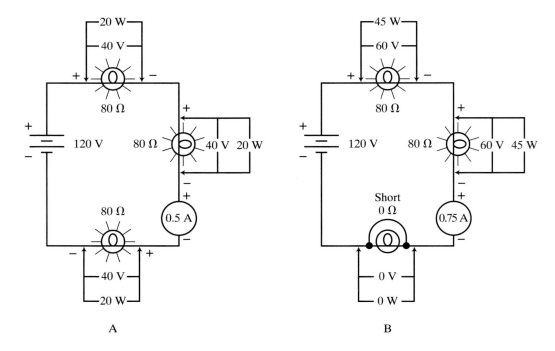

Figure 6-26. Normal circuit calculations are shown on the left. Calculations in a circuit with a short are shown on the right. A short in one part of a series circuit affects the calculations for other parts of the circuit.

In part B of figure 6-26, one load is shorted (the wires connecting it are touching). The total current increases because there is a lower total resistance in the circuit and the voltage remains the same. The voltage distributes equally to the remaining two light bulbs. Each bulb gets a higher voltage, making them much brighter than normal.

An increase in voltage, current, and wattage through the bulbs may be more than they can take. This can result in one or both of the bulbs burning out, resulting in an open circuit. Replacement of the burned light bulb does not correct the short circuit.

6.9 SERIES CIRCUIT GUIDELINES

The primary formulas to use in solving circuits are Ohm's law and the power formulas. The chart in figure 6-27 shows how the use of subscripts describes *which* resistance, voltage, or power is used in the formula. Current is the same throughout a series circuit, therefore it does not use a subscript.

Figure 6-27. Summary of series circuit formulas.

Resistance

Total Resistance:

$$R_T = R_1 + R_2 + R_3 + \dots R_N$$

$$R_T = \frac{V_A}{I} \qquad R_T = \frac{V_A^2}{P_T} \qquad R_T = \frac{P_T}{I^2}$$

Individual Resistors:

$$R_N = \frac{V_{R_N}}{I} \qquad R_N = \frac{P_{R_N}}{I^2} \qquad R_N = \frac{V_{R_N}^2}{P_{R_N}}$$

Current

Using Totals:

$$I = \frac{V_A}{R_T} \qquad I = \frac{P_T}{V_A} \qquad I = \sqrt{\frac{P_T}{R_T}}$$

Individual Currents:

$$I = \frac{V_{R_N}}{R_N} \qquad I = \frac{P_{R_N}}{V_{R_N}} \qquad I = \sqrt{\frac{P_{R_N}}{R_N}}$$

Voltage

Using Totals:

$$V_A = I \times R_T \qquad V_A = \sqrt{P_T \times R_T} \qquad V_A = \frac{P_T}{I}$$

Individual Voltage Drops:

$$V_A = V_{R_1} + V_{R_2} + V_{R_3} + \dots V_{R_N}$$

$$V_A = I \times R_T \qquad V_A = \sqrt{P_T \times R_T} \qquad V_A = \frac{P_T}{I}$$

Power

Using Totals:

$$P_T = V_A \times I \qquad P_T = I^2 \times R_T \qquad P_T = \frac{V_A^2}{R_T}$$

Individual Powers:

$$P_T = P_{R_1} + P_{R_2} + P_{R_3} + \dots P_{R_N}$$

$$P_{R_N} = V_{R_N} \times I \qquad P_{R_N} = I^2 \times R_N \qquad P_{R_N} = \frac{V_{R_N}^2}{R_N}$$

Total resistance, applied voltage, and total power can be found using either the Ohm's law/power formulas or by adding the individual resistances, voltage drops, and powers.

When faced with a circuit to solve and you have no idea where to begin, examine the quantities that are given. Consider each of the four basic circuit parameters (resistance, current, voltage, and power). Be careful to observe the subscripts.

SUMMARY

- A series circuit has only one current path.
- Voltage is dropped across each resistance.
- The total resistance is the sum of the individual resistances.
- The current is the same at all points in the circuit.
- A resistor much larger than the others dominates the circuit, dropping most of the voltage and setting the value of current.
- A resistor much smaller than the others appears to have no effect on the circuit.
- A voltage divider circuit has voltage taps measured to a common reference point.
- Power in a series circuit is the sum of the individual powers.

KEY WORDS AND TERMS GLOSSARY

open circuit: A point in a circuit where the current path is supposed to be that has an infinite amount of resistance. An open circuit will not allow a current.

schematic diagram: A picture that shows how circuit components are electrically connected.

series circuit: An electrical circuit with only one path for the current to flow. Voltage drops across each resistor. Total resistance is the sum of the individual resistances.

short circuit: A defect in a circuit that causes the current to bypass a portion or the load or the entire circuit. If the entire load is shorted, there is no limit to the amount of current flowing.

voltage divider: Reduces the voltage from a power supply as needed by a load.

voltage drop: The difference in potential from one side of a resistor to the other. The amount of voltage drop is directly related to the amount of current and size of resistor. The voltage drop will have a polarity (+/−) equivalent to the direction of the current.

zero volt reference point: The reference point on a voltage divider.

KEY FORMULAS

Formula 6.A

$$R_T = R_1 + R_2 + R_3 + \dots R_N$$

Formula 6.B

$$V_{R_1} = V_A \times \frac{R_1}{R_1 + R_2}$$

Formula 6.C

$$V_{R_2} = V_A \times \frac{R_2}{R_1 + R_2}$$

Formula 6.D

$$V_X = V_A \times \frac{R_X}{R_T}$$

TEST YOUR KNOWLEDGE

Do not write in this text. Please use a separate sheet of paper.

1. List three examples of a series circuit.
2. State the four basic characteristics that identify a series circuit.
3. Write the formula for total resistance in a series circuit when given the individual resistance values.
4. In a certain series circuit with four resistors, the values for total resistance and three of the individual resistances are given. Write an equation that describes how to find the value of the unknown resistance.
5. Current in a series circuit _____.
   a.   drops across each resistor.
   b.   is the same at all points.
   c.   flows from positive to negative.
   d.   is higher near the battery and less at the loads.
6. Voltage in a series circuit _____.
   a.   drops across each resistor.
   b.   is the same at all points.
   c.   flows from positive to negative.
   d.   none of the above.
7. Power in a series circuit _____.
   a.   is the same at all points.
   b.   is the sum of the individual powers.
   c.   flows from positive to negative.
   d.   none of the above.

8. What is the effect on a series circuit if one resistor is 10 times larger than any other?

9. What is the effect on a series circuit if one resistor is very small in comparison to the others?

10. Write the formula to find the voltage drop across one of the resistors in a two-resistor voltage divider.

11. In a series circuit, all of the applied voltage is dropped across the fuse. Is the fuse open or shorted?

12. A circuit has five light bulbs connected in series to a 100 volt supply. If one light goes out and the others get bright, what could be the problem?

13. A circuit has five light bulbs connected in series to a 100 volt supply and all of the lights go out. The ammeter reads zero amps. A voltmeter reads 100 volts at the supply and zero volts across four of the lights. The meter reads 100 volts across the fifth bulb. What is the possible problem?

14. Using the values given in the circuit shown in figure 6-28, calculate total resistance.

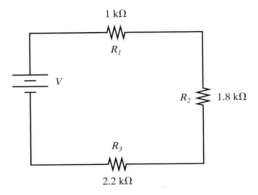

Figure 6-28. Problem #14.

15. In figure 6-29, the total resistance and the values of two resistors are given. Find the value of the missing resistor.

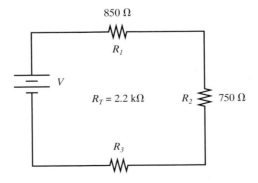

Figure 6-29. Problem #15.

16. The color codes of three resistors are given in figure 6-30. Find:
    a.   Total resistance.
    b.   The current flowing in the circuit.

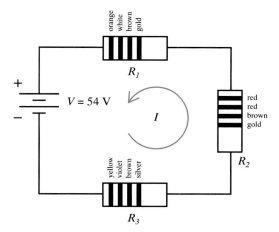

Figure 6-30. Problem #16.

17. The circuit shown in figure 6-31 gives the values of the individual resistances and current. Find:
    a.   Total resistance.
    b.   The applied voltage.

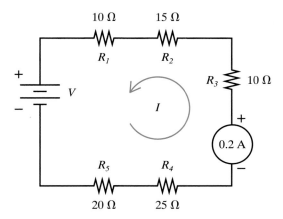

Figure 6-31. Problem #17.

18. Given the values for the applied voltage and current of the circuit shown in figure 6-32, find:
    a.   Total resistance.
    b.   The value of the missing resistor.

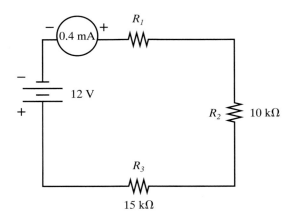

Figure 6-32. Problem #18.

19. Use the color codes of the resistors in figure 6-33 and the applied voltage to find:
    a.   Total resistance.
    b.   Current.
    c.   Voltage drops.

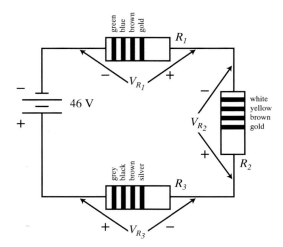

Figure 6-33. Problem #19.

20. Given the voltage drops and resistance values in figure 6-34, find:
    a.   Current.
    b.   Value of R_1.
    c.   Total resistance.
    d.   Applied voltage.

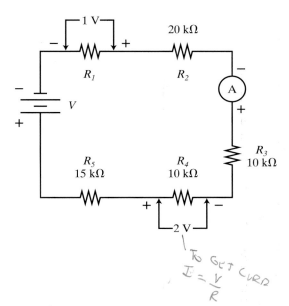

Figure 6-34. Problem #20.

21. Using the values of applied voltage, current, and one voltage drop, shown in figure 6-35, find:
    a. R_T
    b. R_2
    c. R_3

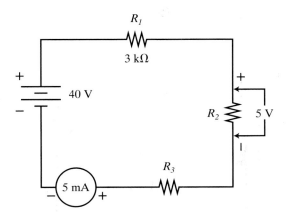

Figure 6-35. Problem #21.

22. Use the color codes of the resistors in the circuit shown in figure 6-36 to find:
    a.   Total resistance.
    b.   Current.
    c.   Power of each resistor.
    d.   Total power.

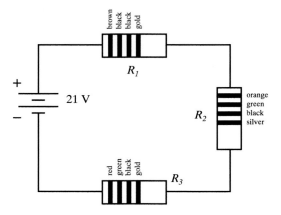

Figure 6-36. Problem #22.

23. Use the two-resistor voltage divider formulas to find the voltages in figure 6-37:
    a. V_1
    b. V_2

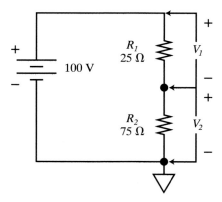

Figure 6-37. Problem #23.

24. Use the multi-tap voltage divider formula to find the following voltages from figure 6-38.
    a. V_1
    b. V_2
    c. V_3
    d. V_4

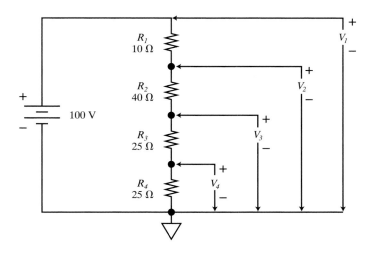

Figure 6-38. Problem #24.

25. Use the values of resistors in figure 6-39 to find:
    a. Total resistance.
    b. Current.
    c. V_{R_1}
    d. P_{R_1}
    e. V_{R_2}
    f. P_{R_2}
    g. V_{R_3}
    h. P_{R_3}
    i. Total power.

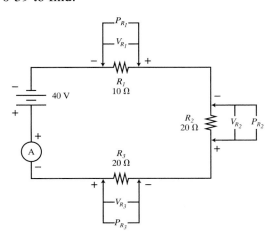

Figure 6-39. Problem #25.

26. Use the voltage drops and powers given in figure 6-40 to find:
    a.   Current.
    b. R_2
    c. R_1
    d. P_{R_1}
    e. V_{R_3}
    f. R_3
    g.   Total resistance.
    h.   Applied voltage.
    i.   Total power.

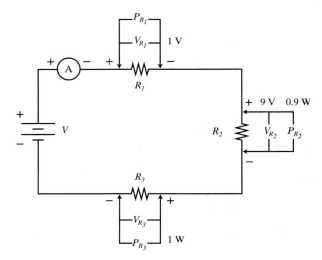

Figure 6-40. Problem #26.

Chapter 7
Parallel Circuits

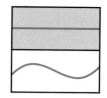

Upon completion of this chapter, you will be able to:
- Recognize and identify loads connected in parallel.
- Calculate the total resistance of a parallel circuit using different formulas.
- Calculate power and current in branch resistances and the total circuit.
- Analyze branch currents to determine mainline currents at various points in a parallel circuit.
- Evaluate symptoms of faulty circuits to determine the problem.

There are two basic types of circuit connections, series and parallel. These connections have very different operating characteristics. Chapter 6 examined series circuits. This chapter covers parallel circuits. In Chapter 8 you will combine the two types.

An example of a parallel circuit is seen in electrical appliances plugged into the wall outlets in a house. Figure 7-1 is a wiring diagram of six rooms in a house. At the top, electrical power enters from the power company. It feeds the main breaker, which is subdivided into individual circuits. These circuits are connected in parallel to each room. Wall outlets appear to be in series because they are connected one after the other. However, due to their internal wiring, they are actually in parallel. When a device is plugged into an outlet, or lights are wired through a switch, the loads are all connected in parallel.

7.1 IDENTIFICATION OF A PARALLEL CIRCUIT

A **parallel circuit** can be identified by these characteristics:
- Voltage is the same throughout the circuit.
- Current splits into each individual branch.
- Total current is the sum of the branch currents.
- Total resistance is smaller than the smallest branch resistance.
- Total power is the sum of the individual powers.

A parallel circuit is used whenever more than one load is to be connected to the same power supply and receive the same voltage. For example, in an automobile, the lights, radio, air conditioner, and computer are all connected such that they receive approximately 12 V dc from the battery/alternator.

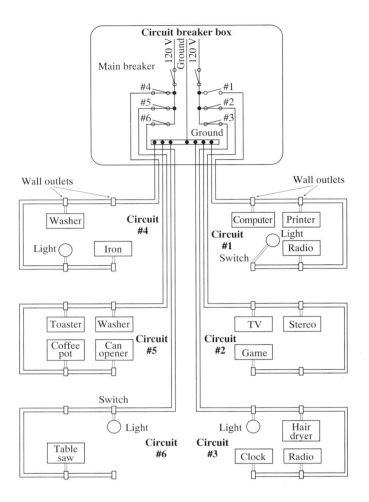

Figure 7-1. House wiring is made up of many parallel circuits.

7.2 TOTAL RESISTANCE OF A PARALLEL CIRCUIT

In a parallel circuit, voltage is the same at all points. The current in an individual resistor is determined by that resistance. Each parallel branch is independent of the others. One branch does not affect the current in the other branches. The total circuit current is the sum of the individual **branch currents.** This is discussed in detail later in this chapter.

The voltage source supplies a current larger than any single branch current. Each branch is only a portion of the total current. The total resistance, therefore, is *less* than the smallest value of the individual branch resistances. This is because the current is inversely related to resistance.

If the total current and applied voltage are both known, Ohm's law is used to calculate the total resistance. When only resistance values are known, the total resistance can be found using four different methods. These methods are:

- The reciprocal formula.
- The conductance method.
- The shortcut formula.
- The equal resistors formula.

Reciprocal Formula

To find the reciprocal of a number, the number is inverted. The reciprocal of 3 is 1/3. The reciprocal of 1/4 would be 4/1 or 4. The reciprocal of some variable R is $1/R$. The reciprocal of the total resistance in a parallel circuit is the sum of the reciprocals of each branch resistance. Remember that the reciprocals, if left as fractions, must be added with a common denominator. With a calculator, it is easy to convert the fractions to decimals before adding.

Note that the left side of the equation, R_T, is also written as a reciprocal. The final step in solving for R_T is to take the reciprocal of the sum of the decimals.

Formula 7.A

$$\frac{1}{R_T} = \frac{1}{R_1} + \frac{1}{R_2} + \frac{1}{R_3} + \cdots \frac{1}{R_N}$$

R_T is the total resistance, measured in ohms.

R_1 through R_N are the individual resistances, measured in ohms.

N is the number of branches in the circuit.

Sample problem 1.

Use the reciprocal formula to find total resistance of the circuit in figure 7-2. Also find total current using Ohm's law. In this figure, the voltmeter across the parallel branches reads the same as the applied voltage.

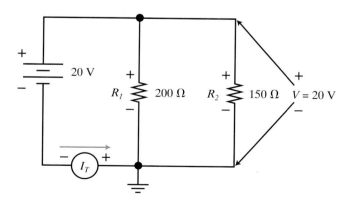

Figure 7-2. Find the total resistance. (Sample problems 1, 2, and 3)

1. Total resistance:

Formula:
$$\frac{1}{R_T} = \frac{1}{R_1} + \frac{1}{R_2} + \frac{1}{R_3} + \dots \frac{1}{R_N}$$

Substitution:
$$\frac{1}{R_T} = \frac{1}{200 \ \Omega} + \frac{1}{150 \ \Omega}$$

Common denominator (600):
$$\frac{1}{R_T} = \frac{3}{600} + \frac{4}{600} = \frac{7}{600}$$

Invert both sides:
$$R_T = \frac{600}{7}$$

Answer:
$$R_T = 85.7 \ \Omega$$

2. Total current:

Formula:
$$I_T = \frac{V}{R_T}$$

Substitution:
$$I_T = \frac{20 \ V}{85.7 \ \Omega}$$

Answer:
$$I_T = 0.233 \ A = 233 \ mA$$

Conductance Method

Conductance (G), measured in siemens (S), is the reciprocal of resistance. Therefore, it is a useful tool for solving the total resistance in parallel circuits. To find total resistance, take the reciprocal of total conductance.

The conductance method uses a calculator to change each reciprocal into its decimal equivalent. The decimals are then added. This results in finding the total conductance. The total resistance is the reciprocal of this number. When the reciprocal formula is used with decimals rather than fractions, it is actually the conductance formula.

Formula 7.B

$$G_T = G_1 + G_2 + G_3 + \dots G_N$$

G_T is the total conductance, measured in siemens.

G_1 through G_N are the individual conductances, measured in siemens.

N is the number of branches in the circuit.

Formula 7.C

$$G_T = \frac{1}{R_1} + \frac{1}{R_2} + \frac{1}{R_3} + \cdots \frac{1}{R_N}$$

G_T is the total conductance, measured in siemens.

R_1 through R_N are the individual resistances, measured in ohms.

N is the number of branches in the circuit.

Formula 7.D

$$R_T = \frac{1}{G_T} \text{ or } G_T = \frac{1}{R_T}$$

R_T is the total resistance, measured in ohms.

G_T is the total conductance, measured in siemens.

Sample problem 2.

Use the conductance method to find total resistance of the circuit in figure 7-2.

Formula:	$G_T = \dfrac{1}{R_1} + \dfrac{1}{R_2} + \dfrac{1}{R_3} + \cdots \dfrac{1}{R_N}$
Substitution:	$G_T = \dfrac{1}{200\ \Omega} + \dfrac{1}{150\ \Omega}$
Conductance step:	$G_T = 5\text{E}{-}03\text{ S} + 6.67\text{E}{-}03\text{ S}$ *Note: The fractions are changed to decimals.*
Total conductance:	$G_T = 11.67\text{E}{-}03\text{ S}$
Total resistance:	$R_T = \dfrac{1}{G_T} = \dfrac{1}{11.67\text{E}{-}03\text{ S}}$
Answer:	$R_T = 85.7\ \Omega$

Shortcut Formula

The shortcut formula can only be used on parallel circuits with only two loads. This formula does not require reciprocals.

Formula 7.E

$$R_T = \frac{R_1 \times R_2}{R_1 + R_2}$$

R_T is the total resistance, measured in ohms.

R_1 is the resistance across resistor R_1, measured in ohms.

R_2 is the resistance across resistor R_2, measured in ohms.

Sample problem 3. _____

Use the shortcut formula to find the total resistance of the circuit in figure 7-2.

Formula: $R_T = \dfrac{R_1 \times R_2}{R_1 + R_2}$

Substitution: $R_T = \dfrac{200\ \Omega \times 150\ \Omega}{200\ \Omega + 150\ \Omega}$

Intermediate step: $R_T = \dfrac{30{,}000}{350}$

Answer: $R_T = 85.7\ \Omega$

More Than Two Resistors in Parallel

The reciprocal and conductance formula is used to find total resistance with any number of resistors. Ohm's law can be used to find total current.

Sample problem 4. _____

Find total resistance and total current of the circuit in figure 7-3.

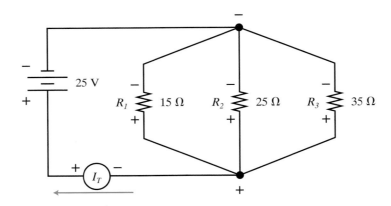

Figure 7-3. Find the total resistance and the total current.
(Sample problem 4)

1. Total resistance:

   Formula:
   $$\frac{1}{R_T} = \frac{1}{R_1} + \frac{1}{R_2} + \frac{1}{R_3} + \cdots \frac{1}{R_N}$$

   Substitution:
   $$\frac{1}{R_T} = \frac{1}{15\ \Omega} + \frac{1}{25\ \Omega} + \frac{1}{35\ \Omega}$$

   Conductance step:
   $$\frac{1}{R_T} = G_T = 0.0667\ S + 0.04\ S + 0.0286\ S$$

   $$\frac{1}{R_T} = 0.1353\ S$$

   Answer: $R_T = 7.39\ \Omega$

2. Total current:

   Formula: $I_T = \dfrac{V}{R_T}$

   Substitution: $I_T = \dfrac{25\ V}{7.39\ \Omega}$

   Answer: $I_T = 3.38\ A$

Equal-Value Resistors in Parallel

When resistances of equal value are connected in parallel their total resistance can be found by dividing the ohmic value by the number of equal-value resistors. The reciprocal formula and conductance methods are also alternate choices.

Formula 7.F

$$R_T = \frac{R}{N}$$

R_T is the total resistance, measured in ohms.

R is the resistance of each resistor, measured in ohms.

N is the total number of resistors.

Sample problem 5.

Use the formula for equal-value resistors in parallel to find the total resistance of the circuit in figure 7-4. This figure uses the ground symbol at the base of each branch. These branches are connected together just as if they were connected with a piece of wire.

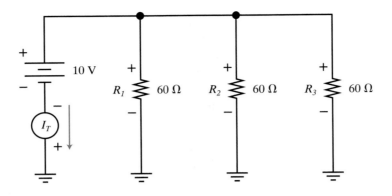

Figure 7-4. Find the total resistance. (Sample problem 5)

Formula: $R_T = \dfrac{R}{N}$

Substitution: $R_T = \dfrac{60\ \Omega}{3}$

Answer: $R_T = 20\ \Omega$

7.3 POWER IN A PARALLEL CIRCUIT

Power calculations are the same in all types of circuits. Individual powers can be found using the power formulas with any two of R, V, or I. Total power can also be found using the power formulas or from the sum of the individual powers. The power formulas listed in Chapters 4 and 6 are repeated here.

$$P_T = P_{R_1} + P_{R_2} + P_{R_3} + \dots P_{R_N}$$

$$P_T = V \times I_T$$

$$P_T = I_T^2 \times R_T$$

$$P_T = \dfrac{V^2}{R}$$

Sample problem 6.

Use the information given in figure 7-5 to find V and P_T.

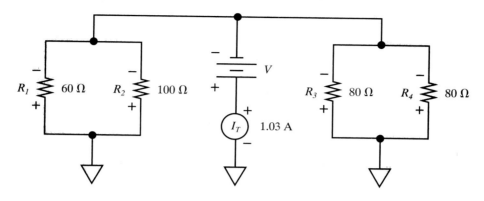

Figure 7-5. Find the voltage and total power. (Sample problem 6)

1. Use the reciprocal formula to find total resistance.

   Formula: $$\frac{1}{R_T} = \frac{1}{R_1} + \frac{1}{R_2} + \frac{1}{R_3} + \dots \frac{1}{R_N}$$

   Substitution: $$\frac{1}{R_T} = \frac{1}{60\ \Omega} + \frac{1}{100\ \Omega} + \frac{1}{80\ \Omega} + \frac{1}{80\ \Omega}$$

   Conductance step: $$\frac{1}{R_T} = G_T = 0.0167\text{ S} + 0.01\text{ S} + 0.0125\text{ S} + 0.0125\text{ S}$$

   $$\frac{1}{R_T} = 0.0517\text{ S}$$

   Answer: $R_T = 19.3\ \Omega$

2. Use total resistance and total current to find total power.

   Formula: $P_T = I_T^2 \times R_T$

   Substitution: $P_T = (1.03\text{ A})^2 \times 19.3\ \Omega$

   Answer: $P_T = 20.5\text{ W}$

3. Use the known values to find applied voltage.

   Formula: $V = I_T \times R_T$

   Substitution: $V = 1.03\text{ A} \times 19.3\ \Omega$

   Answer: $V = 19.9\text{ V}$

Sample problem 7. _____

Use the information given in figure 7-6 to find R_T, R_1, and P_T.

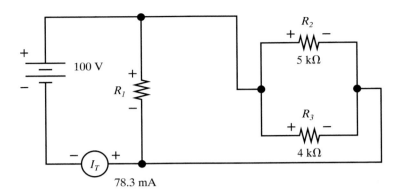

Figure 7-6. Find R_1, the total resistance, and the total power. (Sample problem 7)

1. Use Ohm's law to find total resistance.

    Formula: $R_T = \dfrac{V}{I_T}$

    Substitution: $R_T = \dfrac{100 \text{ V}}{78.3 \text{ mA}}$

    Answer: $R_T = 1277 \ \Omega$

2. Substitute the known values into the reciprocal formula. Add the values on the right, then subtract from the left side of the equation.

    Formula: $\dfrac{1}{R_T} = \dfrac{1}{R_1} + \dfrac{1}{R_2} + \dfrac{1}{R_3} + \dots \dfrac{1}{R_N}$

    Substitution: $\dfrac{1}{1277 \ \Omega} = \dfrac{1}{R_1} + \dfrac{1}{5 \text{ k}\Omega} + \dfrac{1}{4 \text{ k}\Omega}$

    Conductance step: $7.83\text{E}{-}04 \text{ S} = \dfrac{1}{R_1} + 2\text{E}{-}04 \text{ S} + 2.5\text{E}{-}04 \text{ S}$

    After addition: $7.83\text{E}{-}04 \text{ S} = \dfrac{1}{R_1} + 4.5\text{E}{-}04 \text{ S}$

    After subtraction: $3.33\text{E}{-}04 \text{ S} = \dfrac{1}{R_1}$

    Answer: $R_1 = 3 \text{ k}\Omega$

3. Use the known values to find total power.

In this problem, I_T and V are given values. Whenever possible, it is best to use given values rather than calculated. If there is a mistake in the calculated values, it will be transferred to the next calculation. Using given values can help to avoid errors.

Formula: $P_T = I_T \times V$

Substitution: $P_T = 78.3 \text{ mA} \times 100 \text{ V}$

Answer: $P_T = 7.83 \text{ W}$

Sample problem 8. _____

Use the information given in figure 7-7 to find the unknown circuit values.

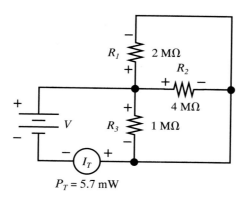

$P_T = 5.7 \text{ mW}$

Figure 7-7. Find the voltage, total resistance, the total current. (Sample problem 8)

1. Use the reciprocal formula to find total resistance.

Formula: $\dfrac{1}{R_T} = \dfrac{1}{R_1} + \dfrac{1}{R_2} + \dfrac{1}{R_3} + \cdots \dfrac{1}{R_N}$

Substitution: $\dfrac{1}{R_T} = \dfrac{1}{2 \text{ M}\Omega} + \dfrac{1}{4 \text{ M}\Omega} + \dfrac{1}{1 \text{ M}\Omega}$

Conductance step: $\dfrac{1}{R_T} = 5\text{E–}07 \text{ S} + 2.5\text{E–}07 \text{ S} + 1\text{E–}06 \text{ S}$

$\dfrac{1}{R_T} = 1.75\text{E–}06 \text{ S}$

Answer: $R_T = .571 \text{ M}\Omega = 571 \text{ k}\Omega$

2. Find voltage using total power and total resistance.

Formula: $V = \sqrt{P_T + R_T}$

Substitution: $V = \sqrt{5.7 \text{ mW} \times 571 \text{ k}\Omega}$

Answer: $V = 57$ V

7.4 BRANCH CURRENT AND POWER

When current leaves a voltage source, it flows through every path available. The primary identification of a parallel circuit is having more than one current path. The amount of current flowing in a particular branch depends on the branch resistance. Power is developed in a resistor with current. Voltage drops are only considered if the branch has more than one resistance.

Total current is the sum of the individual branch currents. Total power is the sum of the individual powers. Examine figure 7-8. The arrows indicating the current show how the total current is the combination of branch currents. Total current can also be calculated using Ohm's law.

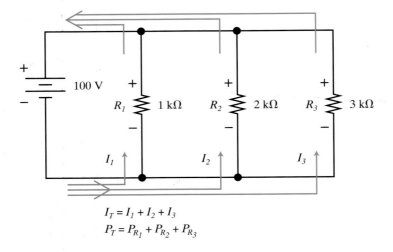

$$I_T = I_1 + I_2 + I_3$$
$$P_T = P_{R_1} + P_{R_2} + P_{R_3}$$

Figure 7-8. Find the total current and total power. (Sample problem 9)

Formula 7.G

$$I_T = I_1 + I_2 + I_3 + ...I_N$$

I_T is the total current, measured in amperes.

I_1 through I_N are the branch currents, measured in amperes.

N is the number of branches in the circuit

Any of the other current formulas derived from Ohm's law or the power formulas can be used as well, provided the proper values are known.

$$I_T = \frac{V}{R_T}$$

$$I_T = \frac{P_T}{V}$$

$$I_T = \sqrt{\frac{P_T}{R_T}}$$

Sample problem 9. _____

Find total current and total power of the circuit in figure 7-8.

1. Branch currents are found using Ohm's law.
    a. Current in R_1:

        Formula: $I_1 = \dfrac{V}{R_1}$

        Substitution: $I_1 = \dfrac{100\text{ V}}{1\text{ k}\Omega}$

        Answer: $I_1 = 0.1\text{ A} = 100\text{ mA}$

    b. Current in R_2:

        Formula: $I_2 = \dfrac{V}{R_2}$

        Substitution: $I_2 = \dfrac{100\text{ V}}{2\text{ k}\Omega}$

        Answer: $I_2 = 0.05\text{ A} = 50\text{ mA}$

    c. Current in R_3:

        Formula: $I_3 = \dfrac{V}{R_3}$

        Substitution: $I_3 = \dfrac{100\text{ V}}{3\text{ k}\Omega}$

        Answer: $I_3 = 0.0333\text{ A} = 33.3\text{ mA}$

2. Total current is found using the sum of the branch currents.

    Formula: $I_T = I_1 + I_2 + I_3 + ...I_N$

    Substitution: $I_T = 100 \text{ mA} + 50 \text{ mA} + 33.3 \text{ mA}$

    Answer: $I_T = 183.3 \text{ mA}$

3. Power in each branch is found using the power formulas. It is best to use the given values of voltage and resistance.
    a. Power of R_1:

        Formula: $P_{R_1} = \dfrac{V^2}{R_1}$

        Substitution: $P_{R_1} = \dfrac{(100 \text{ V})^2}{1 \text{ k}\Omega}$

        Answer: $P_{R_1} = 10 \text{ W}$

    b. Power of R_2:

        Formula: $P_{R_2} = \dfrac{V^2}{R_2}$

        Substitution: $P_{R_2} = \dfrac{(100 \text{ V})^2}{2 \text{ k}\Omega}$

        Answer: $P_{R_2} = 5 \text{ W}$

    c. Power of R_3:

        Formula: $P_{R_3} = \dfrac{V^2}{R_3}$

        Substitution: $P_{R_3} = \dfrac{(100 \text{ V})^2}{3 \text{ k}\Omega}$

        Answer: $P_{R_3} = 3.33 \text{ W}$

4. Total power is found using the sum of the individual powers:

    Formula: $P_T = P_{R_1} + P_{R_2} + P_{R_3} + ... P_{R_N}$

    Substitution: $P_T = 10 \text{ W} + 5 \text{ W} + 3.33 \text{ W}$

    Answer: $P_T = 18.33 \text{ W}$

Sample problem 10.

Use the information given in figure 7-9 to find voltage, branch and total currents, branch and total powers, and R_3. Also, use Ohm's law to find total resistance. Notice in this drawing, the use of the common connection symbol. Treat loads connected at this point as if connected to a conductor.

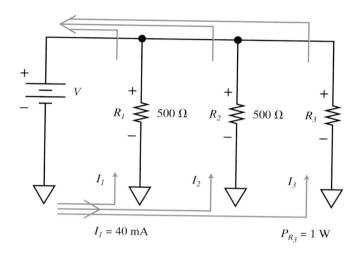

Figure 7-9. Find the unknown circuit parameters. (Sample problem 10)

1. Use Ohm's law to find voltage using I_1 and R_1:

    Formula: $V = I_1 \times R_1$

    Substitution: $V = 40 \text{ mA} \times 500 \text{ }\Omega$

    Answer: $V = 20 \text{ V}$

2. R_3 can be found with Ohm's law, using P_{R_3} and V:

    Formula: $R_3 = \dfrac{V^2}{P_{R_3}}$

    Substitution: $R_3 = \dfrac{(20 \text{ V})^2}{1 \text{ W}}$

    Answer: $R_3 = 400 \text{ }\Omega$

3. Branch currents are found with Ohm's law:

    a. Current in R_1:

        Answer: I_1 is given as 40 mA

b. Current in R_2:

   Formula: $I_2 = \dfrac{V}{R_2}$

   Substitution: $I_2 = \dfrac{200 \text{ V}}{500 \text{ }\Omega}$

   Answer: $I_2 = 0.04 \text{ A} = 40 \text{ mA}$

c. Current in R_3:

   Formula: $I_3 = \dfrac{V}{R_3}$

   Substitution: $I_3 = \dfrac{20 \text{ V}}{400 \text{ }\Omega}$

   Answer: $I_3 = 0.05 \text{ A} = 50 \text{ mA}$

4. Total current is the sum of the branch currents:

   Formula: $I_T = I_1 + I_2 + I_3 + \ldots I_N$

   Substitution: $I_T = 40 \text{ mA} + 40 \text{ mA} + 50 \text{ mA}$

   Answer: $I_T = 130 \text{ mA}$

5. Branch powers are found using Ohm's law:

   a. Power of R_1:

      Formula: $P_{R_1} = \dfrac{V^2}{R_1}$

      Substitution: $P_{R_1} = \dfrac{(20 \text{ V})^2}{500 \text{ }\Omega}$

      Answer: $P_{R_1} = 0.8 \text{ W} = 800 \text{ mW}$

   b. Power of R_2:

      Formula: $P_{R_2} = \dfrac{V^2}{R_2}$

      Substitution: $P_{R_2} = \dfrac{(20 \text{ V})^2}{500 \text{ }\Omega}$

      Answer: $P_{R_2} = 0.8 \text{ W} = 800 \text{ mW}$

   c. Power of R_3:

      Answer: P_{R_3} is given as 1 W

6. Total power is found using the sum of the individual powers:

   Formula: $P_T = P_{R_1} + P_{R_2} + P_{R_3} + \ldots P_{R_N}$

Substitution: $P_T = 0.8 \text{ W} + 0.8 \text{ W} + 1 \text{ W}$

Answer: $P_T = 2.6 \text{ W}$

7. Total resistance is found using Ohm's law, although the reciprocal formula could also be used:

Formula: $R_T = \dfrac{V}{I_T}$

Substitution: $R_T = \dfrac{20 \text{ V}}{130 \text{ mA}}$

Answer: $R_T = 154 \ \Omega$

Sample problem 11. _____

Use the information given in figure 7-10 to find voltage, branch currents, R_1, and branch powers. This circuit drawing indicates that all points are connected to a common chassis ground.

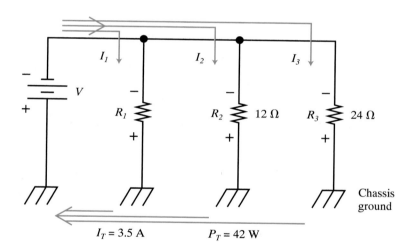

Figure 7-10. Find the unknown circuit parameters. (Sample problem 11)

1. Voltage is found using total current and total power:

Formula: $V = \dfrac{P_T}{I_T}$

Substitution: $V = \dfrac{42 \text{ W}}{3.5 \text{ A}}$

Answer: $V = 12 \text{ V}$

2. Branch currents with known resistors are calculated with Ohm's law:
   a. Current in R_2:

   Formula: $I_2 = \dfrac{V}{R_2}$

   Substitution: $I_2 = \dfrac{12\ \text{V}}{12\ \Omega}$

   Answer: $I_2 = 1\ \text{A}$

   b. Current in R_3:

   Formula: $I_3 = \dfrac{V}{R_3}$

   Substitution: $I_3 = \dfrac{12\ \text{V}}{24\ \Omega}$

   Answer: $I_3 = 0.5\ \text{A}$

3. Current in R_1 is found using the sum of the branch currents:

   Formula: $I_T = I_1 + I_2 + I_3 + \ldots I_N$

   Substitution: $3.5\ \text{A} = I_1 + 1\ \text{A} + 0.5\ \text{A}$

   Intermediate step: $3.5\ \text{A} - 1.5\ \text{A} = I_1$

   Answer: $I_1 = 2\ \text{A}$

4. Unknown resistor R_1 is found using Ohm's law with branch current and voltage:

   Formula: $R_1 = \dfrac{V}{I_1}$

   Substitution: $R_1 = \dfrac{12\ \text{V}}{2\ \text{A}}$

   Answer: $R_1 = 6\ \Omega$

5. Branch powers are found with the power formulas:
   a. Power in R_1:

   Formula: $P_{R_1} = I_{R_1}^2 \times R_1$

   Substitution: $P_{R_1} = (2\ \text{A})^2 \times 6\ \Omega$

   Answer: $P_{R_1} = 24\ \text{W}$

b. Power in R_2:

Formula: $P_{R_2} = I_{R_2}^{\ 2} \times R_2$

Substitution: $P_{R_2} = (1\ A)^2 \times 12\ \Omega$

Answer: $P_{R_2} = 12\ W$

c. Power in R_3:

Formula: $P_{R_3} = I_{R_3}^{\ 2} \times R_3$

Substitution: $P_{R_3} = (0.5\ A)^2 \times 24\ \Omega$

Answer: $P_{R_3} = 6\ W$

7.5 MAINLINE CURRENT

Current flows to and from the branches along wires common to all branches. These positive and negative common wires are referred to as the **mainline buss-bar** wires. The current at a particular point along the mainline buss-bar is dependent on the manner in which each branch is connected.

Current is largest near the power supply. The current measured at any point is the sum of the branch currents that are farther from the voltage source. Refer to figure 7-11 and sample problem 12. In terms of voltage, it makes no difference which loads are closer to the power supply. In some cases, fuses that trigger at different currents are used for the various loads. In these instances, it makes a difference where the loads are placed.

Sample problem 12.

Calculate branch and mainline buss-bar currents of the circuit shown in figure 7-11.

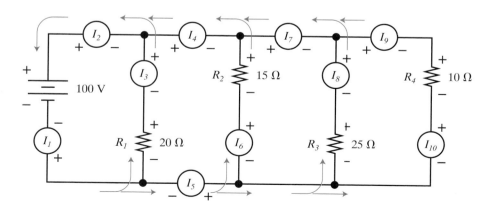

**Figure 7-11. Calculate the branch and mainline buss-bar currents.
(Sample problem 11)**

1. Branch currents are found with Ohm's law:
   a. Current in R_1, represented by ammeter I_3:

      Formula: $I_3 = \dfrac{V}{R_1}$

      Substitution: $I_3 = \dfrac{100 \text{ V}}{20 \text{ }\Omega}$

      Answer: $I_3 = 5 \text{ A}$

   b. Current in R_2, represented by ammeter I_6:

      Formula: $I_6 = \dfrac{V}{R_2}$

      Substitution: $I_6 = \dfrac{100 \text{ V}}{15 \text{ }\Omega}$

      Answer: $I_6 = 6.67 \text{ A}$

   c. Current in R_3, represented by ammeter I_8:

      Formula: $I_8 = \dfrac{V}{R_3}$

      Substitution: $I_8 = \dfrac{100 \text{ V}}{25 \text{ }\Omega}$

      Answer: $I_8 = 4 \text{ A}$

   d. Current in R_4, represented by ammeter I_{10}:

      Formula: $I_{10} = \dfrac{V}{R_4}$

      Substitution: $I_{10} = \dfrac{100 \text{ V}}{10 \text{ }\Omega}$

      Answer: $I_{10} = 10 \text{ A}$

2. Mainline buss-bar currents are found by starting at a point furthest from the power supply:
   Current in ammeter I_9 measures the current in R_4.

   $$I_9 = I_{10} = 10 \text{ A}$$

Current in ammeter I_7 measures the current through I_8 and I_9.

$$I_7 = I_8 + I_9$$
$$I_7 = 4 \text{ A} + 10 \text{ A}$$
$$I_7 = 14 \text{ A}$$

Current in ammeter I_4 measures the current through I_6 (I_8 plus I_9) and I_7.

$$I_4 = I_6 + I_7$$
$$I_4 = 6.67 \text{ A} + 14 \text{ A}$$
$$I_4 = 20.67 \text{ A}$$

Ammeter I_5 is at the same point in the circuit as I_4.

$$I_5 = I_4 = 20.67 \text{ A}$$

Current in ammeter I_2 measures the current through I_3 and I_4.

$$I_2 = I_3 + I_4$$
$$I_2 = 5 \text{ A} + 20.67 \text{ A}$$
$$I_2 = 25.67 \text{ A}$$

Ammeter I_1 is at the same point in the circuit as I_2.

$$I_1 = I_2 = 25.67 \text{ A}$$

Ammeters I_1 and I_2 are each in the position to measure total current, I_T, which is the sum of the branch currents.

Formula: $I_T = I_3 + I_6 + I_8 + I_{10}$

Substitution: $I_T = 5 \text{ A} + 6.67 \text{ A} + 4 \text{ A} + 10 \text{ A}$

Answer: $I_T = 25.67 \text{ A}$

7.6 TROUBLESHOOTING PARALLEL CIRCUITS

Like a series circuit, defects in parallel circuits come in two categories, shorts and opens. A short circuit has the load resistance bypassed. An open disconnects the path for current to flow.

A parallel circuit has its resistance connected directly between the positive and negative terminals of a power supply. If a short circuit happens, there is a zero resistance path between the positive and negative terminals. In figure 7-12, R_2 contains a short circuit. This problem could result from a component failure or from the wires on either side of the load touching. With no resistance to limit current, the current increases until it blows a fuse or trips a circuit breaker. If there is no overcurrent protection, the circuit will be damaged. A wire could melt or the power supply might burn out.

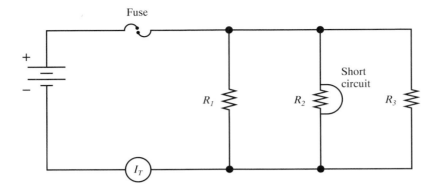

Figure 7-12. Short circuit across R_2 blows the fuse. A short across a branch in a parallel circuit allows an unlimited amount of current to flow.

The symptoms of open circuits in parallel vary depending on the location of the open. Figure 7-13 shows the effects of a burned-out light bulb. With the light bulb's filament open, there is no current in that branch. Voltage will be measured as normal at the wires to the bulb. The other branches are not affected.

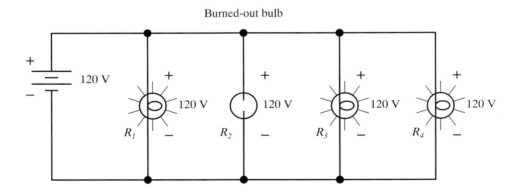

Figure 7-13. Parallel branches are not affected by an open in another branch. The burned-out bulb in this circuit does not change the voltage or current across R_1, R_3, or R_4.

If either the positive or negative side of the mainline buss-wire is open, the effect on the circuit depends on its location. Figure 7-14 shows the connection at the top of R_2 has been disconnected. R_1 and R_2 operate normally. R_3 and R_4 do not have a complete circuit and can not operate. No voltage will be measured across R_3 or R_4. A voltmeter placed at the open measures the supply voltage.

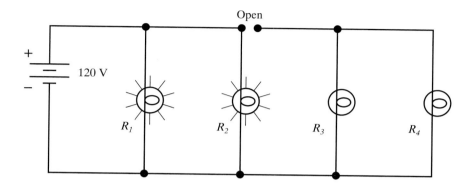

Figure 7-14. Opens in the mainline buss-bar can affect some branches and not others in a parallel circuit. In this figure, R_3 and R_4 have zero volts being dropped across them.

7.7 SIMPLE PARALLEL CIRCUIT GUIDELINES

Total resistance can be found using several different formulas, depending on the situation in the circuit. Ohm's law can be used if any two of the following are given: voltage, total current, or total power. The reciprocal formula and the conductance method can be used provided all of the individual resistors are known. When only two resistors are in parallel, the shortcut formula is available. If the circuit has resistors of the same value, their equivalent can be found by dividing the resistance of one load by the number of resistors.

Total current is found using the Ohm's law/power formulas when the other totals are known. Total current is also the sum of the individual branch currents.

Power is found across individual resistors, regardless if they are connected in series or parallel in a circuit. Total power in all circuits is the sum of the individual powers. Total power can also be calculated using the power formulas.

Voltage is the same throughout the parallel circuit. Any set of two circuit values can be combined in formulas to calculate voltage.

Figure 7-15 is a summary of the formulas used in parallel circuits. You should learn to use all of the formulas. Chapter 8 combines series and parallel connections in the same circuit. It is necessary to be able to determine which resistors are in series and which are in parallel. The formulas for each type of connection cannot be combined.

Figure 7-15. Summary of parallel circuit formulas.

Total Resistance

Reciprocal Formula:

$$\frac{1}{R_T} = \frac{1}{R_1} + \frac{1}{R_2} + \frac{1}{R_3} + \dots \frac{1}{R_N}$$

Conductance Formulas:

$$G = \frac{1}{R} \qquad R = \frac{1}{G} \qquad G_T = G_1 + G_2 + G_3 + \dots G_N$$

Shortcut Formula:

$$R_T = \frac{R_1 \times R_2}{R_1 + R_2}$$

Equal Value Resistors:

$$R_T = \frac{R}{N}$$

Ohm's Law/Power Formulas:

$$R_T = \frac{V}{I_T} \qquad\qquad R_T = \frac{V^2}{P_T} \qquad\qquad R_T = \frac{P_T}{I^2}$$

Total Current

Individual Branches:

$$I_T = I_1 + I_2 + I_3 + \dots I_N$$

Ohm's Law/Power Formulas:

$$I_T = \frac{V}{R_T} \qquad\qquad I_T = \frac{P_T}{V} \qquad\qquad I_T = \sqrt{\frac{P_T}{R_T}}$$

Voltage

Same at all points:

$$V = I \times R \qquad\qquad V = \sqrt{P \times R} \qquad\qquad V = \frac{P}{I}$$

Power

Individual Powers:

$$P_T = P_{R_1} + P_{R_2} + P_{R_3} + \dots P_{R_N}$$

Ohm's Law/Power Formulas:

$$P_T = V \times I_T \qquad\qquad P_T = I_T^2 \times R_T \qquad\qquad P_T = \frac{V^2}{R}$$

SUMMARY

- Parallel circuits have more than one current path.
- Total resistance is less than the smallest branch resistance.
- Total resistance can be calculated using several different formulas.
- Power in all circuits is the sum of the individual powers.
- Total current is the sum of the individual branch currents.
- Each parallel branch is independent of the other branches.
- Mainline current is the sum of the branch currents from that point away from the supply.

KEY WORDS AND TERMS GLOSSARY

branch current: Current in the independent branches of a parallel circuit.

conductance (G): The ease with which a conductor allows a current to flow. It is the reciprocal of resistance. The unit of measure is the siemen (S) or mho.

mainline buss-bar: The wires in electrical equipment connecting all of the positive connections or all of the negative connections.

parallel circuit: Circuit characterized by: the same voltage throughout the circuit, current splitting into branches, total circuit current equaling the sum of the branches, the total resistance being smaller than the smaller branch resistor, the total power equaling the sum of the branch powers.

KEY FORMULAS

Formula 7.A

$$\frac{1}{R_T} = \frac{1}{R_1} + \frac{1}{R_2} + \frac{1}{R_3} + \dots \frac{1}{R_N}$$

Formula 7.B

$$G_T = G_1 + G_2 + G_3 + \dots G_N$$

Formula 7.C

$$G_T = \frac{1}{R_1} + \frac{1}{R_2} + \frac{1}{R_3} + \dots \frac{1}{R_N}$$

Formula 7.D

$$R_T = \frac{1}{G_T}$$

Formula 7.E

$$R_T = \frac{R_1 \times R_2}{R_1 + R_2}$$

Formula 7.F

$$R_T = \frac{R}{N}$$

Formula 7.G

$$I_T = I_1 + I_2 + I_3 + \ldots I_N$$

TEST YOUR KNOWLEDGE

Do not write in this text. Please use a separate sheet of paper.

1. List three examples of a parallel circuit.
2. List five characteristics of a parallel circuit.
3. Write four formulas, not including Ohm's law or power formulas, that are used to calculate total resistance in parallel circuits.
4. Make a statement that describes the size of total resistance in comparison to the branch resistances.
5. What is the unit of measure for conductance?
6. Under what conditions can the shortcut formula be used to solve for R_T?
7. Under what conditions can the equal resistors formula be used to solve for R_T?
8. Make a statement that describes total current in comparison to branch currents.
9. Make a statement that describes mainline current.
10. In a parallel circuit, if one branch is shorted, what is the effect on the remainder of the circuit?
11. In a parallel circuit, if one branch has an open, what is the effect on the remainder of the circuit?
12. Using the information given in the schematic diagram in figure 7-16, find the value of total resistance using:
    a. the reciprocal (or conductance) formula.
    b. the shortcut formula.

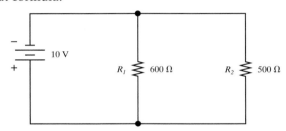

Figure 7-16. Problem #12.

13. Find the total resistance and total current of the circuit shown in figure 7-17.

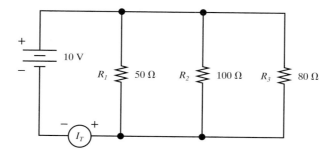

Figure 7-17. Problem #13.

14. Using figure 7-18:
   a. Find the value of total resistance using the reciprocal (or conductance) formula.
   b. Find the value of total resistance using the equal-value resistors formula.
   c. Find the value of total current.

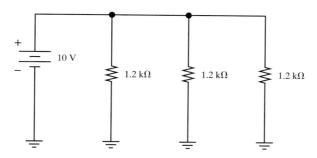

Figure 7-18. Problem #14.

15. Using the information given in the circuit of figure 7-19, find:
   a. the total resistance.
   b. the applied voltage.
   c. total power.

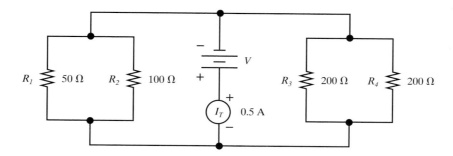

Figure 7-19. Problem #15.

16. With the values given in figure 7-20, find:
    a. total resistance.
    b. total power.
    c. the value of R_2.

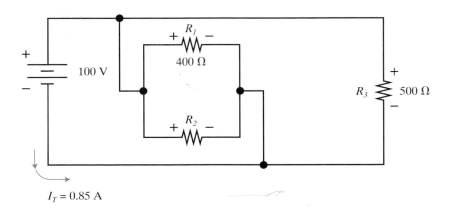

$I_T = 0.85$ A

Figure 7-20. Problem #16.

17. In the parallel circuit shown in figure 7-21, total power and branch resistances are given. Find:
    a. total resistance.
    b. total current.
    c. applied voltage.

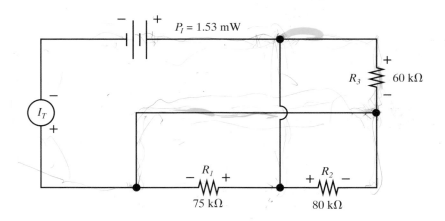

Figure 7-21. Problem #17.

18. Use the given values of the circuit in figure 7-22 to find the unknown values of:
    a. applied voltage.
    b. total current.
    c. total resistance
    d. all branch currents.

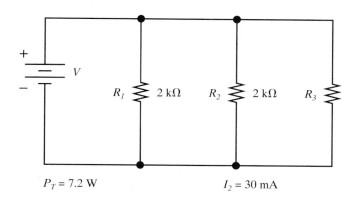

$P_T = 7.2$ W $I_2 = 30$ mA

Figure 7-22. Problem #18.

19. Using the values given in figure 7-23, find the values of:
    a. applied voltage.
    b. total resistance.
    c. resistor R_3.
    d. all branch currents.

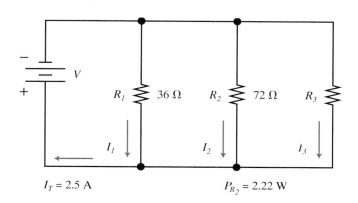

$I_T = 2.5$ A $P_{R_2} = 2.22$ W

Figure 7-23. Problem #19.

20. Calculate the current measured in each of the ammeters shown in the circuit of figure 7-24.

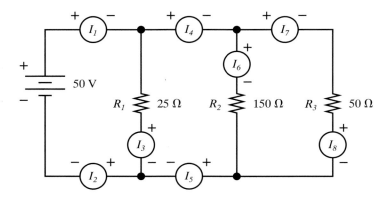

Figure 7-24. Problem #20.

Chapter 8
Series-Parallel
Circuits

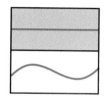

Upon completion of this chapter, you will be able to:
- Differentiate between series connected resistors and parallel connected resistors.
- Apply the rules and circuit formulas to combination circuits.
- Calculate the circuit parameters of increasingly difficult combinations.
- Examine the range resistors connected internally in a voltmeter, ammeter, and ohmmeter.

Series-parallel combination circuits combine connections made in series with those made in parallel. When a component of a certain value is not available, it is a common practice to create that resistance with whatever resistors are available. Also, in certain types of circuits, such as those with transistors and other semiconductors, resistors must be connected in certain formats. These formats contain combinations of series and parallel resistors. It is necessary to be able to analyze these circuit's behavior to be sure they will perform as desired.

This chapter explains how to calculate circuit parameters in several circuit connections. Covered are:
- Resistors connected in parallel, with series resistors in the branches.
- Parallel combinations forming a series circuit.
- Mainline series resistor in a parallel circuit.
- Combination circuits with many mainline resistors and parallel branches.
    Also discussed are the effects of connecting a voltmeter in a circuit and the design of a practical voltage divider circuit.

8.1 SERIES RESISTANCES IN PARALLEL BRANCHES

A unique characteristic of the parallel circuit is that each branch is independent from the other branches in terms of how much current flows through it. When more than one resistor is connected in a branch, it does not affect the other branches. *Each branch can be treated as an individual circuit.*

Equivalent Circuits

To solve a series-parallel circuit, an equivalent circuit is created. Each branch of a parallel circuit can be reduced to its equivalent resistance. As shown in figure 8-1, a new circuit is drawn with the equivalent resistors. Branch A has three resistors connected in series. These three resistors are added together, as a series circuit, to form an equivalent resistor. The equivalent resistor is labeled R_A. The same procedure is used on each of the other branches. Using equivalent circuits simplifies calculations of the total circuit resistance and the total current. The total power in the circuit is the sum of the individual powers in all circuit configurations.

Sample problem 1. _____

Calculate the equivalent resistances of each branch of the circuit in figure 8-1.

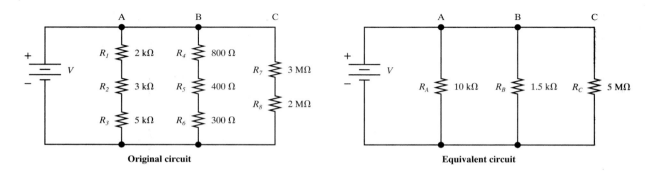

Original circuit   Equivalent circuit

Figure 8-1. Equivalent circuits are used to simplify calculations. An original circuit is on the left. Its equivalent circuit is on the right. (Sample problem 1)

1. Resistance of branch A:

    Formula: $R_A = R_1 + R_2 + R_3$

    Substitution: $R_A = 2 \text{ k}\Omega + 3 \text{ k}\Omega + 5 \text{ k}\Omega$

    Answer: $R_A = 10 \text{ k}\Omega$

2. Resistance of branch B:

    Formula: $R_B = R_4 + R_5 + R_6$

    Substitution: $R_B = 800 \text{ }\Omega + 400 \text{ }\Omega + 300 \text{ }\Omega$

    Answer: $R_B = 1.5 \text{ k}\Omega$

3. Resistance of branch C:

Formula: $R_C = R_7 + R_8$

Substitution: $R_C = 3 \text{ M}\Omega + 2 \text{ M}\Omega$

Answer: $R_C = 5 \text{ M}\Omega$

Voltage Drops in Parallel Branches

Voltage drops are calculated when resistors are connected in series. All the rules learned for series circuits apply to the series resistors in each parallel branch. Recall the basic procedure: first find total resistance for the series circuit, then find current and voltage drops.

Sample problem 2. _____

With the circuit shown in figure 8-2, find the equivalent branch resistances, branch currents, and voltage drops across the individual resistors.

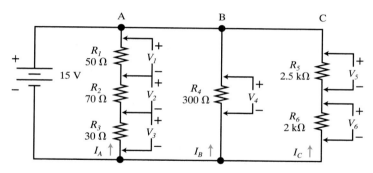

Figure 8-2. Voltage drops for individual branches are like those of individual series circuits. (Sample problem 2)

1. Equivalent branch resistances:
    a. Resistance of branch A:

    Formula: $R_A = R_1 + R_2 + R_3$

    Substitution: $R_A = 50 \text{ }\Omega + 70 \text{ }\Omega + 30 \text{ }\Omega$

    Answer: $R_A = 150 \text{ }\Omega$

    b. Resistance of branch B:

    Answer: $R_B = R_4 = 300 \text{ }\Omega$

    c. Resistance of branch C:

    Formula: $R_C = R_5 + R_6$

    Substitution: $R_C = 2.5 \text{ k}\Omega + 2 \text{ k}\Omega$

    Answer: $R_C = 4.5 \text{ k}\Omega$

2. Branch currents:
   a. Current in branch A:

   Formula: $I_A = \dfrac{V}{R_A}$

   Substitution: $I_A = \dfrac{15\text{ V}}{150\ \Omega}$

   Answer: $I_A = 0.1\text{ A} = 100\text{ mA}$

   b. Current in branch B:

   Formula: $I_B = \dfrac{V}{R_B}$

   Substitution: $I_B = \dfrac{15\text{ V}}{300\ \Omega}$

   Answer: $I_B = 0.05\text{ A} = 50\text{ mA}$

   c. Current in branch C:

   Formula: $I_C = \dfrac{V}{R_C}$

   Substitution: $I_C = \dfrac{15\text{ V}}{4.5\text{ k}\Omega}$

   Answer: $I_C = 3.33\text{ mA}$

3. Voltage drops in branch A ($I_A = 100$ mA):
   a. Voltage across R_1:

   Formula: $V_{R_1} = I_A \times R_1$

   Substitution: $V_{R_1} = 100\text{ mA} \times 50\ \Omega$

   Answer: $V_{R_1} = 5\text{ V}$

   b. Voltage across R_2:

   Formula: $V_{R_2} = I_A \times R_2$

   Substitution: $V_{R_2} = 100\text{ mA} \times 70\ \Omega$

   Answer: $V_{R_2} = 7\text{ V}$

   c. Voltage across R_3:

   Formula: $V_{R_3} = I_A \times R_3$

Substitution: $V_{R_3} = 100 \text{ mA} \times 30 \text{ } \Omega$

Answer: $V_{R_3} = 3 \text{ V}$

4. Voltage drop in branch B ($I_B = 50$ mA):
    Voltage across R_4:

    Formula: $V_{R_4} = I_B \times R_4$

    Substitution: $V_{R_4} = 50 \text{ mA} \times 300 \text{ } \Omega$

    Answer: $V_{R_4} = 15 \text{ V}$

    Notice that with no other resistors in this branch, the voltage drop is equal to the applied voltage.

5. Voltage drops in branch C ($I_C = 3.33$ mA):
    a. Voltage across R_5:

        Formula: $V_{R_5} = I_C \times R_5$

        Substitution: $V_{R_5} = 3.33 \text{ mA} \times 2.5 \text{ k}\Omega$

        Answer: $V_{R_5} = 8.33 \text{ V}$

    b. Voltage across R_6:

        Formula: $V_{R_6} = I_C \times R_6$

        Substitution: $V_{R_6} = 3.33 \text{ mA} \times 2 \text{ k}\Omega$

        Answer: $V_{R_6} = 6.66 \text{ V}$

8.2 PARALLEL COMBINATIONS IN A SERIES CIRCUIT

In a series circuit, current is the same at all points. In a parallel circuit, current splits to each branch. Figure 8-3 is a series circuit with the series resistors made up of parallel combinations.

To find the current through each resistor, it is necessary to first find the equivalent resistances of each of the parallel combinations. Use these equivalent resistances to create a series circuit. Using this series circuit, calculate the voltage drop across each of the parallel combinations. The voltage is the same across all paths of a parallel circuit. Thus, the voltage drop across the entire parallel combination is the voltage drop across each branch of the combination. The current through each resistor is then calculated by dividing its voltage drop by its ohmic value.

Sample problem 3. _____

Use the values given in the top circuit of figure 8-3 to find the equivalent resistances, total resistance, total circuit current, and voltage drops that are given to you in the equivalent circuit.

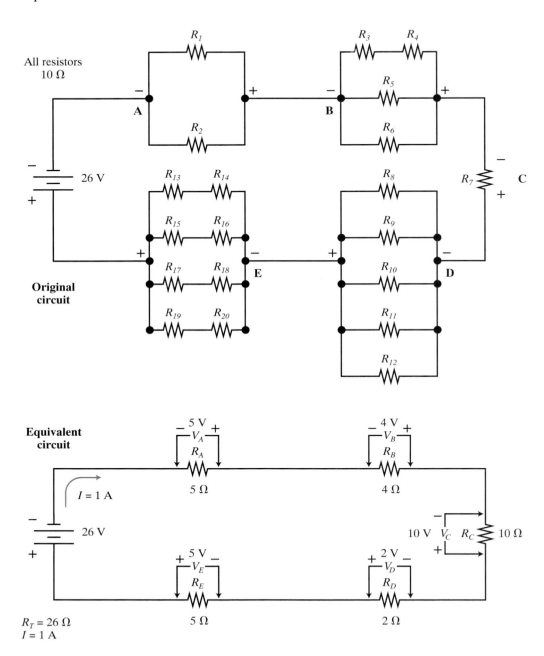

Figure 8-3. Series circuit with parallel combinations. (Sample problem 3)

1. Equivalent resistances:
   a. Parallel combination A:
      (The equal resistance formula can be used here.)

      Formula: $R_A = \dfrac{R}{N}$

      Substitution: $R_A = \dfrac{10\ \Omega}{2}$

      Answer: $R_A = 5\ \Omega$

   b. Parallel combination B:
      (Reciprocal formula is used because not all branches have the same resistance value.)

      Formula: $\dfrac{1}{R_B} = \dfrac{1}{R_3 + R_4} + \dfrac{1}{R_5} + \dfrac{1}{R_6}$

      Substitution: $\dfrac{1}{R_B} = \dfrac{1}{10\ \Omega + 10\ \Omega} + \dfrac{1}{10\ \Omega} + \dfrac{1}{10\ \Omega}$

      Decimals: $\dfrac{1}{R_B} = 0.05 + 0.1 + 0.1$

                        $\dfrac{1}{R_B} = 0.25$

      Answer: $R_B = 4\ \Omega$

   c. Parallel combination D:
      (The equal resistance formula can be used here.)

      Formula: $R_D = \dfrac{R}{N}$

      Substitution: $R_D = \dfrac{10\ \Omega}{5}$

      Answer: $R_D = 2\ \Omega$

   d. Parallel combination E:
      (The equal resistance formula can be used here with 20 Ω in each branch.)

      Formula: $R_E = \dfrac{R}{N}$

      Substitution: $R_E = \dfrac{20\ \Omega}{4}$

      Answer: $R_E = 5\ \Omega$

2. Total circuit resistance:

    Formula: $R_T = R_A + R_B + R_C + R_D + R_E$

    Substitution: $R_T = 5\ \Omega + 4\ \Omega + 10\ \Omega + 2\ \Omega + 5\ \Omega$

    Answer: $R_T = 26\ \Omega$

3. Total circuit current:

    Formula: $I = \dfrac{V}{R_T}$

    Substitution: $I = \dfrac{26\ V}{26\ \Omega}$

    Answer: $I = 1\ A$

4. Equivalent circuit voltage drops:
    (Only R_A is worked out here. Calculate the other voltage drops using the same procedure.)

    a. Voltage drop across R_A:

        Formula: $V_{R_A} = I \times R_A$

        Substitution: $V_{R_A} = 1\ A \times 5\ \Omega$

        Answer: $V_{R_A} = 5\ V$

    b. Voltage drop across R_B:

        Answer: $V_{R_B} = 4\ V$

    c. Voltage drop across R_C:

        Answer: $V_{R_C} = 10\ V$

    d. Voltage drop across R_D:

        Answer: $V_{R_D} = 2\ V$

    e. Voltage drop across R_E:

        Answer: $V_{R_E} = 5\ V$

Sample problem 4. _____

Using the values found in sample problem 3, calculate each of the branch currents. The results are shown in figure 8-4. The formula and substitutions are shown for the current in R_1. The calculations for all of the other resistors use the same procedure. The voltage for each branch is shown for reference.

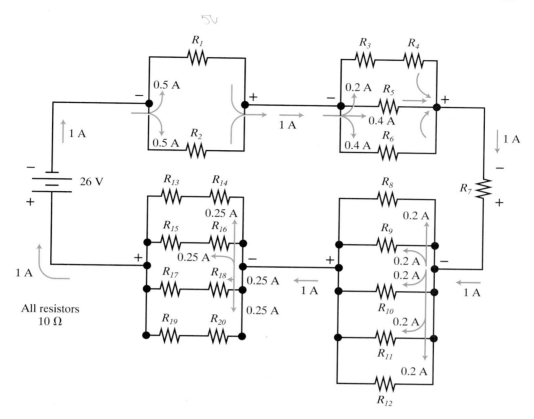

Figure 8-4. Branch currents for circuit in figure 8-3. (Sample problem 4)

a. Current in R_1:

Formula: $I_{R_1} = \dfrac{V_A}{R_1}$

Substitution: $I_{R_1} = \dfrac{5\ V}{10\ \Omega}$

Answer: $I_{R_1} = 0.5\ A$

b. Current in R_2:

$V_A = 5\ V$

Answer: $I_{R_2} = 0.5\ A$

c. Current in R_3 and R_4:

$V_B = 4\ V$

Answer: $I_{R_3} = I_{R_4} = 0.2\ A$

d. Current in R_5:

$$V_B = 4 \text{ V}$$

Answer: $I_{R_5} = 0.4 \text{ A}$

e. Current in R_6:

$$V_B = 4 \text{ V}$$

Answer: $I_{R_6} = 0.4 \text{ A}$

f. Current in R_7:

$$V_C = 10 \text{ V}$$

Answer: $I_{R_7} = 1 \text{ A}$

g. Currents in R_8 through R_{12} are all equal:

$$V_D = 2 \text{ V}$$

Answer: $I_{R_8} = I_{R_9} = I_{R_{10}} = I_{R_{11}} = I_{R_{12}} = 0.2 \text{ A}$

h. Currents in R_{13} through R_{20} are all equal:

$$V_E = 5 \text{ V}$$

Answer: $I_{R_{13}}$ through $I_{R_{20}} = 0.25 \text{ A}$

8.3 MAINLINE SERIES RESISTORS

The mainline is usually associated with a parallel circuit. It is the buss-bar that brings voltage to the parallel branches. However, if a resistor is in the mainline, all of the current to the parallel branches must flow through it. If there is an increase in current in the parallel section (from its resistance changing to a lower value) the increased current causes a larger voltage drop in the mainline resistor.

This type of series-parallel combination circuit is used in power supply circuits in combination with voltage regulator components. The following example demonstrates how a mainline series resistor interacts with the circuit.

Sample problem 5. _____

Using the original circuit of figure 8-5, calculate current and voltage for each resistor.

1. Find the equivalent circuit:

a. Combine parallel resistors R_2 and R_3 (figure B):

Formula: $R_{2\text{-}3} = \dfrac{R_2 \times R_3}{R_2 + R_3}$

Substitution: $R_{2\text{-}3} = \dfrac{20\ \Omega \times 30\ \Omega}{20\ \Omega + 30\ \Omega}$

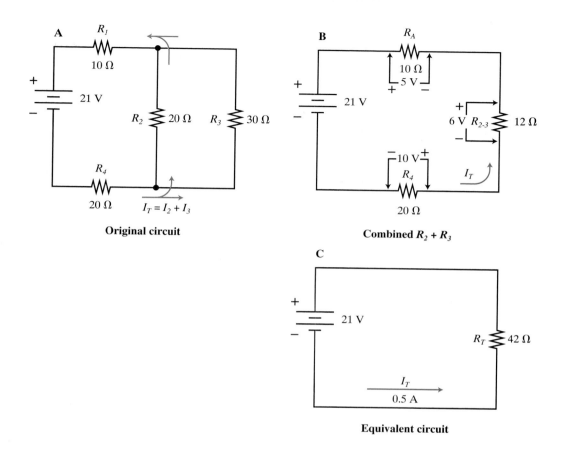

Figure 8-5. Mainline series resistor circuit. First, find the equivalent circuit, then calculate the current and voltage for each resistor. (Sample problem 5)

Intermediate step: $R_{2-3} = \dfrac{600}{50}$

Answer: $R_{2-3} = 12 \ \Omega$

b. Find total resistance (figure C):

Formula: $R_T = R_1 + R_{2-3} + R_4$

Substitution: $R_T = 10 \ \Omega + 12 \ \Omega + 20 \ \Omega$

Answer: $R_T = 42 \ \Omega$

2. Use total resistance to find total current (figure C):

Formula: $I_T = \dfrac{V}{R_T}$

Substitution: $I_T = \dfrac{21 \ V}{42 \ \Omega}$

Answer: $I_T = 0.5 \ A$

3. Use I_T to find the current through I_{R_1}, $I_{R_{2-3}}$, and I_{R_4}.

I_T is the current through the mainline resistors as well as the combined current through the parallel resistors (figure B).

Answer: $I_T = I_{R_1} = I_{R_4} = I_{R_{2-3}} = 0.5$ A

4. Calculate voltage drop across each resistor in the equivalent series circuit (figure B).

   a. Voltage across R_1:

   Only the calculations for R_1 are shown here. Solve the others using the same procedure.

   Formula: $V_{R_1} = I_T \times R_1$

   Substitution: $V_{R_1} = 0.5$ A $\times$ 10 Ω

   Answer: $V_{R_1} = 5$ V

   b. Voltage across R_{2-3}:

   Answer: $V_{R_{2-3}} = 6$ V

   c. Voltage across R_4:

   Answer: $V_{R_4} = 10$ V

5. With the voltage known across the parallel branches, calculate current in each branch (figure A).

   a. Current in R_2:

   Formula: $I_{R_2} = \dfrac{V_{R_{2-3}}}{R_2}$

   Substitution: $I_{R_2} = \dfrac{6 \text{ V}}{20 \text{ }\Omega}$

   Answer: $I_{R_2} = 0.3$ A

   b. Current in R_3:

   Answer: $I_{R_3} = I_T - I_{R_2} = 0.2$ A

8.4 COMBINATION SERIES-PARALLEL CIRCUIT

Combinations of series and parallel resistors can come in an infinite number of forms. There is no pattern or formula to use for every circuit combination. When solving a combination circuit, it is best to start at the point furthest from the power supply. Wherever there is a junction, there is a parallel circuit. Each of these branches must be examined to see if there is more than one resistor forming a series circuit. After finding an equivalent resistance for each of the branches with series circuits, combine all of the parallel branches. Then, redraw the circuit.

Redraw the circuit after each step, showing the new equivalent resistance and its position in the circuit. Redrawing a circuit takes time and patience. However, the work pays off. You have a much higher probability of finding the correct answers.

Sample problem 6.

Find the total resistance, starting with the original circuit shown in figure 8-6.

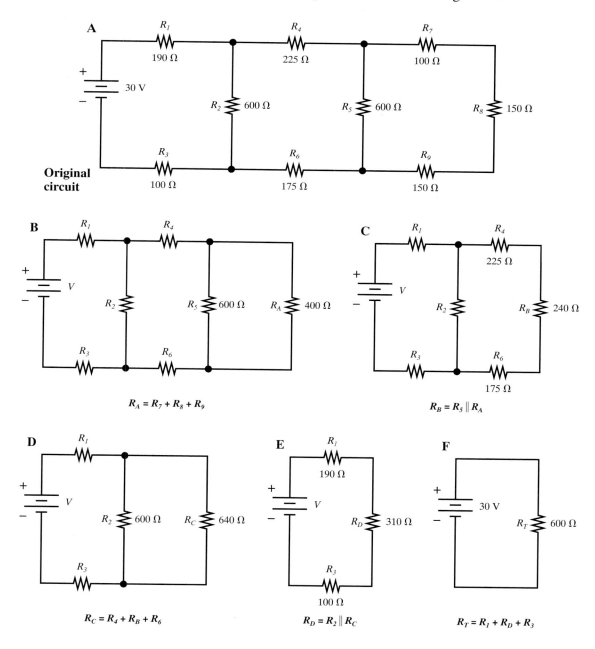

Figure 8-6. Steps in finding total resistance in a combination circuit. (Sample problem 6)

1. Resistors R_7, R_8, and R_9 are in series (figure A).

    Formula: $R_A = R_7 + R_8 + R_9$

    Substitution: $R_A = 100\ \Omega + 150\ \Omega + 150\ \Omega$

    Answer: $R_A = 400\ \Omega$ (redrawn in figure B)

2. Resistors R_5 and R_A are in parallel (figure B).

    Formula: $R_B = \dfrac{R_5 \times R_A}{R_5 + R_A}$

    Substitution: $R_B = \dfrac{600\ \Omega \times 400\ \Omega}{600\ \Omega + 400\ \Omega}$

    Intermediate step: $R_B = \dfrac{24{,}000}{1000}$

    Answer: $R_B = 240\ \Omega$ (redrawn in figure C)

    Note: in the drawing, two slash marks, $\parallel$, indicate "in parallel with."

3. Resistors R_4, R_B, and R_6 are in series (figure C).

    Formula: $R_C = R_4 + R_B + R_6$

    Substitution: $R_C = 225\ \Omega + 240\ \Omega + 175\ \Omega$

    Answer: $R_C = 640\ \Omega$ (redrawn in figure D)

4. Resistors R_2 and R_C are in parallel (figure D).

    Formula: $\dfrac{1}{R_D} = \dfrac{1}{R_2} + \dfrac{1}{R_C}$

    Substitution: $\dfrac{1}{R_D} = \dfrac{1}{600\ \Omega} + \dfrac{1}{640\ \Omega}$

    Decimals: $\dfrac{1}{R_D} = 1.67\text{E}{-03}\ \text{S} + 1.56\text{E}{-03}\ \text{S}$

                       $\dfrac{1}{R_D} = 3.23\text{E}{-03}\ \text{S}$

    Answer: $R_D = 310\ \Omega$ (redrawn in figure E)

5. Total circuit resistance is found by the series combination of resistors R_1, R_D, and R_3 (figure E).

Formula: $R_T = R_1 + R_D + R_3$

Substitution: $R_T = 190 \ \Omega + 310 \ \Omega + 100 \ \Omega$

Answer: $R_T = 600 \ \Omega$ (redrawn in figure F)

Sample problem 7.

Follow the steps shown in figure 8-7 to find the current and voltage drops of each resistor of the circuit in sample problem 6.

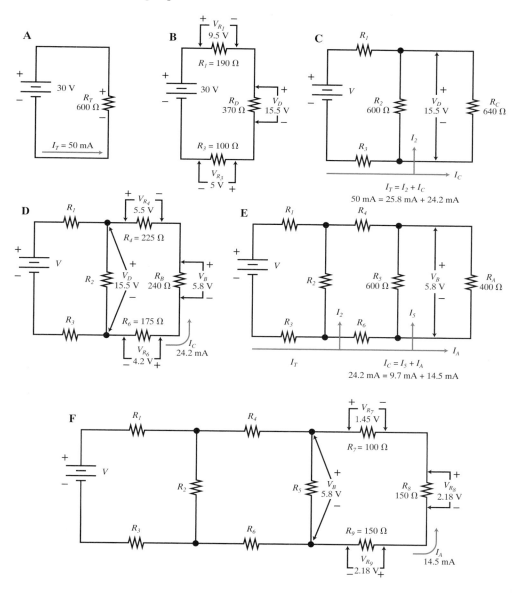

Figure 8-7. Steps in finding the voltage drops and current in the circuit in figure 8-6. (Sample problem 7)

Currents and voltage drops are found using equivalent circuits, in the reverse order. Starting with R_T, and return to the original circuit.

1. Total current is found using applied voltage and total resistance (figure A).

    Formula: $I_T = \dfrac{V}{R_T}$

    Substitution: $I_T = \dfrac{30 \text{ V}}{600 \text{ }\Omega}$

    Answer: $I_T = 0.05 \text{ A} = 50 \text{ mA}$

2. The equivalent circuit in figure B is a series circuit. Calculations with I_T produce voltage drops across the two mainline series resistors, R_1 and R_3. Voltage developed across R_D is the voltage remaining for the rest of the circuit.

    a. Voltage across R_1:

        Formula: $V_{R_1} = I_T \times R_1$

        Substitution: $V_{R_1} = 50 \text{ mA} \times 190 \text{ }\Omega$

        Answer: $V_{R_1} = 9.5 \text{ V}$

    b. Voltage across R_D:

        Answer: $V_D = 15.5 \text{ V}$

    c. Voltage across R_3:

        Answer: $V_{R_3} = 5 \text{ V}$

    As a check, these voltage drops should add up to equal the voltage applied to the series circuit.

    Formula: $V = V_{R_1} + V_D + V_{R_3}$

    Substitution: $V = 9.5 \text{ V} + 15.5 \text{ V} + 5 \text{ V}$

    Answer: $V = 30 \text{ V}$ (equals applied voltage)

3. Figure C expands R_D into its parallel components of R_2 and R_C. The voltage developed across R_D is the same voltage across these parallel components. As shown in this figure, I_T splits between R_2 and R_C. These currents can be calculated using the parallel voltage.

    a. Current through R_2:

        Formula: $I_2 = \dfrac{V_D}{R_2}$

        Substitution: $I_2 = \dfrac{15.5 \text{ V}}{600 \text{ }\Omega}$

        Answer: $I_2 = 25.8 \text{ mA}$

b. Current through R_C:

Formula: $I_C = \dfrac{V_C}{R_C}$

Substitution: $I_C = \dfrac{15.5\ \text{V}}{640\ \Omega}$

Answer: $I_C = 24.2\ \text{mA}$

As a check, the sum of these branch currents should equal the current, I_T, entering the branches.

Formula: $I_T = I_2 + I_C$

Substitution: $I_T = 25.8\ \text{mA} + 24.2\ \text{mA}$

Answer: $I_T = 50\ \text{mA}$ (checks)

4. Figure D expands R_C into its three series components. Use the current, I_C, to find voltage drops.
   a. Voltage across R_4:

   Answer: $V_{R_4} = 5.5\ \text{V}$

   b. Voltage across R_B:

   Answer: $V_B = 5.8\ \text{V}$

   c. Voltage across R_6:

   Answer: $V_{R_6} = 4.2\ \text{V}$

The sum of these voltage drops should equal the voltage applied to this portion of the circuit, V_D.

Formula: $V_D = V_{R_4} + V_B + V_{R_6}$

Substitution: $V_D = 5.5\ \text{V} + 5.8\ \text{V} + 4.2\ \text{V}$

Answer: $V_D = 15.5\ \text{V}$ (checks)

5. Figure E expands R_B into its parallel components of R_5 and R_A. Calculate branch currents using V_B, the voltage applied to this parallel portion of the circuit.
   a. Current through R_5:

   Answer: $I_5 = 9.7\ \text{mA}$

   b. Current through R_A:

   Answer: $I_A = 14.5\ \text{mA}$

The sum of these branch currents should equal the current entering into the branches, I_C.

Formula: $I_C = I_5 + I_A$

Substitution: $I_C = 9.7$ mA + 14.5 mA

Answer: $I_C = 24.2$ mA (checks)

6. In figure F, the original circuit, R_A is expanded into its three series resistors. Use I_C to find voltage drops for R_7, R_8, and R_9.

   a. Voltage across R_7:

      Answer: $V_{R_7} = 1.45$ V

   b. Voltage across R_8:

      Answer: $V_{R_8} = 2.18$ V

   c. Voltage across R_9:

      Answer: $V_{R_9} = 2.18$ V

   The sum of these voltage drops should equal the voltage applied to this section, V_B.

   Formula: $V_B = V_{R_7} + V_{R_8} + V_{R_9}$

   Substitution: $V_B = 1.45$ V + 2.18 V + 2.18 V

   Answer: $V_B = 5.81$ V (checks)

Current and Voltage Summary

Sample problem 6 developed the equivalent circuits and total resistance of figure 8-6. Sample problem 7 used the equivalent circuits to determine current and voltage for each resistor. A summary of the results is shown in figure 8-8 and in the table below.

Summary of results:

$I_{R_1} = 50$ mA	$V_{R_1} = 9.5$ V
$I_{R_2} = 25.8$ mA	$V_{R_2} = 15.5$ V
$I_{R_3} = 50$ mA	$V_{R_3} = 5$ V
$I_{R_4} = 24.2$ mA	$V_{R_4} = 5.5$ V
$I_{R_5} = 9.7$ mA	$V_{R_5} = 5.8$ V
$I_{R_6} = 24.2$ mA	$V_{R_6} = 4.2$ V
$I_{R_7} = 14.5$ mA	$V_{R_7} = 1.45$ V
$I_{R_8} = 14.5$ mA	$V_{R_8} = 2.18$ V
$I_{R_9} = 14.5$ mA	$V_{R_9} = 2.18$ V

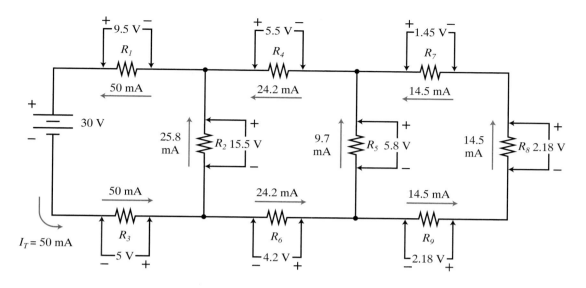

Figure 8-8. Summary of currents and voltages for all of the resistors from sample problems 6 and 7.

8.5 EFFECT OF VOLTMETER LOADING

When a voltmeter is connected in a circuit to measure a voltage, it is connected in parallel with the portion of the circuit it is measuring. Consequently, the internal resistors in a voltmeter, the **multiplier resistors,** are placed in parallel with the circuit being measured. Adding a resistance in parallel to a circuit lowers the effective resistance of that circuit. However, the internal resistance of a voltmeter is very large, large enough that the effect on the circuit is usually considered negligible.

On an analog meter, the input resistance is typically 10,000 ohms per volt. The Ω/V rating is for full scale. This means a meter with 10 kΩ/V on the 25 volt scale would have an input resistance of 25 V × 10 kΩ/V = 250 kΩ.

Normally, this resistance is high enough that when it is used to measure a voltage, it will have no effect on the circuit. Digital meters have an even higher input resistance, typically one megohm or more. This high of a resistance would have no effect on most circuits.

Two examples are given for voltmeter loading, figures 8-9 and 8-10. The figures use three equal value resistors to simplify voltage drop calculations. The voltmeter in the examples has an input resistance of 25 kilohms.

In figure 8-9, the voltmeter is used to measure the voltage across a resistor with a value more than 10 times smaller than the meter's resistance. When calculations are made on the circuit's performance without the meter connected, a 10 volt drop is predicted across each resistor. When the meter is connected, some current is diverted through it. Yet, the current level is so low in comparison to the current traveling through the circuit, there is no change in the voltage drop. The meter will read 10 volts.

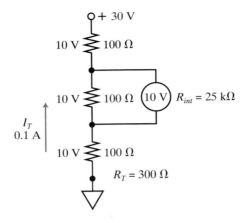

Figure 8-9. When a voltmeter is connected across a low resistance, the effect of the voltmeter is negligible.

Figure 8-10 has a circuit with resistance values equal to that of the meter. When the circuit is calculated without the meter, each resistor receives 10 volts. When the meter is connected, it forms a parallel circuit with R_2. The meter is a 25 kilohm resistor in parallel with a 25 kilohm resistor, creating an equivalent resistance of 12.5 kilohms. With the meter connected, the center resistor drops only 6 volts, while the other two drop 12 volts each. The voltmeter significantly changes this circuit's parameters.

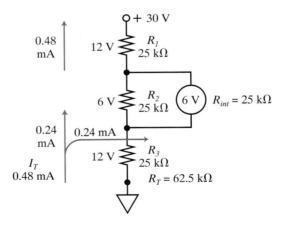

Figure 8-10. When a voltmeter is connected across a large resistance, it can dramatically alter the circuit. In this circuit, resistor R_2 effectively has its value cut in half.

8.6 DESIGNING A VOLTAGE DIVIDER

A **voltage divider** is designed to reduce the voltage from a power supply as needed by a load. For a load to operate properly, it must be supplied with the proper amount of power.

Load specifications usually include a voltage rating and a current rating. For example, the load in figure 8-11 is rated for 12 volts and 0.5 amps. The load's resistance can be calculated to be 24 ohms. In the figure, the power supply has an output of 24 volts. The resistor in parallel with the load, R_B, is also 24 ohms. R_B and the load must receive 0.5 amps, resulting in a need for a total current of 1 amp. R_B and the load drop 12 volts from the 24 volt source. R_A must drop the remaining voltage, 12 volts, and pass the total current, one amp. This calls for a 12 ohm resistor.

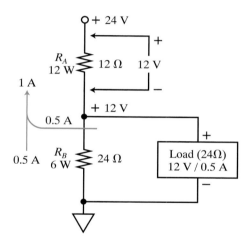

Figure 8-11. Schematic of a voltage divider. The divider created by resistors R_A and R_B allows a 24 V source to deliver the 12 V and 0.5 A required by the load.

With the resistance values known, the only calculations needed now are for the power dissipated across each resistor of the voltage divider.

Power of R_A:

Formula: $P_{R_A} = I_T \times V_{R_A}$

Substitution: $P_{R_A} = 1\,A \times 12\,V$

Answer: $P_{R_A} = 12\,W$

Power of R_B:

Formula: $P_{R_B} = I_B \times V_{R_B}$

Substitution: $P_{R_B} = 0.5 \times 12 \text{ V}$

Answer: $P_{R_B} = 6 \text{ W}$

8.7 INTERNAL CONNECTIONS OF METERS

The standard multimeter used to test electronic circuits contains three separate meters: an ammeter, a voltmeter, and an ohmmeter. Each of these meters has several ranges from which to select. An analog-type multimeter uses an assortment of resistors for range selection. This assortment of resistors is discussed in this section. Digital meters use more complex electronic circuits for range selection.

Ammeter

The needle of an analog meter is operated when a certain amount of current travels through the meter movement. A typical meter allows full-scale deflection with 50 microamps. If a current larger than 50 microamps is to be measured, a resistor must be placed in parallel with the meter movement to provide another path for the current.

A simplified schematic of an ammeter shown in figure 8-12 has a **shunt resistor** connected in parallel with the meter movement of the ammeter. When an ammeter is used, it is connected in series with the current path. If the circuit has one amp of current passing through the meter, 50 microamps or less will go through the meter movement. The rest takes the shunt path. The amount of current traveling through the meter movement is determined by which range is selected.

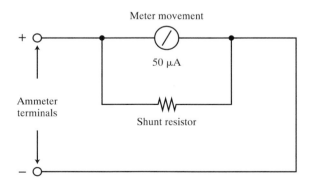

Figure 8-12. Simplified schematic diagram of an ammeter shows the shunt resistor in parallel.

A multirange ammeter, shown in figure 8-13, selects a different shunt resistor for each range chosen. The values of the shunt resistors are low to have as little an effect on the circuit as possible. *The ideal ammeter has zero resistance when placed in the circuit.*

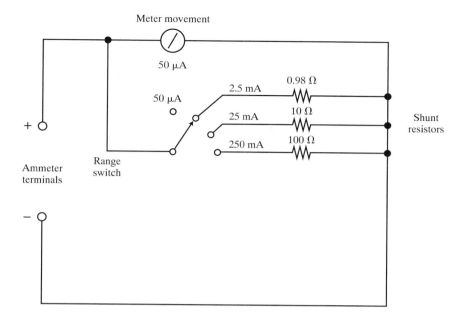

Figure 8-13. Multirange ammeters have several shunt resistors. The shunt resistors chosen for this ammeter are typical values.

Voltmeter

A voltmeter is placed in parallel with a voltage to make a measurement. The resistance of the voltmeter, unlike the ammeter, must be very high. A very large parallel resistor has little current passing through it. With very little current passing, the voltmeter will have only a minor effect on the circuit.

The internal resistance of the meter, the multiplier resistor, is placed in series with the meter movement. This is shown in the simplified schematic of figure 8-14.

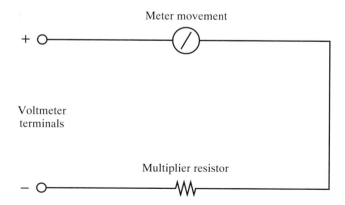

Figure 8-14. Simplified schematic of a voltmeter shows the multiplier resistor in series.

Voltmeter loading is an important consideration when the resistance of the load is high. A typical analog meter with an input resistance of 10,000 ohms per volt is shown in figure 8-15. To find the value of the multiplier resistor, multiply the full-scale value by the ohms per volt rating. For example, in the 0.5 volt range, the multiplier resistor is 5000 ohms.

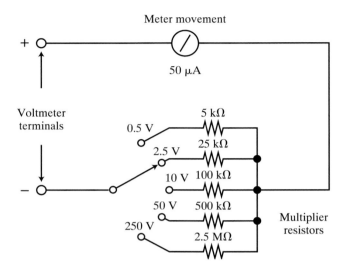

Figure 8-15. Multirange voltmeters have several multiplier resistors. The multiplier resistors chosen for this multimeter are typical values.

Ohmmeter

When an ohmmeter is used, there *cannot* be any voltage in the circuit to be measured. The ohmmeter provides the voltage it needs for a measurement using an internal battery. As shown in figure 8-16, when a resistance is attached to the ohmmeter terminals, the circuit is complete.

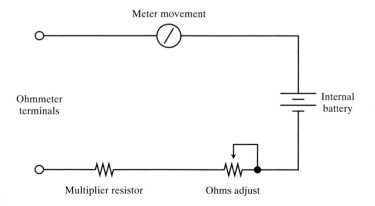

Figure 8-16. Simplified schematic of an ohmmeter shows the internal battery.

The **ohms adjust potentiometer** adjusts the meter to read zero resistance when no resistance is being measured (the leads of the ohmmeter are touching). This adjustment corrects for any changes in the battery voltage.

A multirange ohmmeter, shown in figure 8-17, uses both shunt resistors and multiplier resistors, selected with ganged rotary switches. Some meters use a second battery for the higher resistance ranges to boost the current through the large value resistor being tested.

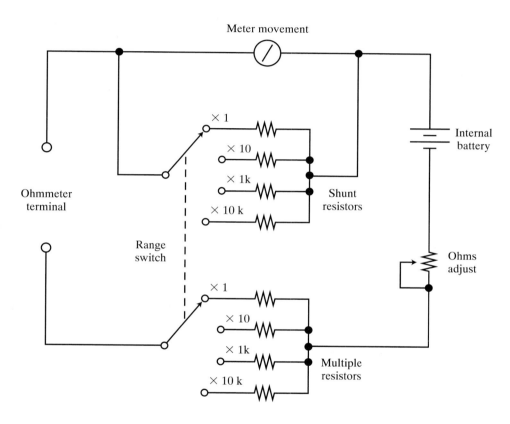

Figure 8-17. Multirange ohmmeter uses both shunt resistors and multiplier resistors.

SUMMARY

- Series-parallel circuits combine both types of circuits, requiring the use of series and parallel formulas.
- Circuits contained within a parallel branch are independent of other branches.
- An equivalent circuit simplifies analysis by replacing complex circuits with a single resistor.
- All current to a circuit flows through a mainline series resistor located in the buss-bar, bringing voltage to parallel branches.

- The internal resistance of a voltmeter can affect a circuit if the circuit's resistance is large.
- Shunt resistors in an ammeter are connected in parallel with the meter movement to provide an alternate current path for large amounts of current.
- Ammeters should have an internal resistance as close to zero as possible.
- Multiplier resistors in a voltmeter are connected in series with the meter movement to drop voltage and to reduce the current through the meter.
- An ohmmeter has its own internal battery and should not be used in a live circuit.

KEY WORDS AND TERMS GLOSSARY

multiplier resistor: A resistor connected internally in a voltmeter. It is connected in series with the meter movement to allow selection of different ranges.

ohms adjust potentiometer: Adjusts a meter to read zero when no resistance is being measured.

shunt resistor: A resistor in parallel.

voltage divider: Reduces the voltage from a power supply as needed by a load.

TEST YOUR KNOWLEDGE

Do not write in this text. Please use a separate sheet of paper.

1. What procedure should be used to solve a circuit with resistors connected in series within a parallel branch?
2. What are the advantages of drawing an equivalent circuit?
3. How does a mainline series resistor affect the rest of the circuit?
4. Under what condition does voltmeter loading have the most effect?
5. How can the effects of voltmeter loading be reduced?
6. When calculating the input resistance of a voltmeter, how is the ohms per volt rating used?
7. How is a shunt resistor is connected? In what type of meter is the shunt resistor used?
8. What is the purpose of a shunt resistor?
9. How is a multiplier resistor connected? In what type of meter is the multiplier resistor used?
10. What is the purpose of a multiplier resistor?

11. Using the information given in the parallel circuit shown in figure 8-18, calculate branch resistances, branch currents, total current, total resistance, and the voltage drops across each resistor.

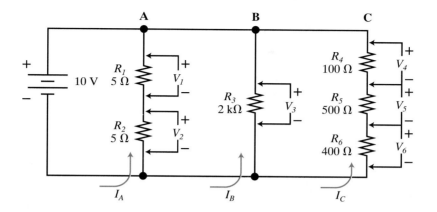

Figure 8-18. Problem #11.

12. With the series circuit containing parallel branches shown in figure 8-19, find total resistance, total current, and voltage drops across each resistor.

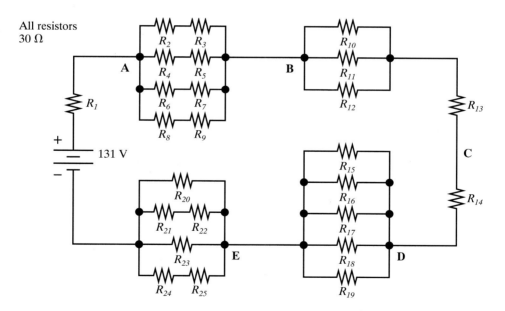

Figure 8-19. Problem #12.

13. Using the resistance values given in the series-parallel combination circuit shown in figure 8-20, calculate total resistance, total current, current through each resistor, and all voltage drops.

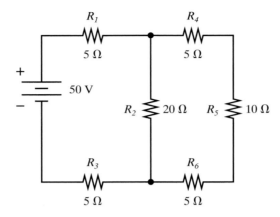

Figure 8-20. Problem #13.

14. Calculate total resistance, total current, current through each resistor, and the voltage drops of the series-parallel combination circuit in figure 8-21.

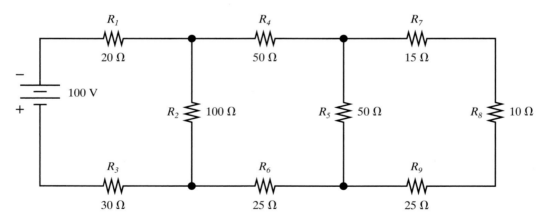

Figure 8-21. Problem #14.

15. Determine the effect of voltmeter loading in the circuit shown in figure 8-22.

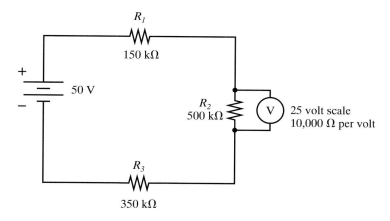

Figure 8-22. Problem #15.

16. Calculate the values of R_1 and R_2 in the voltage divider circuit of figure 8-23. The value of R_2 should equal the load.

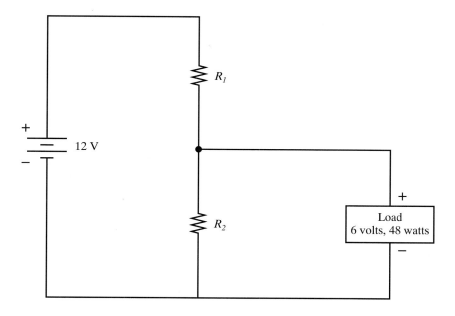

Figure 8-23. Problem #16.

Opening the case of a computer reveals a number of circuit boards. Each circuit board contains many components in a variety of series and parallel relationships.

Chapter 9
DC Circuit Theorems

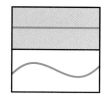

Upon completion of this chapter, you will be able to:
- Solve dc circuits using Kirchhoff's voltage and current laws, the superposition theorem, Thevenin's theorem, and Norton's theorem.
- Solve dc circuits with more than one voltage source using Kirchhoff's laws and the superposition theorem.
- Calculate the values of a Wheatstone bridge.

Solving dc circuits develops the foundation from which all other understanding of electronic circuits is developed. Circuit theorems are used to solve these dc circuits. **Circuit theorems** are mathematical tools that are intended to make it easier to solve complicated circuits. Learning the circuit theorems also aids in the understanding of how circuits perform.

9.1 A BRIEF HISTORY

Chapter 4 included a brief history of the people who helped develop mathematical relationships and whose names are used for units of measure. This chapter includes a history of the people who developed a further understanding of some basic electrical principles.

Gustav R. Kirchhoff was a German physicist who developed two laws concerning current and voltage in electrical networks. With his laws, complicated circuits could be solved and understood much better than with previous techniques.

M. L. Thevenin was a French engineer who developed a theorem that simplified a circuit into an equivalent voltage and series resistance of a network as seen by the load.

E. L. Norton was an American scientist working for the Bell Telephone Laboratories. Norton developed a method of representing a network by a current source in parallel with an equivalent circuit resistance as seen by the load.

Sir Charles Wheatstone was a British physicist who developed the Wheatstone bridge. The Wheatstone bridge is useful for making an exact measurement of an unknown resistance.

9.2 KIRCHHOFF'S CURRENT LAW

In the chapters explaining how to solve dc circuits, Kirchhoff's laws were used to aid in the solutions. The laws were introduced as formulas that just seemed to fit into the natural order of circuit behavior. In this chapter, circuit theorems are also used to find solutions when no other method will work.

Kirchhoff's current law was used in Chapter 8 with parallel circuits. It was used as a check when calculating branch currents. Kirchhoff's current law can be stated as:

The algebraic sum of the currents entering and leaving a point will equal zero.

Figure 9-1 is an example of Kirchhoff's current law. The equation for the circuit is written as: $I_1 + I_2 - I_3 = 0$. The polarity of the currents is determined by the direction of the current. If the current is entering the junction, it is positive. If the current is leaving the junction, it is negative.

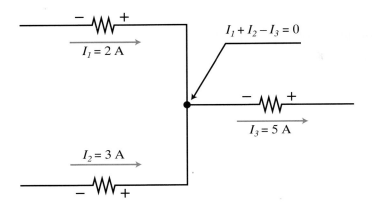

Figure 9-1. The algebraic sum of currents entering and leaving a junction equals zero.

Using Kirchhoff's Current Law

Frequently, when investigating a circuit, currents can be measured at some points but not at others. If it is necessary to find the current for an unknown branch, Kirchhoff's current law can be used.

Sample problem 1. _____

In figure 9-2, the current for three of the four branches is known. Determine the value of the unknown current.

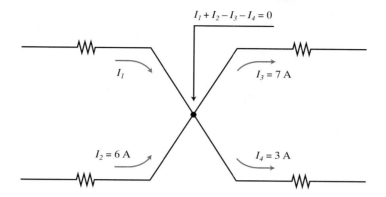

Figure 9-2. Kirchhoff's current law can be used to find an unknown current. (Sample problem 1)

Equation: $I_1 + I_2 - I_3 - I_4 = 0$

Substitution: $I_1 + 6\,A - 7\,A - 3\,A = 0$

Rearranging: $I_1 = -6\,A + 10\,A$

Answer: $I_1 = 4\,A$

Sample problem 2. _____

In figure 9-3, there are seven possible current values. Find the four unknown quantities and their direction of flow. Note, to find the direction of unknown current, simply

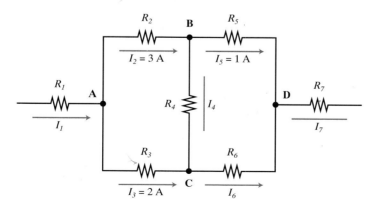

Figure 9-3. Find all of the unknown currents. (Sample problem 2)

choose a direction for that unknown current for the purposes of creating the equation. If the answer for the solved current is positive, the assumed direction was correct. If the answer is negative, the current is traveling in the opposite direction.

1. Find unknown current I_1 at junction A. See figure 9-4. The direction of the current is assumed to be into the junction.

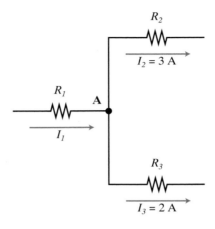

Figure 9-4. Finding the unknown current, I_1, at junction A.

Equation: $I_1 - I_2 - I_3 = 0$

Substitution: $I_1 - 3\,\text{A} - 2\,\text{A} = 0$

Answer: $I_1 = 5\,\text{A}$ (the direction is correct)

2. Find unknown current I_4 at junction B. See figure 9-5. The direction of the current is assumed to be into the junction.

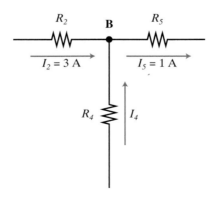

Figure 9-5. Finding the unknown current, I_4, at junction B.

Equation: $I_2 + I_4 - I_5 = 0$

Substitution: $3\,\text{A} + I_4 - 1\,\text{A} = 0$

Answer: $I_4 = -2\,\text{A}$ (the direction is reversed)

$I_4 = 2\,\text{A}$ away from junction

3. Find unknown current I_6 at junction C. See figure 9-6. The direction of the current is assumed to be away from junction.

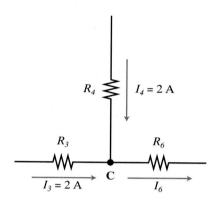

Figure 9-6. Finding the unknown current, I_6, at junction C.

Equation: $I_3 + I_4 - I_6 = 0$

Substitution: $2\,\text{A} + 2\,\text{A} - I_6 = 0$

Answer: $I_6 = 4\,\text{A}$ (the direction is correct)

4. Find unknown current I_7 at junction D. See figure 9-7. The direction of the current is assumed to be away from junction.

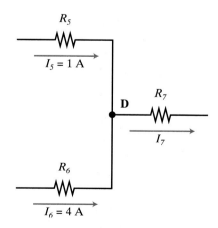

Figure 9-7. Finding the unknown current, I_7, at junction D.

Equation: $I_5 + I_6 - I_7 = 0$

Substitution: $1\,A + 4\,A - I_7 = 0$

Answer: $I_7 = 5\,A$

Figure 9-8 shows all of the currents with their correct directions.

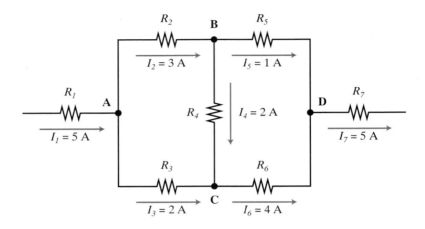

Figure 9-8. Complete circuit with all currents solved. (Sample problem 2)

Observations on Figure 9-8

Listed below are some important observations on figure 9-8. These observations will help in your understanding of how the current travels through this circuit.

- Resistors R_1 and R_7 are series resistors. They should have the same current value.
- Junction A splits two ways. These two currents should add to equal the current entering the junction.
- Junction B splits two ways. These currents should add to equal the current entering the junction.
- Junction C has two currents entering. These currents should add to equal the current out.
- Junction D has two currents entering. These currents should add up to equal the total current entering the series resistor.

9.3 KIRCHHOFF'S VOLTAGE LAW

The principle of Kirchhoff's voltage law was used in Chapter 6 with series circuits to check that all of the voltage drops made in calculations added to equal the applied voltage. Kirchhoff's voltage law can be summed up as:

The algebraic sum of the voltages around a loop will equal zero.

Kirchhoff's voltage law is demonstrated in figure 9-9. The voltage drops around the loop must be labeled with a polarity across each resistor. The polarity is labeled in the direction selected for that loop. The loop is assumed to follow the direction of the current, but this is not critical. What is important is that all the resistors are labeled in one direction following the loop. The negative side of the resistor is where the current enters.

Using Kirchhoff's Voltage Law

Write a loop equation starting at one point and continuing around the loop in the direction of the assumed current. The positive or negative sign for the voltage drop is taken from the sign of polarity as the current first enters the component.

Sample problem 3. _____

All circuit values are given in figure 9-9 to aid in demonstrating how Kirchhoff's voltage law works. Start the loop at point A and continue in the direction following the current arrow.

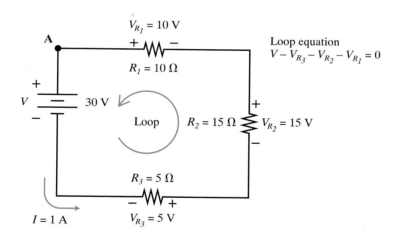

Figure 9-9. Series circuit demonstrating Kirchhoff's voltage law. (Sample problem 3)

Equation:	$V - V_{R_3} - V_{R_2} - V_{R_1} = 0$
Substitution:	$30\text{ V} - 5\text{ V} - 15\text{ V} - 10\text{ V} = 0$
Combining like signs:	$30\text{ V} - 30\text{ V} = 0$

Kirchhoff's law is proven.

Circuits with One Loop

When making voltage measurements around a circuit, the situation frequently arises where a certain voltage must be verified without knowing resistor and current values. An unknown voltage can be found by using Kirchhoff's voltage law.

Sample problem 4. _____

In figure 9-10, the voltage drop across R_2 is unknown. Use Kirchhoff's voltage law to find its value.

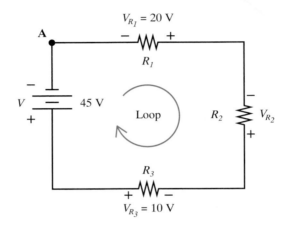

Figure 9-10. Kirchhoff's voltage law can be used to find an unknown voltage in a series circuit. (Sample problem 4)

Equation: $-V_{R_1} - V_{R_2} - V_{R_3} + V = 0$

Substitution: $-20\,V - V_{R_2} - 10\,V + 45\,V = 0$

Answer: $V_{R_2} = 15\,V$

Circuits with More than One Loop

Circuits with parallel branches require the use of more than one loop equation. Kirchhoff's laws often simplify the solving of parallel and combination series-parallel circuits.

Sample problem 5.

As shown in figure 9-11, a circuit can contain more than one loop. Kirchhoff's voltage law is used even though there is no voltage source in loop B.

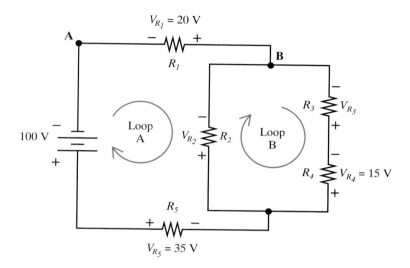

Figure 9-11. More than one equation is needed when solving for unknown voltages in circuits with more than one loop. (Sample problem 5)

Find the unknown voltages in this circuit.

1. Loop A:

   Equation: $-V_{R_1} - V_{R_2} - V_{R_3} + V = 0$

   Substitution: $-20 \text{ V} - V_{R_2} - 35 \text{ V} + 100 \text{ V} = 0$

   Answer: $V_{R_2} = 45 \text{ V}$

2. Loop B:

   Equation: $-V_{R_3} - V_{R_4} + V_{R_2} = 0$

   Substitution: $-V_{R_3} - 15 \text{ V} + 45 \text{ V} = 0$

   Answer: $V_{R_3} = 30 \text{ V}$

One Loop with Two Voltage Sources

Kirchhoff's voltage laws also apply to circuits containing more than one voltage source. To solve these circuits, make sure that all polarities are labeled correctly. Then write the equation following the loop arrow. Treat voltage sources and voltage drops the same.

Sample problem 6. _____

Figure 9-12 has two circuits. One has series-aiding voltage sources. The other has series-opposing voltage sources. All of the circuit values have been given to demonstrate Kirchhoff's voltage law.

1. Aiding voltage sources:

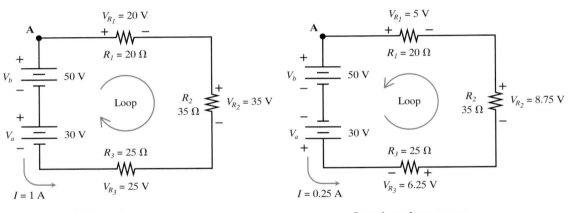

Aiding voltage sources   **Opposing voltage sources**

Figure 9-12. Kirchhoff's voltage law can be used when there is more than one voltage source in the loop. (Sample problem 6)

Equation: $V_{R_1} + V_{R_2} + V_{R_3} - V_a - V_b = 0$

Substitution: 20 V + 35 V + 25 V – 30 V – 50 V = 0

Combining like signs: 80 V – 80 V = 0

2. Opposing voltage sources:

Equation: $V_{R_1} + V_{R_2} + V_{R_3} + V_a - V_b = 0$

Substitution: 5 V + 8.75 V + 6.25 V + 30 V – 50 V = 0

Combining like signs: 50 V – 50 V = 0

Two Loops with Two Voltage Sources

The circuit shown in figure 9-13 cannot be solved using conventional techniques. Resistor R_3 is shared by both voltage sources. Kirchhoff's laws are used to calculate voltage drops and current flow.

When writing the loop equations for this type or circuit, with resistor values and the values of the voltage sources given, the voltage drops must be expressed in terms of Ohm's law: $V_{R_N} = I \times R_N$. Usually, the multiplication sign is removed and the formula is written as $V_{R_N} = IR_N$. Written in loop equations, the voltage drop is called the **IR drop.**

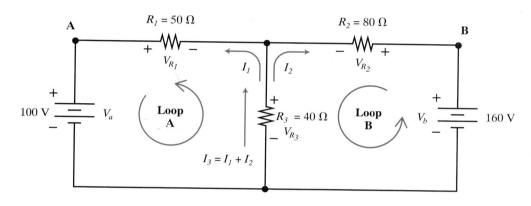

Figure 9-13. Using Kirchhoff's law to solve a two-loop circuit with two voltage sources. (Sample problem 7)

Sample problem 7. _____

Find the current and voltage drops for each resistor in the circuit shown in figure 9-13.

1. Expressing voltage drops:

$$V_{R_1} = I_1 R_1 = I_1 \times 50\ \Omega = 50I_1\ V$$

$$V_{R_2} = I_2 R_2 = I_2 \times 80\ \Omega = 80I_2\ V$$

$$V_{R_3} = (I_1 + I_2)R_3 = (I_1 + I_2) \times 40\ \Omega = 40(I_1 + I_2)\ V$$

2. Loop A: (units of measure are not shown)

   Equation: $V_{R_1} + V_{R_2} - V_a = 0$

   Substitution: $50I_1 + 40(I_1 + I_2) - 100 = 0$

   Simplification: $50I_1 + 40I_1 + 40I_2 = 100$

   $$90I_1 + 40I_2 = 100$$

3. Loop B: (units of measure are not shown)

   Equation: $V_{R_2} + V_{R_3} - V_b = 0$

   Substitution: $80I_2 + 40(I_1 + I_2) - 160 = 0$

   Simplification: $80I_2 + 40I_1 + 40I_2 = 160$

                        $40I_1 + 120I_2 = 160$

Notice that there are now two unknowns and two equations. The solutions to two equations with two unknowns must be solved using the rules of simultaneous equations. There are several different methods available. The *elimination method* is used here.

- Each term of one equation is multiplied by a factor that makes the coefficient of one variable (I_1 or I_2) the same as its corresponding variable in the other equation.
- The two equations are added or subtracted together to eliminate one variable.
- The remaining variable is solved.
- Substitute this value into either original equation and solve for the second variable.

4. Solving loop equations using elimination method:

   Loop A: $90I_1 + 40I_2 = 100$

   Loop B: $40I_1 + 120I_2 = 160$

   a. Multiply all terms of loop A by 3:

      Loop A: $270I_1 + 120I_2 = 300$

   b. Subtract loop B from loop A:

Loop A:	$270 I_1 + 120 I_2$	$=$	300
Loop B:	$-40 I_1 - 120 I_2$	$=$	-160
	$230 I_1$	$=$	140

   c. Isolate the variable by dividing (use decimals rather than fractions):

      $$I_1 = 0.609 \text{ A}$$

   d. Substitute this value into either original equation and solve:

      Loop A: $90I_1 + 40I_2 = 100$

      Substitution: $90(0.609) + 40I_2 = 100$

      Multiply: $54.81 + 40I_2 = 100$

      Subtract from both sides: $40I_2 = 45.19$

      Isolate variable by dividing: $I_2 = 1.13 \text{ A}$

5. Summary of currents: (see figure 9-14)

$$I_1 = 0.609 \text{ A}$$

$$I_2 = 1.13 \text{ A}$$

$$I_3 = I_1 + I_2 = 1.739 \text{ A}$$

6. Voltage drops across each resistor: (see figure 9-14)
    a. Voltage across R_1 (sample calculation):

    Formula: $V_{R_1} = I_1 \times R_1$

    Substitution: $V_{R_1} = 0.609 \text{ A} \times 50 \text{ }\Omega$

    Answer: $V_{R_1} = 30.45 \text{ V}$

    b. Voltage across R_2:

    Answer: $V_{R_2} = 90.4 \text{ V}$

    c. Voltage across R_3:

    Answer: $V_{R_3} = 69.56 \text{ V}$

7. The voltage dropped across R_3 can be verified using Kirchhoff's law: (There are slight differences due to rounding.)
    a. Voltage from supply V_a:

    Formula: $V_{R_3} = V_a - V_{R_1}$

    Substitution: $V_{R_3} = 100 \text{ V} - 30.45 \text{ V}$

    Answer: $V_{R_3} = 69.55 \text{ V}$

    b. Voltage from supply V_b:

    Formula: $V_{R_3} = V_b - V_{R_2}$

    Substitution: $V_{R_3} = 160 \text{ V} - 90.4 \text{ V}$

    Answer: $V_{R_3} = 69.6 \text{ V}$

Notice that even though the currents add through R_3, in this two-voltage circuit, the voltages do not. According to Kirchhoff's law, the algebraic sum of the voltages around a loop must equal zero.

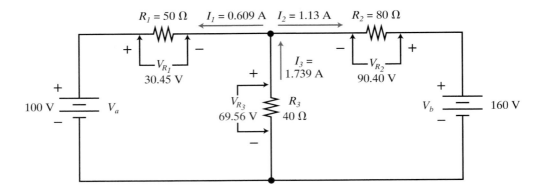

Figure 9-14. Summary of results. (Sample problem 7)

9.4 SUPERPOSITION THEOREM

The superposition theorem allows the use of Ohm's law when solving circuits containing more than one voltage source. Calculations are made to find the effects on current and voltage from each source. These independent effects are then superimposed, which means the results are combined, to determine the complete circuit.

The effects of an individual supply are found by replacing the other source with a short circuit. The circuit may then be solved using conventional methods. The circuit is solved in this fashion for both power supplies. The superposition theorem states:

In a circuit containing more than one voltage source, the current or voltage of individual components is the algebraic sum of the sources acting separately.

Sample problem 8. _____

Use the superposition theorem to find the currents and voltages in the circuit in figure 9-15.

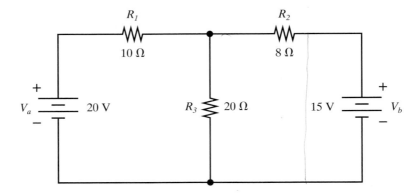

Figure 9-15. Circuit with two voltage sources demonstrates the superposition theorem. (Sample problem 8)

1. Voltage source V_b is replaced with a wire, as shown in figure 9-16. Solve this circuit as viewed from V_a.

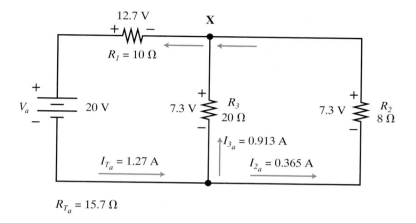

Figure 9-16. Equivalent circuit for figure 9-15 with V_b replaced.

    a. Total resistance from V_a:

        Parallel resistance of : $R_2 \parallel R_3$

        Any parallel formula can be used.

        Formula: $R_{2\text{-}3} = \dfrac{R_2 \times R_3}{R_2 + R_3}$

        Substitution: $R_{2\text{-}3} = \dfrac{8\,\Omega \times 20\,\Omega}{8\,\Omega + 20\,\Omega}$

        Answer: $R_{2\text{-}3} = 5.71\,\Omega$

        Total resistance:

        Formula: $R_{T_a} = R_1 + R_{2\text{-}3}$

        Substitution: $R_{T_a} = 10\,\Omega + 5.71\,\Omega$

        Answer: $R_{T_a} = 15.7\,\Omega$

    b. Total current from V_a:

        Formula: $I_{T_a} = \dfrac{V_a}{R_{T_a}}$

        Substitution: $I_{T_a} = \dfrac{20\,\text{V}}{15.7\,\Omega}$

        Answer: $I_{T_a} = 1.27\,\text{A}$

c.   Series voltage drop across R_1 from V_a:

Formula: $V_{R_{1_a}} = I_{T_a} \times R_1$

Substitution: $V_{R_{1_a}} = 1.27\text{ A} \times 10\ \Omega$

Answer: $V_{R_{1_a}} = 12.7\text{ V}$

d.   Parallel voltage drop across $R_{2\text{-}3}$ from V_a:

Formula: $V_{R_{2\text{-}3_a}} = V_a - V_{R_{1_a}}$

Substitution: $V_{R_{2\text{-}3_a}} = 20\text{ V} - 12.7\text{ V}$

Answer: $V_{R_{2\text{-}3_a}} = 7.3\text{ V}$

e.   Currents in R_2 and R_3 from V_a:

Formula: $I_{R_{2_a}} = \dfrac{V_a}{R_2}$

Substitution: $I_{R_{2_a}} = \dfrac{7.3\text{ V}}{20\ \Omega}$

Answer: $I_{R_{2_a}} = 0.365\text{ A}$

f.   Current in R_3 from V_a:

Answer: $I_{R_{3_a}} = 0.913\text{ A}$

2.   Voltage source V_a is replaced with a wire, as shown in figure 9-17. Solve this circuit as viewed from V_b.

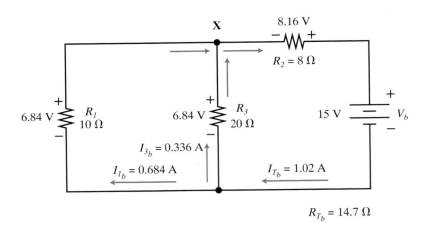

$$R_{T_b} = 14.7\ \Omega$$

Figure 9-17. Equivalent circuit for figure 9-15 with V_a replaced.

a. Total resistance from V_b:

   Parallel resistance of $R_1 \parallel R_3$:

   Formula: $\dfrac{1}{R_{1\text{-}3}} = \dfrac{1}{R_1} + \dfrac{1}{R_3}$

   Substitution: $\dfrac{1}{R_{1\text{-}3}} = \dfrac{1}{10\ \Omega} + \dfrac{1}{20\ \Omega}$

   Decimals: $\dfrac{1}{R_{1\text{-}3}} = 0.1 + 0.05$

   Answer: $R_{1\text{-}3} = 6.67\ \Omega$

   Total resistance:

   Formula: $R_{T_b} = R_2 + R_{1\text{-}3}$

   Substitution: $R_{T_b} = 8\ \Omega + 6.67\ \Omega$

   Answer: $R_{T_b} = 14.7\ \Omega$

b. Total current from V_b:

   Formula: $I_{T_b} = \dfrac{V_b}{R_{T_b}}$

   Substitution: $I_{T_b} = \dfrac{15\ \text{V}}{14.7\ \Omega}$

   Answer: $I_{T_b} = 1.02\ \text{A}$

c. Series voltage drop across R_2 from V_b:

   Formula: $V_{R_{2_b}} = I_{T_b} \times R_2$

   Substitution: $V_{R_{2_b}} = 1.02\ \text{A} \times 8\ \Omega$

   Answer: $V_{R_{2_b}} = 8.16\ \text{V}$

d. Parallel voltage drop across $R_{1\text{-}3}$ from V_b:

   Answer: $V_{R_{1\text{-}3_b}} = 6.84\ \text{V}$

e. Currents in R_1 from V_b:

   Formula: $I_{R_{1_b}} = \dfrac{V_{R_{1_b}}}{R_1}$

   Substitution: $I_{R_{1_b}} = \dfrac{6.84\ \text{V}}{10\ \Omega}$

   Answer: $I_{R_{1_b}} = 0.684\ \text{A}$

f.   Current in R_3 from V_b:

   Answer: $I_{R_{3_b}} = 0.336$ A

3. Superimpose the results from each voltage source algebraically using the polarities of the currents and voltage drops. The results of this superimposing are the values measured with meters in an active circuit. The results are shown in figure 9-18.

All of the polarities are made in reference to point X. Arbitrary assignments are made to the polarities for the purpose of performing the arithmetic. Actual polarities are marked on the schematics, figures 9-16 and 9-17.

   + towards point X has been assigned positive.

   − towards point X has been assigned negative.

a.   Voltage across R_1:

$$V_{R_{1_a}} = -12.7 \text{ V}$$
$$V_{R_{1_b}} = +6.84 \text{ V}$$

Combined: $V_{R_1} = -5.86$ V

b.   Voltage across R_2:

$$V_{R_{2_a}} = +7.3 \text{ V}$$
$$V_{R_{2_b}} = -8.16 \text{ V}$$

Combined: $V_{R_2} = -0.86$ V

c.   Voltage across R_3:

$$V_{R_{3_a}} = +7.3 \text{ V}$$
$$V_{R_{3_b}} = +6.84 \text{ V}$$

Combined: $V_{R_3} = +14.14$ V

Examine the voltages around the first loop and confirm that they correspond to Kirchhoff's law.

d.   Current through R_1:

$$I_{R_{1_a}} = -1.27 \text{ A}$$
$$I_{R_{1_b}} = +0.684 \text{ A}$$

Combined: $I_{R_1} = -0.586$ A

e.   Current through R_2:

$$I_{R_{2_a}} = +0.365 \text{ A}$$
$$I_{R_{2_b}} = -1.02 \text{ A}$$

Combined: $I_{R_2} = -0.655$ A

f.   Current through R_3:

$$I_{R_{3_a}} = + 0.913 \text{ A}$$

$$I_{R_{3_b}} = + 0.336 \text{ A}$$

Combined: $I_{R_3} = + 1.249 \text{ A}$

Examine the currents entering and leaving point X to confirm that Kirchhoff's law applies.

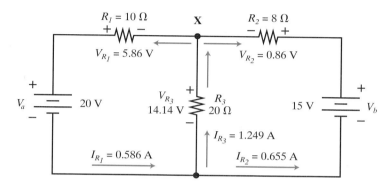

Figure 9-18. Summary of voltage and currents from sample problem 8.

9.5 THEVENIN'S THEOREM

Thevenin's theorem analyzes a circuit's relationship to the load. The entire circuit, except for the load, becomes the voltage source. This circuit includes a resistance equal to the internal resistance of the power supply.

Figure 9-19 demonstrates Thevenin's theorem. The dashed line represents the circuit inside of the power supply. The voltage applied to the entire circuit is not the

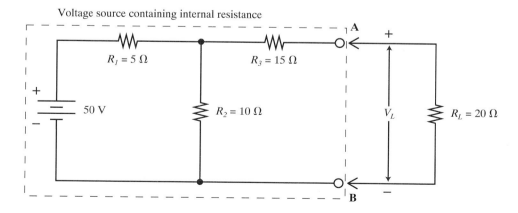

Figure 9-19. In a Thevenin equivalent circuit, everything except the load becomes part of the voltage source.

same as the voltage at the load. Also, the load resistor is not the only resistor connected to the voltage.

A distinct advantage of "Thevenizing" a circuit is its simplicity when there is a
change in the load. With a **Thevenin equivalent circuit,** the bulk of the calculations
have already been performed. New load voltages and currents can be calculated quickly
using a simple series circuit.

Finding the Thevenin Equivalent Voltage

The **Thevenin equivalent voltage** is found by removing the load resistor and finding what voltage would appear at the load terminals. This leaves an open circuit where
the load was connected. A voltmeter has such a large resistance it will measure the open
circuit voltage.

Sample problem 9. _____

In figure 9-20, the load resistor has been removed from terminals A and B. Calculate
the Thevenin equivalent voltage measured by the voltmeter.

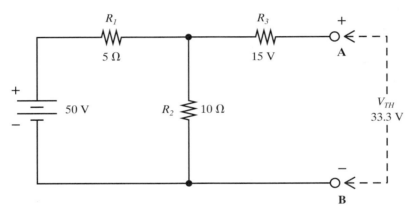

**Figure 9-20. Thevenin voltage is found at the open circuit where the load is
removed. (Sample problem 9)**

To calculate the Thevenin equivalent voltage, notice that with the load removed
from this circuit there is no current through resistor R_3. When there is no current, there
is no voltage drop. Therefore, V_{TH} will be equivalent to the voltage measured across
resistor R_2.

R_2 can be seen as a voltage divider with R_1.

Formula: $V_{R_2} = V_A \times \dfrac{R_2}{R_1 + R_2}$

Substituting: $V_{R_2} = 50 \times \dfrac{10}{5 + 10}$

Answer: $V_{R_2} = V_{TH} = 33.3$ V

Calculating the Thevenin Equivalent Resistance

The **Thevenin equivalent resistance** is the resistance that would be measured at the load terminals, with the voltage removed and replaced by a short. Thevenin resistance is the internal resistance of the voltage source.

Sample problem 10. _____

Find the Thevenin equivalent resistance, as shown in figure 9-21.

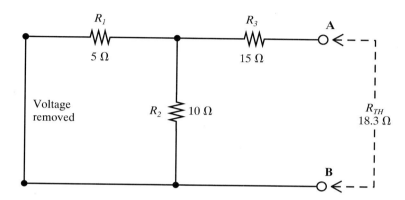

Figure 9-21. R_{TH} **looks at the circuit with the voltage removed. (Sample problem 10)**

1. Parallel combination of $R_1 \parallel R_2$:

   Formula: $R_{1-2} = \dfrac{R_1 \times R_2}{R_1 + R_2}$

   Substitution: $R_{1-2} = \dfrac{5\ \Omega \times 10\ \Omega}{5\ \Omega + 10\ \Omega}$

   Answer: $R_{1-2} = 3.33\ \Omega$

2. Add in the mainline resistor R_3:

   Formula: $R_{TH} = R_{1-2} + R_3$

   Substitution: $R_{TH} = 3.33\ \Omega + 15\ \Omega$

   Answer: $R_{TH} = 18.3\ \Omega$

Thevenin Equivalent Circuit

The Thevenin equivalent circuit is drawn by placing R_{TH} in series with V_{TH} and the load is reconnected to terminals A and B. Figure 9-22 is the equivalent circuit. Notice that V_{TH} is not the same as V_L. Current traveling in this equivalent circuit causes a voltage drop across R_{TH}.

Sample problem 11. _____

Using the Thevenin equivalent circuit shown in figure 9-22, find load current and voltage.

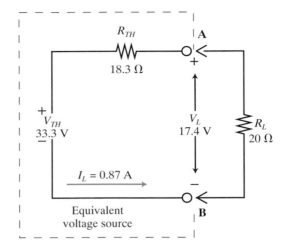

Figure 9-22. Thevenin equivalent circuit. (Sample problem 11)

1. Total resistance of equivalent circuit:

   Formula: $R_T = R_{TH} + R_L$

   Substitution: $R_T = 18.3 \ \Omega + 20 \ \Omega$

   Answer: $R_T = 38.3 \ \Omega$

2. Equivalent circuit current (load current):

   Formula: $I_L = \dfrac{V_{TH}}{R_T}$

   Substitution: $I_L = \dfrac{33.3 \ V}{38.3 \ \Omega}$

   Answer: $I_L = 0.87 \ A$

3. Voltage drop across load resistor:

   Formula: $V_L = I_L \times R_L$

   Substitution: $V_L = 0.87 \ A \times 20 \ \Omega$

   Answer: $V_L = 17.4 \ \Omega$

Thevenin Compared to Conventional Solutions

The Thevenin circuit is calculated independent of the load. Consequently, once it is calculated, it remains the same for any load. When the load is changed, the equivalent circuit remains unchanged. Since each load current will be different, there will be a different voltage drop across R_{TH} and the load voltage will change. The advantage of using Thevenin's theorem is the simplified calculations when working with the equivalent circuit.

As a comparison, sample problem 12 shows calculations of load voltage and current made using conventional techniques. This will help you better understand Thevenin's equivalent circuit, and it proves the theorem is valid.

Sample problem 12.

Find the voltage and current across R_L of the circuit shown in figure 9-23, using conventional circuit solving techniques.

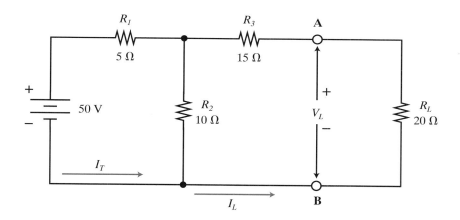

Figure 9-23. Solving this circuit with conventional methods proves Thevenin's theorem. (Sample problem 12)

1. Total resistance (including R_L):
   a. Series combination of R_3 and R_L:

   Formula: $R_{3\text{-}L} = R_3 + R_L$

   Substitution: $R_{3\text{-}L} = 15\ \Omega + 20\ \Omega$

   Answer: $R_{3\text{-}L} = 35\ \Omega$

   b. Parallel combination of $R_2 \parallel R_{3\text{-}L}$:

   Formula: $R_{2\text{-}3\text{-}L} = \dfrac{R_2 \times R_{3\text{-}L}}{R_2 + R_{3\text{-}L}}$

Substitution: $R_{2\text{-}3\text{-}L} = \dfrac{10\ \Omega \times 35\ \Omega}{10\ \Omega + 35\ \Omega}$

Answer: $R_{2\text{-}3\text{-}L} = 7.78\ \Omega$

c. Combining with the mainline resistor:

Formula: $R_T = R_1 + R_{2\text{-}3\text{-}L}$

Substitution: $R_T = 5\ \Omega + 7.78\ \Omega$

Answer: $R_T = 12.78\ \Omega$

2. Total current:

Formula: $I_T = \dfrac{V}{R_T}$

Substitution: $I_T = \dfrac{50\ \text{V}}{12.78\ \Omega}$

Answer: $I_T = 3.91\ \text{A}$

3. Voltage drop across mainline resistor R_1:

Formula: $V_{R_1} = I_T \times R_1$

Substitution: $V_{R_1} = 3.91\ \text{A} \times 5\ \Omega$

Answer: $V_{R_1} = 19.6\ \text{V}$

4. Voltage across parallel combination:

Answer: $V_{R_{2\text{-}3\text{-}L}} = 30.4\ \text{V}$

5. Current through load branch (including R_3):

Formula: $I_{R_L} = \dfrac{V_{R_{2\text{-}3\text{-}L}}}{R_L}$

Substitution: $I_{R_L} = \dfrac{30.4\ \text{V}}{20\ \Omega}$

Answer: $I_{R_L} = 0.87\ \text{A}$

(Checks with Thevenin's equivalent circuit.)

6. Voltage drop across load resistor:

Formula: $V_{R_L} = I_{R_L} \times R_L$

Substitution: $V_{R_L} = 0.87 \text{ A} \times 20 \text{ } \Omega$

Answer: $V_{R_L} = 17.4 \text{ V}$

(Checks with Thevenin's equivalent circuit.)

9.6 SOLVING A WHEATSTONE BRIDGE USING THEVENIN'S THEOREM

A *Wheatstone bridge* connects resistors in such a way that the load is placed between two parallel branches, refer to figure 9-24. The voltage to the load is regulated by two voltage dividers, R_1–R_3 and R_2–R_4. Any of the four resistors in the voltage divider could be replaced with a variable resistor. Using variable resistors allows the voltage, and current, to be adjusted to whatever situation is needed.

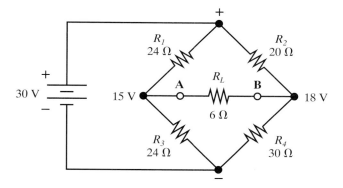

Figure 9-24. Typical Wheatstone bridge circuit with a load resistor.

Using fixed resistors in the bridge is an ideal application of Thevenin's equivalent circuit. The load resistor can be changed as needed, and the calculations to determine the change in load voltage and current are minimized.

Finding the Thevenin Voltage of a Wheatstone Bridge

Thevenin voltage in the bridge circuit, V_{TH}, is measured between the two load terminals, A and B. Voltages at the load terminals are in reference to ground. Calculations to find the voltages can be made in two ways. One way is by finding the branch current and then the voltage drop across the resistor. The other way is by using the voltage divider formulas.

Sample problem 13. _____

Figure 9-24 is redrawn in figure 9-25 to calculate Thevenin voltage. The load resistor has been removed and replaced with a voltmeter.

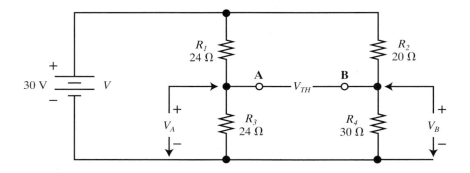

Figure 9-25. With the load removed, V_{TH} is the difference in the two voltage dividers. (Sample problem 13)

1. Voltage at terminal A equals the voltage drop across R_3.
    a. Branch current through R_1 and R_3:

    Formula: $I_{1-3} = \dfrac{V}{R_1 + R_3}$

    Substitution: $I_{1-3} = \dfrac{30 \text{ V}}{24 \text{ }\Omega + 24 \text{ }\Omega}$

    Answer: $I_{1-3} = 0.625 \text{ A}$

    b. Voltage drop across R_3:

    Formula: $V_{R_3} = I_{1-3} \times R_3$

    Substitution: $V_{R_3} = 0.625 \text{ A} \times 24 \text{ }\Omega$

    Answer: $V_{R_3} = V_A = 15 \text{ V}$

2. Voltage at terminal B using the voltage divider formula:

    Formula: $V_{R_4} = V \times \dfrac{R_4}{R_2 + R_4}$

    Substitution: $V_{R_4} = 30 \text{ V} \times \dfrac{30 \text{ }\Omega}{20 \text{ }\Omega + 30 \text{ }\Omega}$

    Answer: $V_{R_4} = V_B = 18 \text{ V}$

3. Thevenin voltage is the difference between points A and B.

    $V_A = 15 \text{ V and } V_B = 18 \text{ V}$

    $V_{TH} = 18 - 15 = 3 \text{ V}$

Finding Thevenin Resistance of a Wheatstone Bridge

To find the Thevenin equivalent resistance, the power supply is removed and replaced with a short. The Thevenin resistance is equal to the resistance that would be seen by the load. Therefore, it is the resistance that would be measured with an ohm-meter between terminals A and B of the bridge circuit.

In figure 9-26 the bridge circuit has been redrawn to allow a better visualization of the relationship of the resistors. With conventional methods of solving circuits, total resistance is viewed from the voltage source. This is not the case with the Thevenin resistance. As a result, the positioning of the resistors is different.

Sample problem 14. _____

Use figure 9-26 to find the Thevenin equivalent resistance.

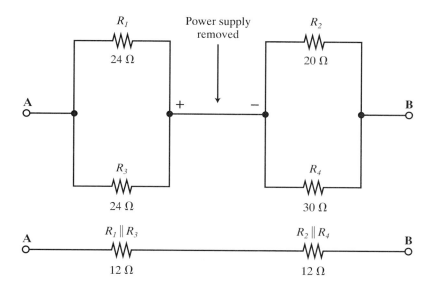

Figure 9-26. Bridge circuit is redrawn (top) with R_L and V removed. The parallel resistors can be simplified (bottom). (Sample problem 14)

1. Find the parallel equivalents.

$$R_{1-3} = R_1 \parallel R_3 = 12 \ \Omega$$

$$R_{2-4} = R_2 \parallel R_4 = 12 \ \Omega$$

2. Find the series combination.

$$R_{TH} = R_{1-3} + R_{2-4}$$

$$R_{TH} = 24 \ \Omega$$

Examination of the Thevenin Equivalent Circuit

Even though the Thevenin voltage is measured at the load terminals, it is measured with the load removed. Having an open circuit at the load terminals results in a voltage somewhat different from the complete circuit.

When the load is attached to the terminals, if there is a difference in potential (voltage), current travels through the load resistor. That current will alter the current through the voltage divider resistors. This changes their voltage drops. As a result, the voltage that is measured with the load attached is less than Thevenin's voltage.

Sample problem 15. _____

Using the equivalent circuit in figure 9-27, determine the voltage across the load and the load current.

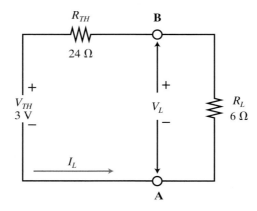

Figure 9-27. Thevenin equivalent circuit of the bridge circuit. (from figure 9-24)

1. Total resistance of Thevenin's equivalent circuit:

   Formula: $R_T = R_{TH} + R_L$

   Substitution: $R_T = 24\ \Omega + 6\ \Omega$

   Answer: $R_T = 30\ \Omega$

2. Current to the load:

   Formula: $I_L = \dfrac{V_{TH}}{R_T}$

   Substitution: $I_L = \dfrac{3\ V}{30\ \Omega}$

   Answer: $I_L = 0.1\ A$

3. Voltage across the load:

   Formula: $V_L = I_L \times R_L$

   Substitution: $V_L = 0.1 \text{ A} \times 6 \text{ }\Omega$

   Answer: $V_L = 0.6 \text{ V}$

9.7 NORTON'S THEOREM

Norton's theorem is similar to Thevenin's theorem in that it reduces the circuit as seen by the load. However, rather than finding an equivalent voltage source with a series resistor, Norton's theorem finds an equivalent current source with a parallel resistor.

Frequently, when designing a circuit, it is necessary to know the amount of current within the power supply's circuitry. In the **Norton equivalent circuit,** the calculated current remains constant, regardless of changes made in the load. As the current to the load changes, the current flowing through the Norton equivalent resistor also changes. The total current produced by the power supply remains constant, however.

Finding the Norton Equivalent Current

The **Norton equivalent current** is found by removing the load and replacing it with a short. Compare figures 9-28 and 9-29. By shorting the output terminals, the current produced is with a load of zero ohms.

Sample problem 16. _____

Find the Norton equivalent current for the circuit drawn in figure 9-28. This circuit is redrawn in figure 9-29 showing the load replaced by a short.

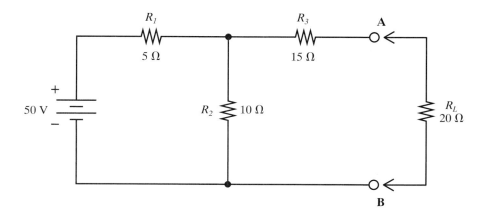

Figure 9-28. Find the Norton equivalent of this circuit.

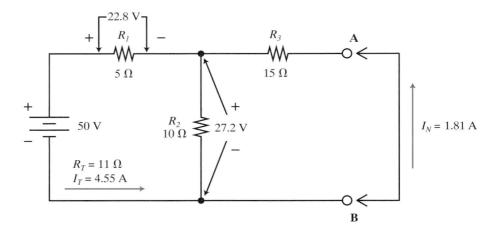

Figure 9-29. Norton current is found by replacing the load resistor with a short circuit. (Sample problem 16)

1. Total circuit resistance (with the load replaced by a short).
    a. Parallel combination of $R_2 \parallel R_3$:

    Formula: $\dfrac{1}{R_{2\text{-}3}} = \dfrac{1}{R_2} + \dfrac{1}{R_3}$

    Substitution: $\dfrac{1}{R_{2\text{-}3}} = \dfrac{1}{10\ \Omega} + \dfrac{1}{15\ \Omega}$

    Decimals: $\dfrac{1}{R_{2\text{-}3}} = 0.1 + 0.0667$

    Answer: $R_{2\text{-}3} = 6\ \Omega$

    b. Combine with mainline series resistor:

    Formula: $R_T = R_1 + R_{2\text{-}3}$

    Substitution: $R_T = 5\ \Omega + 6\ \Omega$

    Answer: $R_T = 11\ \Omega$

2. Find total current:

    Formula: $I_T = \dfrac{V}{R_T}$

Substitution: $I_T = \dfrac{50\text{ V}}{11\ \Omega}$

Answer: $I_T = 4.55\text{ A}$

3. Voltage drop across mainline resistor:

Formula: $V_{R_1} = I_T \times R_1$

Substitution: $V_{R_1} = 4.55\text{ A} \times 5\ \Omega$

Answer: $V_{R_1} = 22.8\text{ V}$

4. Voltage drop across the parallel combination:

Formula: $V_{R_{2\text{-}3}} = V - V_{R_1}$

Substitution: $V_{R_{2\text{-}3}} = 50\text{ V} - 22.8\text{ V}$

Answer: $V_{R_{2\text{-}3}} = 27.2\text{ V}$

5. Norton's equivalent current equals current through R_2:

Formula: $I_N = \dfrac{V_{R_{2\text{-}3}}}{R_3}$

Substitution: $I_N = \dfrac{27.2\text{ V}}{15\ \Omega}$

Answer: $I_N = 1.81\text{ A}$

Finding the Norton Equivalent Resistance

The next step in creating the Norton equivalent circuit is to find the **Norton equivalent resistance.** Do not confuse the Norton resistance with the total circuit resistance that was calculated in step 1 of sample problem 16. Total resistance is seen by the voltage source. Norton resistance is seen by the load.

To calculate the Norton resistance, the voltage source is removed and replaced with a short. Resistance is calculated at the load terminals, with the load removed, as if there were an ohmmeter in its place.

The procedure to calculate the Norton equivalent resistance is exactly the same as calculating the Thevenin equivalent resistance. Therefore, $R_N = R_{TH}$.

Sample problem 17. _____

   Calculate the Norton equivalent resistance for the circuit in figure 9-30.

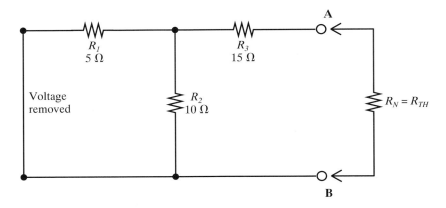

Figure 9-30. Norton resistance is found using the same procedure used to find the Thevenin resistance. (Sample problem 17)

1. Parallel combination of $R_1 \parallel R_2$:

   Formula: $R_{1\text{-}2} = \dfrac{R_1 \times R_2}{R_1 + R_2}$

   Substitution: $R_{1\text{-}2} = \dfrac{5\ \Omega \times 10\ \Omega}{5\ \Omega + 10\ \Omega}$

   Answer: $R_{1\text{-}2} = 3.33\ \Omega$

2. Combine with series resistor:

   Formula: $R_N = R_{1\text{-}2} + R_3$

   Substitution: $R_N = 3.33\ \Omega + 15\ \Omega$

   Answer: $R_N = R_{TH} = 18.33\ \Omega$

Drawing the Norton Equivalent Circuit

   The Norton equivalent circuit is drawn with a constant current source, shown as a circle with an arrow in it. The arrow points in the direction the current travels around the circuit, from negative to positive.

   The Norton equivalent resistor is drawn in parallel with the current source and the load is returned to the load terminals, as shown in figure 9-31. The current splits to the load and to the Norton resistor. Current to the load can be calculated by first finding total resistance, then using Ohm's law.

Sample problem 18.

Determine the current and voltage to the load resistor in the Norton equivalent circuit shown in figure 9-31.

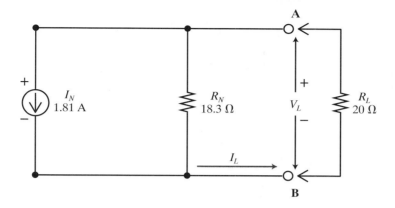

Figure 9-31. Find the current and the voltage to the load in this Norton equivalent circuit. (Sample problem 18)

1. Calculate total resistance of the equivalent circuit.

    Formula: $R_T = \dfrac{R_N \times R_L}{R_N + R_L}$

    Substitution: $R_T = \dfrac{18.3 \ \Omega \times 20 \ \Omega}{18.3 \ \Omega + 20 \ \Omega}$

    Answer: $R_T = 9.56 \ \Omega$

2. Calculate circuit voltage:

    Formula: $V = I_N \times R_T$

    Substitution: $V = 1.81 \ A \times 9.56 \ \Omega$

    Answer: $V = 17.3 \ V$

3. Calculate load current:

    Formula: $I_L = \dfrac{V}{R_L}$

    Substitution: $I_L = \dfrac{17.3 \ V}{20 \ \Omega}$

    Answer: $I_L = 0.865 \ A$

9.8 COMPARING THEVENIN WITH NORTON CIRCUITS

Both Thevenin and Norton circuits are concerned with what takes place at the load. It is the load current and load voltage that has significance when designing a circuit.

Figure 9-32 shows the equivalent circuits from both theorems. The Thevenin equivalent circuit is from sample problem 12. The Norton equivalent circuit is from sample problem 18. The results of calculations in the sample problems are written on each circuit for comparison. The slight variance in the numbers is due to rounding at various stages of the calculations.

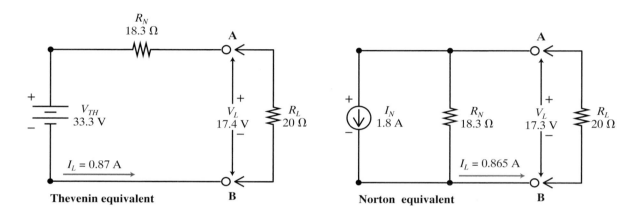

Figure 9-32. Comparing Thevenin and Norton equivalent circuits.

9.9 BALANCED WHEATSTONE BRIDGE

A balanced Wheatstone bridge has equal voltages between terminals A and B. The two parallel branches of the bridge are each voltage dividers. If the ratio of the voltage dividers is equal, there will be no difference in potential between the two terminals.

A galvanometer is connected between the terminals. A **galvanometer** is a voltmeter with the zero in the center. If there is a difference in voltages between the meter's terminals, the needle swings toward the higher voltage. With a galvanometer connected, a Wheatstone bridge is used as an instrument for measuring resistance values. The voltage applied to the circuit has no effect on accuracy since both voltage dividers receive the same applied voltage. In figure 9-33, an unknown resistor is connected between two test terminals. Variable resistor R_A is adjusted until the galvanometer reads exactly zero, a balanced condition. When balanced, the unknown resistance can be calculated. The formula for calculating the resistance is:

Formula 9.A

$$\frac{R_X}{R_A} = \frac{R_1}{R_2}$$

R_X is the unknown resistance, measured in ohms.

R_1 and R_2 are the resistances of fixed resistors, measured in ohms.

R_A is the resistance of variable resistor R_A, measured in ohms.

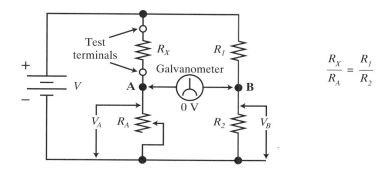

Figure 9-33. A balanced Wheatstone bridge (with a zero volt difference between points A and B) occurs when the ratio of the resistors is equal.

Sample problem 19. _____

What is the value of R_X needed to balance the Wheatstone bridge in figure 9-34?

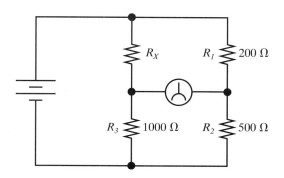

Figure 9-34. Find the value of R_X needed to balance this Wheatstone bridge. (Sample problem 19)

Formula:	$$\dfrac{R_X}{R_A} = \dfrac{R_1}{R_2}$$
Substitution:	$$\dfrac{R_X}{1000\ \Omega} = \dfrac{200\ \Omega}{500\ \Omega}$$
Multiply both sides by 1000 Ω:	$$R_X = \dfrac{200\ \Omega}{500\ \Omega} \times 1000\ \Omega$$
Answer:	$$R_X = 400\ \Omega$$

SUMMARY

- Kirchhoff's current law states that the current leaving a junction equals the current entering the junction.
- Kirchhoff's voltage law states that the voltages around a loop add to equal zero.
- Circuits with two voltage sources can be solved using Kirchhoff's laws or the superposition theorem.
- The superposition theorem states that either the current or voltage of a component are the algebraic sum of the sources acting separately.
- Thevenin's theorem reduces a circuit to a power source consisting of an equivalent voltage with a series resistor as seen by a load.
- Norton's theorem reduces a circuit to a power source consisting of an equivalent current source with a parallel resistor as seen by a load.
- Thevenin resistance and Norton resistance are the same value in a circuit.
- A balanced Wheatstone bridge has no difference in potential between the two voltage dividers.

KEY WORDS AND TERMS GLOSSARY

circuit theorem: Mathematical tools that make it easier to solve complicated circuits.

galvanometer: A voltmeter with the zero in the center. If there is a difference in voltages between its terminals, the needle swings toward the higher voltage.

IR drop: A name for the voltage drop in loop equations.

Norton equivalent circuit: Circuit where all but the load is turned into an equivalent current source and parallel resistor.

Norton equivalent current: Current that would be measured by removing the load and replacing it with a short.

Norton equivalent resistance: Resistance that would be measured across the load terminals if the load resistor was removed and the voltage source was replaced with a short.

Thevenin equivalent circuit: Circuit where all but the load is turned into an equivalent voltage source and series resistor.

Thevenin equivalent resistance: Resistance that would be measured at the load terminals if the load resistor was removed and the voltage source was removed and replaced with a short.

Thevenin equivalent voltage: Found by removing the load resistor and measuring the voltage at the load terminals.

Wheatstone bridge: A circuit that places the load between two parallel resistor branches.

KEY FORMULA

Formula 9.A

$$\frac{R_X}{R_A} = \frac{R_1}{R_2}$$

TEST YOUR KNOWLEDGE

Do not write in this text. Please use a separate sheet of paper.
1. State Kirchhoff's current law.
2. State Kirchhoff's voltage law.
3. Describe how to write a loop equation.
4. State the superposition theorem.
5. Describe how to find the Thevenin voltage in a circuit.
6. Describe how to find the Thevenin equivalent resistance.
7. Draw a Thevenin equivalent circuit.
8. How is the load voltage and current found in a Thevenin equivalent circuit?
9. After the Thevenin equivalent circuit has been found, what values will change from one load to another?
10. Describe how to find the Norton current.
11. Describe how to find the Norton equivalent resistance.
12. Draw a Norton equivalent circuit.
13. How does the Norton resistance compare to the Thevenin resistance.
14. Use Kirchhoff's current law to find current I_2 in figure 9-35.

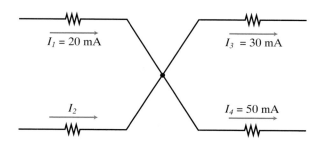

Figure 9-35. Problem #14.

15. Use Kirchhoff's current law to find the unknown currents and their directions in the circuit in figure 9-36.

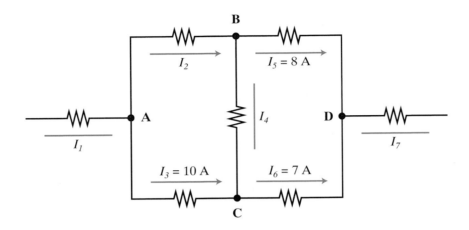

Figure 9-36. Problem #15.

16. Use Kirchhoff's voltage law to find the missing voltage V_{R_2} in figure 9-37.

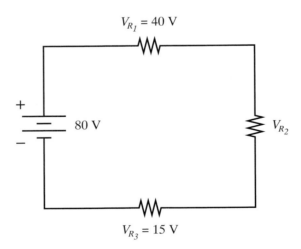

Figure 9-37. Problem #16.

17. Use Kirchhoff's voltage law to find the unknown voltages in figure 9-38.

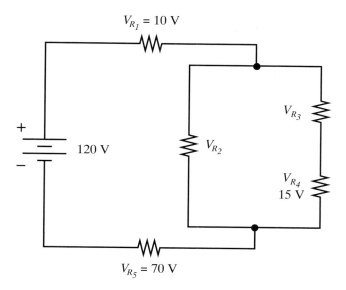

Figure 9-38. Problem #17.

18. Use Kirchhoff's laws to find the current and voltage of each resistor of the circuit in figure 9-39.

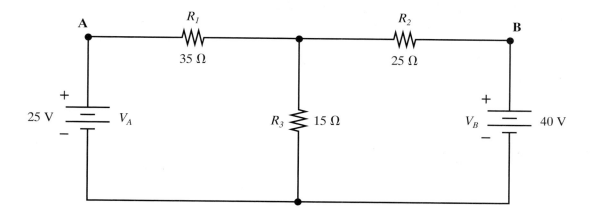

Figure 9-39. Problem #18.

19. Use the superposition theorem to find the current and voltage of each resistor in figure 9-40.

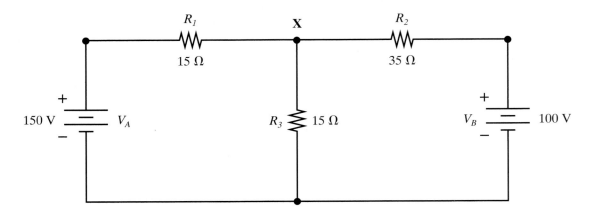

Figure 9-40. Problem #19.

20. Find the Thevenin equivalent circuit of figure 9-41.

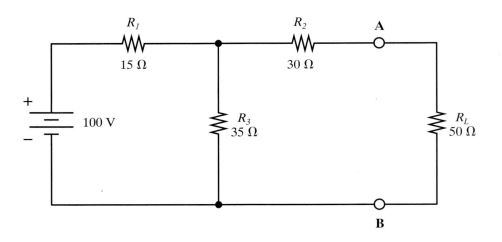

Figure 9-41. Problem #20.

21. Find the Thevenin equivalent circuit of the Wheatstone bridge in figure 9-42.

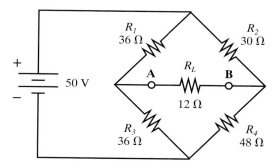

Figure 9-42. Problem #21.

22. Find the Norton equivalent circuit of figure 9-43.

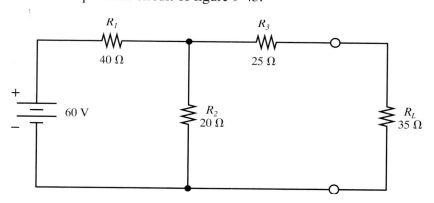

Figure 9-43. Problem #22.

23. Find the value of R_X needed to balance the Wheatstone bridge in figure 9-44.

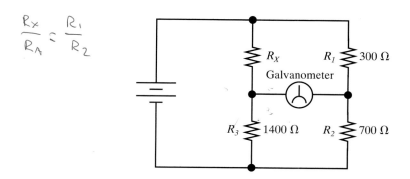

Figure 9-44. Problem #23 .

**An understanding of the basic electrical principles
leads to a better grasp of more complex ideas.
(DeVry Institutes)**

Chapter 10
Producing DC Voltages

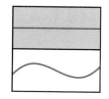

Upon completion of this chapter, you will be able to:
- Relate several individuals to important inventions and developments in producing electricity.
- Describe the characteristics and behavior of static electricity.
- Identify examples of the piezoelectric effect.
- Give examples where the effects of heat can be used.
- Describe applications of light-sensitive devices.
- Recognize devices using magnetism to develop electrical signals.
- Describe the differences between different batteries.
- Calculate the results of voltage sources connected in series and parallel.
- Predict the effect of a voltage source's internal resistance.
- Identify the maximum transfer of power.

There are six ways to produce electricity:
- Static/electric charges.
- Pressure.
- Heat.
- Light.
- Magnetism.
- Chemical reaction.

Each of these has a place in electronics technology. This chapter discusses how electricity is produced and some of its applications.

There have been very few recent developments in producing electricity. However, advances have been made in the miniaturization of electronics. This has helped to make many new and better applications for electricity. A student preparing to enter a career in electronics technology can be confident that a sound foundation in dc/ac electronics will enable the understanding of future developments.

10.1 A BRIEF HISTORY

There have been many people involved in the discovery and invention of electricity and electrical products. The following are some of the most important individuals and their impact on the matter contained in this chapter.

Thomas Edison, an American inventor in the early 1900s developed a secondary cell, also called a nickel-iron battery, that could be recharged. The principles of these early cells are still in use today.

Benjamin Franklin, an American statesman and inventor, experimented with proving that lightning is an electrical phenomenon and invented the lightning rod in the 1750s. Franklin developed the *one-fluid theory* to explain the two kinds of electricity, positive and negative.

Luigi Galvani, an Italian physiologist, was noted for his studies with frogs in the 1790s. Galvani found electricity activates animal nerves and muscles.

Georges Leclanché, a French chemist, invented the primary cell, called the Leclanché cell, in the 1860s. This is a type of dry cell. The modern dry cell is based on the Leclanché cell.

Gaston Planté, a French physicist, developed a secondary cell in 1859. This cell was a rechargeable lead-acid wet cell battery.

Thomas Seebeck, a German physicist, in 1821 observed that if two dissimilar metals are joined, electrical current is created when the joint is heated. This is known as the **Seebeck effect.**

Robert Jemison Van de Graaff, an American physicist, developed the Van de Graaff generator. The **Van de Graaff generator,** an electrostatic machine, produces extremely high voltages.

Alessandro Volta, an Italian physicist, developed the earliest battery called the Volta cell. This cell produced a steady stream of electrons.

10.2 STATIC/ELECTRIC CHARGES

Static electricity is a familiar sight, usually caused by friction. **Static electricity** is a buildup of electrons, or a shortage of electrons, that is stored until a discharge path is provided. For example, a person walking across carpeting in dry weather becomes charged. If that person then touches someone else, a tiny spark occurs, giving the person a "shock." Atmospheric lightning is also caused by the buildup of static electricity. It is the result of clouds building a static charge and then discharging either to the ground or to another cloud.

A charge on an object is an unbalanced electrical condition. This condition is determined by the amount of excess electrons (on negatively charged objects) or the amount of holes (on positively charged objects). A **hole** is an atom that is seeking an electron. When some item is charged, it is either negative or positive. The item either has too many free electrons that are looking for holes, or it has too many holes and is looking to attract some electrons. Static electricity, by its nature, is a dc voltage. It has a definite polarity.

In figure 10-1, three types of charges are shown. A balanced condition results when there is an equal number of electrons and holes. A negative charge has an excess number of electrons. A larger excess, produces a larger negative charge. A positive charge has a shortage of electrons. The greater the shortage, the higher the positive charge.

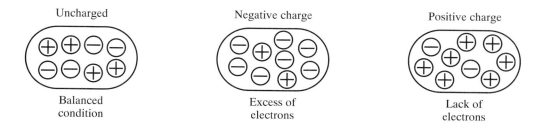

Figure 10-1. Static charge is determined by the relative amount of electrons.

Do not confuse the excess of electrons associated with a negative charge with the *free electrons* in a conductor. In a conductor, the electrons are loosely bound in orbit and easily move along the conductor. A static charge develops in *insulators,* such as glass or rubber, by gathering electrons in a material that allows little movement. Other materials, such as nylon, easily give up electrons when contact is made.

Discharging a Static Charge

A charged object will neutralize its charge when given the chance. The charged object does not need to find an object with the opposite charge, only an object with a *lesser charge.* A negative charge needs to find something less negative to accept the excess electrons. A positive charge needs to find an object less positive to supply electrons.

A problem with discharging static electricity is that it is not always a controlled situation. For example, the information stored on computer disks can be lost if the disk is exposed to a static discharge.

Figure 10-2 shows the three ways in which static electricity can be discharged. These ways are:
• Through a wire.
• Through touching.
• Through arcing.

Each of these methods of discharge can happen by accident, or when intended.

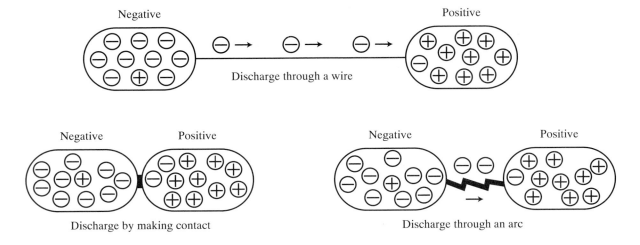

Figure 10-2. Static electric charges can discharge through three methods.

Discharge by Touching or with a Wire

Touching or using a wire to eliminate a static charge buildup is generally a safe way. In electronics equipment manufacturing, workers involved with the assembly of static-sensitive components wear a wristband with a wire attached to electrical ground. Any possible static buildup is released harmlessly.

Discharge through an Arc

The voltage necessary to cause an **arc,** electricity jumping through the air, is approximately 25,000 volts per inch. The static electricity shock that is heard, and felt, from a person walking on a carpet is a small arc.

Lightning jumping from the clouds has the potential of many millions of volts. With voltages that large, even a short duration of exposure can cause severe shock and burns.

Sometimes, when lightning strikes, the discharge reaches into the electrical power lines. This causes an electrical **surge,** a sudden burst of high voltage. A surge entering electrical equipment can cause severe damage. Surge protectors are often used as a means of protection. Surge protectors are plugged into the electrical line and the equipment is plugged into them.

10.3 PRESSURE

Certain types of crystals produce a voltage when a mechanical pressure is applied. This is called the **piezoelectric effect.** The voltage that is produced is very small, but it is large enough to be used in applications such as microphones and radio crystals.

The voltage that is produced by applying pressure to a crystal is a dc voltage with a polarity that does not change. However, the amount of voltage generated fluctuates as the amount of pressure changes. If a constant pressure is maintained on the crystal, the voltage produced drops to zero. In other words, the crystal responds to the *change in pressure.*

Piezoelectric Microphones

Figure 10-3 shows a microphone, which is used to change sound waves into electricity. The electrical waves are then amplified in an audio amplifier. This amplifier may consist of several amplification steps. Next, the amplified signal is sent where it can be translated into sound waves once again. This could be through a radio transmission, magnetic tape recording, or simply though a loudspeaker, as shown in the figure.

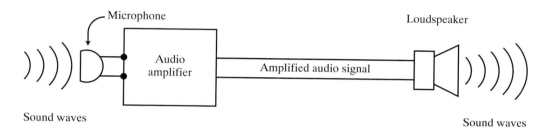

Figure 10-3. Sound waves are converted into electricity, amplified, and then converted back into sound waves.

A crystal microphone works when sound waves strike a diaphragm. This puts pressure on the crystal, as shown in figure 10-4. Sound waves follow a sine wave pattern (sine waves are explained in detail in Chapter 15). The sine wave pattern goes from zero to a maximum positive voltage, back to zero, to a maximum negative voltage, back to zero, and then repeats. The voltage of the microphone follows the changing sound waves producing a form in the shape of a sine wave. However, the voltage of the piezoelectric crystal does not go negative. The crystal produces a zero voltage to correspond to the maximum negative of the sound wave and increases to a maximum positive voltage. The fluctuations all remain positive.

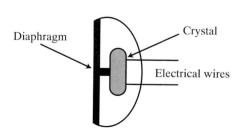

Figure 10-4. Structure of a crystal microphone.

Microphones are available in types other than with the crystal. Figure 10-5 shows microphones in a variety of sizes and shapes. Applications determine which type is best. How much voltage needs to be produced by the microphone, how rugged it is, and how sensitive it is are all important.

Figure 10-5. Microphones come in a variety of shapes and sizes. (Sima, Cobra Electronics Corp., Lonestar Technologies Ltd.)

Radio Frequency Crystals

Piezoelectric crystals are used in radio frequency transmitters and receivers because of their ability to accurately produce a desired frequency. Applying electrical pressure to a crystal causes it to vibrate. The crystal produces a sine wave voltage with a frequency equal to the vibrations.

The circuit shown in figure 10-6 is a simplification of a crystal oscillator used in a radio frequency application. The two coils drawn side-by-side represent a transformer. The radio frequency oscillations produced by the crystal are coupled through the transformer to an amplifier circuit. Transformers respond only to a changing voltage. This makes them ideal for use with the piezoelectric crystals.

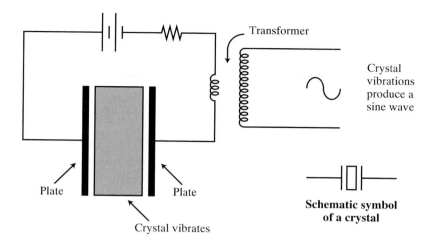

Figure 10-6. Electricity applied to a crystal causes vibrations in the crystal. These vibrations produce a constant electrical frequency.

10.4 HEAT

Electricity can be affected three different ways by heat. These three ways are:
- The resistance can be varied.
- A voltage can be produced.
- A switch can be activated.

Each of these are applied in the field of electronics. Some applications require miniature components and others use much larger devices.

Thermistor

A **thermistor** is a resistor that varies its resistance with changes in temperature. A thermistor's resistance varies over a wide temperature range. A very popular application of the thermistor is measuring actual temperature. The electronic thermometer uses a thermistor.

The thermistor as a component, shown in figure 10-7, must be connected to supporting circuitry. The circuitry is responsible for interpreting the changes in resistance. The circuitry may sense a change in current or a change in a voltage drop.

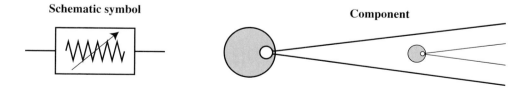

Figure 10-7. Thermistors change resistance with changes in temperature.

Thermistors come with ratings. The ratings are:
* Working temperature range. (Example: –80° to +75°C.)
* Nominal resistance at a specified temperature, usually stated at +25°C.
* Resistance tolerance.
* Temperature accuracy.

Thermocouple

A **thermocouple** is two pieces of metal, made of different materials, joined at one end. This is also called a **bimetallic strip.** When the junction is heated, a dc voltage is produced at the other end. This voltage remains constant when constant heat is applied and changes with a changing heat. The voltage output drops to zero when there is insufficient heat.

A thermocouple can be as simple as two dissimilar wires welded together or it can be specially selected bimetallic strips. Figure 10-8 shows the metal strips, their junction, and a voltmeter to register a voltage. No external circuitry is needed to measure the voltage. However, some circuitry is needed if the voltage is used to do work.

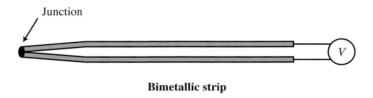

Figure 10-8. When heat is applied to the junction of two dissimilar metals, a voltage is produced.

A thermocouple is useful in applications where heat needs to be monitored. One use is in an alarm system checking for a furnace to reach a certain temperature. Thermocouples can also give feedback concerning combustion. It is used in this manor in electronically controlled burners.

Thermostat

A **thermostat** is a heat-activated switch. The schematic symbols for a thermostat are shown in figure 10-9. Thermostats are made in many different ways. One of the popular methods of production is with bimetallic strips. When a bimetallic strip is

Figure 10-9. There are several schematic symbols used for a thermostat.

heated, the two metals expand at different rates. Consequently, the strip bends when heated. A switch is created in which the bimetallic strip is one of the contacts. When enough current flows to heat the metal, it bends and opens the switch. Current stops flowing, the strip cools, and contact is made again. These circuits work well as flashers.

A thermostat can be used to operate a separate circuit, such as the circuit that controls the heat and air conditioning in buildings. Adjustments can be made to the thermostat. These adjustments change the temperature at which the switch operates.

10.5 LIGHT

The use of light has found many applications in electronics technology. The brightness can be varied. This allows it to be used in the communications industry, such as with fiber optics. Light can be switched on and off at an almost unlimited rate of speed. Also, some materials are highly sensitive to light.

Photosensitive materials respond when light strikes. Some photosensitive materials respond by producing a voltage. Other materials change their resistance. See figure 10-10 for a picture of light striking the material and the schematic symbols.

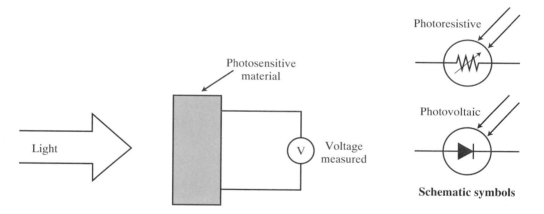

Figure 10-10. Voltage is produced or resistance changes when light strikes a photosensitive substance.

The output of photosensitive material is a dc voltage. With increasing light, there is an increasing voltage. If the light does not reach some critical level, there is no voltage produced. When the light source oscillates, the voltage oscillates in a similar pattern.

Optoelectric Devices

Optoelectric is the name given to electronic components that use light to operate. Optoelectric devices are sensitive to different types of light, either visible or infrared. An optoelectric device must be connected to supporting circuitry to perform a useful

Figure 10-11. Shown are just a sample of the many shapes and sizes of optoelectric devices. The variety of shapes is necessary because of their many different applications.

operation. As shown in figure 10-11, there are so many shapes, sizes, and applications that it is not possible to describe them all.

The packaging of the device determines the type of applications it can be used for. Some contain only a light receiver. Some devices contain only a light source. Some devices contain both receiver and source within the same package.

Figure 10-12 shows a familiar application. The bar codes on packages are easily *read* by an optoelectric device. The sensor detects the reflected light from the white bars and no light from the black bars. The on and off light flashes are translated into a numerical code. This code is sent to a computer. If the computer recognizes the information, it sends a signal to tell the operator the reading was successful. What the

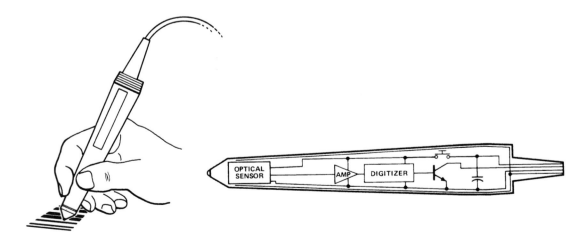

Figure 10-12. Most items sold in retail stores come with a bar code on the package. A pass with a bar code wand rings up the price for the consumer and tracks inventory for the store.

computer does with the code is entirely up to the computer programmer. The information is often used for item identification, purchase price, and inventory.

Another type of optically coded message is received through the use of a wheel with holes passing in front of an optical sensor. One of these sensors is shown in figure 10-13. One application of this system is measuring the speed and number of turns of a motor shaft. Robots use this type of device to determine the speed and position of the motors operating the wheels, gripper, and arm.

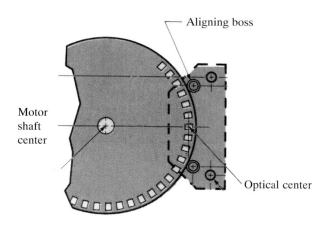

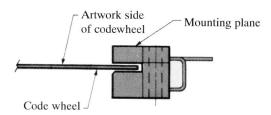

Figure 10-13. An optical sensor tracks the light passing though this code wheel. The mouse for a computer uses this system to move your cursor.

10.6 MAGNETISM

Magnetism finds many uses in the electrical and electronics fields. When a conductor passes through a magnetic field, a voltage is produced. This is diagrammed in figure 10-14. Inversely, when electric current flows through a conductor, a magnetic field is produced. Magnetism, its principles, and applications are explored in greater detail in Chapter 11.

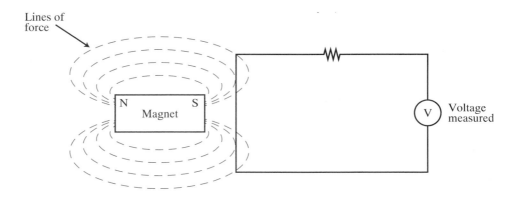

Figure 10-14. Passing a wire through a magnetic field produces electricity.

Generators and Motors

Strong magnetic fields are used to produce voltages in generators, such as those used by the power companies to supply electricity to homes and businesses. Large amounts of current can be used to produce very strong magnetic fields. These electromagnets can be made strong enough to pick up an automobile. Electric motors are another example of using electricity to produce a magnetic field.

Magnetic Pickups

Electrical information is coded magnetically in a number of ways. One common form of coded magnetic information is magnetic tape. Magnetic tape is coded by arranging tiny magnets on the tape into a pattern that matches the electrical signal. Examples of magnetic storage include: audio cassette tapes, VCR tapes, computer disks, and coded information on the back of credit cards.

Another form of coded electrical information comes from magnetic pulses produced by placing a magnet on a spinning shaft. These are used on the tachometer on an engine and the sensor used with a vehicle's cruise control.

The reading of this information stored on magnetic tapes, or magnetic pulses, requires a **magnetic pickup.** The magnetic pickup, commonly called a **head**, is shown in figure 10-15. It is a coil of wire wrapped around a core made of a metal that easily conducts magnetic lines of force (flux).

When the tape moves past the head, the magnetic particles on the tape induce a magnetic flux into the core. This magnetic field passes by the wire, which induces a voltage in the wire. The voltage in the wire is very small and needs to be amplified for use in the electronic circuitry. Then, in the case of an audio signal, the voltage is applied to a speaker, allowing people to listen.

To induce a voltage in a magnetic head, the tape must be moving, or in the case of a VCR, both the tape and the head are moving. The voltage produced in the head is an ac sine wave signal. It changes from zero, to a maximum positive voltage, back through zero, to a maximum negative voltage.

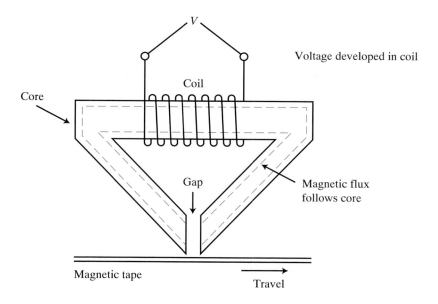

Figure 10-15. Construction of a magnetic tape head. A changing magnetic field from the tape passing below induces a voltage in the coil.

10.7 BATTERIES

Two dissimilar metals can be made to produce a voltage from a chemical reaction using an electrolyte. An **electrolyte** conducts electricity using ionic conduction. The electrolyte can be the acid in a wet cell battery, the paste or powder in dry cell batteries, the gel used in gel-cell batteries, or even the pulpy liquid of a grapefruit.

One type of metal plate is selected to produce negatively charged ions when it comes in contact with the electrolyte. The other plate is selected to produce positively charged ions. An **ion** is an atom or molecule (group of atoms) that is electrically charged. Positive ions are lacking electrons. Negative ions have excess electrons.

When an external circuit is connected to the cell, electrons from the negative terminal are provided with a path to the positive terminal. The electron current is the result of an attempt to equalize the difference in potential between the plates. As the electrons enter the positive plate, the ions are neutralized. More ions are produced and the process continues. The electrolyte weakens during this conduction and eventually the process will stop. The cell, then, has lost its charge.

Recharging a Battery

To recharge a battery, an outside voltage forces current through the electrolyte in the *reverse* direction. The recharge process returns the ions onto their original plates. The charging unit voltage must be slightly higher than that of the battery. The charging unit must also have the capability to supply enough current to run the recharging process.

Not all batteries can be recharged. An automobile battery is a commonly recharged battery. A car battery usually recharges from the car's alternator while the engine is running. However, sometimes a car battery "dies" and the car will not start. The battery can be recharged from another car battery or from a battery charger working off household electricity.

The gel-cell battery and the nickel-cadmium type are other kinds of batteries that can be recharged. Both of these types are limited in the number of times the battery can be recharged. Although, many improvements have been made in recent years to lengthen the life of these batteries.

Wet Cell Battery

The most common type of wet cell battery is the lead-acid type used in automobiles. Figure 10-16 demonstrates the principles of a basic wet cell battery. Two dissimilar metals are placed in the electrolyte. Positive ions are collected on one plate. Negative ions are collected on the other. The collecting of ions continues until a certain potential difference is developed. At that time, the ions are no longer attracted due to the repelling effect of like charges.

A typical wet cell battery has a potential of two volts per cell. A car battery would be made of six of these cells.

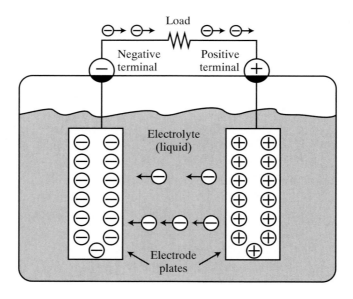

Figure 10-16. Construction of a wet cell. A car battery is made up of several wet cells connected together.

Dry Cell Battery

The dry cell battery is made with a powder or paste substance as the electrolyte, figure 10-17. The advantage of the dry cell is its portability. One common application is in the flashlight.

Dry cell batteries have two disadvantages. One is that they can not be recharged easily. In addition, dry cells do not produce large amounts of current.

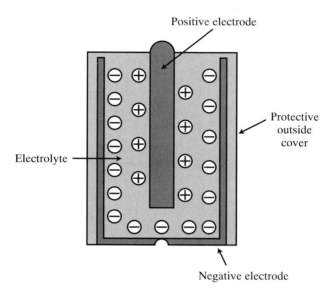

Figure 10-17. Dry cell batteries get their name because they use a powder or paste as their electrolyte.

Rating a Battery

There are many different types of batteries available. The different types are good for certain applications and not good for others. Imagine carrying a car battery to operate a flashlight. Or, think of trying to use enough AA size batteries to start a car. See figure 10-18 for pictures of some popular battery sizes.

The electrolyte and construction of the battery is the major deciding factor for the other ratings. Each type has its own advantages and disadvantages. There is no single battery that would be ideal in every application.

The different electrolytes include: lead-acid (wet cell), lead-dioxide (gel-cell), nickel-cadmium, lithium, alkaline, mercury, silver oxide, and carbon-zinc (all dry cell).

A battery has ratings in the following categories.
• Per-cell voltage.
• Primary/secondary cell.
• Current capacity.
• Applications.

Automotive

C cells

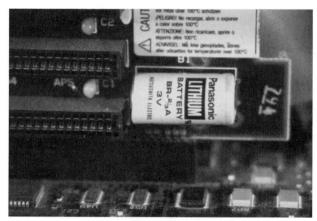

Circuit board mounted

Figure 10-18. Different size batteries are needed for different applications. Some popular sizes are shown here.

Per-Cell Voltage Rating

A battery can consist of two cells, such as the two D cells in a flashlight. A battery also can be made of many cells connected together to obtain a higher voltage, such as those that make a car battery.

The per-cell rating is determined by the type of electrolyte, not the number of cells in the package. Figure 10-19 shows a chart comparing battery electrolytes, typical voltages, and their applications. The nominal voltage rating is for a new cell with no load attached. Later in this chapter, the effect a load has on the output voltage is discussed in detail.

Primary/Secondary Cells

In the process of drawing current, the chemical reaction in each cell weakens. This and other factors lead to the battery wearing. Eventually the battery is no longer able to produce sufficient current to be useful. After a charge has been expended, certain electrolytes easily accept recharging. Others are very difficult, if not impossible to recharge.

Figure 10-19. Comparison of battery electrolytes, voltage output, and applications.

Dry Cell Batteries		
Electrolyte	*Voltage*	*Applications*
carbon-zinc	1.5	general purpose
alkaline	1.5	general purpose, premium service
mercury	1.35	watches, calculators, cameras, medical, hearing aids
mercury	1.4	medical, hearing aids, cameras
silver-oxide	1.55	watches, cameras, medical
lithium	3.0	backup memory, watches
nickel-cadmium	1.2	rechargeable, general use

A **primary cell** is a cell that cannot be easily recharged. The chemical making up the electrolyte is consumed when changed into ions. A slight amount of recharging *may* be obtained by breaking down some of the internal resistance and gaining a short burst of energy.

A **secondary cell** is a cell that can be easily recharged. Different factors are involved in determining if recharging a battery is possible. Some batteries can only be recharged a limited number of times. Other batteries can be recharged almost indefinitely. With some batteries, if they are drained too low, they will no longer accept a charge. Other batteries retain a *memory* of the point at which they were recharged. In future use, these batteries will drain no lower than the point at which they were recharged. Consequently, when using these batteries they should be completely drained before they are recharged.

Current Capacity

Each battery has a voltage rating based on the type of electrolyte. A battery also has a current capacity rating, which is based on its size. Generally speaking, the larger the battery, the higher the capacity.

As the name suggests, a battery's current capacity is the amount of current that can be supplied by the battery. This translates directly into how many electrons are made available through the ionization process.

The unit of measure for current capacity is **ampere-hour (Ah).** The ampere-hour rating is given as a specific number, such as 200 mAh (milliampere-hour). If 200 mA of current is drawn, the battery will have a life of one hour. A smaller current drain results in a longer battery life. For example, 20 mA of load current can be drained for 10 hours.

Ratings such as these must be used only as a comparison of one battery to another. The battery in the example above will not supply a constant 20 mA for 10 hours, then abruptly stop. Rather, the battery will supply a fairly constant current for most of the time. Then, the voltage drops off and the current will be less. The drop-off is gradual, not abrupt. Battery manufacturers produce charts showing how fast the voltage drops off.

Applications

When a battery is manufactured and marketed, it is intended to meet certain needs. Certain electrolytes supply a sudden burst of current, while others are better suited to a constant drain. To some degree, you should be careful and try to match the application to the battery's intended use.

10.8 SERIES AND PARALLEL VOLTAGE SOURCES

If a higher voltage is needed than one battery can supply, more cells must be added together in series. If an electronic circuit needs a larger current capacity, with the same voltage, cells are connected in parallel.

Characteristics of Voltage Sources Connected in Series

Dry cell batteries are made with one cell. When more voltage is necessary, the cells are connected in series. A flashlight is an example of batteries connected in series for a higher voltage. In figure 10-20, two dry cell batteries (each 1.5 V) are used for the needed three volts.

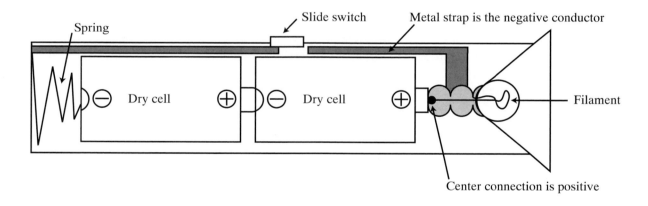

Figure 10-20. Diagram of a two-cell flashlight. Two 1.5 volt batteries connected in series produce three volts.

The schematic diagram for this simple circuit is shown in figure 10-21. If it is necessary to show independent cells, the symbols are drawn separately, as in this figure. For all other situations, the standard symbol for a voltage source is drawn.

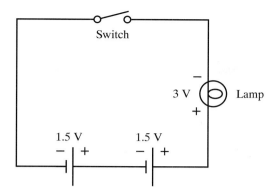

Figure 10-21. Schematic of two-cell flashlight from figure 10-20. Voltages of batteries connected in series add.

Some rules to remember when connecting voltage sources in series are:
1. Voltages connected in series *add*.

Formula 10.A

$$V_T = V_1 + V_2 + \ldots V_N$$

V_T is the total voltage, measured in volts.

V_1 through V_N are the individual voltages, measured in volts.

N is the number of voltage sources in series.

If all of the cells have the same voltage value, you can also use:

Formula 10.B

$$V_T = V \times N$$

V_T is the total voltage, measured in volts.

V is the voltage value, measured in volts.

N is the number of voltage sources in series.

2. Current capacity is equal to the capacity of the lowest capacity cell.

There are design decisions to be made when creating an electric appliance that will be powered by a series of batteries. A flashlight provides a good example. The brilliance of a light bulb is rated electrically with the unit of watts. A higher wattage produces more light, but it also requires more electricity. Higher wattage can be obtained by increasing either the voltage or current supplied.

An increase in current shortens the life of the battery. To increase the voltage, more cells must be used, which increases the weight and size of the flashlight. Manufacturers offer a wide selection of sizes and styles to allow the customer to make a decision as to what is the best.

One of the problems with connecting dry cell batteries in series occurs when one of the batteries goes dead. One dead cell prevents the good batteries from powering the circuit. This is because a dead dry cell has an extremely high internal resistance.

Sample problem 1. _____

Most 12 volt automobile batteries are made with wet cells with a nominal voltage of 2 V. How many cells must be connected in series to complete the battery?

Formula: $V_T = V \times N$

Substitution: $12\ V = 2\ V \times N$

Answer: $N = 6$ cells connected in series

Characteristics of Voltage Sources Connected in Parallel

All points in a parallel circuit have the same voltage. This also applies to voltage sources connected in parallel. The major advantage of connecting voltage sources in parallel is an increase in current capacity.

Some rules to remember when connecting voltage sources in parallel are:
1. The voltage produced is the same for all cells.
2. Current capacity is the sum of the individual capacities.

Figure 10-22 depicts a stalled car with a dead battery being jump started by another car. The batteries are connected in parallel by the jumper cables. The first positive terminal is connected to the other positive terminal, and the first negative terminal is connected to the other negative terminal. The parallel connection allows the good battery to boost the voltage of the dead battery to 12 volts. The current needed to start the stalled car is drawn from the running car's battery. The parallel connection increases the effective current capacity.

The schematic diagram of the stalled car being jump started is shown in figure 10-23. The electrical systems of both cars will draw current from both batteries.

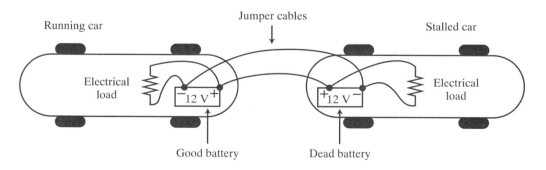

Figure 10-22. Batteries are connected in parallel when starting a stalled car. This increases the current capacity.

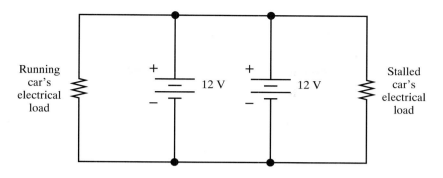

Figure 10-23. Schematic of the circuit made by two cars in figure 10-22.

10.9 INTERNAL RESISTANCE OF A VOLTAGE SOURCE

If a voltage source were perfect, it would produce its rated voltage with no losses internally. However, there is some resistance within a voltage source. This **internal resistance** (r_i) is due to electrons traveling through the electrolyte.

As with any resistance, when current flows there is a voltage drop. The lower the internal resistance, the smaller the voltage lost within the source will be. As a cell weakens, its internal resistance builds. This causes more voltage to be lost, and less current is delivered to the load.

Charging Resistance Is Different from Internal Resistance

Do not confuse the internal resistance of a voltage source with the **charging resistance.** To charge a cell, a voltage is applied in the reverse direction forcing electrons through the electrolyte. A dry cell exhibits extremely high resistance in the reverse direction. This makes them non-rechargeable.

The resistance of a wet cell to charging current is dependent on its charge. If a wet cell is fully charged, its reverse resistance is high, and the forward internal resistance is low. However, when the charge in the cell is low, the reverse resistance is low and the forward resistance is high. High current flows when charging a dead wet cell. Low current flows to charge a wet cell that is almost full.

No Load Voltage vs Loaded Voltage

The internal resistance of a battery varies depending on whether it is connected to a load or in an open circuit. When a voltmeter is connected to an unloaded battery, the internal resistance has no voltage drop. The voltmeter reads the battery voltage at its highest potential output voltage.

If a battery is fully charged, it will have a very low internal resistance. The open circuit voltage for the battery should be equal to its nominal voltage. When a load is connected, the fully charged battery supplies voltage at a value so close to nominal, it may be difficult to tell the difference. In figure 10-24, a 1.5 V dry cell battery is tested no

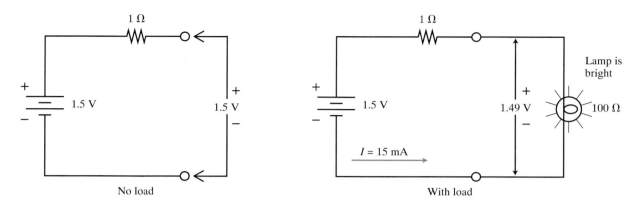

Figure 10-24. New batteries have very low internal resistances. Testing this battery with or without a load makes little difference.

load and with a 100 ohm load. With an estimated internal resistance of one ohm, there is virtually no drop in voltage when the load is connected.

As the battery weakens, the internal resistance begins to increase. The no load voltage shows only a slight drop from a fully charged battery. However, when the load is connected, there is a significant drop in output voltage, figure 10-25.

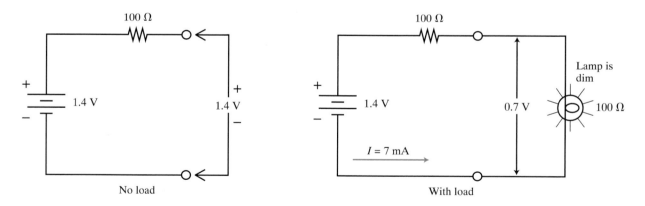

Figure 10-25. A weak battery may test near its full voltage (left) when tested without a load. However, when the battery is connected across a load (right), a significant drop in the voltage can be measured.

This demonstrates that the proper way to test the condition of a battery is with a load. Frequently, inexperienced technicians test a battery with only a voltmeter connected. The results can be misleading.

10.10 MAXIMUM TRANSFER OF POWER

In certain applications, it is necessary to obtain the maximum power possible from a source. In order to accomplish this, the load resistance must be equal to the internal resistance, r_i. When the load resistance is smaller than the internal resistance, the voltage drop across the load is less than the voltage drop across r_i. When the load resistance is larger than r_i there is more voltage dropped across the load, but less current can flow.

Figure 10-26 is a test circuit to prove the maximum transfer of power theory. In this circuit, the internal resistance of the power supply is 60 ohms. A variable resistor is connected as the load to allow measuring of both current and voltage with a changing load.

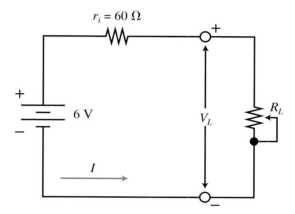

Figure 10-26. Schematic of a circuit designed to measure load power.

Figure 10-27 is a curve of load resistance vs load power for the circuit in figure 10-26. Notice that maximum power occurs when it is equal to r_i. The results of calculations for this graph are shown in the table of figure 10-28.

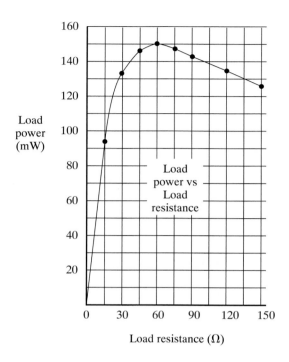

Load resistance (Ω)

Figure 10-27. Maximum transfer of power occurs when the load resistance is equal to the internal resistance at 60 ohms.

Figure 10-28. Table of the calculations for the circuit in figure 10-26.

R_L (Ω)	15	30	45	60	75	90	120	150
I (mA)	80	67	57	50	44	40	33	29
V_{R_L} (V)	1.2	2.0	2.6	3	3.3	3.6	4.0	4.3
P_{R_L} (mW)	96	133	147	150	147	144	138	124

SUMMARY

- Many individuals have contributed important inventions and developments related to electricity.
- Static electricity is a buildup of charges on an insulator caused by friction.
- The piezoelectric effect produces electricity by applying pressure to a crystal.
- Heat is used in electrical applications in three ways: changing the resistance of a material, producing a voltage, and operating a switch.

- Light produces a voltage in photosensitive materials.
- Magnetism is used to generate electricity and as a means of storing information to be recovered with electrical pulses.
- Chemical reaction produces electricity in a battery.
- Voltage sources are connected in series to increase voltage.
- Voltage sources are connected in parallel to increase current.
- Maximum transfer of power occurs when the load equals the internal resistance.

KEY WORDS AND TERMS GLOSSARY

ampere-hour (Ah): The unit of measure for current capacity of a battery. The ampere-hour rates the life of the battery for an amount of current for a period of time. For example, if a battery has a rating of 200 mAh (milliampere-hour), 200 mA can be drawn for one hour. A smaller current drain results in a longer time. For example, 20 mA of load current can be drained for 10 hours.

arc: A discharge of electricity through the air, approximately 25,000 volts per inch.

bimetallic strip: Two pieces of metal, made of different materials, joined at one end. It is also called a thermocouple. When the junction is heated, a voltage is produced at the other end.

charging resistance: A voltage applied to a battery in the reverse direction forcing electrons through the electrolyte.

electrolyte: A substance which produces ions when it conducts electricity.

head: Device that reads the magnetic pulses stored on tapes or disks. Also called a magnetic pickup.

hole: An atom missing an electron in its valence band. It carries a positive charge. A hole is the majority current carrier in a P-type semiconductor.

internal resistance: Resistance within a voltage source.

ion: An atom or molecule (group of atoms) which is electrically charged. Positive ions are lacking electrons. Negative ions have excess electrons.

magnetic pickup: Device the reads the magnetic pulses stored on tapes or disks. Also called a head.

optoelectric: Electronic components that use light to operate.

photosensitive: Materials responding when light strikes by either producing a voltage or changing resistance.

piezoelectric effect: Voltage produced in certain types of crystals when a mechanical pressure is applied.

primary cell: A single-cell of a battery which cannot be recharged. The chemical making up the electrolyte is consumed when changed into ions.

secondary cell: A single-cell battery which can be recharged easily.

Seebeck effect: Effect by which a current is created when the joint between two dissimilar metals is heated.

static electricity: A buildup of electrons or a shortage of electrons creating a difference in potential until a discharge path can be provided.

surge: A sudden burst of high voltage.

thermistor: A resistor that varies its resistance with changes in temperature.

thermocouple: Two pieces of metal, made of different materials, joined at one end. It is also called a bimetalic strip. When the junction is heated, a voltage is produced at the other end.

thermostat: A heat-activated switch.

Van de Graaff generator: An electrostatic machine that produces extremely high voltages.

KEY FORMULAS

Formula 10.A

$$V_T = V_1 + V_2 + ...V_N$$

Formula 10.B

$$V_T = V \times N$$

TEST YOUR KNOWLEDGE

Do not write in this text. Please use a separate sheet of paper.

1. Give a brief statement of the contribution by the following individuals, as related to this chapter:
   a. Thomas Edison
   b. Benjamin Franklin
   c. Luigi Galvani
   d. Georges Leclanché
   e. Gaston Planté
   f. Thomas Seebeck
   g. Robert Van de Graaff
   h. Alessandro Volta
2. How is a static charge developed?
3. On what type of material does static electricity accumulate?
4. Name three ways to discharge static electricity.
5. What is it called when pressure is applied to a crystal to produce a voltage?
6. Describe how a piezoelectric microphone changes sound waves into electrical impulses.
7. What is the purpose of crystals in a radio?
8. Name three ways in which heat is used with electricity.
9. Describe the operation of the three basic heat-activated electrical devices.
10. What is the name of a material that responds to light?
11. What is the name of electronic components that use light to operate?
12. Describe how a code reader is able to read the coded bars.

13. How does a magnetic pick-up change the information stored on the magnetic tape to electrical signals?

14. Make a list of several devices that can change magnetism to electrical signals.

15. What is the name of the substance that produces ions in a battery?

16. How do the ions produce electricity?

17. What is the difference between a wet cell and a dry cell battery?

18. Describe how a battery is recharged.

19. What is the voltage produced in a portable stereo system with six D-size batteries?

20. A 12 volt emergency light requires two amps to operate properly. It is connected to a rechargeable 12 volt battery rated for 20 ampere-hours. What is the estimated time of usefulness in an emergency?

21. A second battery of the same type is connected to the light in problem #20. Should it be connected in series or parallel? What would be the estimated time of usefulness?

22. What is the difference between internal resistance and charging resistance?

23. As a battery weakens, how does the internal resistance change?

24. How does the internal resistance of a voltage source affect the load?

25. How is a voltage source able to achieve the maximum transfer of power?

The power supply in a computer steps down the 120 volts ac coming from an outlet. The circuitry in the computer operates on a much lower dc voltage.

Chapter 11
Magnetic Principles
and Devices

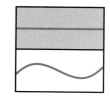

Upon completion of this chapter, you will be able to:
- Relate the names of selected individuals to magnetic principles they helped to develop.
- Define technical terms used with magnetism.
- Describe how certain materials are affected by magnetism.
- Explain how electricity produces magnetism.
- Identify units of measure related to magnetism.
- Perform calculations using formulas related to magnetism.
- Explain the operation of basic electromagnetic devices and give their applications.
- Interpret the ratings of electromagnetic devices.

Magnetism was discovered many thousands of years ago. Since then, magnetism has been used for many applications and in many devices. One of the most familiar is the compass.

Magnetism can be used to produce electricity, and electricity can be used to produce magnetism. These reciprocal properties make magnetism especially useful in modern technology.

In DC/AC FOUNDATIONS OF ELECTRONICS, magnetism is discussed in more than one chapter. Chapter 10 introduced magnetic sensors. Chapter 12 introduces inductance and how a magnetic field is used as an electronic component. Chapter 15 examines the sine wave and how it is produced from magnetism. Chapter 16 discusses the transfer of magnetic fields through transformers and the associated electricity. Chapter 17 uses magnetism to power a motor. In this chapter, details of magnetic principles are investigated as well as producing magnetism with electricity and magnetic devices.

11.1 A BRIEF HISTORY

Following the example of previous chapters, it is important to mention the names of some of the people involved in developments in magnetism and magnetic forces. Frequently, principles of operation and/or units of measure are named after technology pioneers.

Michael Faraday, a British physicist and chemist, is best known for his discoveries of electromagnetic induction and the laws of electrolysis. In 1821 he plotted the magnetic field around a conductor carrying an electric current. The farad, the unit of measure of capacitance in the SI mks system, was named in his honor.

Karl F. Gauss, a German mathematician, worked with magnetism around 1800. The gauss, the unit of measure of flux density in the cgs system, was named in his honor.

William Gilbert, an English physicist, experimented in the late 1500s with static electricity and magnetism. He introduced terms for force, attraction, and magnetic poles. The gilbert, the unit of measure of magnetomotive force in the cgs system, was named after him.

James Clark Maxwell, a British physicist, performed extensive experiments with the strength of magnetic fields around 1870. The maxwell, the unit of measure of magnetic flux in the cgs system, was named in his honor.

Hans Christian Oersted, a Danish physicist, discovered the existence of a magnetic field around an electrical conductor in 1819. The oersted, the unit of measure of magnetizing force in the cgs system, was named in his honor.

Nikola Tesla, a Yugoslav-born American inventor, experimented with electricity and magnetism in the early 1900s. The tesla, the unit of measure for flux density in the SI mks system, was named in his honor.

Wilhelm Weber, a German physicist, worked with magnetic fields in the mid to late 1800s. The weber, the unit of measure for magnetic field strength in the mks system, was named in his honor.

11.2 PRINCIPLES OF MAGNETISM

A **magnet** has the ability to attract certain types of materials. These materials are called ferromagnetic. The most common are iron, nickel, cobalt, and their alloys. These materials hold magnetic properties for a period of time, some much longer than others. A **lodestone** is a material appearing with magnetic properties in its natural state. A lodestone is the only natural magnet. Others are called artificial magnets. Artificial magnets are produced by being exposed to a magnetic field.

Definitions of Important Magnetic Terms

There are a number of terms and concepts that are important when working with magnetism. The definitions that follow are pictured in figure 11-1.

The **magnetic field** is the total area influenced by a magnet. The strength of the magnetic field is strongest at the poles and weakest near the center. The magnetic field strength, made up of lines of force, weakens with distance from the magnet. The weakening of the field is caused by the lines of force becoming farther apart.

The **poles** are what appear to be the sources of the magnetic field as it extends from a magnet. The poles of a magnet are north and south. These terms come from the north and south poles of the earth. The north and south poles of a magnet can be compared to the positive and negative terminals of a battery.

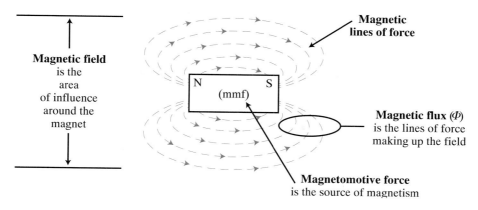

Figure 11-1. Definitions of magnetism shown in relation to the field around a bar magnet.

The **magnetomotive force (mmf)** is the strength of the source of the magnetism. The unit of measure of mmf is the **gilbert (Gb)** in the cgs system. In the SI mks system, mmf is measured in **ampere-turns (At or NI).** Magnetomotive force can be compared to the voltage of a battery.

Magnetic lines of force are the invisible lines making up the magnetic field. In experiments with iron filings sprinkled on a piece of plastic above a magnet, the iron particles form a definite pattern of lines surrounding the magnet. Magnetic lines of force flow from north to south. This can be compared to electron current flowing from negative to positive.

Magnetic flux (Φ), is the total lines of force, as a group, making up the magnetic field. The term *flux* can be interchanged with *lines of force*. The unit of measure in the cgs system is the **maxwell (Mx),** which is equal to one line of force. In the SI mks system, the unit is the **weber (Wb),** which is equal to 10^8 lines of force.

Interaction with a Nearby Magnetic Field

When two magnets are brought close to each other, there is an interaction between the fields. Holding the two magnets and feeling their interaction demonstrates the magnetic field to be a physical entity, even though it is invisible.

As shown in figure 11-2, unlike poles attract. Examination of the drawing explains why. The magnetic fields join in a manner that allows the field strength to increase. As the unlike poles get closer together, the field gets stronger. The increasing strength of the field pulls the magnets together. The field increases until it reaches a maximum equal to the sum of the two individual fields.

Like poles repel, as shown in figure 11-3. When like poles are brought close together, their fields push on each other. There is a distortion in the shape of the field due to the repulsion. When holding strong magnets by hand and trying to push them together, the field strength can become so strong that it is impossible to hold the magnets together.

The repulsion of magnetic fields can be used to do some powerful things. Large trains have been produced that operate on a single track system using magnetic levitation. For

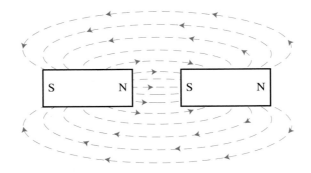

Figure 11-2. Unlike poles attract. The magnetic fields of the two magnets shown add.

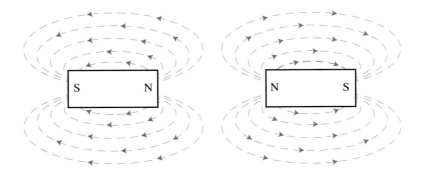

Figure 11-3. Like poles repel. The magnetic field of each magnet becomes distorted when the two like poles are brought near.

these **maglev trains,** the track is energized with electric current, creating a strong magnetic field. The train contains large magnets with their fields set to oppose those of the track. The repelling strength of the magnets is strong enough to hold the train above the track, giving it practically zero frictional resistance.

Field Strength, Flux Density, and Magnetic Field Intensity

Field strength, flux (Φ), and flux density (B) are closely related. Figure 11-4 shows that **field strength** is a quantity that can be measured with test equipment. Flux density is a mathematical relationship of the concentration of flux in a certain area. **Flux density** is a measure of the strength of the magnetic field in a particular area, such as one square inch. Flux density is highest at the poles where there is the greatest concentration of lines of flux.

Magnetic field intensity (H) is a measure of the strength of the magnetic field inside of the magnet. This is a mathematical calculation with units of measure of **oersted (Oe)**, in the cgs system, and **ampere-turns per meter (At/m)** in the SI mks system.

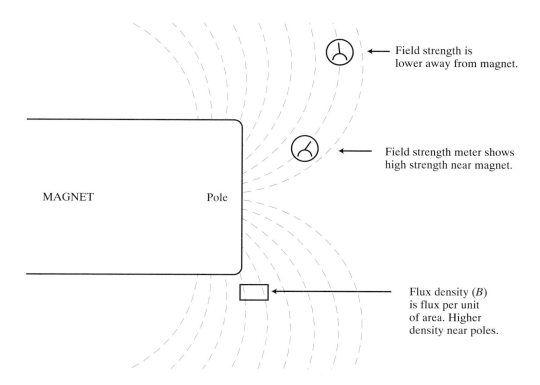

Figure 11-4. Field strength meters are used to measure the strength of a magnetic field.

11.3 MATERIALS

All materials respond in some way to a magnetic force and can be classified into one of four groups:
*   Ferromagnetic.
*   Ferrites.
*   Paramagnetic.
*   Diamagnetic.

Ferromagnetic materials and ferrites are considered to be magnetic materials. They are strongly attracted to magnets and easily conduct magnetic flux. Paramagnetic and diamagnetic materials have such a slight response they are considered nonmagnetic.

Magnetic and Nonmagnetic Classifications

Ferromagnetic materials are strongly attracted to a magnetic field, and they are also good conductors of electricity. Some examples are: iron, nickel, cobalt, and certain alloys such as steel and alnico.

Ferrites are materials that are strongly attracted to a magnetic field but do not conduct electricity. These are chemical compounds made with a magnetic material combined with a ceramic material.

Paramagnetic materials are classified as nonmagnetic. Examples are: aluminum, platinum, oxygen, and copper sulfate. These materials display a very slight magnetic attraction.

Diamagnetic materials are also classified as nonmagnetic. However, diamagnetic materials are actually slightly repelled by a magnetic field. Examples include: copper, lead, gold, antimony, bismuth, and mercury.

Magnetic Properties

The classification of materials is determined by their relative properties in several areas. These areas are:

- Reluctance.
- Reluctivity.
- Permeance.
- Permeability.
- Retentivity.

These properties describe how a material responds to a magnetic field.

Reluctance ($\mathcal{R}$) is the opposition to the flow of magnetic flux. Reluctance corresponds to the term resistance in electricity. Reluctance for a magnetic material is not a constant. It varies with flux density. The reluctance of a nonmagnetic material is a constant.

Reluctivity is the specific reluctance or the reluctance per cubic centimeter. Reluctivity corresponds to the electrical characteristic resistivity, which is the resistance of a material in a specific volume.

Permeance is the ability of a material to carry magnetic lines. Permeance is the reciprocal of reluctance. It corresponds to the electrical term conductance.

Permeability (μ), is the measure of ease with which magnetic lines can flow through a material. It is the reciprocal of reluctivity. Permeability is actually the ratio of lines of force passing through the material as compared to the lines of force passing through the air. Generally, permeability is used as a means of comparing the quality of a magnetic material. A perfectly nonmagnetic material has a permeability of one gauss per oersted. Typical values for iron and steel are 100 to 9000 G/Oe. Permeability's units of measure are **gauss per oersted (G/Oe)** in the cgs system and **henry per meter** in the SI mks system.

Retentivity is the ability of a material to retain magnetism after the magnetizing field has been removed. Permanent magnets have a high retentivity. Temporary magnets have a low retentivity. **Residual magnetism** is the magnetism that remains after the magnetizing force is removed.

Magnetic Domains

A magnetic material is made up of tiny magnets, called **magnetic domains.** Each of these tiny magnets has a north and south pole. In figure 11-5, a magnetic material that is not exposed to a magnetic field has its magnetic domains in a random order. If this material is exposed to a magnetic field, the domains arrange themselves in alignment with the lines of force, as shown in figure 11-6.

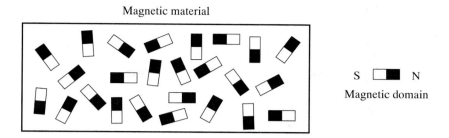

Magnetic material

Magnetic domain

Figure 11-5. Magnetic domains are in a random order when not exposed to a magnetic field.

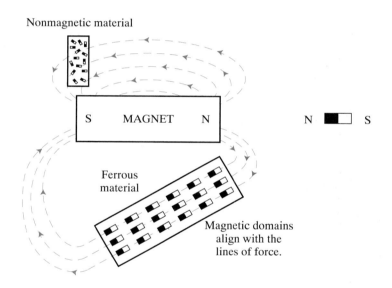

Nonmagnetic material

S   MAGNET   N

Ferrous material

Magnetic domains align with the lines of force.

Figure 11-6. Magnetic materials easily conduct magnetic flux. This causes the field from the magnet to bend.

Also shown in figure 11-6 is the manner in which a magnetic material conducts the magnetic flux much easier than air. As a result the field is distorted. The lines of flux take the path of least resistance, through the magnetic material. Notice that the arrows show the direction of the field is from north to south.

The domains in a magnetic material will remain aligned for a period of time. The length of time relates to the material's retentivity.

Magnetic Shield

A **magnetic shield** is a magnetic material that conducts a magnetic field around an area that should not be exposed to magnetic flux. The magnetic field is drawn into the magnetic material and away from the space being protected. In figure 11-7, a

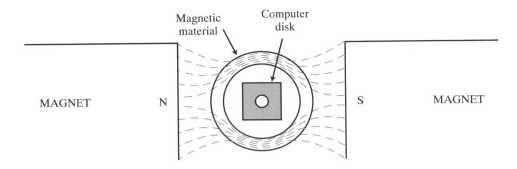

Figure 11-7. Items that are sensitive to magnetism can be protected with a magnetic shield.

computer disk, which stores information on a magnetic sheet, is enclosed in a shield. If a computer disk is exposed to a magnetic field, the information stored on the disk could be lost.

11.4 PRODUCING MAGNETISM WITH ELECTRICITY

Current flowing through a conductor produces a magnetic field in a circular pattern around the conductor. A larger current produces a stronger magnetic field. This principle can be demonstrated with an experiment such as the one shown in figure 11-8. A

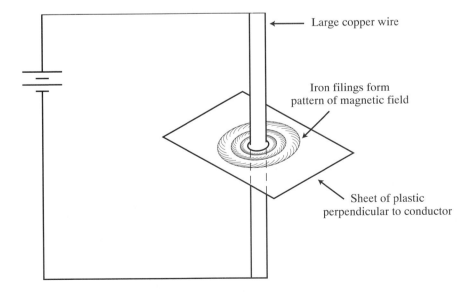

Figure 11-8. The magnetic field around a current carrying wire is strongest near the wire. The metal filings will demonstrate this.

wire is passed through a hole in a sheet of plastic. A battery is connected and iron filings are sprinkled on the sheet. The iron filings will form a pattern of concentric circles. This indicates the presence of magnetic lines of flux.

Electromagnets are produced with electricity. The magnetic strength of an electromagnet is much stronger than that of a permanent magnet of the same size, as shown in figure 11-9. The strength of the field can also be adjusted quite easily. Simply adjust the amount of current in the magnet.

The type of wire used in making electromagnet coils has a special coating to insulate it electrically. The coating is extremely thin, taking up very little space in the coil. This type of wire is called **magnet wire.** A weak acid is used to remove the coating so that solder can stick to it.

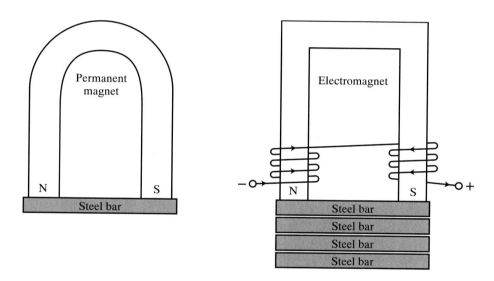

Figure 11-9. Electromagnets have a higher strength to size ratio than permanent magnets.

Increasing the Strength of an Electromagnet

The strength of an electromagnet can be changed, either with its design or by changing the current. At first, it might be thought that the strongest magnet is the best. However, there are compromises to consider. Changing the design of an electromagnet also changes the size and cost of manufacture. Too large an increase in current requires a different size wire and increases the operating cost of the magnet.

There are three easy ways to increase the strength of an electromagnet. The number of turns can be increased, the current can be increased, and an iron core can be added.

An increase in the number of turns on an electromagnet increases the magnetic field. In figure 11-10, the electromagnet on the left has three turns, and the one on the right has six turns. The magnetic field is approximately double in the magnet on the

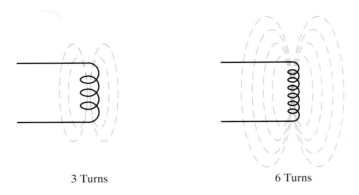

3 Turns   6 Turns

Figure 11-10. Increasing the number of turns increases the field strength of an electromagnet.

right, not considering any increase in losses. In actual practice, a coil may have several hundred turns. The number of turns of a coil is determined during the design and manufacturing process.

Changes to the amount of current traveling through a coil are made easily. In figure 11-11, the current is increased and results in a stronger magnetic field. Caution must be taken in making changes to the amount of current in a coil. Each coil is made of wire with a current rating that must not be exceeded.

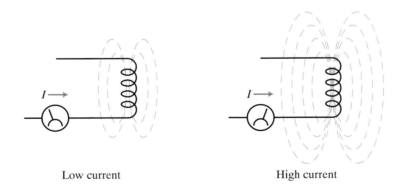

Low current   High current

Figure 11-11. Increasing the current increases the field strength of an electromagnet.

Wrapping a coil around an iron core greatly increases the strength of an electromagnet. The iron core concentrates the flux density. The core is a flux conductor and there is little loss due to stray magnetic lines. Figure 11-12 shows how the magnetic field radiates at the poles from the concentrated area of the core.

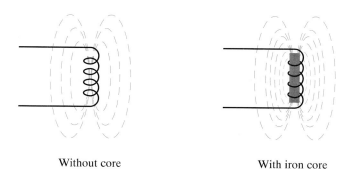

Without core   With iron core

Figure 11-12. Adding an iron core concentrates the flux. Increasing the flux density produces a stronger electromagnet.

How a Core Increases Magnetic Strength

How cores in electromagnets work is an important concept and one that needs to be examined further. Ferromagnetic materials and ferrites are composed of tiny magnetic domains. When the material is not magnetized, the magnetic domains are arranged in a random order. When exposed to a magnetic field, they arrange themselves in an order that follows the lines of force. Refer to figure 11-13.

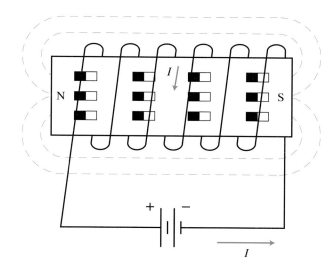

Figure 11-13. Magnetic domains in an iron core align with a magnetic field. They add to the field produced by the coil.

When a ferrous core is used with an electromagnet, it serves two purposes. First, it is a conductor of the magnetic lines of force. This concentrates the flux. Second, the magnetic domains add to the flux produced by the electricity. As a result, the flux is made up of both magnetism from the coil *and* the magnetism from the core.

Materials used for the core of an electromagnet are selected according to their retentivity. As stated earlier, retentivity is the ability to retain magnetic properties after the magnetizing field has been removed. Leftover magnetism, the residual magnetism, is not a desirable quality in electromagnets. If a core has residual magnetism, it is much more difficult to control the strength of the field using changes in the current. In figure 11-14, the electromagnet used to lift a car must be able to instantly release its magnetic holding power at the moment the current is shut off.

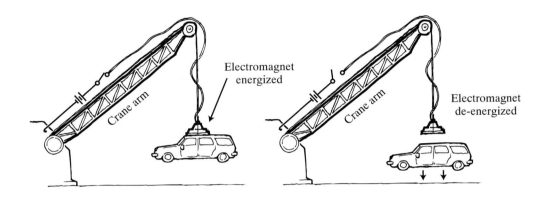

Figure 11-14. Energizing the electromagnet allows it to lift the car. When the power is shut off to the magnet, it should release the car immediately.

Schematic Symbols for Electromagnets

The symbol used for an electromagnet is the same as used for any coil of wire. Figure 11-15 shows two symbols. The coil symbol on the left is used for components that produce a magnetic field without a core. The symbol on the right has two lines drawn alongside the coil to indicate the use of an iron core.

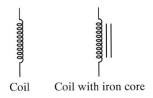

Figure 11-15. Schematic symbols of electromagnets.

These symbols are used again in Chapter 12 as the symbol for an inductor. As you will discover, the inductor is an electromagnet used as an electronic component for its ability to produce magnetic energy.

Left Hand Rule

The left hand rule has two parts. The first part is for finding the direction of magnetic lines when the direction of the current is known. The second part is for locating the north pole when the direction of the current is known.

For the left hand rule part one, refer to figure 11-16. To find the direction of magnetic lines of flux, point the thumb of your left hand in the direction the current is traveling. Your fingers will curl in the direction of magnetic lines of flux.

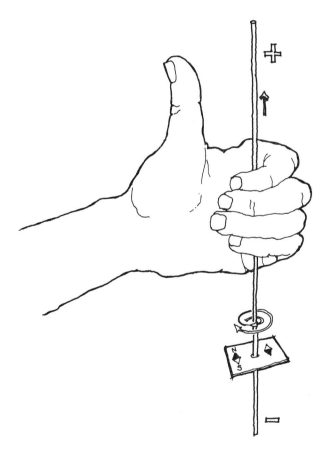

Figure 11-16. Left hand rule, part 1. Used for finding the direction of magnetic lines of flux.

For the left hand rule part two, refer to figure 11-17. Wrap your fingers around a coil in the direction of the current. Your thumb will point in the direction of the north pole of the magnet.

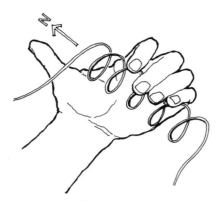

Figure 11-17. Left hand rule, part 2. Used to find the north pole on a electromagnet.

Aiding and Opposing Magnetic Fields

The direction of a coil's winding is critical in certain applications. Figure 11-18 uses a horseshoe-shaped core to demonstrate the effects of coil winding direction. Use the left hand rule to evaluate the location of the north and south poles.

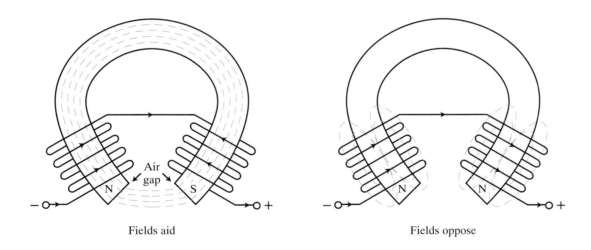

Fields aid   Fields oppose

Figure 11-18. Current in coils must produce opposite poles for a magnet to produce a strong field. A smaller air gap also increase magnetic strength.

In the figure on the left, the winding closest to the negative terminal is wound to create a north pole. The winding closest to the positive is wound to create a south pole. This horseshoe magnet has its poles in a position to allow a circulation of flux lines around the magnet. This is the best possible condition.

In the figure on the right, the winding closest to positive is in the wrong direction. Both ends of the horseshoe have a north pole. The fields oppose and there is no circulation of flux within the core.

Having the coils wound in the proper direction is one way to increase field strength. Another way to increase the field strength is to use a smaller air gap between the poles.

11.5 MAGNETIC UNITS AND MATHEMATICAL RELATIONSHIPS

The theory of how magnets and magnetic fields perform is mostly straightforward and easy to understand. Part of the reason for this is the physical evidence that can be observed by handling magnets.

In the study of electronics technology, in addition to understanding theory and new technical terms, it is also important to use the mathematical tools available. Theory, technical terms, and the mathematics of foundation subjects are all needed to fully understand the concepts discussed in more advanced studies.

Units of measure for magnetism are given in two different systems, cgs and SI mks. The abbreviations stand for **centimeter-gram-second (cgs)** and **standard international meter-kilogram-second (SI mks)**. Figure 11-19 summarizes the units of measure used in this chapter. The SI mks is generally considered the preferred system of units. The units are demonstrated in both systems, with the exception of force, which is expressed in cgs only.

Figure 11-19. Magnetic units of measure.

Term	*Symbol*	*cgs system*	*SI mks system*
Magnetic force	f	dyne and unit pole	
Area	A	square centimeter (cm^2)	square meters (m^2)
Flux	Φ	maxwell (Mx) or lines	webers (Wb) 1 Wb = 10^8 lines
Flux density	B	gauss (G) or Mx/cm^2	tesla (T) or Wb/m^2
Magnetomotive force	mmf	gilbert (Gb)	ampere-turns (At) or (NI)
Magnetic field intensity	H	oersted (Oe) or Gb/cm	ampere-turns per meter (At/m)
Permeability	μ	gauss per oersted (G/Oe)	henry per meter (H/m)
Reluctance	$\mathcal{R}$	gilbert per maxwell (Gb/Mx)	ampere-turns per weber (At/Wb)

Force of Attraction and Repulsion (*f*)

Coulomb's law states the force between two magnetic poles is directly proportional to the strengths of the poles and inversely proportional to the square of the distance between the poles. This means that if the distance is increased by a factor of 2, the force is decreased by a factor of 4 (2^2).

Formula 11.A

$$f = \frac{m_1 \times m_2}{d^2}$$

f is force, measured in dynes.

m_1 is strength of first pole, measured in unit poles.

m_2 is strength of second pole, measured in unit poles.

d is the distance between the poles, measured in cm.

Notes on units of measure (cgs system):
- One **dyne** = 2.248×10^{-6} pounds (U.S. conventional).
- One **unit pole** = force of one dyne repelling a pole of similar strength, in one second, when placed one centimeter apart.
- Unit poles are expressed as positive for north poles and negative for south poles.
- If the result of the calculation is negative, the force is repelling. If the result is positive, the force is attracting.

Sample problem 1. _____

Calculate the force exerted if a north pole of 20 unit poles is facing a south pole of 25 unit poles. They are separated by a distance of 10 cm.

Formula: $f = \dfrac{m_1 \times m_2}{d^2}$

Substitution: $f = \dfrac{20 \times 25}{10^2} = \dfrac{500}{100}$

Answer: $f = 5$ dynes (attracting)

Flux Density (*B*)

Flux density (*B*) is the number of magnetic lines in a plane perpendicular to the direction of the magnetic field.

Formula 11.B (mks units)

$$B = \frac{\Phi}{A}$$

B is flux density, measured in teslas (T).

Φ is total flux, measured in webers (Wb).

A is area, measured in square meters (m²).

Formula 11.C (cgs units)

$$B = \frac{\Phi}{A}$$

B is flux density, measured in gauss (G).

Φ is total flux, measured in maxwells (Mx).

A is area, measured in square centimeters (cm²).

Notes on units of measure:
- One gauss (G) = one line per square centimeter.
- One **tesla (T)** = 10^4 gauss = 10^4 lines/cm².
- One maxwell (Mx) = one magnetic line.
- One weber (Wb) = 10^8 maxwells = 10^8 lines.
- 100 Mx = 1 μWb.

Sample problem 2. (flux density, mks system)

What is the flux density of a magnet with 600 microwebers in an area of 0.00032 square meters?

Formula: $B = \dfrac{\Phi}{A}$

Substitution: $B = \dfrac{600 \ \mu Wb}{.00032 m^2}$

Answer: $B = 1.875$ T

Sample problem 3. (flux density, cgs system)

The north pole of a certain bar magnet has a total flux of 600,000 maxwells. What is the flux density if the dimensions of the pole are 6 centimeters by 10 centimeters?

Formula: $B = \dfrac{\Phi}{A}$

Substitution: $B = \dfrac{600{,}000 \text{ Mx}}{6 \text{ cm} \times 10 \text{ cm}}$

Answer: $B = 10{,}000 \text{ G} = 1 \text{ T}$

Magnetomotive Force (mmf)

Magnetomotive force (mmf) is the measure of the magnetizing force, as it applies to electromagnets.

Formula 11.D (mks units)

$$mmf = N \times I.$$

mmf is magnetomotive force, measured in ampere-turns (At).

I is current in coil, measured in amps.

N is number of turns.

Formula 11.E (cgs units)

$$mmf = N \times I.$$

mmf is magnetomotive force, measured in gilberts (Gb).

I is current flow in coil, measured in amps.

N is number of turns.

Notes on units of measure:
- One ampere turn (At) = 1.26 gilberts (Gb).
- One gilbert (Gb) = 0.794 ampere turns.

Sample problem 4. (magnetomotive force, cgs units)

Calculate the mmf of a solenoid with 250 turns and three amps of current.

Formula: $mmf = I \times N$

Substitution: $mmf = 3 \text{ A} \times 250 \text{ T}$

Answer: $mmf = 750 \text{ At} = 945 \text{ Gb}$

Field Intensity (*H*)

The field intensity (*H*) is the measure of the strength of the field, inside of the magnet. The formula compares the strength of the magnetic flux to the size of the magnet.

The first two formulas, in mks and cgs systems, give the field intensity for electromagnets. The third formula, given only in cgs units, is a measure of field intensity for permanent magnets.

Formula 11.F (electromagnets in mks units)

$$H = \frac{\text{mmf}}{l}$$

H is field intensity, measured in ampere-turns per meter (At/m).

mmf is magnetomotive force, measured in ampere-turns (At).

l is length, measured in meters (m).

Formula 11.G (electromagnets in cgs units)

$$H = \frac{\text{mmf}}{l}$$

H is field intensity, measured in oersteds (Oe).

mmf is magnetomotive force, measured in gilberts (Gb).

l is length, measured in centimeters (cm).

Formula 11.H (permanent magnets in cgs units)

$$H = \frac{\Phi}{A}$$

H is field intensity, measured in oersteds.

Φ is total flux, measured in maxwells.

A is area of magnet, measured in cm².

Notes on units of measure:
- One oersted = 79.37 At/m.
- One ampere-turn per meter (At/m) = 0.0126 oersteds (Oe).

Sample problem 5. (field intensity, mks units)

What is the field intensity of a solenoid with mmf of 500 ampere-turns and a core length of 0.1 meters?

Formula: $H = \frac{\text{mmf}}{l}$

Substitution: $H = \frac{500 \text{ At}}{0.1 \text{ m}}$

Answer: $H = 5000 \text{ At/m}$

Sample problem 6. (magnetomotive force, cgs units)

In a particular electromagnet with a core length of 10 centimeters, the field intensity is 20 oersteds. Calculate the magnetomotive force.

Formula: $mmf = H \times l$

Substitution: $mmf = 20 \text{ Oe} \times 10 \text{ cm}$

Answer: $mmf = 200 \text{ Gb}$

Sample problem 7. (field intensity, cgs units)

A permanent magnet with the dimensions of four centimeters by six centimeters has a magnetic field of 72,000 lines (maxwells). Determine the field intensity.

Formula: $H = \dfrac{\Phi}{A}$

Substitution: $H = \dfrac{72,000 \text{ Mx}}{4 \text{ cm} \times 6 \text{ cm}}$

Answer: $H = 3000 \text{ Oe}$

Permeability (μ)

The permeability of a material is a ratio of its flux density, *B*, to its field intensity, *H*. The value of permeability for the material is compared to the permeability of air. A high value of permeability indicates that the material easily conducts and concentrates magnetic lines of force. Permeability is a value of the magnetic field that can actually be produced by the material.

Formula 11.I (mks units)

$$\mu = \frac{B}{H}$$

μ is permeability, measured in henry per meter (H/m).

B is flux density, measured in teslas (T).

H is field intensity, measured in ampere-turns per meter (At/m).

Formula 11.J (cgs units)

$$\mu = \frac{B}{H}$$

μ is permeability, measured in gauss per oersted (G/Oe).

B is flux density, measured in gauss (G).

H is field intensity, measured in oersted (Oe).

Sample problem 8. (flux density, mks units)

Through experimental testing, the permeability of a material is found to be 220 microhenry per meter. With a field intensity is 800 ampere-turns per meter, what is the flux density?

Formula: $B = \mu \times H$

Substitution: $B = 220~\mu\text{H/m} \times 800~\text{At/m}$

Answer: $B = 0.176~\text{T}$

Sample problem 9. (permeability, cgs units)

What is the permeability of a certain piece of steel with a flux density of 13,000 gauss and a magnetizing force of 26 oersteds?

Formula: $\mu = \dfrac{B}{H}$

Substitution: $\mu = \dfrac{13,000~\text{G}}{26~\text{Oe}}$

Answer: $\mu = 500~\text{G/Oe}$

Reluctance ($\mathcal{R}$)

Reluctance is the opposition to the flow of magnetic lines of force. It is similar to resistance in an electrical circuit. It can be calculated as a comparison of the magnetic force to the flow of magnetic flux. The formula for reluctance is also referred to as "Ohm's law for magnetic circuits."

Following are a list of the magnetic/electrical comparisons:

mmf to V

Φ to I

$\mathcal{R}$ to R

Formula 11.K (mks units)

$$\mathcal{R} = \frac{\text{mmf}}{\Phi}$$

$\mathcal{R}$ is reluctance, measured in ampere-turns per weber.

mmf is magnetomotive force, measured in ampere-turns.

Φ is total flux, measured in webers.

Formula 11.L (cgs units)

$$\mathcal{R} = \frac{mmf}{\Phi}$$

$\mathcal{R}$ is reluctance, measured in gilberts per maxwell (Gb/Mx).

mmf is magnetomotive force, measured in gilberts (Gb).

Φ is total flux, measured in maxwells (Mx).

Sample problem 10. (reluctance, mks units)

A relay has a total flux of 250 microwebers. What is its reluctance with a magnetomotive force of 50 ampere-turns?

Formula: $\mathcal{R} = \dfrac{mmf}{\Phi}$

Substitution: $\mathcal{R} = \dfrac{50 \text{ At}}{250 \text{ } \mu\text{Wb}}$

Answer: $\mathcal{R} = 200,000$ At/Wb

Sample problem 11. (magnetomotive force, cgs units)

What is the magnetomotive force required for a flux density of 36,000 maxwells in a material with a reluctance of 0.002 gilberts per maxwell?

Formula: $mmf = \mathcal{R} \times \Phi$

Substitution: $mmf = 0.002$ Gb/Mx $\times$ 36,000 Mx

Answer: $mmf = 72$ Gb

11.6 APPLICATIONS OF ELECTROMAGNETS

There are many applications for the use of electromagnets. This chapter examines a few of the more common devices. Electromagnets are used in many more areas, but the focus of this chapter is on understanding how magnetic devices operate.

Moving Coil Galvanometer

Many electrical instruments use magnetism to move the pointer of the meter. Figure 11-20 shows the principles of operation of the moving coil galvanometer. The leads of the meter are connected to the coil. When there is no current, the needle rests on the left side. Some galvanometers have the resting place in the center. When current is applied, the coil develops a polarity that is the same as the nearby permanent magnet. Since like

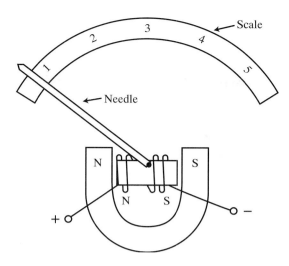

Figure 11-20. Current flowing through the meter drives the needle to the right. A moving coil galvanometer is used in some meters as the meter movement.

poles repel, the electromagnet rotates away from its resting position. A stronger current will rotate the coil farther.

Even though it is the current that develops the magnetic field, meters can be used to measure voltage, resistance, power, or a variety of other quantities. It is a matter of the circuitry within the meter being wired to allow whatever current is necessary to pass through the meter movement.

Speakers

To produce sound in a speaker, an audio signal is applied to a coil. The coil develops a magnetic field that vibrates a cone. The speaker in figure 11-21 uses a fairly large core, which is typical of speakers used for low frequencies. These speakers are called base speakers or woofers. Frequency describes the cycle at which an ac waveform changes from its maximum positive to maximum negative. The unit of measure of frequency is hertz (Hz). AC waveforms are discussed in detail in Chapter 15.

The magnetic field in the coil expands and collapses to follow the current from the audio signal. The magnetism pulls on the diaphragm, which in turn pulls on the attached cone. The diaphragm can be pulled and released quite rapidly, allowing the cone to vibrate. The vibrations of the cone move the air. This air movement is heard in the form of sound waves.

Some speakers can be operated in the reverse fashion. Many intercoms use a speaker that acts as both a speaker *and* as a microphone. When the cone of the speaker is vibrated from sound waves, the diaphragm varies its distance from the core. This movement develops a magnetic field in the coil. The coil produces a voltage proportional to the developed magnetic field. This induced voltage travels to an audio amplifier where

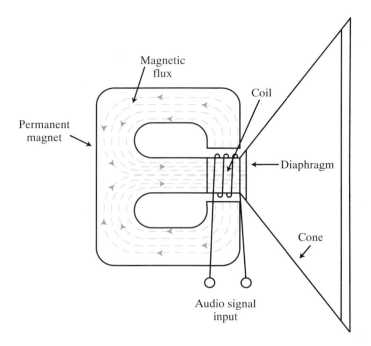

Figure 11-21. Speakers use an electromagnet to vibrate the speaker cone.

it is applied to a speaker at the other end. All speakers can be used as microphones to some degree, though with limited results. The speakers for the intercoms are made specially for the application.

Figure 11-22 shows the shapes of some speakers. The shape of the speaker is more to make it look good in the finished box than it is for function. When deciding what is the best speaker for a particular application, there are several ratings to be considered:
- Frequency response.
- Maximum wattage.
- Ohmic value of the voice coil.
- Weight of the magnet.

Frequency Response of Speakers

The **frequency response** of a speaker is its ability to follow the sounds over a range of frequencies. To best judge the range of frequencies required in a certain application, it is helpful to be aware of the frequencies of sounds. There are basically two types of sounds listened to on speakers, voice and music.

Frequencies to consider:
- The human ear hears sounds from 16 hertz (Hz) to approximately 20,000 Hz.
- Human voice ranges from 80 Hz to 1600 Hz.
- Musical instruments range from 30 Hz to 16,000 Hz.

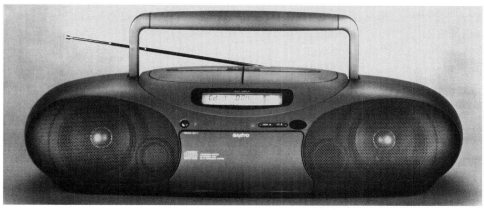

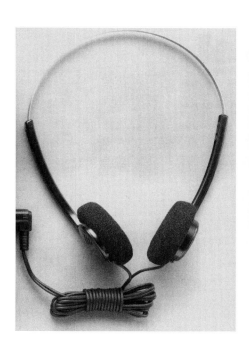

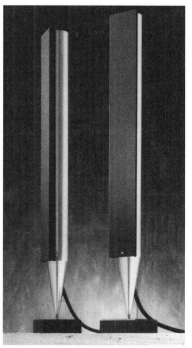

Figure 11-22. There are several ratings to consider when choosing speakers. There are other factors than simply the size and shape. (Kenwood, Sanyo, Bang & Olufsen)

Speakers that are intended for only reproducing voice, such as the speakers in aircraft, portable and mobile two-way radios, and telephones are made more rugged with a lower frequency response. They only reproduce sounds up to 5000 Hz. Woofers in stereo systems range from 15 Hz to 4000 Hz. Mid-range speakers range from 450 Hz to 7000 Hz. Tweeters, which reproduce the high notes, range from 2000 Hz to 30,000 Hz. As you can see, no single speaker covers the entire range of frequencies. If a single speaker is used, usually a mid-range is selected.

Crossover filters, discussed in Chapter 24, are used to electronically select which speaker is used in systems that have more than one speaker.

Wattage Rating of Speakers

Most electrical and electronic devices have a wattage rating. Generally, the **wattage rating** is the maximum wattage that can be dissipated without causing damage from excessive heat.

Some speakers have two wattage ratings, a maximum and minimum. The maximum rating is the same as for any device. The minimum wattage rating is associated with large speakers, especially woofers. The rating indicates that the speaker needs a certain amount of wattage just to make a sound.

It is not unusual to find speakers with maximum wattage ratings of 40 or 50 watts, and higher. Some small speakers, like those used in small portable radios have maximum ratings of one watt or less. Aircraft and other mobile two-way radios use speakers of up to 10 watts.

Is a higher wattage better? Generally, higher wattage speakers are used where the volume is expected to be high. In these cases, a higher wattage is needed. However, higher wattage speakers are usually more costly, they are heavier, and they put a heavier drain on a battery.

Speaker Coil Ohmic Value

Speakers are given an ohmic rating of the voice coil. Eight ohms is common for many audio applications. Four ohm coils are used for two-way radios. Sixteen ohms or 45 ohms are common for intercoms.

The **ohmic rating** is a resistance value. However, it is not the dc resistance of the coil. In other words, if an ohmmeter were used to read the dc resistance, it would not read the rated value. The ohmic rating is the ac resistance, called impedance. This impedance is stated at a specific frequency, usually 1000 Hz. Impedance is examined in detail in Chapter 18.

Weight of the Magnet

Heavier magnets are used in woofers to produce the very low frequencies. Heavier magnets are also used for higher power applications. They can produce a much deeper and fuller sounding load.

Solenoids

In a **solenoid,** a portion of the core is movable, see figure 11-23. The movable portion is called the **plunger.** This plunger is attached to something that needs to be moved. Solenoids are used to operate sliding locks, count mechanisms on counters, and the gear on the starter motor of a car.

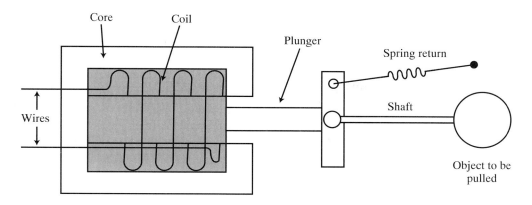

Figure 11-23. When the coil is energized, the plunger on a solenoid is pulled in. The plunger operates as a mechanical trigger.

When current flows through the coil, the magnetic force pulls the plunger fully into the coil. When released, a spring returns the plunger to the full out position.

Figure 11-24 shows some of the sizes and shapes of solenoids. They are available in very small forms, soldered to a circuit board, all the way to solenoids ample enough to slide large steel bars.

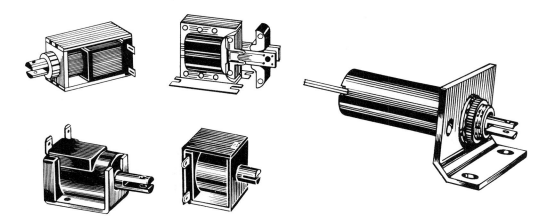

Figure 11-24. There are numerous applications for solenoids, so solenoids come in a variety of shapes and sizes.

Ratings of a Solenoid
Solenoids have four ratings to consider. These ratings are:
- Repeated operations.
- Coil voltage.
- Ohmic value of coil.
- Pull-in strength.

How often the solenoid is operated makes a difference in the heat it is able to dissipate. Solenoids are rated as either intermittent or continuous. **Intermittent** means that the solenoid will be turned off long enough to let the coil cool. **Continuous operation** means that the solenoid will be on for such long periods of time that it will not have time to cool. These solenoids must be able to dissipate the heat buildup.

Coil voltages are stated as either dc or ac, depending on the application. The voltage rating also includes a numerical value of the operating voltage, such as 12 volts dc.

The ohmic value given for a solenoid is a dc resistance, which can be measured with an ohmmeter. This rating is used to determine the amount of current needed to operate the coil.

Pull-in strength is measure of how much the solenoid can move when it is activated. Pull-in ratings can be from a few ounces to several pounds.

Relays

A **relay** is an electrically operated switch. An electronic circuit can be used to apply voltage to the relay coil, such as might be used in an electronic timer.

Refer to figure 11-25. When voltage is applied to the coil, the magnetic field pulls the armature in tight against the coil. The common of a switch is attached to the armature. The switch contacts alternate with the moving armature. The contacts of the switch can be either normally open or normally closed. The *normal position* for the switch is when there is no voltage applied. Also note in figure 11-25 the schematic diagram of a relay. Dashed lines are used between the coil and switch to indicate that they are connected through magnetism.

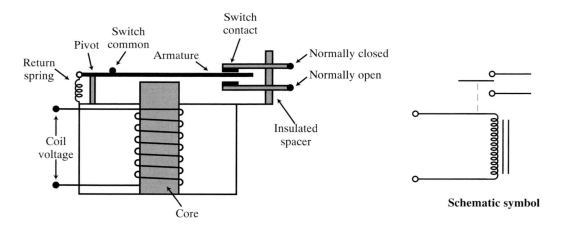

Figure 11-25. Relays are electrically operated switches. The parts of a relay and its schematic symbol are shown here.

Relays are available in a wide range of sizes and shapes as shown in figure 11-26. They can be small enough to mount on a circuit board or big enough to operate large electrical transfer stations. The advantage of using a relay is that the coil can be operated from one circuit, while the switch contacts connect to a separate circuit.

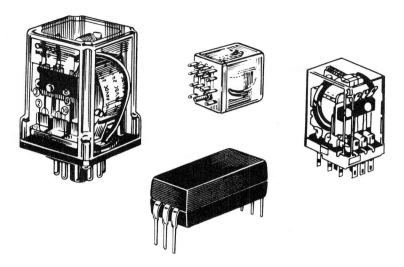

Figure 11-26. Like solenoids, relays come in a variety of shapes and sizes.

Buzzers and Bells

Buzzers and bells are electrical devices that use a relay to vibrate a striker arm to ring a bell. Examine figure 11-27 to see both a drawing and a schematic of a typical bell.

The voltage to the relay coil is through the relay switch contacts. The contacts are in the normally closed position. When a voltage is applied, the coil receives the voltage

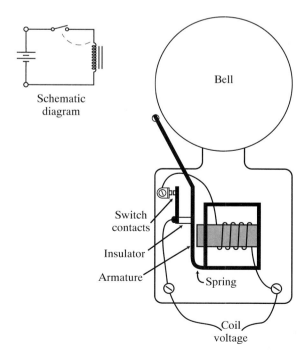

Figure 11-27. Schematic and diagram of a simple bell.

through the contacts. As soon as the coil develops a strong enough magnetic field, it pulls the armature. This breaks the switch contact. When the switch opens, the magnetic field collapses and releases the armature.

The armature is attached to a striker that rings the bell. Adjustments can be made to the time the bell needs to complete its on/off cycle.

Different Voltages for Coil and Contacts

Frequently, a relay circuit is used to operate a circuit with a different voltage. Figure 11-28 shows an alarm clock, which operates with a nine volt battery, operating a light connected to household voltage.

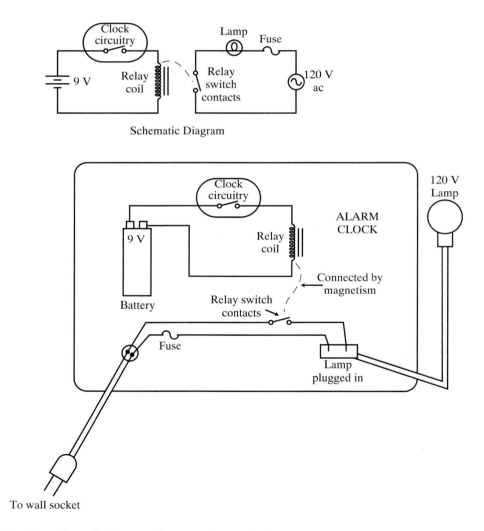

Figure 11-28. This schematic and diagram show an alarm clock, powered by a nine volt battery, operating a 120 volt ac circuit using a relay.

The nine volt battery operates the clock circuitry and the relay coil. The clock circuitry can be any combination of circuits that act like a switch. When the clock is at the correct time, it energizes the relay coil.

The household line voltage is fed through the relay switch contacts to a place where a light is plugged in to a wall socket. When the relay activates, the switch contacts close and apply 120 volts to the light. This type of two-voltage circuit is a common application of relays.

Relay Ratings

Switch contacts are available in single-pole or double-pole configurations. Single-pole is either on or off. It can be normal in either position. Double-pole has a common and can be connected to a circuit in either, or both, the normally open and closed positions. A relay can have one or several sets of switch contacts.

There are three ratings for a relay: coil resistance, switch contact current, and coil voltage. Relays, like solenoids, can be operated with either dc or ac voltages. The coil resistance is the value measured with an ohmmeter. The switch contacts are rated in the same manner as any switch would be. The contacts have a maximum current rating. Frequently, relays are also given a voltage rating. This states the maximum safe voltage that can be applied to the switch contacts.

SUMMARY

- A magnet has the ability to attract ferromagnetic materials.
- When magnets are brought close together, like poles oppose and opposite poles attract.
- All materials can be classified by how they react to magnetism. Some materials slightly repel (diamagnetic). Some have virtually no effect (paramagnetic). Some materials conduct magnetism but not electricity (ferrites). Some conduct magnetism and electricity (ferromagnetic).
- Electricity can be used produce an electromagnet.
- The left hand rule is used to tell the direction of a current and the location of the north pole of an electromagnet.
- Two systems of units are used for measuring magnetic quantities, cgs and SI mks.
- Electromagnets are used in applications such as: electrical instruments, speakers, solenoids, relays, and buzzers.

KEY WORDS AND TERMS GLOSSARY

ampere-turn (At or NI): The unit of measure in the mks system for magnetomotive force.

ampere-turns per meter (At/m): The unit of measure in the mks system for magnetic field intensity.

centimeter-gram-second (cgs): A system of measurement.

diamagnetic: Classified as nonmagnetic. However, diamagnetic materials are actually very slightly repelled by a magnetic field. Examples are: copper, lead, gold, antimony, bismuth, mercury.

dyne: The unit of measure in the cgs system for magnetic force.

electromagnet: Magnet produced with the aid of electricity.

ferrites: Materials that are strongly attracted to a magnetic field but will not conduct electricity. These are chemical compounds, made with a magnetic material combined with a ceramic material.

ferromagnetic: Materials that are strongly attracted to a magnetic field and are also good conductors of electricity. Examples include: iron, nickel, cobalt.

field strength: The strength of a magnetic field.

flux density: The intensity of a magnetic field. Symbol is B.

frequency response: The ability to respond to a range of frequencies.

gauss per oersted (G/Oe): Units for measuring permeability in the cgs system.

gilbert: Unit of measure in the cgs system for magnetomotive force.

henry per meter: Units for measuring permeability in the mks system.

lodestone: A material with magnetic properties in its natural state.

magnet: A substance which produces a magnetic field.

magnet wire: The type of wire used to make magnetic coils.

magnetic domain: Tiny magnetic particles making up a magnet. Each of these tiny magnets has a north and south pole.

magnetic field: The area influenced by a magnet.

magnetic field intensity (H): The measure of the magnetic field inside of the magnet.

magnetic flux (Φ): The lines of force, as a group, making up the magnetic field.

magnetic lines of force: The invisible lines making up the magnetic field. Magnetic lines of force flow from north to south. This is synonymous to electron current flowing from negative to positive.

magnetic shield: A material that conducts a magnetic field around an area that should not be exposed to magnetic flux.

magnetomotive force (mmf): The strength of the source of magnetism. The unit of measure is the gilbert (Gb). Magnetomotive force is synonymous to the voltage of a battery.

maxwell (Mx): Unit of measure in the cgs system for magnetic flux.

oersted (Oe): Unit of measure in the cgs system for magnetic field intensity.

ohmic rating: The ac resistance in a speaker.

paramagnetic: Materials that are nonmagnetic. Examples are: aluminum, platinum, oxygen, copper sulfate. These materials actually display a very slight magnetic attraction.

permeability: Measure of ease with which magnetic lines can flow through a material. It is the reciprocal of reluctivity. It is actually the ratio of lines of force passing through the material as compared to the lines of force passing through the air. Generally, permeability is used as means of comparing the quality of a magnetic material. A perfectly nonmagnetic material has a permeability of one. Typical values for iron and steel are 100 to 9000. Letter symbol for permeability is μ.

permeance: The ability of a material to carry magnetic lines of force. It is the reciprocal of reluctance and corresponds to the electrical term conductance.

poles: The origins of the north and south of a magnet. The north and south poles are synonymous with the positive and negative of a battery.

plunger: Moveable portion of a solenoid.

relay: A electromagnetically operated switch.

reluctance: The opposition to the flow of magnetic flux. It corresponds to the term resistance in electricity. Reluctance for a magnetic material is not a constant but varies with flux density. The reluctance of a nonmagnetic material is a constant.

reluctivity: The specific reluctance or the reluctance per cubic centimeter. Reluctivity corresponds to the electrical characteristic resistivity.

residual magnetism: Magnetism remaining in a temporary magnet after the magnetizing force has been removed.

retentivity: The ability of a material to retain magnetism after the magnetizing field has been removed. Permanent magnets have a high retentivity. Temporary magnets have a low retentivity.

solenoid: An electromagnetic with a moveable core.

standard international meter-kilogram-second (SI mks): A system of measurement.

tesla (T): Unit of measure in the cgs system for flux density.

unit pole: Unit of measure in the cgs system for magnetic force.

weber (Wb): Unit of measure in the mks system for magnetic flux.

KEY FORMULAS

Formula 11.A

$$f = \frac{m_1 \times m_2}{d^2}$$

Formulas 11.B and C

$$B = \frac{\Phi}{A}$$

Formulas 11.D and E

$$mmf = N \times I$$

Formulas 11.F and G

$$H = \frac{mmf}{l}$$

Formula 11.H

$$H = \frac{\Phi}{A}$$

Formulas 11.I and J

$$\mu = \frac{B}{H}$$

Formulas 11.K and L

$$\mathcal{R} = \frac{mmf}{\Phi}$$

TEST YOUR KNOWLEDGE

Do not write in this text. Please use a separate sheet of paper.

1. What is a material that has magnetic properties in its natural state called?
2. Which magnetic term, and in which system, are each of the following units of measure associated?
   a. ampere-turns (At or NI)
   b. gilbert per maxwell (Gb/Mx)
   c. ampere-turns per meter (At/m)
   d. maxwell (Mx)
   e. ampere-turns per weber (At/Wb)
   f. henry per meter (H/m)
   g. dyne
   h. oersted (Oe)
   i. gauss (G)
   j. tesla (T)
   k. gauss per oersted (G/Oe)
   l. unit pole
   m. gilbert (Gb)
   n. webers (Wb)
3. Briefly list the contributions in the understanding of magnetism made by each of these technology pioneers.
   a. Michael Faraday
   b. Hans Christian Oersted
   c. Karl Gauss
   d. Nikola Tesla
   e. William Gilbert
   f. Wilhelm Weber
   g. James Maxwell

4. A magnet has the ability to attract what types of materials?
5. When two magnets are brought close together, what is the effect on the magnets and their fields?
6. Describe the characteristics of the construction of an electromagnet.
7. List three ways to change the strength of an electromagnet.
8. Describe how to use the left hand rule to find the direction the current is traveling.
9. Describe how to use the left hand rule to locate the north pole.
10. Calculate the force exerted if a north pole of 40 unit poles is facing a south pole of 65 unit poles from a distance of five cm.
11. Calculate the flux density (mks system) of a magnet with 720 microwebers in an area of 0.004 square meters.
12. A selected bar magnet has a north pole with a total flux of 500,000 maxwells. Determine the flux density (cgs system) if the dimensions of the pole are four centimeters by 20 centimeters.
13. Calculate the magnetomotive force of a solenoid with 1500 turns of wire and a current of five amps. Give answer in both mks and cgs systems.
14. Find the field intensity (mks system) of an electromagnet with mmf of 1000 ampere-turns and a core length of 0.08 meters.
15. Calculate the field intensity (cgs system) of a permanent magnet with the dimensions of 5 cm × 10 cm and a magnetic field of 25,000 lines.
16. Determine the flux density (mks system) when a field intensity of 500 ampere-turns per meter is applied to a material with a permeability of 250 microhenry per meter.
17. What is the reluctance (mks system) of a relay with an mmf of 80 ampere-turns and a total flux of 250 microwebers?
18. Calculate the magnetomotive force (cgs system) required to produce a flux density of 40,000 maxwells in a material with a reluctance of 0.005 gilberts per maxwell.
19. List four ratings of a speaker.
20. How does the frequency response of a speaker determine the design of a crossover filter?
21. If an ohmmeter is used to test the resistance of a speaker, will it read the rated ohmic value?
22. To make a solenoid useful, which part is movable?

Most read/write computer storage media hold information using magnetism. Disks like these 3 1/2 inch floppies are the cheapest and most common form.

Chapter 12
Inductance

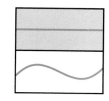

Upon completion of this chapter, you will be able to:
- Explain how a voltage is induced when a conductor is moved through a magnetic field.
- Describe how a counterelectromotive force is developed.
- Describe how a counterelectromotive force opposes a change in the induced magnetic field.
- Analyze how inductance opposes a changing current.
- Estimate the effects on a circuit with a very high resistance discharge path.
- Identify the construction characteristics of an inductor.
- Recognize the ratings of an inductor.
- Calculate the resultant when inductors are connected in series or parallel.
- Calculate the effects of mutual inductance.

Chapter 11 investigated magnetism with both permanent magnets and electromagnets. This chapter uses the electromagnet as an electronic component, taking advantage of the magnetic energy produced by electrical current.

When electricity produces magnetism, the conversion from one energy form to another is not instantaneous. **Inductance** is the name given to the opposition to *change* when converting from electrical to magnetic energy and from magnetic to electrical energy.

Inductors are used as electronic components, making use of their opposition to change. However, when there is no change in the current, an inductor acts just like a piece of wire.

12.1 A BRIEF HISTORY

The individuals whose names follow are noted for their work with electromagnetism. They discovered behavioral laws, used in this chapter, that help us better understand and predict the operation of an inductor.

Michael Faraday, mentioned in Chapter 11, discovered electromagnetic induction and the induction of one electric current by another in 1831. The laws of behavior named in his honor predict the amount of voltage that is developed. The unit measure of capacitance (Chapter 13), the farad (F), is also named in his honor.

Joseph Henry, an American physicist and professor of mathematics and natural philosophy at Albany Academy, discovered self induction in 1832. Henry also devised and constructed the first electric motor. The unit measure of inductance, the henry (H), was named in his honor.

Heinrich Lenz, a German physicist working during the middle 1800s, discovered an opposition to change when a conductor is moved through a magnetic field. The behavioral law predicting the polarity of induced voltage is named in his honor.

12.2 INDUCING A VOLTAGE

A voltage can be induced when the relationship between a conductor and a magnetic field is changed in one of three methods.
- The conductor is moved past a magnetic field.
- The magnetic field is moved past the conductor.
- The strength of the magnetic field is changed.

To induce a voltage, there must be a change. If the relationship between the conductor and magnetic field remains the same, there will not be an induced voltage.

Induced voltage is demonstrated in figure 12-1. A voltage is measured by the voltmeter when the conductor is moved through the magnetic field. The faster the movement, the larger the voltage will be.

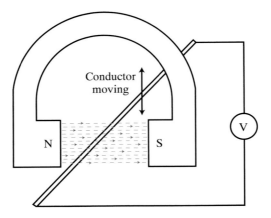

Figure 12-1. Voltage is induced into a conductor when it is moved though a magnetic field.

This process of converting magnetic to electrical energy requires the use of force. Figure 12-2 shows that as a wire is pressed through a magnetic flux, three things take place. First, current starts to flow in the conductor. Second, the current develops a magnetic field around the conductor. Third, these two magnetic fields oppose each other.

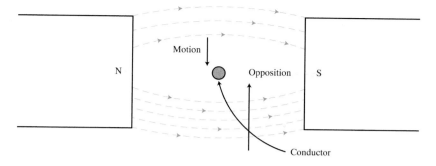

Figure 12-2. When the conductor is moved, the induced voltage opposes the magnetic field.

If the conductor is moved faster, more current is produced, which produces a stronger magnetic field. This, in turn, develops a stronger opposition. The opposition created by inducing a voltage in a wire is a property of inductance called **counterelectromotive force (cemf).** It can also be called **back emf.**

Counterelectromotive Force

When current is applied to a purely resistive circuit, the effect is instantaneous. It happens at the speed of light. When a coil of wire is placed in the circuit, the conversion of energy from electrical current to a magnetic field is *not* instantaneous.

Counterelectromotive force opposes any change in the building of a magnetic field. A faster change results in a larger cemf, which gives stronger opposition to building the magnetic field. The polarity of the cemf is opposite to the applied voltage and to the *changing* current flowing through the conductor.

Laws of a Magnetic Circuit

There are three laws of primary concern in understanding how a magnetic circuit works. These laws are:
- Kirchhoff's voltage law.
- Faraday's law.
- Lenz's law.

Kirchhoff's voltage law states that the algebraic sum of the voltages around a loop is zero. For review and further details, refer to Chapter 9.

Faraday's and Lenz's laws use the term *induced emf.* By combining the two laws, it is clear that the *induced voltage is the counterelectromotive force.*

Faraday's Law

The magnitude of the induced emf is directly proportional to the rate of change of current through the conductor.

Lenz's Law

Current from an induced emf will develop a flux that opposes the original change in the magnetic field that produced it.

12.3 INDUCTANCE IN A DEMONSTRATION CIRCUIT

Figures 12-3 through 12-8 are a demonstration circuit to explore the process of build-ing a magnetic field and producing an opposing voltage. Figure 12-9 shows a timing diagram. This shows the relationships between applied voltage, cemf, resistive voltage drop, and the current.

Inductance in a circuit opposes any change in current by producing a counterelec-tromotive force. The result of this is that it takes a period of time for the current to reach its full value. The inductor used in this demonstration circuit is not given a specific value for simplicity. Chapter 14 uses this circuit again, only with values to calculate the actual time required to induce the magnetic field.

First Instant Voltage Is Applied

The voltage is applied for the first instant in figure 12-3. The inductor has no mag-netic field at the start. At this instant, there is no current. A 20 ohm resistor in a circuit with 20 volts should allow a current of one amp according to Ohm's law. The inductor is the only other opposition to current in this circuit. The inductor opposes the change in current, from zero amps to one amp.

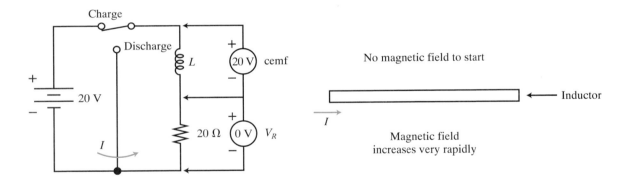

Figure 12-3. Right after the switch is closed, the current is increasing rapidly and the opposing cemf is at a maximum.

At this first instant, the rate of change is at a maximum. The current is trying to increase from zero to one amp instantly. Faraday's law says the cemf is at its maximum value when the rate of change in the current is at a maximum. Lenz's law says the cemf has a polarity opposite the applied voltage.

When there is no current, there is no voltage drop across the resistor. The cemf acts like an opposing voltage source, at least for a short period of time. Applying Kirchhoff's voltage law, the algebraic sum of the voltages around the loop equals zero. Consequently, the cemf of the inductor must equal the applied voltage.

Magnetic Field Partly Developed

The cemf will not hold back the current for very long. Remember, cemf depends on a changing current. The rapidly rising current develops the magnetic field quickly. As the magnetic field develops, the current rises, and the rate of change begins to slow.

The magnetic field is developed in proportion to the size of the current. If the current is one half of its maximum, for example, the magnetic field will be one half of its maximum, for that instant.

Figure 12-4 shows a current of 0.5 amps and a voltage drop across the 20 ohm resistor of 10 volts. The cemf is, therefore, producing 10 volts, which indicates a slowing in the rate of change of the current.

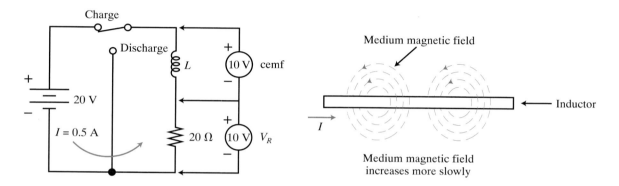

Figure 12-4. Here the magnetic field has partially developed. The opposing cemf is less, with less change in current.

Fully Developed Magnetic Field

When the current reaches its maximum value, the magnetic field cannot increase any further. The current in figure 12-5 is at its full value of one amp. At this time, the *rate of change* in the current drops to zero, along with the cemf.

As long as the current remains a steady value, the inductor maintains its magnetic field and offers resistance to the circuit only equal to a length of wire. An inductor is a coil of wire with some small value of dc resistance.

With inductors there is a difference between the dc, or **steady state resistance,** and the ac, or **dynamic resistance.** The dc resistance can be measured with an ohmmeter. It is the resistance of the wire. The ac resistance is the opposition to a changing current. Chapter 18 discusses how to calculate the ac resistance, called *inductive reactance.*

First Instant of Discharge

In figure 12-6, the switch is moved to the discharge position. This removes the battery from the circuit and provides a discharge path for the inductor.

The inductor once again opposes a change in current. This time the change is from one volt to zero volts. The inductor tries to keep the same amount of current traveling

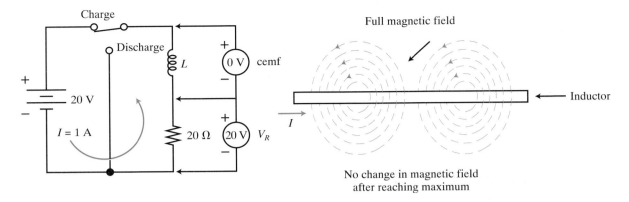

**Figure 12-5. The magnetic field and current are at a maximum.
There is no cemf as there is no change in current.**

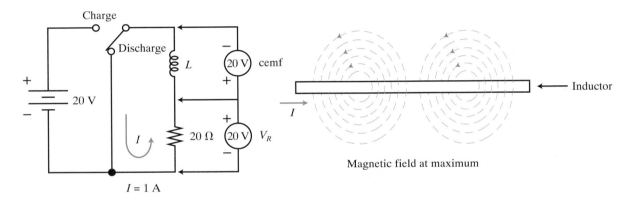

**Figure 12-6. When the switch is shut off, the cemf changes polarity
to oppose the change in current.**

in the same direction by acting like a voltage source. The inductor's magnetic field begins a rapid collapse as it converts its magnetic field back into electrical current.

The polarity of the inductor voltage switches instantly to supply current. The polarity of the voltage across the resistor remains the same, in the direction of the current.

During the Discharge

The voltage drop across the resistor is equal to the cemf of the inductor. In figure 12-7, the magnetic field has collapsed to approximately one half its full value. This can be seen by the value of current and the resistor voltage drop.

The rapidly changing magnetic field continues to induce a voltage, which continues to supply current. As the magnetic field weakens, the rate of change slows, producing less current.

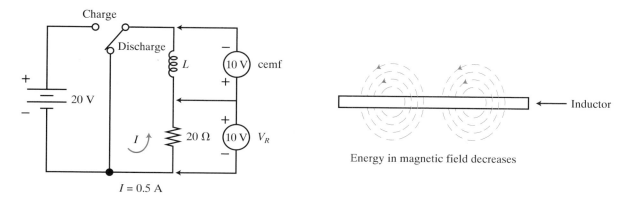

Figure 12-7. The collapsing magnetic field supplies current to the circuit.

Fully Collapsed Magnetic Field

In figure 12-8, the switch remains in the discharge position. The magnetic field is exhausted and the current ceases. This circuit is ready for another charge cycle. With no magnetic field, the inductor is only a coil of wire.

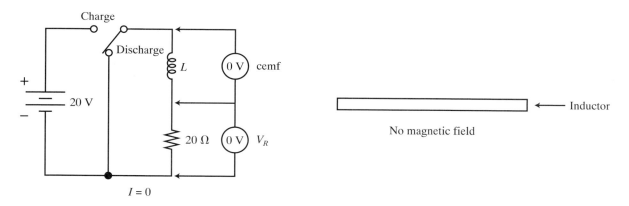

Figure 12-8. The magnetic field has no more energy to supply current to the circuit.

Charge/Discharge Timing Cycle

Figure 12-9 shows the relationships of the voltages and current of the demonstration circuit. Each waveform starts at zero, goes through its charge cycle, holds at a steady value, and then goes through a discharge cycle.

The top waveform, V_A, represents the applied voltage. When the switch is in the on position, voltage is applied to the circuit. The applied voltage is either zero or a full 20 volts.

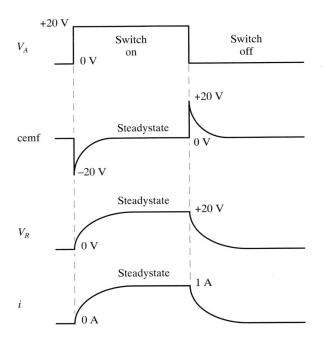

Figure 12-9. Shown is the relationship of counterelectromotive force to applied voltage, resistor voltage drop, and current.

The cemf waveform shows it instantly jumping to –20 volts at the same time the switch is moved to the on position. The negative voltage represents a polarity opposite that of the applied voltage. The cemf voltage quickly moves toward zero as the current increases.

The resistor voltage is always the difference between the applied voltage and the cemf. Kirchhoff's law states the voltages around the loop have an algebraic sum of zero. Notice how this statement holds true.

Here, current is represented by the lowercase *i*, rather than the usual uppercase *I*. The lowercase *i* represents a changing current. Notice that the resistor and the current curves have exactly the same shape. The voltage drop across a resistor is directly proportional to the amount of current flowing.

During the steady state portion of the graphs, the circuit behavior is exactly the same as a dc circuit with no inductor. The current is not changing, therefore there is no cemf.

When the voltage is switched to the off position, the polarity of the cemf instantly changes to a positive value. It must do this to supply current in the same direction.

12.4 AN INDUCTIVE CIRCUIT WITH A HIGH RESISTANCE DISCHARGE PATH

In the previous demonstration circuit, a path was provided for both charge and discharge. The circuit in figure 12-10 does not provide a discharge path.

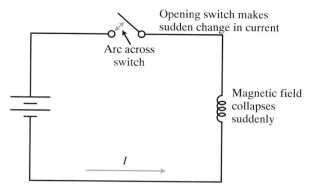

Figure 12-10. When the switch opens, the magnetic field will release its energy with an arc at the switch contacts.

When the switch is opened, the inductor does not allow the current to instantly drop to zero. The open switch acts as a very large resistor and the cemf builds to a large enough value to cause an arc across the switch contacts.

According to Ohm's law, with a given value of current, if resistance is low, voltage will be low. If resistance is very high, voltage will be very high.

12.5 FACTORS AFFECTING INDUCTANCE

Self inductance is the property of a conductor to induce voltage within itself. If some of the flux from the magnetic field around a conductor crosses a portion of the coil, it develops an opposing voltage within the conductor. How much counterelectromotive force is developed depends on the amount of self inductance. Self inductance is usually simply called inductance. The principle of self inductance can be seen in figures 12-11 and 12-12.

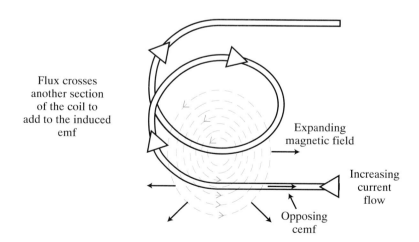

Figure 12-11. An expanding magnetic field opposes the increasing current.

In figure 12-11, voltage is applied to the coil, current is increasing, and the magnetic field is building. As the magnetic field builds, some of the lines of force cross over from one turn of the coil to a nearby turn. This interaction increases the cemf. The more turns involved in the interaction, the larger the cemf for the same amount of change in current.

In figure 12-12, the collapsing magnetic field is aided by the interaction of magnetic lines. A stronger magnetic field results in a slower collapse and more aid to the current.

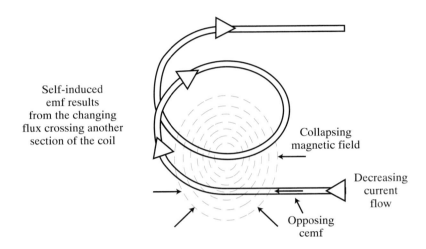

Self-induced emf results from the changing flux crossing another section of the coil

Collapsing magnetic field

Decreasing current flow

Opposing cemf

Figure 12-12. A collapsing magnetic field opposes a decrease by adding to the current flow.

12.6 PHYSICAL CONSTRUCTION OF AN INDUCTOR

The physical construction of an inductor determines how much magnetic flux will induce voltage within the coil. In figure 12-13, the various construction techniques are shown for discussion. *L* is the letter symbol used to represent an inductor. Magnetic terms can be reviewed in Chapter 11, if necessary.

• Number of turns: the more turns a coil has, the higher its inductance will be.
• Core material: a ferrous material is the best material to use for a magnetic core. A diamagnetic core, such as copper, will weaken the magnetic field, lowering the inductance.
• Spacing between turns: a closer spacing between turns results in a higher flux density, increasing the inductance.
• Wire size: the size of the wire has two effects. A larger wire allows more current, producing a stronger magnetic field. A larger wire also makes the spacing between the turns closer. Therefore, a larger wire size results in more inductance.

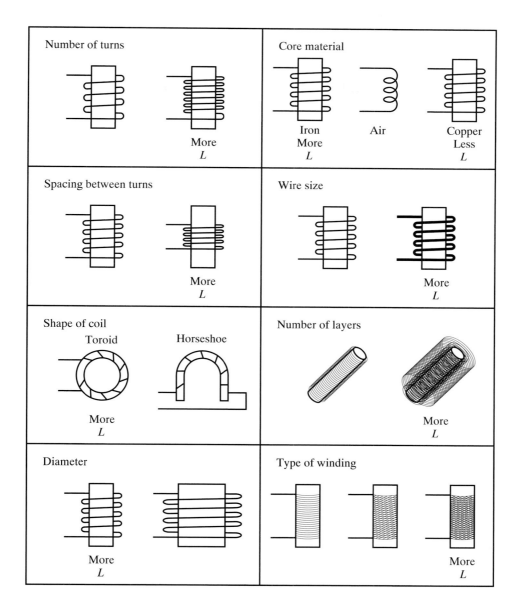

Figure 12-13. Many factors of physical construction affect the amount of inductance in an inductor.

- Shape of the coil: although a bar-shaped coil is often drawn, it is actually the poorest shape. The best shape for a coil is one where the north and south poles have the smallest possible air gap, allowing a circulation of flux within the coil. The horseshoe shape is an excellent design and is used in many applications. However, the best shape for an inductor is the **toroid,** which looks like a doughnut. The toroid has the highest concentration of magnetic flux and the least leakage. The toroid coil is most popular when large values of inductance are needed for filter circuits.

- Number of layers: when a coil is wound in layers, the spacing is made tighter since the wires are on top of each other. Consequently, a higher value of inductance results from more layers.
- Diameter: the diameter of the coil is closely related to the spacing of the turns. Inductance is increased with a more concentrated magnetic field.
- Type of winding: crisscrossing the windings improves the angle of the magnetic field. A right angle, 90°, induces the highest voltage. The crisscrossed coil is popular with miniature inductors.

12.7 INDUCTANCE AS A COMPONENT

The symbol for inductance is *L* and the unit of measure is the henry (H). For most electronic circuits, the millihenry (mH) and microhenry (μH) are practical sizes of inductance. Very large electromagnets are measured in henrys.

The schematic symbol for an inductor represents a coil of wire. Figure 12-14 shows four symbols. The inductor can have either a fixed value or variable value. The variable inductor has an arrow either through it or pointing at it, in a manner similar to the symbols for variable resistors. When an iron core inductor is used, it is shown in the schematic using two parallel lines alongside the coil.

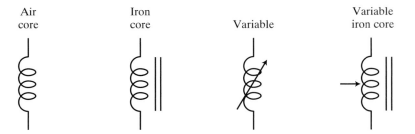

Figure 12-14. Schematic symbols used for inductors.

Figure 12-15 shows pictures of some shapes and sizes of inductors. As with most components, the size and shape selected depend on the application in the circuit.

Inductor Ratings

Inductors have applications in both dc and ac circuits. A typical application of an inductor in a dc circuit is as a filter to help regulate fluctuations in the current. With some dc power supplies, such as the supplies for computer circuits, regulation is critical. A radio frequency filter circuit is a typical ac application. The inductor is used to select a specific frequency or to select a group of frequencies.

Inductors have four basic ratings:

- Value in henrys, usually μH or mH.
- DC resistance in ohms, the value measured with an ohmmeter. Generally speaking, dc resistance should be as low as possible. Values of less than one ohm are common.

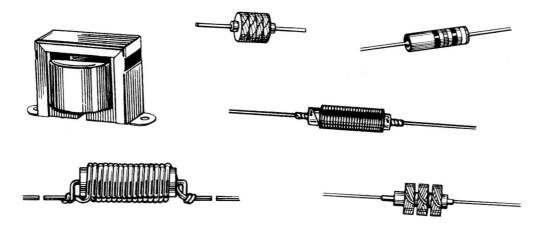

Figure 12-15. The inductor as a component.

- Maximum current, for which almost every electronic component has a rating.
- *Q*, which stands for quality. *Q* is a ratio of ac resistance (inductive reactance) to dc resistance. *Q* is measured at a particular frequency. Circuit *Q* is discussed in Chapter 23 as a factor of a resonant circuit.

Many catalogs, used to order components, refer to the inductor as a **choke.** The word choke has no significance, other than another name for inductor.

12.8 INDUCTANCE CONNECTED IN SERIES AND PARALLEL

As with other electronic components, inductors can be connected in a circuit in series or in parallel with other inductors. Formulas are used to calculate the total inductance in a manner similar to resistance. The formula for inductors in series is:

Formula 12.A

$$L_T = L_1 + L_2 + L_3 + \ldots L_N$$

L_T is the total inductance, measured in henrys.

L_1 through L_N are the individual inductances, measured in henrys.

N is the number of inductors in series.

The formula for inductors in parallel is:

Formula 12.B

$$\frac{1}{L_T} = \frac{1}{L_1} + \frac{1}{L_2} + \frac{1}{L_3} + \dots \frac{1}{L_N}$$

L_T is the total inductance, measured in henrys.

L_1 through L_N are the individual inductances, measured in henrys.

N is the number of inductors in parallel.

The shortcut formulas available for parallel resistance may be used with inductance.

Sample problem 1.

Determine the total inductance of three inductors connected in series with the values: 35 mH, 50 mH, and 75 mH.

Formula: $L_T = L_1 + L_2 + L_3 + \dots L_N$

Substitution: $L_T = 35 \text{ mH} + 50 \text{ mH} + 75 \text{ mH}$

Answer: $L_T = 160 \text{ mH}$

Sample problem 2.

A 40 µH inductor is connected in parallel with a 60 µH inductor. What is their combined inductance?

Formula: $\frac{1}{L_T} = \frac{1}{L_1} + \frac{1}{L_2} + \frac{1}{L_3} + \dots \frac{1}{L_N}$

Substitution: $\frac{1}{L_T} = \frac{1}{40 \text{ µH}} + \frac{1}{60 \text{ µH}}$

Decimals: $\frac{1}{L_T} = 25 \text{ E03} + 16.7 \text{ E03}$

Answer $L_T = 24 \text{ µH}$

12.9 MUTUAL INDUCTANCE

One aspect of an inductor that must be taken into consideration is its physical proximity to another inductor. With each inductor producing a magnetic field, it is possible to have interaction between lines of force. **Mutual inductance (L_M)** is the effect of a magnetic field from one inductor crossing the turns of a different inductor.

Depending on the locations of the north and south poles, one inductor can either aid the field of the other inductor, or it can oppose the other inductor's field. **Phasing dots** are used to indicate the direction in which the coil is wound. Figure 12-16 shows the use of phasing dots and two different wiring connections. Mutual inductance is the principle used when designing and constructing transformers. Transformers are discussed in detail in Chapter 16.

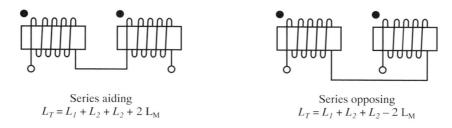

Series aiding
$L_T = L_1 + L_2 + L_2 + 2\,L_M$

Series opposing
$L_T = L_1 + L_2 + L_2 - 2\,L_M$

Figure 12-16. Phasing dots indicate polarity on inductors.

Formulas are given below for mutual inductance in series, but not parallel. Parallel inductance uses the reciprocal formula and is not a practical consideration.

$$\text{Aiding: } L_T = L_1 + L_2 + 2L_M$$

$$\text{Opposing: } L_T = L_1 + L_2 - 2L_M$$

Sample problem 3. _____

Using the values given in figure 12-17, find total inductance.

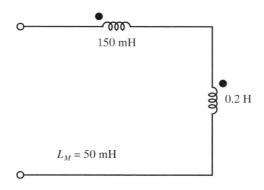

150 mH

0.2 H

$L_M = 50$ mH

Figure 12-17. Series aiding inductors. (Sample problem 3)

Formula: $L_T = L_1 + L_2 + 2L_M$

Substitution: $L_T = 150$ mH + 200 mH + 50 mH
(Note: 0.2 H = 200 mH)

Answer: $L_T = 400$ mH

Sample problem 4. _____

Find the total inductance of the circuit shown in figure 12-18.

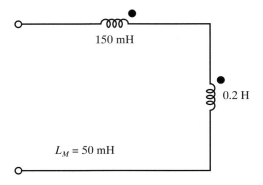

Figure 12-18. Series opposing inductors. (Sample problem 4)

Formula: $L_T = L_1 + L_2 - 2L_M$

Substitution: $L_T = 150 \text{ mH} + 200 \text{ mH} - 50 \text{ mH}$

Answer: $L_T = 300 \text{ mH}$

SUMMARY

- Inductance is the opposition to change when converting from electrical to magnetic energy and magnetic to electrical energy.
- Voltage is induced when a conductor is moved through a magnetic field.
- Faraday's law states that the amount of induced emf is proportional to the rate of change of current through the conductor.
- Lenz's law states that the polarity of induced emf opposes the direction of the current.
- At the first instant of charge, all of the voltage is across the inductor.
- After the inductor has a full charge, the voltage drop is minimal across the inductor.
- During discharge, the inductor supplies current to the circuit in the same direction as the charging current.
- The physical construction of the inductor affects the amount of inductance.
- Calculations for the resultant of series and parallel inductor connections use the same type of formulas as used for resistance.

KEY WORDS AND TERMS GLOSSARY

back emf: Counterelectromotive force.

choke: An inductor placed in series as a filter to pass dc voltages while stopping ac signals.

counterelectromotive force (cemf): The property of an inductor to oppose any change in the instantaneous building of the magnetic field. It also opposes a change in the current.

dynamic resistance: The ac resistance of a circuit.

inductance: The property of a circuit to oppose a change in current due to a counterelectromotive force. It is the result of converting electrical energy to magnetic energy or magnetic energy to electrical energy.

mutual inductance: The effect of a magnetic field from one inductor crossing the turns of a different inductor.

phasing dots: Dots on a schematic symbol used to indicate the direction in which the coil is wound.

self inductance: The property of a conductor to induce voltage within itself. This is a result of the magnetic field developed from current crossing the conductor to reverse the energy conversion. The amount of self inductance determines how much counterelectromotive force is developed. Self inductance is usually just called inductance.

steady state resistance: The dc resistance of a circuit.

toriod: The best shape for an inductor. Shaped like a doughnut.

KEY FORMULAS

Formula 12.A

$$L_T = L_1 + L_2 + L_3 + ...L_N$$

Formula 12.B

$$\frac{1}{L_T} = \frac{1}{L_1} + \frac{1}{L_2} + \frac{1}{L_3} + ...\frac{1}{L_N}$$

TEST YOUR KNOWLEDGE

Do not write in this text. Please use a separate sheet of paper.

1. Who is credited with discovering self induction?
2. List three ways an induced voltage is developed.
3. Describe how a counterelectromotive force is developed and how to determine its magnitude and direction.
4. What is the letter symbol of inductance? What is the unit of measure? What is the letter symbol for the unit of measure?
5. Use a demonstration circuit to analyze the performance of the inductance as it opposes a changing current for both the charge and discharge cycles. Include a timing diagram as part of the discussion.

6. What is the effect on an inductive circuit with a very high resistance discharge path?

7. Identify the construction characteristics of an inductor.

8. List the ratings of an inductor.

9. Calculate the resultant of three inductors connected in series with the following values: 150 μH, 250 μH, and 300 μH.

10. Calculate the resultant of three inductors connected in parallel with the following values: 150 μH, 250 μH, and 300 μH.

11. Calculate the resultant of two inductors connected in series with the values of 150 μH and 250 μH with 50 μH aiding mutual inductance.

12. Calculate the resultant of two inductors connected in series with the values of 25 mH and 75 mH with 5 mH opposing mutual inductance.

Chapter 13
Capacitance

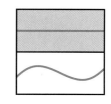

Upon completion of this chapter, you will be able to:
- Define capacitance.
- Describe the construction of a capacitor.
- Explain how a capacitor works.
- Identify the factors affecting capacitance.
- Relate capacitor ratings to a catalog listing.
- List types of capacitors.
- Calculate the resultant of capacitors connected in series and parallel.
- Calculate the voltages of a capacitive voltage divider.
- Test a capacitor with an ohmmeter.

Capacitance is the ability of a circuit to store an electrical charge. In building, storing, and releasing a charge, capacitance *opposes a change in voltage.*

Capacitors have many applications. In dc power supplies, capacitors are used as filters to remove fluctuations in voltage. This is discussed in detail in Chapter 26. Capacitors also make effective electronic timers to control activities in circuits. This use is discussed in Chapter 14.

The capacitor also finds many applications with ac voltages. Capacitors offer an ac resistance, called *reactance*, which is addressed in Chapter 19. Capacitors also block dc voltages while passing ac voltages. This is an important characteristic discussed in Chapter 24. In addition, when capacitors are combined with an inductor in an ac circuit, the result is a filtering action with a certain amount of cancellation of the ac resistance. You will learn about this in Chapter 23.

13.1 CONSTRUCTION OF A CAPACITOR

A capacitor is made with two conductors that are separated by an insulator. This description fits virtually all wiring where two wires run side by side. Although there is a certain amount of capacitance in wires, for most applications it is so small it is not of any consideration. In a similar manner, inductance is present in all wires.

To develop a capacitor with enough capacitance to be used as a circuit component, it must be made more efficient. The two conductors are formed into plates with a specific surface area. The insulator between the plates can be made of different materials. The choice of material depends upon the application of the capacitor.

Definitions of Capacitor Parts

Figure 13-1 shows the parts of a capacitor, the plates and the dielectric. Electrons gather on the plates. This produces an electrostatic field through the dielectric material.

A capacitor's **plate** is conductive. The plate is often made of a foil-type material, similar to aluminum foil used in cooking. The function of the negative plate is to collect electrons. The function of the positive plate is to give up electrons.

The **dielectric** is an insulator between the two plates. Some capacitors use air as the dielectric. Other capacitors use more efficient materials, such as mica or ceramics. A more *efficient* dielectric material produces a greater capacitance.

An electrostatic field is developed through the dielectric, from the negative plate to the positive plate. The **electrostatic field** is the attraction between the negative and the positive charges. Do not confuse this with current. Electrons do not actually flow.

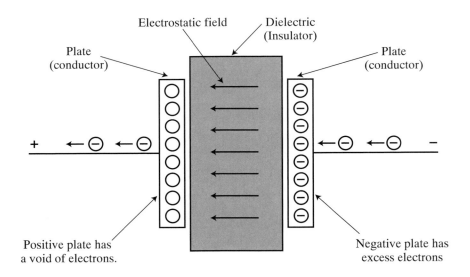

Figure 13-1. Capacitors are two conductors separated by an insulator.

13.2 HOW A CAPACITOR WORKS

The charge stored in a capacitor is potential energy. The electrons, gathered on the negative plate and removed from the positive plate, are the stored energy available for future use. This is much like the potential energy stored in a battery. The difference between a battery and a capacitor is the storage capacity. A battery has a much larger supply of electrons.

First Instant of Charge

At the first instant of charge, figure 13-2, the capacitor has no voltage across its plates. With no energy stored, the capacitor offers no resistance to the current. At this first instant, current is limited only by the resistor, just as if the capacitor was not there. The current is one amp, as calculated by Ohm's law. The full amount of the supply voltage is dropped across the resistor.

The current travels through the resistor to the capacitor. It cannot pass through the capacitor because of the dielectric. To have a current, a number of electrons equal to the amount leaving the negative side of the voltage supply must return to the positive side of the voltage supply. That quantity of electrons is taken from the positive side of the capacitor, which had a neutral charge until this point. The electrons traveling from the positive side of the capacitor leave holes on the plate. Notice that even though electrons do not actually travel through the capacitor, the current is the same at any point in the series circuit.

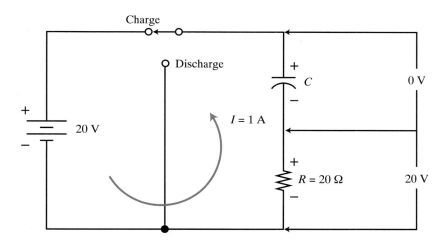

Figure 13-2. During the first instant of the charge cycle, the circuit has its maximum current.

Part of the Way through the Charge

As electrons gather on the negative side and leave the positive side, the electrostatic field in the capacitor gets stronger. The capacitor acts like an opposing voltage source. As shown in figure 13-3, the capacitor voltage subtracts from the supply voltage. Current decreases, and the voltage drop across the resistor is the difference between the capacitor voltage and the supply voltage. Kirchhoff's law states the voltages around the loop must add to equal zero.

It is important to note the polarity of the voltages in the circuit. The voltage drop across the resistor is in the direction of the current. The voltage developed across the

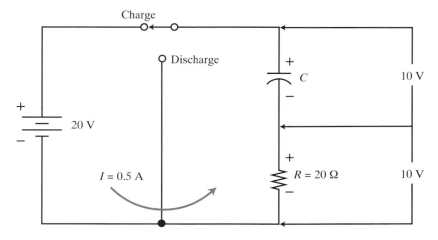

Figure 13-3. Part of the way through the cycle, the capacitor is acting as an opposing voltage source. Current is reduced.

capacitor is also in the direction of current. These voltages oppose the supply voltage. The polarity across the resistor will change during the discharge time, as shown in figure 13-6.

Fully Charged Capacitor

The capacitor is fully charged when it has a voltage equal to the supply voltage. The length of time it takes to become fully charged is dependent on the values of the capacitance and circuit resistance. Charge and discharge times are discussed in detail in Chapter 14.

With opposing voltages equal, there is no current traveling in the circuit. If there is no current, there is no voltage dropped across the resistor. The switch in figure 13-4 can remain in the charge position indefinitely with no change in the circuit. There will be a very small amount of current in the circuit due to leakage of electrons across the dielectric. However, the leakage current is usually too small to measure.

Capacitor Holds Charge

The capacitor can now be removed from the circuit and still hold its charge. If the capacitor were charged to 20 volts, a voltmeter would measure that value even after some period of time had passed. In some equipment, such as televisions, capacitors with large values are used in the power supply. Even if the television is unplugged, it can still contain voltages in its capacitors that are large enough to cause a severe shock.

Figure 13-5 shows the switch in the off position. The battery is removed from the circuit, but there is no path for the capacitor to discharge. If a voltmeter were to monitor the capacitor's voltage, there would be a very slow discharge through the voltmeter. Although the voltmeter has a high input resistance, current passes through it in order to create the measurement.

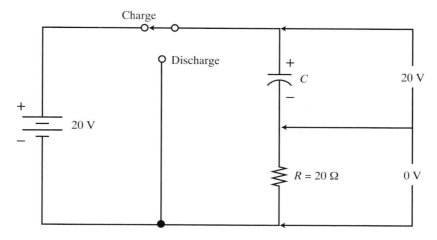

Figure 13-4. The capacitor is now fully charged. There is no current flowing.

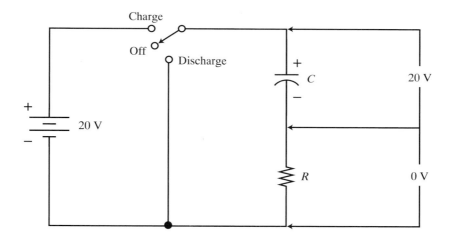

Figure 13-5. So long as there is no discharge path, the capacitor will hold its charge indefinitely.

First Instant of Discharge

In figure 13-6, the switch is placed in the discharge position. The capacitor is allowed to equalize the electrostatic field. The current is limited by the resistor.

During discharge, the capacitor becomes the voltage source for the circuit. Notice the polarity of the resistor in this figure. Compare it to figure 13-3. The current is traveling in the opposite direction during charge. The voltage drop across the resistor is also opposite, as it must follow the direction of the current.

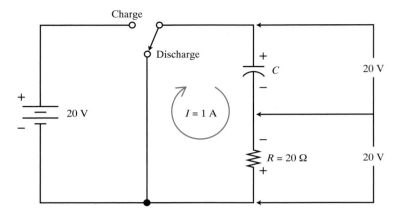

Figure 13-6. At the first instant of discharge, current changes direction and the polarity of the resistor changes with it. The capacitor acts as a voltage source.

Part of the Way through the Discharge

The size of the current is determined by the circuit resistance. The length of time a capacitor can supply current is determined by the value of capacitance.

In figure 13-7, observe that as the capacitor loses its charge, the voltage drop across the resistor also decreases. With the capacitor as the only supply of voltage, the resistor has a voltage equal to the capacitor's voltage at any moment.

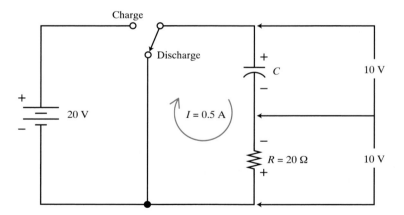

Figure 13-7. As the capacitor discharges, the voltage on the resistor is reduced.

Timing Diagram of Charge/Discharge

Figure 13-8 is a timing diagram of the demonstration circuit shown in figures 13-2 through 13-7. V_A is the voltage applied to the circuit. It is shown as either +20 volts or zero volts. V_C is the voltage across the capacitor. V_R is the voltage across the resistor.

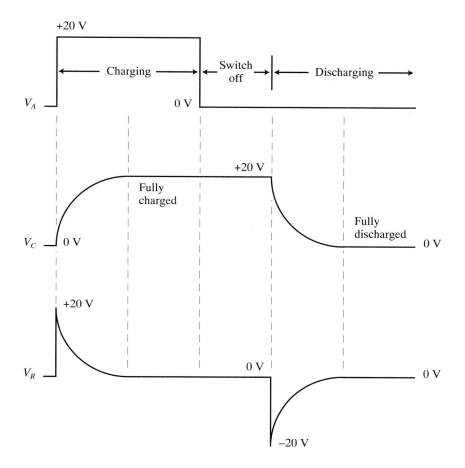

Figure 13-8. Timing diagram of one charge/discharge cycle of the capacitor circuit in figures 13-2 through 13-7.

During the charge time, the capacitor voltage rises quickly, but not instantly, toward the applied voltage. Notice that the sum of V_C and V_R add to equal V_A. When the capacitor is fully charged, the resistor voltage is zero.

When the switch is in the off position, there is no change in circuit conditions. The capacitor remains fully charged. This condition can remain for long periods of time. It could last indefinitely if the capacitor were perfect and had no leakage current.

During the discharge time, the resistor instantly switches its polarity. It jumps to –20 volts, then follows the capacitor as it discharges.

13.3 FACTORS AFFECTING CAPACITANCE

Capacitance is the property of a circuit that opposes a change in voltage. This is performed by producing an electrostatic field. An improvement in the efficiency of the capacitor allows it to hold a charge for a longer period of time.

Figure 13-9 shows there are three factors that determine the amount of capacitance. These factors are:

- Plate surface area.
- Distance between plates.
- Dielectric material.

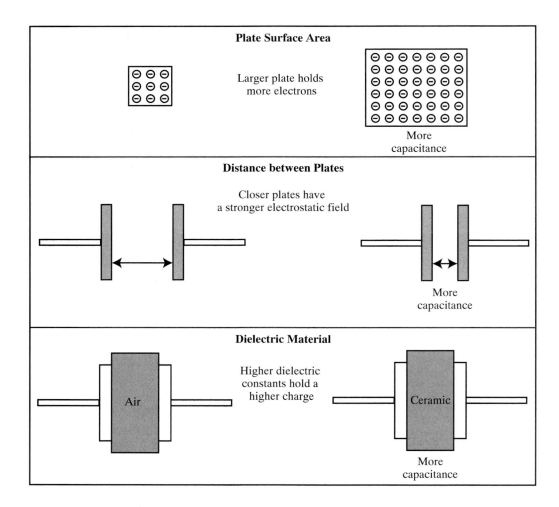

Figure 13-9. Factors affecting capacitance.

Plate Surface Area

The surface of the plates holds the electrons and the holes that produce the electric field. A larger surface area holds more electrons and holes, therefore it holds more electric charge. A capacitor charges to the applied voltage. A larger capacitance takes longer to reach full charge and longer to discharge.

In the construction of some types of capacitors, the plates are made of a foil material. Many layers of this foil can be wrapped together with an insulator in between. This produces the maximum surface area in the smallest space possible.

Distance between Plates

An electric field displays physical characteristics similar to those of a magnetic field. The plates are like the poles of a magnet. The closer they come, without touching, the stronger the field.

Dielectric Material

The plates are held apart by the dielectric. The dielectric should be as thin as possible while still maintaining good insulating characteristics.

The materials used as dielectrics in capacitors are given a rating, called the **dielectric constant.** This value reflects how many times better it is than a vacuum. For example, figure 13-10, air has a value of 1.0006. This is effectively the same as a vacuum. Mica has a dielectric constant of five. Therefore, mica increases the capacitance five times larger than air or a vacuum.

Figure 13-10. Dielectric constants of selected materials.

Material	Constant
Vacuum	1
Air	1.0006
Oil	2
Rubber	2.5—35
Wax paper	3.5
Mica	5—6
Glass	5—10
Pure water	81
Ceramic	up to 7500

Along with the values given for dielectric constants, consideration must be given to the quality of the dielectric material. As seen in figure 13-11, leakage current is a result of electrons going across the dielectric. Any leakage decreases the electric field and decreases the capacitance.

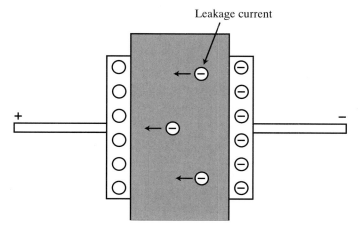

Figure 13-11. Capacitors have a leakage current between their plates. A stronger dielectric produces a smaller leakage.

13.4 CAPACITOR RATINGS

Capacitors have two ratings, the dielectric strength and the capacitance value. Usually, these two values are stamped on the body of the capacitor. Most electronic components have ratings for current and/or power. However, neither of these are necessary for a capacitor, because current does not travel through it.

Dielectric Strength

The **dielectric strength** states the maximum voltage that can be applied to the capacitor without a destructive breakdown. The dielectric strength is given as a voltage rating. Many small disc capacitors have voltage ratings of over 600 volts. Electrolytic capacitors have much smaller voltage ratings. Some as low as 10 volts.

Voltage ratings are usually stated as a dc voltage or sometimes as DCWV, standing for dc working voltage. In some cases, ratings are given for a surge voltage. This is a voltage that the capacitor can survive, but only if it lasts for a brief period of time. This rating is usually much higher. In addition, a rating may be given for ac voltages.

Unit of Measure of Capacitance

The unit of measure of capacitance is the **farad, F.** This unit is so large that it is always preceded by a multiplier. Capacitors are rated in microfarads, μF (10^{-6}) and picofarads, pF (10^{-12}). The multipliers milli- (10^{-3}) and nano- (10^{-9}) are not used.

The multiplier of micro-microfarad, $\mu\mu F$ ($10^{-6} \times 10^{-6} = 10^{-12}$) is obsolete, having been replaced by picofarad. Even though $\mu\mu F$ is obsolete, it is still found because there are many old components and old schematic diagrams in use.

13.5 SCHEMATIC SYMBOLS

Figure 13-12 shows six symbols in three groups: general, electrolytic, and variable. The general capacitor symbol is used for capacitors that are not polarized, such as those used with ac. The electrolytic type capacitor must show polarity signs. Variable capacitors use the arrow through the symbol.

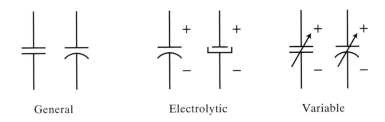

General   Electrolytic   Variable

Figure 13-12. Schematic symbols representing capacitors.

The symbols with one flat line and one curved line are the most common. However, be aware of the others as they do appear on schematic diagrams.

13.6 TYPES OF CAPACITORS

The construction of capacitors varies widely. Each type of capacitor has its own benefits. Figure 13-13 shows a variety of nonpolarized capacitors. Nonpolarized capacitors are used in applications with ac voltages. Most of these capacitors have very high voltage ratings.

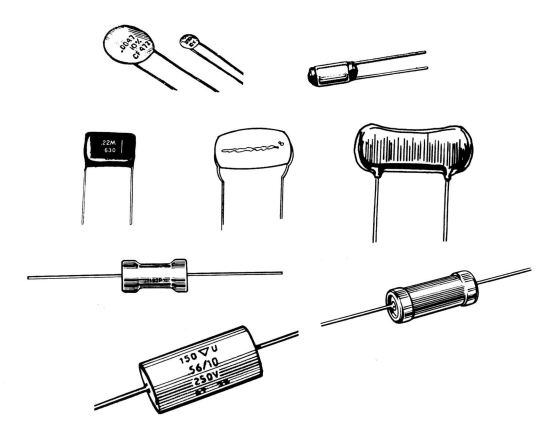

Figure 13-13. Nonpolarized capacitors come in a large variety of shapes and sizes.

Figure 13-14 shows several variable capacitors. These are used in tuning filters. One use of the variable capacitor is in the tuning dial on a portable AM/FM radio. Air dielectric capacitors are usually used as variables. The distance between the plates is adjusted to vary the capacitance.

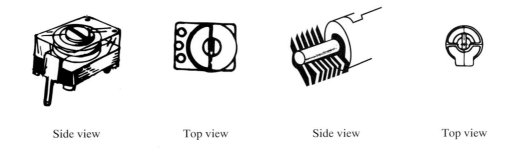

| Side view | Top view | Side view | Top view |

Figure 13-14. Variable capacitors are used to tune filter circuits.

Figure 13-15 is a collection of polarized capacitors. Most of these are **electrolytic capacitors.** It is critical to connect a polarized capacitor with the proper polarity. If connected in reverse polarity, the capacitor will get very hot and explode. Electrolytic capacitors are constructed to have the highest capacitance value with the least amount of leakage in a small package. They range from about the size of the eraser on a pencil to a canister several inches high. Due to electrolytic capacitor's very high capacitance ratings, they are used in dc power supplies and other applications.

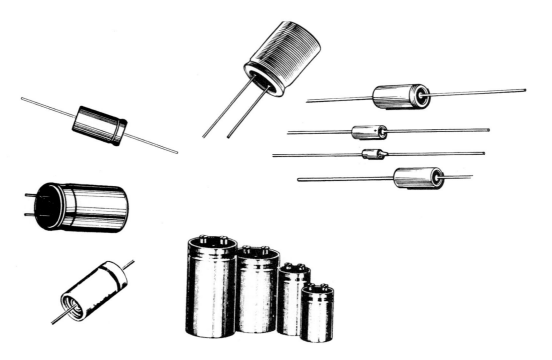

Figure 13-15. Polarized capacitors can only be used with dc voltages.

13.7 CAPACITORS IN SERIES AND PARALLEL

Capacitors are connected in series and parallel in the same manner as resistors, inductors, and other components.

Capacitors in Series

When capacitors are connected in series, the effect is the same as increasing the thickness of the dielectric, as shown in figure 13-16. A thicker dielectric inversely affects the net capacitance. The reciprocal formula is used for total capacitance in series. It is the same type of formula used for resistors and inductors in parallel. Shortcut formulas can be used when they apply. The formulas for capacitors in series are:

Formula 13.A

$$\frac{1}{C_T} = \frac{1}{C_1} + \frac{1}{C_2} + \frac{1}{C_3} + \dots \frac{1}{C_N}$$

C_T is the total capacitance, measured in farads.

C_1 through C_N are the individual capacitances, measured in farads.

N is the number of capacitors in series.

Formula 13.B

$$C_T = \frac{C_1 \times C_2}{C_1 + C_2}$$

C_T is the total capacitance, measured in farads.

C_1 and C_2 are the individual capacitances, measured in farads.

Formula 13.C

$$C_T = \frac{C}{N}$$

C_T is the total capacitance, measured in farads.

C is the capacitance value, measured in farads.

N is the number of capacitors in series.

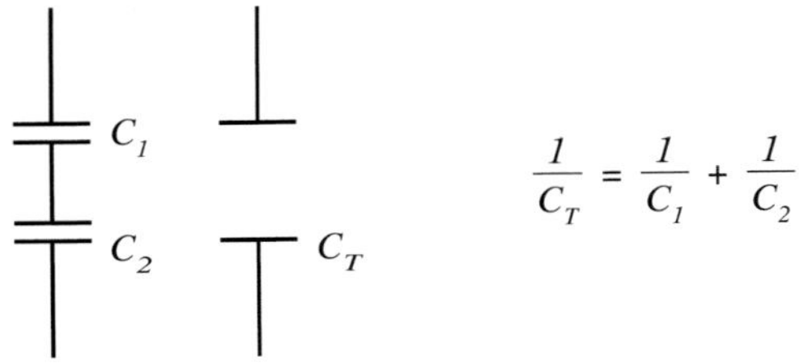

$$\frac{1}{C_T} = \frac{1}{C_1} + \frac{1}{C_2}$$

Figure 13-16. Putting capacitors in series reduces the total capacitance. It has the effect of increasing the thickness of the dielectric.

Sample problem 1.

Determine the total capacitance of two capacitors connected in series with values of 250 μF and 125 μF.

Formula: $C_T = \dfrac{C_1 \times C_2}{C_1 + C_2}$

Substitution: $C_T = \dfrac{250 \ \mu F \times 125 \ \mu F}{250 \ \mu F + 125 \ \mu F}$

Answer: $C_T = 83.3 \ \mu F$

Capacitors Connected in Parallel

Capacitors connected in parallel have the effect of increasing the plate area, as shown in figure 13-17. An increased plate area increases the capacitance, with the resultant being the sum of the capacitors connected in parallel. This formula is of the same type as the formula used for resistors and inductors in series. The formula for capacitors in parallel is:

Formula 13.D

$$C_T = C_1 + C_2 + C_3 + \ldots C_N$$

C_T is the total capacitance, measured in farads.
C_1 through C_N are individual capacitances, measured in farads.
N is the number of capacitors in parallel.

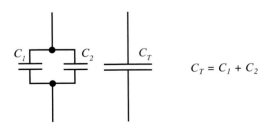

$$C_T = C_1 + C_2$$

Figure 13-17. Putting capacitors is parallel increases the total capacitance. It has the effect of increasing the plate surface area.

Sample problem 2. _____

Four 47 µF capacitors are connected in parallel. What is the total capacitance?

Formula: $C_T = C_1 + C_2 + C_3 + ...C_N$

Substitution: $C_T = 47 \text{ µF} + 47 \text{ µF} + 47 \text{ µF} + 47 \text{ µF}$

Answer: $C_T = 188 \text{ µF}$

13.8 CAPACITIVE VOLTAGE DIVIDERS

A capacitive voltage divider results from capacitors being connected in series, see figure 13-18. The amount of voltage across the capacitors is a ratio of the opposite capacitance to the total multiplied by the applied voltage. In equation form:

Formula 13.E

Voltage across C_1:

$$V_{C_1} = \frac{C_2}{C_1 + C_2} \times V_A$$

V_{C_1} is the voltage across C_1, measured in volts.
C_1 and C_2 are individual capacitances, measured in farads.
V_A is the applied voltage, measured in volts.

Formula 13.F

Voltage across C_2:

$$V_{C_2} = \frac{C_1}{C_1 + C_2} \times V_A$$

V_{C_2} is the voltage across C_2, measured in volts.
C_1 and C_2 are individual capacitances, measured in farads.
V_A is the applied voltage, measured in volts.

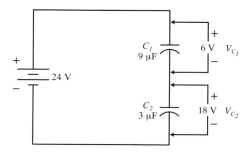

Figure 13-18. In a capacitive voltage divider, the smaller capacitance value has the largest voltage drop.

Sample problem 3. _____

    Verify the voltage readings in figure 13-18.

1. Voltage across C_1:

    Formula: $V_{C_1} = \dfrac{C_2}{C_1 + C_2} \times V_A$

    Substitution: $V_{C_1} = \dfrac{3\ \mu F}{9\ \mu F + 3\ \mu F} \times 24\ V$

    Answer: $V_{C_1} = 6\ V$

2. Voltage across C_2:

    Formula: $V_{C_2} = \dfrac{C_1}{C_1 + C_2} \times V_A$

    Substitution: $V_{C_2} = \dfrac{9\ \mu F}{9\ \mu F + 3\ \mu F} \times 24\ V$

    Answer: $V_{C_2} = 18\ V$

Note: Kirchhoff's voltage law should be used as a check. The sum of the voltages should equal the applied voltage.

13.9 TESTING A CAPACITOR WITH AN OHMMETER

An ohmmeter has its own dc voltage source and can be used to make a rough check to determine if a capacitor is good or faulty. An analog ohmmeter should be used for these measurements. Different ohms scales should be tried to find which works the best with a particular meter.

    First, short the leads of the capacitor together. This eliminates any charge that may already be present. If the testing is repeated, be sure to repeat discharging the capacitor. Next, connect the ohmmeter across the leads of the capacitor.

- *If the capacitor is good,* the readings will be as shown in figure 13-19. The needle swings to the right (towards zero). Then, the needle steadily drops back to the left (towards infinity).
- *If the capacitor is leaky (faulty dielectric),* the readings will be as shown in figure 13-20. The needle goes to some point on the scale, which depends on the amount of leakage, and stays there. Try different ohms scales and check the readings again to verify the result.
- *If the capacitor is open,* as with any open circuit, the ohmmeter will read infinity as shown in figure 13-21.

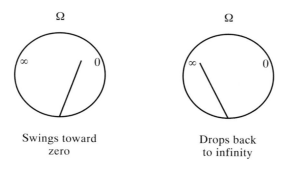

Swings toward
zero

Drops back
to infinity

Figure 13-19. Ohmmeter shows that the capacitor is good.

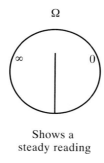

Shows a
steady reading

Figure 13-20. Ohmmeter shows a leaky capacitor.

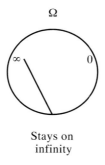

Stays on
infinity

Figure 13-21. Ohmmeter shows an open capacitor.

SUMMARY

- Capacitance is the ability of a circuit to store an electric charge.
- A capacitor is made of two conductors separated by an insulator.
- When first charging, a capacitor circuit has a high current even though the current does not actually flow through the capacitor.
- When fully charged, the capacitor has a voltage equal to the applied voltage and current stops flowing.

- A capacitor holds its charge even after the voltage supply has been removed.
- When discharging, the capacitor supplies voltage to the circuit until its supply of electrons is depleted.
- Larger plate surface area produces larger amounts of capacitance.
- A thinner dielectric, allowing the plates to be closer, produces larger amounts of capacitance.
- A capacitor has a value of capacitance and a voltage rating of the dielectric.

KEY WORDS AND TERMS GLOSSARY

capacitance (*C*): The ability of a device to store an electric charge. Through the storing and discharging of a charge, capacitance opposes a change in voltage. The unit of measure is farad (F). Common units of measure are µF and pF.

dielectric: An electrical insulator between the plates of a capacitor.

dielectric constant: The rating for materials used as dialectrics. The number reflects how many times better the material is than a vacuum.

dielectric strength: States the maximum voltage that can be applied to a capacitor before breakdown occurs.

electrolytic capacitor: A type of capacitor that is polarized.

electrostatic field: The attraction between negative and positive voltages.

farad (F): Unit measure of capacitance.

plate: In reference to capacitors, it is a conductive surface with the functions of collecting electrons on the negative side of the capacitor and give up electrons on the positive side.

KEY FORMULAS

Formula 13.A

$$\frac{1}{C_T} = \frac{1}{C_1} + \frac{1}{C_2} + \frac{1}{C_3} + \dots \frac{1}{C_N}$$

Formula 13.B

$$C_T = \frac{C_1 \times C_2}{C_1 + C_2}$$

Formula 13.C

$$C_T = \frac{C}{N}$$

Formula 13.D

$$C_T = C_1 + C_2 + C_3 + \ldots C_N$$

Formula 13.E

$$V_{C_1} = \frac{C_2}{C_1 + C_2} \times V_A$$

Formula 13.F

$$V_{C_2} = \frac{C_1}{C_1 + C_2} \times V_A$$

TEST YOUR KNOWLEDGE

Do not write in this text. Please use a separate sheet of paper.

1. The charge in a capacitor is stored as _____ energy.
2. What part of an electrolytic capacitor is formed with aluminum foil?
3. What is air called when it is used in some capacitors?
4. What does the dielectric constant measure?
5. Describe the process of charging a capacitor, from zero charge to full charge.
6. Describe the process of discharging a capacitor, from full charge to zero.
7. What is the effect on resistor voltage during the charge of a capacitor?
8. What is the effect on resistor voltage during the discharge of a capacitor?
9. List three factors affecting capacitance.
10. What are the two basic capacitor ratings?
11. List three types of capacitors.
12. Calculate the resultant of two capacitors connected in parallel with values of 150 μF and 200 μF.
13. Calculate the resultant of two capacitors connected in series with values of 100 μF and 300 μF.
14. Calculate the voltages of a capacitive voltage divider. The two capacitors have values of 100 μF and 300 μF. The applied voltage is 12 volts.
15. Describe the procedure to test a capacitor with an ohmmeter.

Electric cables are designed to keep stray inductance and capacitance from interfering with the signals they are relaying. The cables themselves often create the interference. (AMP Inc.)

Chapter 14
Time Constants and Waveshaping

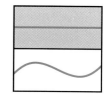

Upon completion of this chapter, you will be able to:
- Define time constants as related to inductors and capacitors.
- Calculate the values of time constants.
- Plot a universal time constant curve and use it to predict the performance of time constant circuits.
- Calculate the performance of time constant circuits.
- Plot the output waveforms of waveshaping circuits with different time constants.
- Use waveshaping circuits in filter applications.
- Apply time constants to circuits requiring high voltage or high current.

In this chapter, the actual time needed for the charging and discharging of inductors and capacitors is examined in detail. Both of these components are shown in simple circuits to consider the effects of different time constants.

Integrator and differentiator waveshaping circuits are introduced, with their waveforms. These terms refer to the mathematical expressions in calculus that approximate these circuits under certain conditions. The calculus is not discussed, but the waveshaping curves are presented in detail. The circuits are used in a wide range of applications, including video signals and timing circuits.

14.1 TIME CONSTANTS

One **time constant** is defined as the length of time to reach 63.2 percent of full charge or discharge. Full charge, or full discharge, is reached in a period of five time constants. In calculations using actual circuit values, the time constant is expressed in seconds, milliseconds, microseconds, or any other unit of time. The formula for calculating the time constant for an inductive circuit can be expressed as:

Formula 14.A

$$\tau = \frac{L}{R}$$

τ is one time constant, measured in seconds.

L is the inductance, measured in henrys.

R is the resistance, measured in ohms.

The formula for calculating the time constant for a capacitive circuit can be expressed as:

Formula 14.B

$$\tau = R \times C$$

τ is one time constant, measured in seconds.

C is the capacitance, measured in farads.

R is the resistance, measured in ohms.

Sample problem 1.

Determine the time constant and the time needed to fully charge or discharge a circuit with a resistor of 250 ohms in series with an inductor of 500 millihenrys.

One Time Constant:

Formula: $\tau = \dfrac{L}{R}$

Substitution: $\tau = \dfrac{500 \text{ mH}}{250 \ \Omega}$

Answer: $\tau = 2$ ms

Full charge or discharge:

Formula: Full $= 5 \times \tau$

Substitution: Full $= 5 \times 2$ ms

Answer: Full $= 10$ ms

Sample problem 2.

Find one time constant and the time to reach full charge of a circuit with a 200 ohm resistor in series with a 100 microfarad capacitor.

One Time Constant:

Formula: $\tau = R \times C$

Substitution: $\tau = 200\ \Omega \times 100\ \mu F$

Answer: $\tau = 20$ ms

Full Charge or Discharge:

Formula: Full = $5 \times \tau$

Substitution: Full = 5×20 ms

Answer: Full = 100 ms

14.2 UNIVERSAL TIME CONSTANT CURVE

The time constant formulas calculate the value of one time constant or (by multiplying by five) find the time needed for a capacitor or inductor to reach full charge or full discharge. For many circuits, this is sufficient. In certain applications, however, it is necessary to know conditions at some other instant in time.

The universal time constant curve, shown in figure 14-1 is a graphical representation of the charge and discharge of an RL or RC circuit. Using this chart, instantaneous values can be found at any point along the curves.

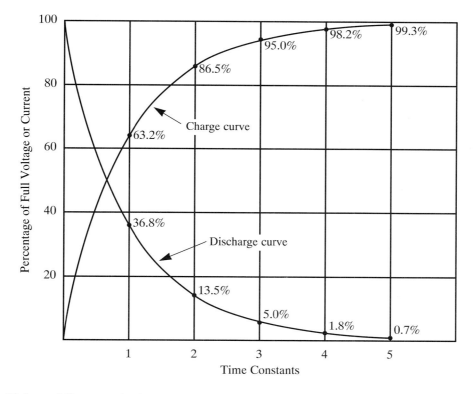

Figure 14-1. Universal time constant curve.

Shape of the Curve

The basic shape of the time constant curve applies to all circuits, regardless of the circuit values. In some situations, the curve may be smaller or larger, depending on how the axis are labeled, but the shape is the same.

The bottom axis of the graph is labeled in time constants. Actual values of time depend on the circuit. The left side is labeled in percentage of full charge or discharge.

Plotting the Charge Curve Using Percentages

The charge curve starts at zero and rises toward 100 percent. During each time constant, 63.2 percent of the remaining charge is accumulated. Starting from zero, the first time constant rises 63.2 percent, which explains the steep initial slope. After the first time constant, 36.8 percent of the charge remains ($100\% - 63.2\% = 36.8\%$).

For the second through fifth time constants, you can determine the portion of charge by multiplying the remaining percentage by 63.2 percent (0.632). Add this figure to the previous percentage to find the current charge. Subtract from 100 percent to find the percentage of charge still remaining.

Second time constant: $36.8 \times 0.632 = 23.3$

$63.2 + 23.3 = 86.5\%$ (charged)

$100 - 86.5 = 13.5\%$ (remaining)

Third time constant: $13.5 \times 0.632 = 8.5$

$86.5 + 8.5 = 95.0\%$ (charged)

$100 - 95.0 = 5.0$ (remaining)

Fourth time constant: $5.0 \times 0.632 = 3.2$

$95.0 + 3.2 = 98.2\%$ (charged)

$100 - 98.2 = 1.8\%$ (remaining)

Fifth time constant: $1.8 \times 0.632 = 1.1$

$98.2 + 1.1 = 99.3\%$ (charged)

$100 - 99.3 = 0.7\%$ (remaining)

Five time constants is considered full charge.

Plotting the Discharge Curve Using Percentages

The discharge curve begins at 100 percent and drops toward zero at the rate of 63.2 percent per time constant. After the first time constant, there is 36.8 percent remaining ($100\% - 63.2\% = 36.8\%$).

For the second through fifth time constants, determine the portion by multiplying the remaining percentage by 63.2 percent (0.632). Subtract from the previous percentage to find the charge still remaining.

Second time constant:	$36.8 \times 0.632 = 23.3\%$
	$36.8 - 23.3 = 13.5\%$ (remaining)
Third time constant:	$13.5 \times 0.632 = 8.5\%$
	$13.5 - 8.5 = 5.0\%$ (remaining)
Fourth time constant:	$5.0 \times 0.632 = 3.2\%$
	$5.0 - 3.2 = 1.8\%$ (remaining)
Fifth time constant:	$1.8 \times 0.632 = 1.1\%$
	$1.8 - 1.1 = 0.7\%$ (remaining)

Five time constants is considered fully discharged.

14.3 GRAPHICAL SOLUTIONS FOR INSTANTANEOUS VALUES

Two types of solutions to a problem can be found using the universal time constant curves, voltage (or current), and time of charge or discharge. Voltage is found for a capacitive circuit. Current is found for an inductive circuit. Instantaneous circuit values of voltage and current are represented by lower case letters v and i.

There are three different time values to be aware of: instantaneous time, the value of one time constant, and the total number of time constants. The formula for converting instantaneous time into time constants is:

Formula 14.C

$$T = \frac{t}{\tau}$$

T = total time constants.

t = instantaneous time.

τ = one time constant.

The following two examples examine the RL time constant.

Sample problem 3. _____

Use the graph shown in figure 14-2 and the circuit in figure 14-3 to find the current after 0.75 milliseconds of charge time.

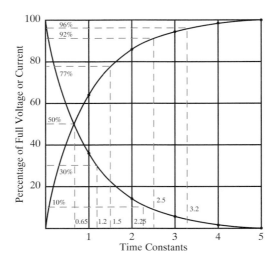

Figure 14-2. The universal time constant curve is used to find solutions to time constant circuits graphically. Use this curve for Sample problems 3 through 8.

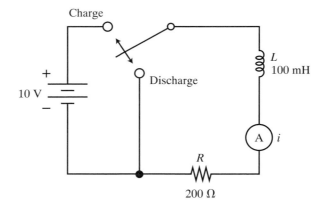

Figure 14-3. RL circuit for Sample problem 3.

1. Find one time constant, using circuit values.

   Formula: $\tau = \dfrac{L}{R}$

   Substitution: $\tau = \dfrac{100 \text{ mH}}{200 \text{ }\Omega}$

   Answer: $\tau = 0.5$ ms

2. Change instantaneous time into time constants.

> **Formula:** $T = \dfrac{t}{\tau}$

> **Substitution:** $T = \dfrac{0.75 \text{ ms}}{0.5 \text{ ms}}$

> **Answer:** $T = 1.5$ time constants

3. Use the graph in figure 14-2 to find the percentage.

> First, locate time constant equivalent along the horizontal axis. Second, draw a line *up* to the curve. Third, draw a line from the curve to the *left* to the percentage. Graphically estimate the current percentage: 77 percent (approximately).

4. Find maximum current using Ohm's law.

> **Formula:** $I_{max} = \dfrac{V}{R}$

> **Substitution:** $I_{max} = \dfrac{10 \text{ V}}{200 \text{ }\Omega}$

> **Answer:** $I_{max} = 50$ mA

5. Find instantaneous current.

> **Formula:** $i = \% \times I_{max}$

> **Substitution:** $i = 77\% \times 50$ mA

> **Answer:** $i = 38.5$ mA

Sample problem 4. _____

The circuit shown in figure 14-4 is fully charged. How long will it take to reach a discharge current of 25 milliamps?

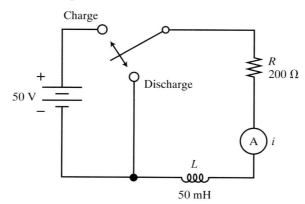

Figure 14-4. RL circuit for Sample problem 4.

1. Use Ohm's law to calculate maximum current.

Formula: $I_{max} = \dfrac{V}{R}$

Substitution: $I_{max} = \dfrac{50 \text{ V}}{200 \text{ }\Omega}$

Answer: $I_{max} = 250$ mA

2. Determine the percentage of maximum current the circuit has discharged to.

Formula: $\% = \dfrac{i}{I_{max}} \times 100\%$

Substitution: $\% = \dfrac{25 \text{ mA}}{250 \text{ mA}} \times 100\%$

Answer: $\% = 10\%$

3. Use the discharge curve in figure 14-2 to find the time constants.

First, find 10 percent on the percent of full current axis. Second, draw a line to the *right* to the discharge curve. Third, draw a line *down* to the time constant axis. Solved graphically: $T = 2.25$ time constants (approximately).

4. Find the value of one time constant.

Formula: $\tau = \dfrac{L}{R}$

Substitution: $\tau = \dfrac{50 \text{ mH}}{200 \text{ }\Omega}$

Answer: $\tau = 0.25$ ms

5. Find the instantaneous time using time constants.

Formula: $t = T \times \tau$

Substitution: $t = 2.25 \times 0.25$ ms

Answer: $t = 0.563$ ms

These next two sample problems examine RC time constant circuits.

Sample problem 5. _____

With the circuit shown in figure 14-5, graphically determine the voltage across the capacitor after a charging time of 125 ms. Use figure 14-2.

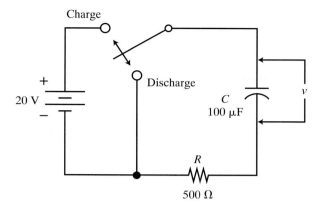

Figure 14-5. RC circuit for Sample problem 5.

1. Find one time constant, using circuit values.

   Formula: $\tau = R \times C$

   Substitution: $\tau = 500 \ \Omega \times 100 \ \mu F$

   Answer: $\tau = 50$ ms

2. Change instantaneous time into time constants.

   Formula: $T = \dfrac{t}{\tau}$

   Substitution: $T = \dfrac{125 \ ms}{50 \ ms}$

   Answer: $T = 2.5$ time constants

3. Use the graph in figure 14-2 to find percentage.

   Solved graphically: % = 92% (approximately)

4. Maximum charging voltage in a capacitor circuit is equal to the applied voltage.

   Answer: $V_{max} = 20$ V

5. Find instantaneous voltage.

   Formula: $v = \% \times V_{max}$

   Substitution: $v = 92\% \times 20$ V

   Answer: $v = 18.4$ V

Sample problem 6. _____

> After reaching full charge, the circuit begins to discharge. How long will it take the circuit in figure 14-6 to reach an instantaneous voltage of 15 volts?

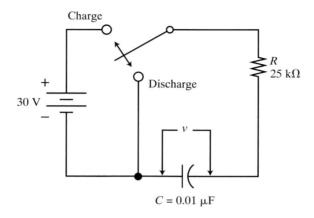

Figure 14-6. RC circuit for Sample problem 6.

1. Determine what percentage of maximum voltage is the instantaneous value.

   Formula: $\% = \dfrac{V}{V_{max}} \times 100\%$

   Substitution: $\% = \dfrac{15\ V}{30\ V} \times 100\%$

   Answer: $\% = 50\%$

2. Use the discharge curve in figure 14-2 to find the time constants.

   Solved graphically: $T = 0.65$ time constants (approximately)

3. Find the value of one time constant.

   Formula: $\tau = R \times C$

   Substitution: $\tau = 25\ k\Omega \times 0.01\ \mu F$

   Answer: $\tau = 0.25$ ms

4. Find the instantaneous time using time constants.

   Formula: $t = T \times \tau$

   Substitution: $t = 0.65 \times 0.25$ ms

   Answer: $t = 0.16$ ms

Different Charge and Discharge Times

The next two sample problems examine circuits that have different charge and discharge times. Notice that in the circuits in figures 14-7 and 14-8 the resistors in series with the inductor and capacitor change during the discharge cycle. This changes the discharge time.

Sample problem 7. _____

Determine the time to reach full charge and full discharge of the circuit shown in figure 14-7. Also, graphically determine the instantaneous charging current after 1.6 ns.

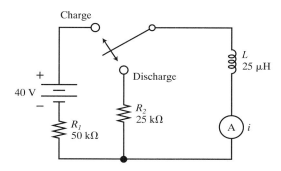

Figure 14-7. RL circuit for Sample problem 7.

1. Find charge time constant.

   Formula: $\tau = \dfrac{L}{R}$

   Substitution: $\tau = \dfrac{25\ \mu H}{50\ k\Omega}$

   Answer: $\tau = 0.5$ ns

   Multiply by five to get full charge time: 2.5 ns

2. Find discharge time constant.

   Formula: $\tau = \dfrac{L}{R}$

   Substitution: $\tau = \dfrac{25\ \mu H}{25\ k\Omega}$

   Answer: $\tau = 1.0$ ns

   Multiply by five to get full discharge time: 5 ns

3. Change instantaneous charge time into time constants.

Formula: $T = \dfrac{t}{\tau}$

Substitution: $T = \dfrac{1.6 \text{ ns}}{0.5 \text{ ns}}$

Answer: $T = 3.2$ time constants

4. Use the charge curve in figure 14-2 to find the percentage of maximum current.

Solved graphically: % = 96% (approximately)

5. Use Ohm's law to calculate maximum charging current. In this circuit, R_1 is the charge resistor.

Formula: $I_{max} = \dfrac{V}{R_1}$

Substitution: $I_{max} = \dfrac{40 \text{ V}}{50 \text{ k}\Omega}$

Answer: $I_{max} = 0.8 \text{ mA}$

6. Use percentage to determine instantaneous current.

Formula: $i = \% \times I_{max}$

Substitution: $i = 96\% \times 0.8 \text{ mA}$

Answer: $i = 0.77 \text{ mA}$

Sample problem 8.

Determine the charge and discharge time constants for the circuit shown in figure 14-8. Also, graphically determine the instantaneous voltage after 0.3 seconds of discharge (assume the circuit was fully charged).

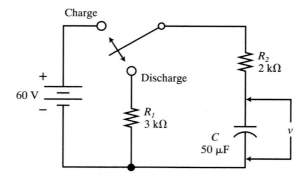

Figure 14-8. RC circuit for Sample problem 8.

1. Find the time constant for charge.

   Formula: $\tau = R \times C$

   Substitution: $\tau = 2\ k\Omega \times 50\ \mu F$

   Answer: $\tau = 0.1\ s$

2. Time constant for discharge.

   Formula: $\tau = R \times C$

   Substitution: $\tau = 5\ k\Omega \times 50\ \mu F$

   Answer: $\tau = 0.25\ s$

3. Change instantaneous time into time constants.

   Formula: $T = \dfrac{t}{\tau}$

   Substitution: $T = \dfrac{0.3\ s}{0.25\ s}$

   Answer: $T = 1.2$ time constants

4. Graphically determine the percentage of voltage remaining.

   Solved graphically: $\% = 30\%$ (approximately)

5. Calculate the instantaneous voltage.

   Formula: $v = \% \times V_{max}$

   Substitution: $v = 30\% \times 60\ V$

   Answer: $v = 18\ V$

14.4 EXPONENTIAL FORMULA FOR INSTANTANEOUS VALUES

The universal time constant curve, shown on page 387, gets its shape from two exponential formulas. One formula is for charging, and one formula is for discharging. These formulas can also be used to find an exact value for the charge or discharge.

These formulas are given to find percent of maximum value. This percent can be used to find the maximum current for inductive circuits or maximum voltage for capacitive circuits. The two exponential formulas are written:

Formula 14.D

Instantaneous charge: $\% = (1 - e^{-T}) \times 100\%$

Note: e is the natural log (e^x key on a calculator)

$$-T = -\frac{t}{\tau} \qquad - \text{Time constant} = -\frac{\text{instantaneous time}}{\text{one time constant}}$$

Formula 14.E

Instantaneous discharge: $\% = (e^{-T}) \times 100\%$

Note: e is the natural log (e^x key on a calculator)

$$-T = -\frac{t}{\tau} \qquad - \text{Time constant} = -\frac{\text{instantaneous time}}{\text{one time constant}}$$

Sample problem 9. _____

Find the instantaneous charge current after four microseconds of the circuit shown in figure 14-9.

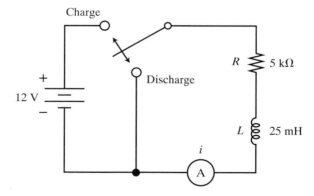

Figure 14-9. Find the instantaneous current. (Sample problem 9)

1. Find the value of one time constant.

    Formula: $\tau = \dfrac{L}{R}$

    Substitution: $\tau = \dfrac{25 \text{ mH}}{5 \text{ k}\Omega}$

    Answer: $\tau = 5 \text{ μs}$

2. Change time into time constants.

    Formula: $T = \dfrac{t}{\tau}$

Substitution: $T = \dfrac{4\ \mu s}{5\ \mu s}$

Answer: $T = 0.8$ time constants

3. Find the instantaneous percent of full charge.

   Formula: $\% = (1 - e^{-T}) \times 100\%$

   Substitution: $\% = (1 - e^{-0.8}) \times 100\%$

   Intermediate step: $e^{-0.8} = 0.449$

   Substitution: $\% = (1 - 0.449) \times 100\%$

   Answer: $\% = 55.1\%$

4. Use Ohm's law to find maximum current.

   Formula: $I_{max} = \dfrac{V}{R}$

   Substitution: $I_{max} = \dfrac{12\ V}{5\ k\Omega}$

   Answer: $I_{max} = 2.4\ mA$

5. Instantaneous charge current is a percentage of maximum.

   Formula: $i = \% \times I_{max}$

   Substitution: $i = 55.1\% \times 2.4\ mA$

   Answer: $i = 1.32\ mA$

Sample problem 10. _____

Assuming the circuit in figure 14-10 has reached full charge, find the instantaneous discharge voltage after 80 milliseconds.

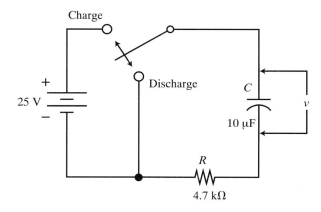

Figure 14-10. Find the instantaneous voltage. (Sample problem 10)

1. Find the value of one time constant.

   Formula: $\tau = R \times C$

   Substitution: $\tau = 4.7 \text{ k}\Omega \times 10 \text{ μF}$

   Answer: $\tau = 47 \text{ ms}$

2. Change time into time constants.

   Formula: $T = \dfrac{t}{\tau}$

   Substitution: $T = \dfrac{80 \text{ ms}}{47 \text{ ms}}$

   Answer: $T = 1.7$ time constants

3. Find the instantaneous percent of full charge, using the discharge formula.

   Formula: $\% = (e^{-T}) \times 100\%$

   Substitution: $\% = (e^{-1.7}) \times 100\%$

   Intermediate step: $e^{-1.7} = 0.183$

   Substitution: $\% = 0.183 \times 100\%$

   Answer: $\% = 18.3\%$

4. Instantaneous discharge voltage is a percentage of maximum voltage. The maximum voltage is the applied voltage.

   Formula: $v = \% \times V$

   Substitution: $v = 18.3\% \times 25 \text{ V}$

   Answer: $v = 4.58 \text{ V}$

14.5 INTEGRATORS AND DIFFERENTIATORS

Integration and differentiation are mathematical terms used in calculus. Time constant circuits, using inductors and capacitors, can be used to produce waveforms that are approximated by these mathematical functions. The outputs of time constant circuits are complementary, with the shape of the differentiator being dependent on the shape of the integrator. These circuits are used in applications such as filters and timers. The shape of the output curve is controlled by varying the time constant.

For discussion purposes, the input to these circuits is an entirely positive square wave. This is often used in real applications. This input can be thought of as either *on* or *off*. The time that the square wave is on or off is called the **pulse width**. A **symmetrical square wave** has the on pulse width equal to the off pulse width.

Short, Medium, and Long Time Constant Circuits

A **short time constant circuit** is classified as a circuit that has the time for a full charge or discharge in much less than the time of one pulse width. The circuit has more than enough time to reach full charge or discharge.

A **medium time constant circuit** has its full charge/discharge time slightly smaller than or equal to the time of one pulse width. The circuit has enough time to reach full charge or discharge.

A **long time constant circuit** has its full charge/discharge time greater than the time of one pulse width. The circuit does not have enough time to reach full charge or full discharge. As a reminder, full charge/discharge is defined as five time constants.

Integrators

An **integrator** is a circuit that when an all positive square wave is applied to the input, the output is all positive. Figure 14-11 shows the range of outputs. Notice, with

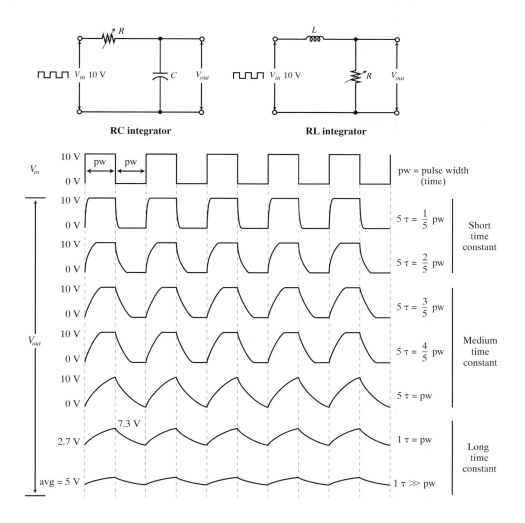

Figure 14-11. RC and RL integrator circuits are shown (top). With a pulse input, the outputs are drawn for short, medium, and long time constants (bottom).

a capacitive circuit, the output is across the capacitor. With an inductive circuit, the output is across the resistor. The output will have the shape of the universal time constant curve. The steepness of charge or discharge curve depends on the time constant.

Sample problem 11. _____

This problem examines a short time constant integrator. Plot the output waveform of the integrator circuit shown in figure 14-12. Pulse width is five ms.

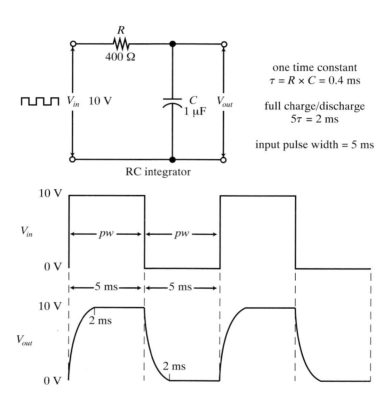

Figure 14-12. With a short time constant, $5\tau <$ pw, there is more than enough time for charging and discharging. (Sample problem 11)

1. Calculate the value of one time constant.

    Formula: $\tau = R \times C$

    Substitution: $\tau = 400\ \Omega \times 1\ \mu F$

    Answer: $\tau = 0.4$ ms

2. Calculate the time for full charge/discharge.

    Formula: $5\tau = 5 \times \tau$

    Substitution: $5\tau = 5 \times 0.4$ ms

    Answer: $5\tau = 2$ ms

3. Compare the full charge/discharge to the pulse width and plot the waveform.

    Answer: $5\tau \ll$ pulse width (5τ is much less than the pulse width)

    See figure 14-12 for the plot

Sample problem 12. _____

This sample problem examines a medium time constant integrator. Plot the output waveform of the integrator circuit shown in figure 14-13. Pulse width is five ms.

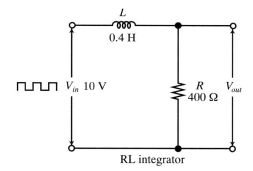

one time constant
$$\tau = \frac{L}{R} = 1 \text{ ms}$$

full charge/discharge
$$5\tau = 5 \text{ ms}$$

input pulse width = 5 ms

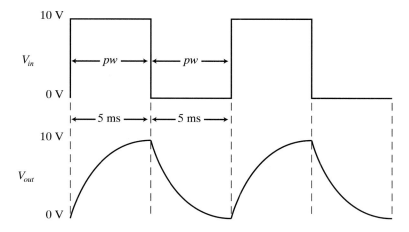

Figure 14-13. With a medium time constant, 5τ = pw, there is enough time for full charging and discharging. (Sample problem 12)

1. Calculate the value of one time constant.

   Formula: $\tau = \dfrac{L}{R}$

   Substitution: $\tau = \dfrac{0.4\ \text{H}}{400\ \Omega}$

   Answer: $\tau = 1$ ms

2. Calculate the time for full charge/discharge.

   Formula: $5\tau = 5 \times \tau$

   Substitution: $5\tau = 5 \times 1$ ms

   Answer: $5\tau = 5$ ms

3. Compare full charge/discharge to the pulse width and plot the waveform.

   Answer: 5τ = pulse width

   See figure 14-13 for the plot.

Sample problem 13. _____

This sample problem examines the long time constant integrator. Plot the output wave-form of the integrator circuit shown in figure 14-14. Pulse width is five ms.

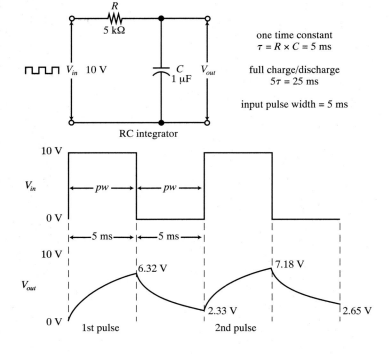

Figure 14-14. With a long time constant, $5\tau <$ pw, there is not enough time for full charging or discharging. (Sample problem 13)

1. Calculate the value of one time constant.

   Formula: $\tau = R \times C$

   Substitution: $\tau = 5 \text{ k}\Omega \times 1 \text{ μF}$

   Answer: $\tau = 5 \text{ ms}$

2. Calculate the time for full charge/discharge.

   Formula: $5\tau = 5 \times \tau$

   Substitution: $5\tau = 5 \times 5 \text{ ms}$

   Answer: $5\tau = 25 \text{ ms}$

3. Compare full charge/discharge to the pulse width.

   Answer: $5\tau \gg$ pulse width (5τ is much greater than the pulse width)

4. To plot this waveform, use the instantaneous formula to determine voltage reached during first pulse charge time. Rather than using percentage, use voltage. First change pulse width time into time constants.

   Formula: $T = \dfrac{t}{\tau}$

   Substitution: $T = \dfrac{5 \text{ ms}}{5 \text{ ms}}$

   Answer: $T = 1$ time constant

   Instantaneous voltage:

   Formula: $v = (1 - e^{-T}) \times V_{max}$

   Substitution: $v = (1 - e^{-1}) \times 10 \text{ V}$

   Intermediate step: $e^{-1} = 0.368$

   Substitution: $v = (1 - 0.368) \times 10 \text{ V}$

   Answer: $v = 6.32 \text{ V}$

5. Use the instantaneous discharge formula to determine the voltage reached during the discharge time between the first and second pulses. The maximum voltage (V_{max}) used in the equation is the value reached at the end of the charge time (6.32 V).

   Formula: $v = (e^{-T}) \times V_{max}$

   Substitution: $v = (e^{-1}) \times 6.32 \text{ V}$

   Intermediate step: $e^{-1} = 0.368$

   Substitution: $v = 0.368 \times 6.32 \text{ V}$

   Answer: $v = 2.33 \text{ V}$

6. To find the voltage reached during the second charge time, you must first subtract the remaining voltage from the applied voltage. This determines the value to use for V_{max}.

    Formula: $V_{max} = V_{applied} - V_{remaining}$

    Substitution: $V_{max} = 10\ V - 2.33\ V$

    Answer: $V_{max} = 7.67\ V$

    Second, use the instantaneous charging formula.

    Formula: $v = (1 - e^{-T}) \times V_{max}$

    Substitution: $v = (1 - e^{-1}) \times 7.67\ V$

    Intermediate step: $e^{-1} = 0.368$

    Substitution: $v = (1 - 0.368) \times 7.67\ V$

    Answer: $v = 4.85\ V$

    Third, add the instantaneous charging voltage to the starting voltage to determine the voltage reached at the end of charge time:

    Formula: $V_{ending} = v + V_{starting}$

    Substitution: $V_{ending} = 4.85\ V + 2.33\ V$

    Answer: $V_{ending} = 7.18\ V$

7. The voltage reached during discharge is found with the instantaneous formula using the previous high voltage as V_{max}. In this problem, $v = 2.65\ V$.

8. When the input square wave has produced pulses with their on times adding to approximately five time constants, a steady state is reached. The center of the steady state waveform is equal to the average of the input square wave. For this particular circuit, the steady state waveform will have a maximum of approximately 7.3 volts and a minimum of 2.7 volts. See plot in figure 14-14.

Differentiators

A **differentiator** has two parts of output to consider, the positive charge portion and the negative discharge portion. The charge portion is the difference between the input voltage and the voltage dropped across the integrator. The discharge portion goes negative, as result of the energy being released in the circuit. The average of the differentiator output is zero.

The output waveform of the differentiator is the complement of the integrator. Kirchhoff's voltage law states that the algebraic sum of the voltages around the loop must add to equal zero. During the charge time, the differentiator has a voltage equal to the difference between the integrator and the input voltage. During the discharge time, the differentiator has a voltage equal to the opposite of the integrator.

Figure 14-15 shows a comparison between an integrator output and a differentiator output. Both of these are taking place in the circuit at the same time. The application determines which output is to be used. The circuit is connected to allow the output to be taken across the component on the bottom.

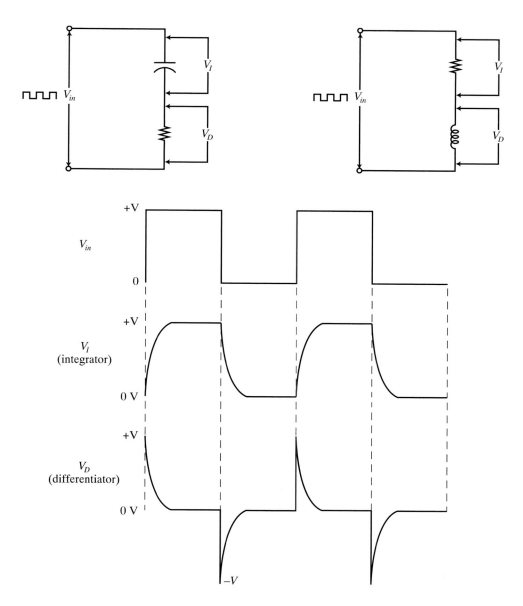

Figure 14-15. Comparison of integrator and differentiator output waveforms.

The negative portion of the differentiator waveform is a result of the circuit releasing the stored energy in a direction that is opposite the direction in which it was stored.

In figure 14-16, differentiator outputs are shown over a range of time constants, from short to very long. The short time constant results in sharp spikes. Medium time constants result in wider spikes. Long time constants result in waveforms that look more like the applied square wave. In all cases, the average of the waveform is zero.

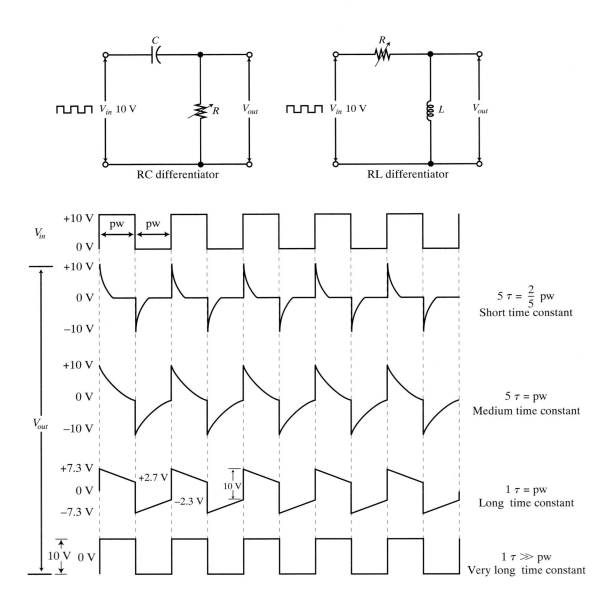

Figure 14-16. RC and RL differentiator circuits are shown (top). With a pulse input, the outputs are drawn for short, medium, long, and very long time constants (bottom).

Sample problem 14.

Figure 14-17 shows a differentiator circuit with a short time constant. Plot the output waveform. Pulse width is five ms.

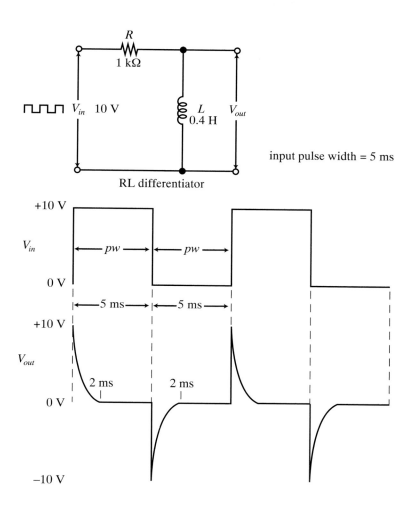

Figure 14-17. With a short time constant, $5\tau <$ pw, there is more than enough time for charging and discharging. (Sample problem 14)

1. Calculate the value of one time constant.

   Formula: $\tau = \dfrac{L}{R}$

   Substitution: $\tau = \dfrac{0.4\ \text{H}}{1\ \text{k}\Omega}$

   Answer: $\tau = 0.4\ \text{ms}$

2. Calculate the time for full charge/discharge.

Formula: $5\tau = 5 \times \tau$

Substitution: $5\tau = 5 \times 0.4$ ms

Answer: $5\tau = 2$ ms

3. Compare full charge/discharge to the pulse width.

Answer: $5\tau <<$ pulse width (5τ is much less than pulse width)

4. Plot the waveform. The negative discharge spike is the exact opposite of the positive charge portion because the pulse widths are the same and the charge and discharge time constants are the same. See plot in figure 14-17.

Sample problem 15. _____

Figure 14-18 shows a differentiator with a medium time constant. Plot the output waveform. Pulse width is five ms.

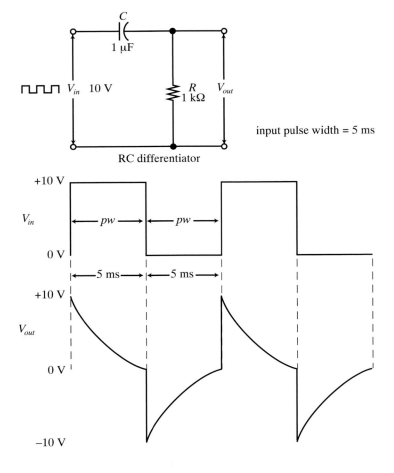

Figure 14-18. With a medium time constant, $5\tau = $ pw, there is enough time for full charging and discharging. (Sample problem 15)

1. Calculate the value of one time constant.

    Formula: $\tau = R \times C$

    Substitution: $\tau = 1\ k\Omega \times 1\ \mu F$

    Answer: $\tau = 1\ ms$

2. Calculate the time for full charge/discharge.

    Formula: $5\tau = 5 \times \tau$

    Substitution: $5\tau = 5 \times 1\ ms$

    Answer: $5\tau = 5\ ms$

3. Compare full charge/discharge to the pulse width and plot the waveform.

    Answer: 5τ = pulse width.

    See plot in figure 14-18.

Sample problem 16. _____

Figure 14-19 show a differentiator with a long time constant. Plot the output waveform. Pulse width is five ms.

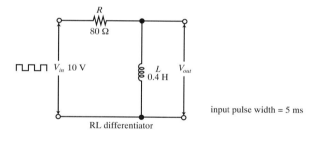

RL differentiator

input pulse width = 5 ms

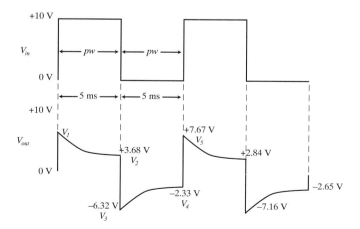

Figure 14-19. With a long time constant, $\tau >$ pw, there is not enough time for full charging or discharging. (Sample problem 16)

1. Calculate the value of one time constant.

    Formula: $\tau = \dfrac{L}{R}$

    Substitution: $\tau = \dfrac{0.4\ \text{H}}{80\ \Omega}$

    Answer: $\tau = 5$ ms

2. Calculate the time for full charge/discharge.

    Formula: $5\tau = 5 \times \tau$

    Substitution: $5\tau = 5 \times 5$ ms

    Answer: $5\tau = 25$ ms

3. Compare full charge/discharge to the pulse width.

    Answer: $5t \gg$ pulse width (5τ is much greater than the pulse width)

4. The first point on the curve, V_1, is equal to the applied voltage because there is no charge on the circuit to start.

5. Use the instantaneous formula to determine voltage reached during first pulse charge time. Rather than using percentage, use voltage. *Note: these formulas are for an integrator circuit, therefore, you must subtract the value found from the applied voltage.* First change pulse width time into time constants.

    Formula: $T = \dfrac{t}{\tau}$

    Substitution: $T = \dfrac{5\ \text{ms}}{5\ \text{ms}}$

    Answer: $T = 1$ time constant

    Instantaneous voltage:

    Formula: $v = (1 - e^{-T}) \times V_{max}$

    Substitution: $v = (1 - e^{-1}) \times 10$ V

    Intermediate step: $e^{-1} = 0.368$

    Substitution: $v = (1 - 0.368) \times 10$ V

    Answer: $v = 6.32$ V

    Subtract from the applied voltage.

    Formula: $V_2 = V_{in} - v$

Substitution: $V_2 = 10\text{ V} - 6.32\text{ V}$

Answer: $V_2 = 3.68\text{ V}$

6. The first portion of the negative spike is found by subtracting the **applied voltage** from the last positive instantaneous voltage.

   Formula: $V_3 = V_2 - V_{in}$

   Substitution: $V_3 = 3.68\text{ V} - 10\text{ V}$

   Answer: $V_3 = -6.32\text{ V}$

7. Use the instantaneous discharge formula to determine the voltage reached during the discharge time between the first and second pulses. Maximum voltage is the value of the first part of the negative spike.

   Formula: $V_4 = (e^{-T}) \times V_{max}$

   Substitution: $V_4 = (e^{-1}) \times -6.32\text{ V}$

   Intermediate step: $e^{-1} = 0.368$

   Substitution: $V_4 = 0.368 \times -6.32\text{ V}$

   Answer: $V_4 = -2.33\text{ V}$

8. When the input switches to the charge portion, the differentiator output instantly jumps up a voltage equal to the input voltage. Add to the last negative voltage, V_4.

   Formula: $V_5 = V_{in} + V_4$

   Substitution: $V_5 = 10\text{ V} + (-2.33\text{ V})$

   Answer: $V_5 = 7.67\text{ V}$

9. When the input voltage has produced enough charge pulses to equal five time constants, the output reaches its steady state value. See plot in figure 14-19.

14.6 SELECTING COMPONENT VALUES

There are many inductor/resistor and resistor/capacitor combinations that produce a given time constant. In certain applications, one of the components may have been selected from a previous design. In other applications, components are selected according the ac resistance, to be discussed in Chapters 18 and 19. Still other circuits use components selected according to what is in stock.

Figure 14-20 shows some possible combinations that will produce a desired time constant. The time constants selected for this table are the ones used in the sample problems of integrators and differentiators.

Figure 14-20. RL and RC combinations for selected time constants.

Time Constant	RL Combinations	RC Combinations
0.4 ms	$L = 0.4$ H $R = 1$ kΩ	$R = 400$ Ω $C = 1$ μF
	$L = 0.2$ H $R = 500$ Ω	$R = 1$ kΩ $C = 0.4$ μF
	$L = 50$ mH $R = 125$ Ω	$R = 40$ kΩ $C = 0.01$ μF
1 ms	$L = 0.4$ H $R = 400$ Ω	$R = 1$ kΩ $C = 1$ μF
	$L = 0.2$ H $R = 200$ Ω	$R = 2500$ Ω $C = 0.4$ μF
	$L = 50$ mH $R = 50$ Ω	$R = 100$ kΩ $C = 0.01$ μF
5 ms	$L = 0.4$ H $R = 80$ Ω	$R = 5$ kΩ $C = 1$ μF
	$L = 0.2$ H $R = 40$ Ω	$R = 12.5$ kΩ $C = 0.4$ μF
	$L = 50$ mH $R = 10$ Ω	$R = 500$ Ω $C = 0.01$ μF

14.7 FILTER CIRCUITS

A **filter** is a circuit that offers varying degrees of opposition to different frequencies. Inductors and capacitors change with the frequency and this characteristic is useful to filter selected frequencies.

Parts of a Square Wave

A square wave is made up of a low frequency component and a high frequency component, as seen in figure 14-21. The low frequency is the flat, horizontal portion, equal to a dc voltage. The high frequency portion is the switching time from zero to maximum, or from maximum to zero.

The square wave also has a leading edge and a trailing edge. The **leading edge** is the portion switching from zero to maximum, also called the *positive-going edge*. The **trailing edge** is the portion switching from maximum to zero, also called the *negative-going edge*.

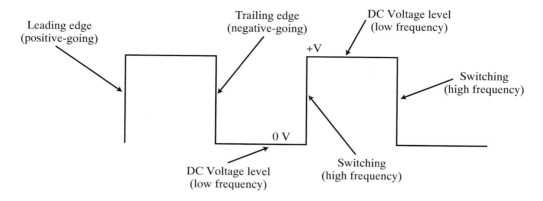

Figure 14-21. Parts of a square wave.

Integrator as a Low-Pass Filter

When an integrator is used as a **low-pass filter,** the low frequency portions of the square wave are passed with little opposition. The high frequencies receive more opposition with longer time constants.

In figure 14-22, the results are seen when a short time constant integrator is used as a low-pass filter. The dc voltages, zero and +V are passed, without distortion. However, the opposition to the switching times changes their shape from a straight line to a curved line. The change results in a slower switching time.

Too much filtering of the high frequencies, results in loss of the low frequencies as well. A longer time constant affects the high frequencies more. However, increasing the

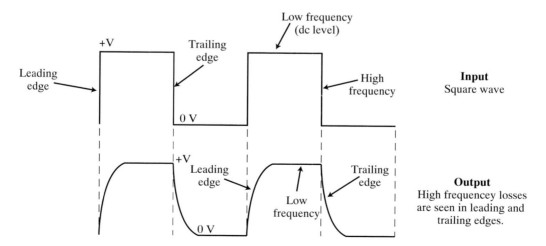

Figure 14-22. Output of an integrator has of an effect on high frequencies with a shorter time constant.

time constant also results in shorter on and off times for the low frequencies. The design of the circuit must be a compromise.

Differentiator as a High-Pass Filter

In a differentiator circuit, the switching time remains unaffected. In figure 14-23, a long time constant differentiator is used as a **high-pass filter.** Losses in the dc portion are seen by the horizontal lines being slanted.

A characteristic of a square wave, and all other ac waveforms, is its average dc level. For example, if the waveform starts at zero and rises to 10 volts, it has an average dc level of plus five volts. The differentiator always produces an output waveform with an average voltage equal to zero. The dc component of the dc waveform has been removed.

If a very long time constant differentiator is used, the output will have a shape very close to the original square wave, with an average value of zero.

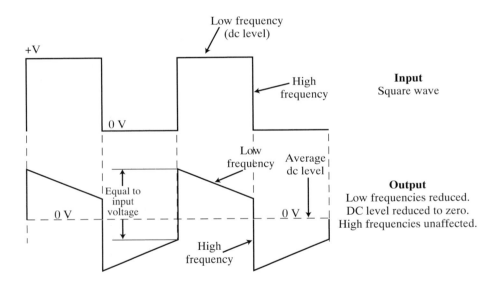

Figure 14-23. Differentiators stop dc voltages. The shape of the output is affected less with a longer time constant.

14.8 PERFORMANCE OF TIME CONSTANT CIRCUITS

The release of stored energy in a time constant circuit can be used in applications that might not be possible with other circuits. For example, an inductor can produce the thousands of volts necessary to make a spark plug operate, even though the voltage source may only be a few volts. A capacitor can produce enough current to fire a flash bulb using a very small battery.

Long time constant circuits can be used when it is desired to hold the voltage, or current, on the load as long as possible.

RL Circuit with a Small Discharge Resistor

Figure 14-24 has its load in the discharge section of the circuit. The load is a light bulb with a filament resistance of 10 ohms. One possible application of this circuit has the inductor being a relay coil. When the switch is in the charge position, the relay coil is activated and the light bulb is off. If the power to the relay is turned off, the light will glow for a period of time. Other applications are possible.

The waveforms compare the relative charge time to the relative discharge time. With the low value of resistance, the discharge time constant is quite long in comparison to the charge time. In this example, the discharge time is 10 times longer than the charge time.

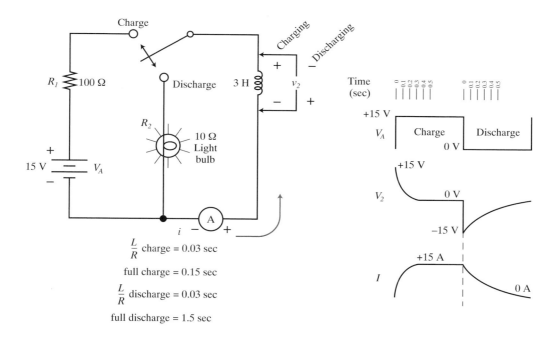

Figure 14-24. Small discharge resistance in RL circuit has long time constant.

High Voltage Released from Inductive Kick

The **inductive kick** is a name given to the high voltage produced when the discharge path for an inductor is an open circuit. Figure 14-25 uses a spark plug in the discharge path. A resistance of 80 kilohms is given for this example.

To calculate the inductive kick, first find the maximum current flowing in the inductor during charge. Remember, an inductor opposes a change in current. It tries to maintain the same current when the switch is moved to discharge. If the circuit is opened, instantaneous discharge current produced the high cemf.

Sample problem 17. _____

Calculate the voltage applied to the spark plug during the discharge time in figure 14-25.

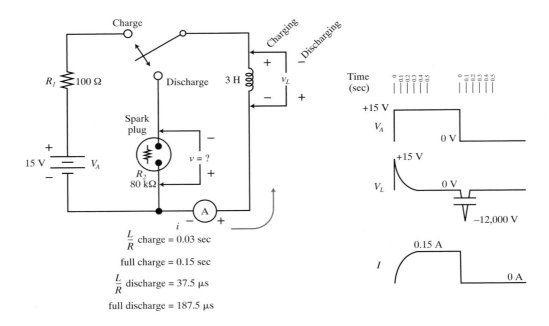

$\dfrac{L}{R}$ charge = 0.03 sec

full charge = 0.15 sec

$\dfrac{L}{R}$ discharge = 37.5 μs

full discharge = 187.5 μs

Figure 14-25. Instantaneous high voltage fires a spark plug when inductor is discharged. (Sample problem 17)

1. Calculate maximum current during the charge time.

   Formula: $I = \dfrac{V}{R_1}$

   Substitution: $I = \dfrac{15\text{ V}}{100\ \Omega}$

   Answer: $I = 0.15$ A

2. Use the maximum current to find the instantaneous voltage developed by the inductor's cemf. The cemf will also be across the load, which is the spark plug.

   Formula: $v = I \times R_2$

Substitution: $v = 0.15 \text{ A} \times 80 \text{ k}\Omega$

Answer: $v = 12 \text{ kV}$

RC Circuit with a Large Discharge Resistance

A capacitor will store voltages for a long period of time, depending on the current drain during discharge. A capacitor can replace a battery for a short period of time. An application is the filter capacitor in a power supply. A power supply converts a sine wave voltage to a dc voltage using rectifier diodes and filters. The diode converts the positive and negative voltages of the ac sine wave to pulsating dc. A capacitor is used to smooth the pulses. The capacitor charges during the positive-going portion and discharges during the negative-going portion. Power supplies are discussed in detail in Chapter 26.

Figure 14-26 shows a circuit with a large value of discharge resistance. The large resistance provides a long time constant, allowing the capacitor to hold its charge for a long period of time, such as would be needed for a power supply application.

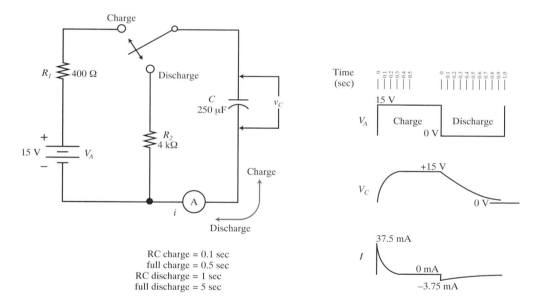

RC charge = 0.1 sec
full charge = 0.5 sec
RC discharge = 1 sec
full discharge = 5 sec

Figure 14-26. Large resistance results in long discharge time.

High Current with Low Resistance Capacitor Discharge

A capacitor has the advantage of discharging very large amounts of current. An application is the flash unit for a camera, which uses a large amount of current for a very short time. The batteries in a portable flash unit are typically AA size. If the AA batteries were to provide a large amount of current, they would have a very short life.

The capacitor charges from the battery voltage. When the flash is fired, the battery is taken out of the circuit and the capacitor does the work.

Sample problem 18. _____

Calculate the discharge current in the circuit in figure 14-27. Assume the charge time is long enough for the capacitor to reach full charge.

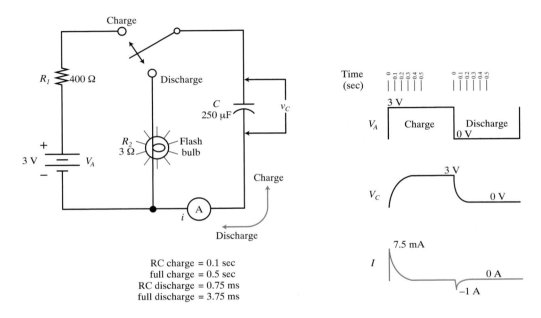

RC charge = 0.1 sec
full charge = 0.5 sec
RC discharge = 0.75 ms
full discharge = 3.75 ms

Figure 14-27. Instantaneous large current spike lights a flash bulb during a capacitor discharge. (Sample problem 18)

1. Voltage at the start of discharge:

   Answer: $V_C = 3$ V

2. Use Ohm's law to calculate the instantaneous discharging current.

   Formula: $i = \dfrac{V}{R}$

   Substitution: $i = \dfrac{3 \text{ V}}{3 \text{ }\Omega}$

   Answer: $i = 1$ A

SUMMARY

- One time constant is equal to 63.2 percent of the full charge or discharge time.
- Full charge or discharge time is equal to five time constants.
- The universal time constant curve is used to graphically predict the percentage of charge or discharge from zero to full.
- The exponential formula can be used to calculate the percentage of charge of discharge for any instantaneous time.
- Time constant circuits can be used to change the shape of a square wave.
- Time constant circuits can be used as filters to pass or reject selected bands of frequencies.
- An inductive circuit produces a large voltage if discharged into a high resistance.
- A capacitive circuit produces a high current if discharged into a low resistance.

KEY WORDS AND TERMS GLOSSARY

differentiator: An electronic circuit whose output waveform has two portions to consider, the positive charge portion and the negative discharge portion. The charge portion is the difference between the input voltage and the voltage dropped across other circuit components. The discharge portion goes negative as result of the energy being released in the circuit. The average of the differentiator output is zero.

filter: A circuit that offers varying degrees of opposition to different frequencies.

high pass filter: A filter that offers little opposition to high frequency signals and greater opposition to low frequency signals.

inductive kick: The high voltage produced when the discharge path for an inductor is an open circuit.

integrator: An application of a time constant circuit. When an all-positive square wave is applied to the input, the output has the shape of the universal time constant curve. The steepness of charge/discharge depends on the value of the time constant.

leading edge: The portion of a wave switching from zero to a maximum.

long time constant circuit: A circuit that has its full charge/discharge time greater than the time of one pulse width.

low pass filter: A filter that offers little opposition to low frequency signals and greater opposition to high frequency signals.

medium time constant circuit: A circuit that has its full charge/discharge time slightly smaller than or equal to the time of one pulse width.

pulse width: The length of time that a square wave is either on or off.

short time constant circuit: A circuit that has its full charge/discharge time much smaller than the time of one pulse width.

symmetrical square wave: A square wave that has the on pulse width equal to the off pulse width.

time constant: The length of time for a resistor-capacitor circuit or resistor-inductor circuit to reach 63.2 percent of full charge or discharge. Full charge is reached in five time constants.

trailing edge: The portion of the wave switching from a maximum to zero.

KEY FORMULAS

Formula 14.A

$$\tau = \frac{L}{R}$$

Formula 14.B

$$\tau = R \times C$$

Formula 14.C

$$T = \frac{t}{\tau}$$

Formula 14.D

Instantaneous charge: $\% = (1 - e^{-T}) \times 100\%$

Formula 14.E

Instantaneous discharge: $\% = (e^{-T}) \times 100\%$

TEST YOUR KNOWLEDGE

Do not write in this text. Please use a separate sheet of paper.
 1. What percent of full charge is reached in a time period equal to three time constants?
 2. What is the difference between actual time and a time constant?
 3. How many time constants to reach full charge?
 4. How many time constants to reach full discharge?
 5. What percentage of full charge or discharge is reached in one time constant?
 6. What percentage of full charge or discharge is reached in two time constants?
 7. What percentage of full charge or discharge is reached in three time constants?
 8. What is the formula to calculate one time constant in an inductive circuit?
 9. What is the formula to calculate one time constant in an capacitive circuit?
10. Determine the time needed for one time constant and the time needed to reach full charge in a circuit with a 500 ohm resistor in series with a 250 mH inductor.
11. Determine the time needed for one time constant and the time needed to reach full charge in a circuit with a 15 kilohm resistor in series with a 20 μF capacitor.

12. Use the universal time constant curve shown in figure 14-2 to calculate the current after 0.65 µs charge time. The circuit is a 350 µH inductor in series with a 1400 ohm resistor with 10 volts applied.

13. A circuit consists of a two H inductor in series with a 10 ohm resistor with 10 volts applied. Use the universal time constant curve shown in figure 14-2 to calculate the current after 0.3 seconds discharge time if the circuit started from full charge.

14. A circuit consists of a 1000 µF capacitor in series with a 500 ohm resistor with 10 volts applied. Use the universal time constant curve shown in figure 14-2 to calculate capacitive voltage after 1.2 seconds charge time.

15. A circuit consists of a 250 µF capacitor in series with a 10 kilohm resistor with 10 volts applied. Use the universal time constant curve shown in figure 14-2 to calculate capacitive voltage after two seconds discharge time if the circuit started from full charge.

16. Use the exponential formula to find the instantaneous charge current after 0.27 milliseconds in a circuit with a 0.5 H inductor in series with a 2.5 kilohm resistor with 10 volts applied.

17. A circuit has a 250 mH inductor in series with a 50 ohm resistor with 10 volts applied. Use the exponential formula to find the instantaneous discharge current after 7.1 ms if the circuit started fully charged.

18. Use the exponential formula to find the instantaneous charge voltage after 10.2 ms in a circuit with a 10 µF capacitor in series with a 300 ohm resistor with 10 volts applied.

19. Use the exponential formula to find the instantaneous discharge voltage after 0.12 seconds in a circuit with a 15 µF capacitor in series with a two kilohm resistor with 10 volts applied.

20. Plot the output waveform of an RC integrator waveshaping circuit with a 5 ms time constant and 25 ms pulse width.

21. Plot the output waveform of an RL integrator waveshaping circuit with a 2 ms time constant and 25 ms pulse width.

22. Plot the output waveform of an RC differentiator waveshaping circuit with a 5 ms time constant and 25 ms pulse width.

23. Plot the output waveform of an RL differentiator waveshaping circuit with a 2 ms time constant and 25 ms pulse width.

24. When an integrator with a short time constant is used with a square wave, what part of the output waveform is affected the most?

25. When a differentiator with a long time constant is used with a square wave, what part of the output waveform is affected the most?

26. An RL time constant circuit having an inductor with 200 ohms internal resistance has 15 volts applied during the charge time. Calculate the voltage developed across the discharge resistor of 100 kilohms.

27. An RC time constant circuit charges the capacitor to 4.8 volts. What is the current during discharge with a load resistor of two ohms?

Electrical waves, like sound waves, come in an infinite variety. (Roland)

Chapter 15
AC Waveforms

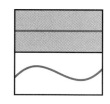

Upon completion of this chapter, you will be able to:
- Distinguish between dc and ac voltages.
- Describe the characteristics of an ac waveform.
- Analyze square waves: positive, negative, and symmetrical.
- Calculate the four basic methods of measuring voltage.
- Calculate the period and frequency of ac waveforms.
- Explain how a sine wave is produced from a generator.
- Calculate the instantaneous voltage values along a sine wave.
- Recognize phase shift, harmonic frequencies, and other characteristics of ac waveforms.
- Analyze basic ac circuits with resistive loads.
- Use an oscilloscope to observe and measure an ac waveform.

Electricity is found in many forms. It falls into three general groupings: dc, ac, and random signals. DC voltages, such as those produced by batteries, have a constant polarity. AC voltage sources vary in polarity in a fixed periodic pattern. Random noise signals have no set pattern.

DC voltages (direct current) are defined as having a continuous polarity. This means the voltage is always either positive or negative. The current from a dc voltage source continually flows in the same direction.

AC voltages (alternating current) are defined as changing polarity. The current periodically alternates its direction of flow. AC voltages repeatedly alternate from a maximum positive value, to a maximum negative value, and back to positive. The cycle then repeats.

15.1 WAVEFORMS

A **waveform** is the shape of a voltage over a period of time. Waveforms are usually thought of as the picture that is obtained if the voltage is viewed with an oscilloscope.

Waveform Characteristics

The characteristics of waveforms fall into two groups, voltage and time.

Voltage describes the vertical (up and down) direction of the waveform. Keep in mind, when voltage is applied to a circuit, current flows and power is produced. With purely resistive circuits, current and power both have the *same* waveform as the voltage, except with a difference in magnitude. For simplicity, this chapter describes the waveforms only in terms of voltage, rather than including current and power.

Time is the horizontal (side-to-side) direction. On waveform drawings, time is from left to right. Terminology related to time are the words period, cycle, and frequency.

Voltage terminology

The **amplitude** of a waveform is the height. Amplitude is described as either peak or peak-to-peak. **Peak** is measured from zero to either the maximum positive value or the maximum negative. **Peak-to-peak** is measured from the maximum positive value to the maximum negative value.

If the voltage is from zero to the maximum positive, it is the positive peak. From zero to the maximum negative is the negative peak. **A symmetrical waveform** has equal values of positive and negative voltages. With a symmetrical waveform, the peak-to-peak value is equal to twice the peak value.

Time terminology

A **cycle** is from one point on a waveform to the point at which it repeats itself. The time to complete one cycle has two terms that are inversely related. These terms, period and frequency, describe the on and off switching.

Period is the length of time required for the waveform to complete one cycle. The unit of measure is seconds, or fractions of a second, such as millisecond or microsecond. **Frequency** is the number of cycles that take place in one second. The unit of measure of frequency is **hertz (Hz).** Frequencies are frequently stated in kilohertz (kHz), megahertz (MHz), and gigahertz (GHz). **Cycles per second (cps)** is an older unit of measure for frequency, generally considered obsolete. Frequency is not expressed using unit multipliers smaller than one, such as milli-, micro-, and nano.

15.2 SQUARE WAVES

Square waves were first presented in Chapter 14 as a means of switching a dc voltage on and off. Square waves are classified as ac waveforms, because they switch on and off at a periodic rate. However, if the waveform does not change *polarity,* it is technically a dc voltage, rather than ac. For the purposes of this chapter however, square waves are considered ac due to their switching characteristics.

Positive Square Waves

Figure 15-1 is an all positive square wave. It is a dc voltage switching from zero volts to +15 volts and back to zero. It is symmetrical, with the time duration at +15 volts equal to the time duration at zero volts. The peak voltage is +15. It would not be proper to refer to it as a peak-to-peak voltage since there is no negative portion.

One cycle in figure 15-1 can be identified between points a and b. At point a, the voltage is switching from zero toward +15. The cycle ends at the next point where the

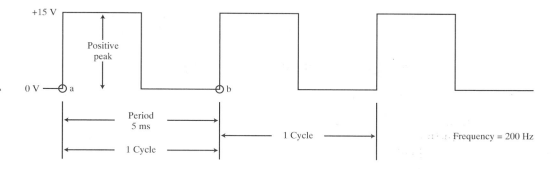

Figure 15-1. This square wave is produced by a positive dc voltage switching on and off at a periodic rate.

voltage repeats the pattern, at point b. A period of five ms is the time it takes to complete the cycle. Frequency can be calculated by taking the reciprocal of the period.

Negative Square Waves

A negative square wave switches from zero to a peak negative voltage, refer to figure 15-2. The peak is –15 volts. Period is measured by selecting a starting point and following the waveform to the next point where the wave repeats itself. It is important to note the direction of the changing voltage. In this figure, the starting point is at the instant where waveform switches to –15 V. The end point is the next time it switches to –15.

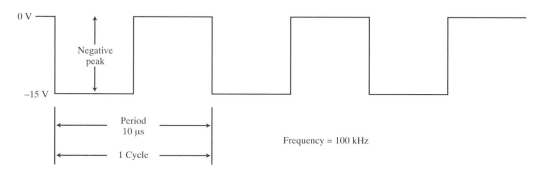

Figure 15-2. This square wave is produced by a negative dc voltage switching on and off at a periodic rate.

Symmetrical Square Wave

A symmetrical waveform has an equal positive and negative voltage. The period of a symmetrical wave also has equal portions of negative and positive. Figure 15-3 is an example of a symmetrical square wave. The peak-to-peak value of this wave can be calculated by adding the negative and positive portions together or by multiplying one peak by two. In equation form:

$$\text{peak-to-peak} = \text{positive peak} + \text{negative peak}$$

or

$$\text{peak-to-peak} = \text{peak} \times 2 \text{ (symmetrical waveforms only)}$$

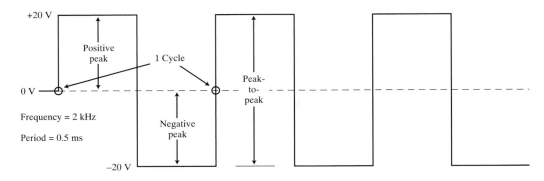

Figure 15-3. Symmetrical waveforms have equal positive and negative voltages. This waveform is 40 V peak-to-peak.

Sample problem 1. _____

Determine the peak-to-peak voltage of the waveform in figure 15-3.

    Formula: p-to-p = peak × 2

    Substitution: p-to-p = 20 V × 2

    Answer: p-to-p = 40 V

15.3 INVERSE RELATIONSHIP OF PERIOD AND FREQUENCY

An inverse relationship means when a number on one side of the equation gets smaller, the number on the other side of the equation gets larger. Period and frequency are inversely related. Therefore, as the period decreases in time, the frequency increases. As the frequency decreases, the period of one cycle increases. This is shown in figure 15-4. Period and frequency are calculated by taking the reciprocal of each other.

THE SHORTER THE CYCLE THE HIGHER THE FREQUENCY

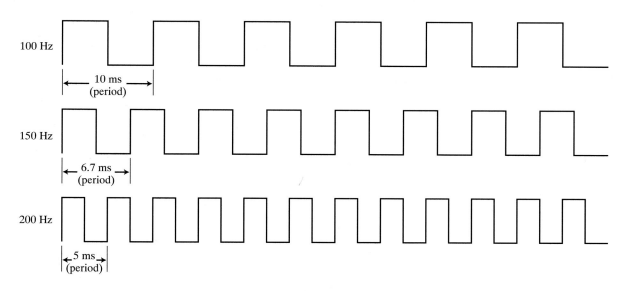

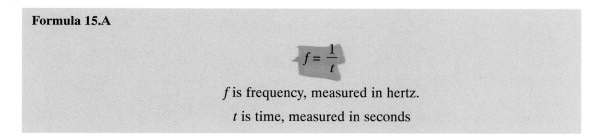

Figure 15-4. Higher frequency waves have shorter periods.

The frequency can be found using:

Formula 15.A

$$f = \frac{1}{t}$$

f is frequency, measured in hertz.

t is time, measured in seconds

The period can be found using:

Formula 15.B

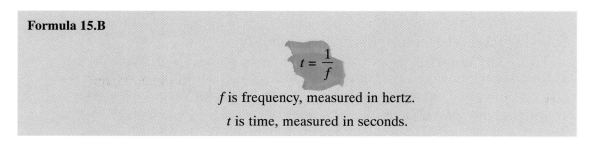

$$t = \frac{1}{f}$$

f is frequency, measured in hertz.

t is time, measured in seconds.

Sample problem 2. _____

With a period of five ms, find the frequency of the waveform in figure 15-1.

Formula: $f = \dfrac{1}{t}$

Substitution: $f = \dfrac{1}{5 \text{ ms}}$

Answer: $f = 200 \text{ Hz}$

Sample problem 3. _____

Calculate the frequency of the square wave in figure 15-2, using the period of 10 μs.

Formula: $f = \dfrac{1}{t}$

Substitution: $f = \dfrac{1}{10 \text{ μs}}$

Answer: $f = 100 \text{ kHz}$

Sample problem 4. _____

Calculate the period of the symmetrical square wave of figure 15-3, with a frequency of two kHz.

Formula: $t = \dfrac{1}{f}$

Substitution: $t = \dfrac{1}{2 \text{ kHz}}$

Answer: $t = 0.5 \text{ ms}$

15.4 OTHER WAVEFORMS

All waveforms with a pattern that periodically repeats have two common characteristics: their amplitude is measured in peak and/or peak-to-peak, and they have a period of one cycle that is measured in time, allowing calculations of frequency.

Unsymmetrical Rectangular Waveforms

Figure 15-5 is a series of pulses, such as might be found in a digital timing circuit. This waveform is classified as an unsymmetrical waveform. This wave does not have equal positive and negative portions. Frequently, timing circuits allow adjustments to vary the time of the positive and negative portions separately. To determine the period of one cycle, add the positive and negative times together. The waveform has a positive peak voltage of 15 V and a negative peak voltage of 5 V. The peak-to-peak is their sum, 20 V.

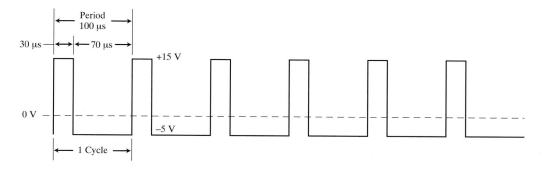

Figure 15-5. Unsymmetrical rectangular waveform. The wave does *not* have equal positive and negative voltages. This pulse has a period of 100 μs and a frequency of 10 kHz.

Ramp Waveforms

The **ramp waveform** in figure 15-6 is also referred to as a **sawtooth wave.** The voltage of this wave is 20 volts peak. With no negative voltage, the peak-to-peak value is equal to the peak value. Keep in mind, waveforms are displayed on an oscilloscope, where the easiest measurements are from peak-to-peak.

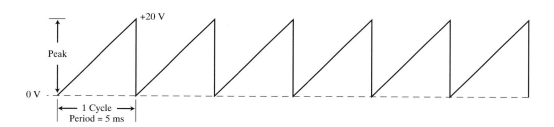

Figure 15-6. Ramp waveform. This wave has a peak of 20 V and a frequency of 200 Hz.

Triangular Waveforms

The **triangular waveform** in figure 15-7 shows the measurements for one cycle taken at the zero center line. The starting point of one cycle can be considered at any point, provided the ending point is taken when the wave repeats. The advantage of using the zero reference line is both convenience and accuracy.

Sine Waves

The **sine wave** is the shape of voltage used in residential electricity. With most electronic equipment being plugged into a wall socket, the sine wave becomes a very familiar shape.

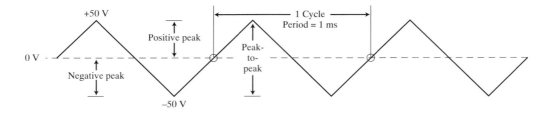

Figure 15-7. Triangular waveform. This wave has a peak-to-peak value of 100 V and a frequency of 1000 Hz.

The sine wave in figure 15-8 is an example of why the best place to measure period is at the center line. That part of the waveform is a straight line. The peaks are curved and make taking an accurate measurement very difficult.

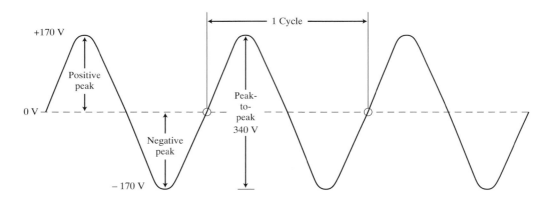

Figure 15-8. A sine wave is the waveform used in residential electricity.

Waveforms from Rectifier Circuits

Figures 15-9 and 15-10 are the output waveforms of rectifier circuits found in electronic power supplies. The full wave, shown in figure 15-9 repeats itself twice as often as the half wave, shown in figure 15-10. The frequency of the full wave is twice that of the half wave.

A full wave rectifier circuit changes the negative portion of a sine wave into a positive voltage. A half wave rectifier circuit chops off the negative portion of a sine wave. In both cases, the voltages are all the same polarity, resulting in what is classified as **pulsating dc.** Development of these waveforms is examined in detail in Chapter 26, Power Supplies.

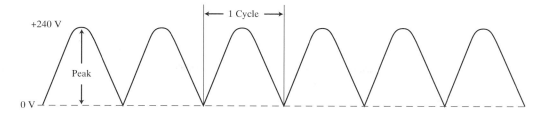

Figure 15-9. Output of a sine wave from a full wave rectifier.

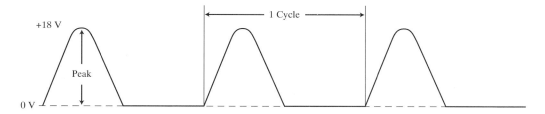

Figure 15-10. Output of a sine wave from a half wave rectifier.

15.5 SINE WAVES

The sine wave is a naturally occurring shape and can be observed in applications other than electricity. For example, the waves of an ocean increase and decrease at a periodic rate in the shape of the sine wave. Light, sound waves, and radio signals travel as sine waves. Other examples can be found in nature.

In electrical applications, the sine wave is the most common type of voltage used. Many electronic circuits use dc voltages that are produced by rectifying a sine wave.

Producing a Sine Wave with an AC Generator

If a wire moves through a magnetic field, or if a wire is stationary and the strength of the magnetic field changes, current is induced in the wire.

Figure 15-11 is a simplified ac generator. In this figure, a single loop of wire rotates through a magnetic field. Slip rings and brushes are used at the ends of the loop to allow the generated voltage to pass to the circuit load. Figure 15-12 traces the loop as it rotates in the generator. Position A is the starting point. When the loop is in line with the lines of flux, there is no voltage produced. The wire must cut through the flux to produce voltage. This point is labeled 0°. This indicates the start of the circle, which will produce one cycle.

Passing from point A to point B, the solid-colored portion of the loop is moving toward the north pole, while the light-colored portion moves toward the south. The voltage is increasing in a positive direction.

At point B, the loop is the closest to the face of the poles. At this point, the voltage is at its maximum value, or peak. The label of 90° indicates 1/4 of the 360° circle is completed.

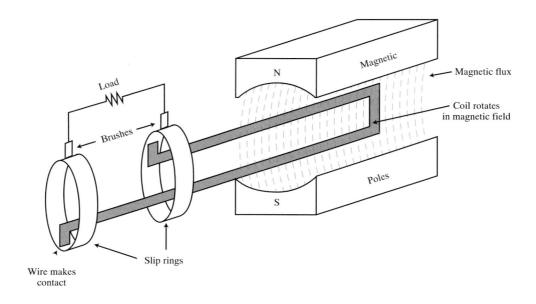

Figure 15-11. AC generator produces a sine wave by rotating a wire loop inside of a magnetic field.

Between points B and C, the voltage is decreasing as the loop moves away from the poles. The voltage reaches zero at a point exactly halfway between the poles. At this point, the loop is opposite of its starting position. Point C is 180° on the circle, which is 1/2 of a complete cycle.

Between points C and D, the solid-colored portion of the wire loop is moving toward the south and the light-colored portion moves toward the north. This produces a negative voltage, because it is opposite the first 1/2 cycle.

At point D, 270°, or 3/4 of a complete circle, the loop reaches its maximum voltage as it passes the center of the poles. The negative peak is equal in amplitude to the positive peak.

Between points D and E, the voltage is decreasing as the loop leaves the poles and reaches a point where the wire does not cross any lines of force.

At point E, the wire loop has completed the full 360° of one complete circle. One cycle is complete and the wire is at its starting point. In a circle, 360° = 0°. The sine wave starts a new cycle as the wire loop continues to rotate.

Factors Affecting Frequency

The voltage developed in an ac generator is determined primarily by the strength of the magnetic field. The frequency, the number of cycles per second, is determined by how often the wire loop passes by the pair of magnetic poles.

With the two-pole generator, shown in figure 15-13, the speed of rotation has been set at 60 rps (revolutions per second) and 120 rps. With a rotation of 60 rps, the sine wave has 60 cps (cycles per second), or 60 Hz. With a rotation speed of 120 rps, the sine wave has 120 cps, or 120 Hz. As indicated, in 1/60 second the faster generator produces twice as many cycles.

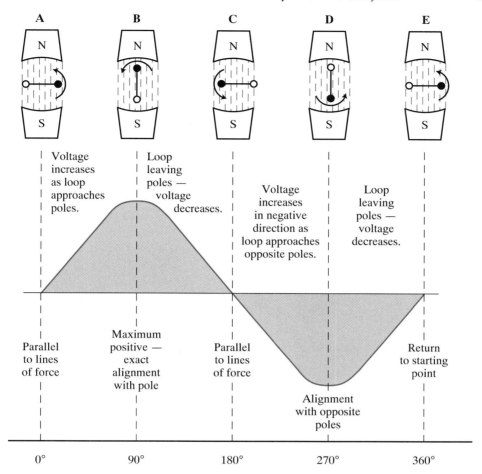

Figure 15-12. Highest voltage is produced when the wire loop crosses perpendicular to the magnetic poles.

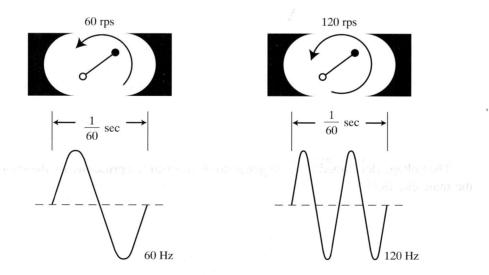

Figure 15-13. A faster turning generator produces a higher frequency.

If more poles are added to the generator, the wire loop will pass more pole pairs in the same amount of time. The four-pole generator in figure 15-14 produces *two* cycles per revolution. If the generator were spinning at 60 rps, the output frequency would be 120 Hz.

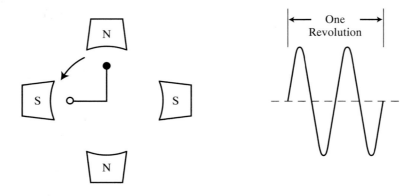

Figure 15-14. A four-pole generator produces two complete cycles per revolution.

Instantaneous Values of a Sine Wave

The **instantaneous value** is a specific point on a sine wave at an instant in time. The sine wave is made up of an infinite number of points, each having an instantaneous value. This value can be calculated using a formula.

Formula 15.C

$$v = V_{max} \times \sin \theta$$

v is instantaneous voltage.

V_{max} is peak voltage.

sin θ is the sine of the angle θ measured in degrees or radians.

Sample problem 5. _____

Find the instantaneous value at 45° of a sine wave with 100 volts peak.

Formula:	$v = V_{max} \times \sin \theta$
Substitution:	$v = 100 \text{ V} \times \sin 45°$
Calculator step:	$\sin 45° = 0.707$
Substitution:	$v = 100 \text{ V} \times 0.707$
Answer:	$v = 70.7 \text{ V}$

Sample problem 6.

Find the angle (in degrees) of an instantaneous voltage of 63.6 on a sine wave having a peak of 100 volts. Use only the angles less than 180°.

Formula: $v = V_{max} \times \sin \theta$

Substitution: $63.6 = 100 \text{ V} \times \sin \theta$

Rearranging: $\sin \theta = 0.636$

Rearranging: $\theta = \sin^{-1} 0.636$

Note: $\sin^{-1}$ means the *"angle whose sine is..."* Use the inverse of sine on a calculator.

Answer: $\theta = 39.5°$

Using instantaneous values to plot a sine wave

The sine wave shown in figure 15-15 was plotted by calculating the instantaneous values every 15 degrees. The results of these calculations are shown in the table in figure 15-16.

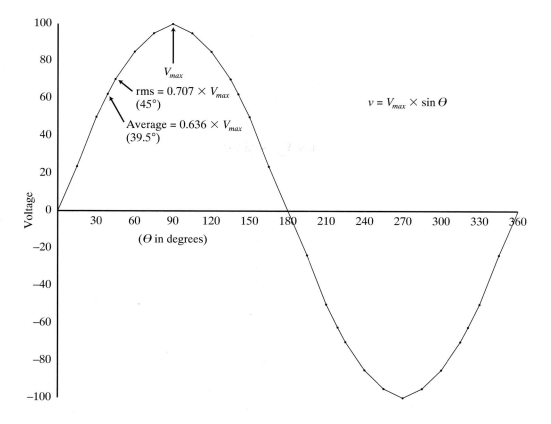

Figure 15-15. Plot of the instantaneous values of a sine wave.

Figure 15-16. Table of the instantaneous values of the sine wave shown in figure 15-15.

Degrees	Voltage	Degrees	Voltage
0	0	180	0
15	25.9	195	−25.9
30	50.0	210	−50.0
45	70.7	225	−70.7
60	86.6	240	−86.6
75	96.6	255	−96.6
90	100.0	270	−100.0
105	96.6	285	−96.6
120	86.6	300	−86.6
135	70.7	315	−70.7
150	50.0	330	−50.0
165	25.9	345	−25.9
180	0	360	0

15.6 ANGULAR MEASURE

The sine wave is generated by a loop of wire rotating in a circle. The distance around the circle, the circumference, can be measured in degrees. A full circle is 360 degrees. The circumference can also be measured in radians, abbreviated rad.

The **radius** is a term from geometry, referring to the distance from the center to the circumference of the circle. The **circumference** is the distance around the circle. The circumference is calculated by the formula: $c = 2 \times \pi \times r$.

A **radian** can be defined as the distance along the circumference of the circle equal to the radius of the circle. Referring to the formula, there are 2π radians in the circumference of a circle. As π is approximately 3.14, one radian equals 57.3 degrees. Refer to figure 15-17.

The instantaneous values along a sine wave can be found using degrees or radians. Certain applications make the use of one or the other more convenient.

Converting Radians and Degrees

The conversion factor between degrees and radians is based on there being 360 degrees or 2π radians in a full circle. Therefore, $360° = 2\pi$ rad. Figure 15-18 is a summary of degrees vs radian measure.

Formula 15.D

$$\text{degrees to radians: rad} = \frac{\pi \text{ rad}}{180°} \times \text{degrees}$$

Formula 15.E

$$\text{radians to degrees: deg} = \frac{180°}{\pi \text{ rad}} \times \text{rad}$$

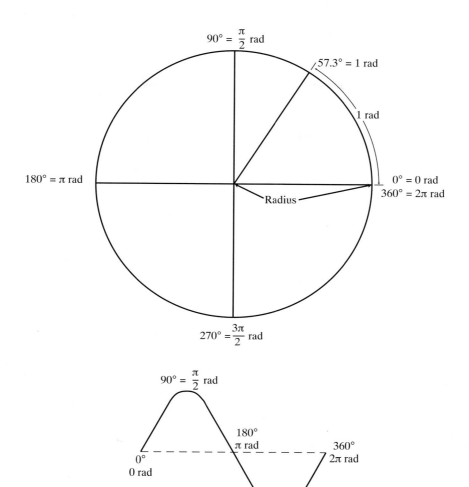

Figure 15-17. Degrees or radians can be used to measure the rotation of a wire to produce a sine wave.

Figure 15-18. Summary of degrees vs radian measure.

Degrees	Radians
0	0
45	$\pi/4$
90	$\pi/2$
135	$3\pi/4$
180	π
225	$5\pi/4$
270	$3\pi/2$
315	$7\pi/4$
360	2π

Sample problem 7. _____

Convert 135 degrees to radians.

Formula: $rad = \dfrac{\pi \; rad}{180°} \times degrees$

Substitution: $rad = \dfrac{\pi \; rad}{180°} \times 135°$

Answer: $\dfrac{3\pi}{4} \; rad$

Note, a fraction of π is the best form for the answer.

_____

Sample problem 8. _____

Convert $7\pi/4$ rad to degrees.

Formula: $deg = \dfrac{180°}{\pi \; rad} \times rad$

Substitution: $deg = \dfrac{180°}{\pi \; rad} \times \dfrac{7\pi}{4}$

Answer: $315°$

_____

15.7 RMS AND AVERAGE VALUES OF SINE WAVES

The useful amount of electrical work performed by a continually changing sine wave can be compared to the steady voltage found in dc voltages. The comparison is made with the mathematical relationships of rms and average.

RMS value stands for **root mean square.** It is the effective value of a sine wave. When a sine wave is applied to a resistive circuit, it produces the same amount of heat as a dc voltage equal to the rms value, as shown in figure 15-19. AC voltmeters read the rms value of a sine wave.

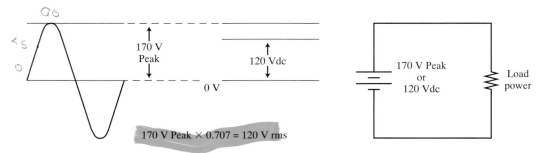

Figure 15-19. The rms value of a sine wave has the same effective power as a dc voltage of the same value.

Average value is determined for the instantaneous values having the same polarity. If both the positive and negative half cycles were to be averaged together, the result would be zero. However, when both half cycles have the same polarity, the equivalent dc voltage is the average value of the sine wave. A full wave rectifier produces this result, as shown in figure 15-20. A dc voltmeter connected across the full wave rectifier reads the average.

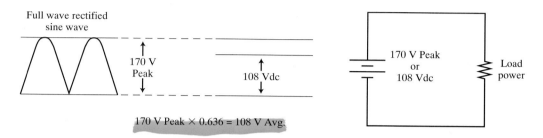

Figure 15-20. The average value of a full wave rectified sine wave is its dc equivalent.

Calculating RMS and Average

To calculate the rms of a sine wave, multiply the peak value by 0.707. This value is derived by squaring the instantaneous values of the sine wave, finding their average (mean), and then taking the square root.

To calculate the average of a sine wave, multiply the peak value by 0.636. This value is derived by taking the average of the instantaneous values of one half cycle. This is also equal to the average of two half cycles with the same polarity. In equation form:

Formula 15.F

$$rms = peak \times 0.707$$

rms is the root mean square value of a wave, measured in volts.

peak is the peak value of a wave, measured involts.

Formula 15.G

$$avg = peak \times 0.636$$

avg is the average value of a wave, measured in volts.

peak is the peak value of a wave, measured in volts.

Figure 15-21 is a summary of results used to derive the factors used to calculate rms and average values. To find the average value: find the sine (sin) of each angle, add all values together, and divide by the number of values. To find the rms value: square the sine of each angle, add all values together, divide by the number of values, then take the square root.

Figure 15-21. Summary of the results used to calculate the rms and average values of a sine wave.

angle ϕ	$(sin\ \phi)^2$	$sin\ \phi$
10°	0.030	0.174
20°	0.117	0.342
30°	0.250	0.500
40°	0.413	0.643
50°	0.587	0.766
60°	0.750	0.866
70°	0.883	0.940
80°	0.970	0.985
90°	1.000	1.000
100°	0.970	0.985
110°	0.883	0.940
120°	0.750	0.866
130°	0.587	0.766
140°	0.413	0.643
150°	0.250	0.500
160°	0.117	0.342
170°	0.030	0.174
180°	0.000	0.000
Addition Total:	9.000	11.432
Average:	$\dfrac{9.000}{18} = 0.5$	$\dfrac{11.432}{18} = 0.636$
Square root:	$\sqrt{0.5} = 0.707$	

Converting Sine Wave Values

The four voltage (or current) values of a sine wave can be converted from one value to the other using the chart in figure 15-22. Use the formulas to first find the peak value. Once the peak is known, use it to find the desired value.

Figure 15-22. Conversion formulas for sine waves.

To Find Peak	*When Peak Is Known*
peak $= \dfrac{\text{peak-to-peak}}{2}$	peak-to-peak $=$ peak $\times$ 2
peak $= \dfrac{\text{rms}}{0.707}$	rms $=$ peak $\times$ 0.707
peak $= \dfrac{\text{average}}{0.636}$	average $=$ peak $\times$ 0.636

Sample problem 9. _____

A certain sine wave has a peak value of 25 volts, find the peak-to-peak, rms, and average values.

1. Find peak-to-peak value:

   Formula: p-to-p = peak × 2

   Substitution: p-to-p = 25 V × 2

   Answer: p-to-p value = 50 V

2. Find rms value:

   Formula: rms = peak × 0.707

   Substitution: rms = 25 V × 0.707

   Answer: rms value = 17.7 V

3. Find average value:

   Formula: average = peak × 0.636

   Substitution: average = 25 V × 0.636

   Answer: average value = 15.9 V

Sample problem 10. _____

A sine wave is measured on an oscilloscope to have a peak-to-peak value of 80 volts. Find the rms value.

1. Convert to peak value:

   Formula: $peak = \dfrac{p\text{-to-p}}{2}$

   Substitution: $peak = \dfrac{80\ V}{2}$

   Answer: peak value = 40 V

2. Find rms value:

   Formula: rms = peak × 0.707

   Substitution: rms = 40 V × 0.707

   Answer: rms value = 28.3 V

Sample problem 11. _____

An ac voltmeter reads 120 volts rms. What is the average value?

1. Convert to peak value:

   Formula: $peak = \dfrac{rms}{0.707}$

   Substitution: $peak = \dfrac{120\ V}{0.707}$

   Answer: peak value = 170 V

2. Find average value:

   Formula: average = peak × 0.636

   Substitution: average = 170 V × 0.636

   Answer: average value = 108 V

15.8 PHASE SHIFT

When two waveforms of the same frequency appear together, it is possible to have a difference in time between their starting points. **Phase shift** is the difference in time, measured in degrees of the sine wave, ranging from 0° to 180°. Figure 15-23 shows four pairs of sine waves, each with different phase shifts.

In an actual circuit, one of the waveforms is selected as the reference. For example, the input to a circuit can be the reference, and the output of the circuit can be compared to it. When comparing a signal to a reference, there can be a leading or lagging phase shift. A **leading phase shift** results from the waveform appearing prior to the reference. A **lagging phase shift** results from the waveform appearing after the reference. Phase shift does not depend on the voltages of either waveform. Phase shift depends entirely on the relationship of one waveform to the other.

Examples of when phase shift is an important consideration can be found in Chapters 18, 19, 25, and 27. In Chapters 18 and 19, the voltage drop across a resistor is compared to the voltage drop across an inductor or capacitor. Inductors and capacitors cause shifts in phase. In Chapter 25, the voltage produced by a generator creates different sine waves at different phases. In Chapter 29 the output of a transistor amplifier is compared to the input signal.

15.9 HARMONIC FREQUENCIES

A **harmonic frequency** is a multiple of a sine wave's fundamental frequency. For example, the second harmonic is twice the fundamental frequency. The third harmonic is three times the fundamental frequency.

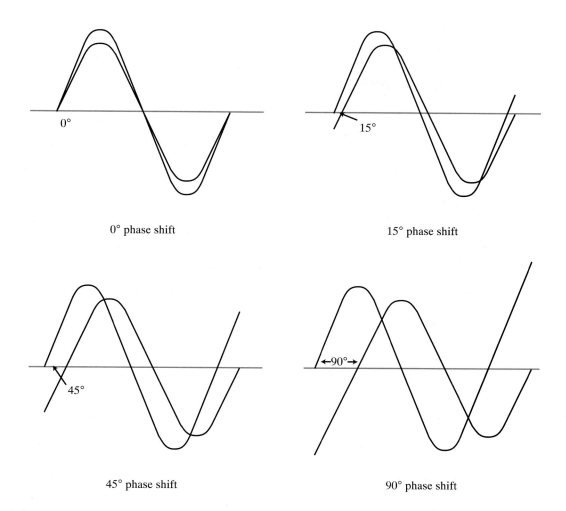

0° phase shift

15° phase shift

45° phase shift

90° phase shift

Figure 15-23. Series of increasing phase shifts.

In certain situations, harmonic frequencies are undesirable and filters are used to eliminate them. Radio transmitters, for example, can be mistuned and result in transmission of harmonic frequencies. These harmonics must be filtered or they can interfere with other radio signals.

Harmonic frequencies can also be used to produce desirable results. They can be used to produce a square wave, as demonstrated in figure 15-24. If a square wave of 1000 Hz were needed, the fundamental frequency of the sine wave would be 1000 Hz, as shown in figure A. Odd harmonics are added to the fundamental. This produces an approximation of a square wave.

Figure B shows the addition of the third harmonic. The sum of the first and third harmonics makes a more square shape. Though, the waveform has a pronounced dip in the center.

Figure C adds the third and fifth harmonics to the fundamental, resulting an even more square wave. The ripples across the top and bottom of the wave are decreased.

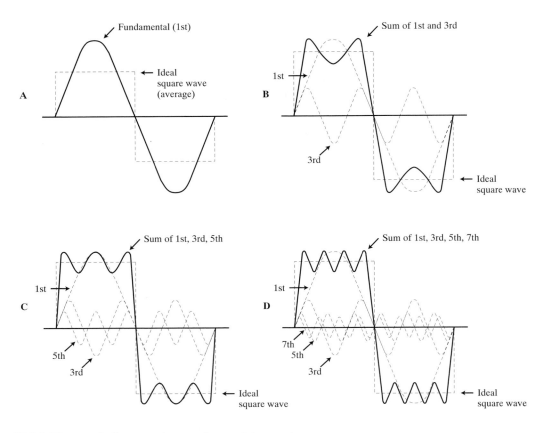

Figure 15-24. Harmonic frequencies can be used to create a square wave. Shown here, the fundamental frequency and the first three odd harmonics have been added to approximate a square wave.

Figure D includes the seventh harmonic. The addition of more harmonic frequencies produces a better square wave. Notice that each successive harmonic has a much lower amplitude than the previous one. The further the frequency is from the fundamental, the smaller its amplitude will be.

15.10 ALTERNATING CURRENT IN RESISTIVE CIRCUITS

An ac voltage applied to a resistive circuit follows all of the basic circuit laws that apply to dc circuits. To calculate current, power, and voltage drops, use Ohm's law and the power formulas.

AC voltages can be described using four different values: peak-to-peak, peak, rms, and average. These same four amplitude values also apply to current and power. Use caution when making calculations. Make sure that all circuit relationships use the same amplitude value.

Notice that the letter symbols used to represent some circuit quantities are lower-case. This is used to show that these values are instantaneous values of an ac circuit: voltage (*v*), current (*i*), power (*p*). Uppercase letters are often used when describing an ac waveform as a whole, since it is no longer an instantaneous value. AC values such as average, rms, and peak-to-peak are not instantaneous values and thus also use upper-case representation. Peak ac values are often shown both ways, depending on how the value is being used. Resistors use an uppercase *R* because the value of a resistor does not change in a circuit.

Current Alternates Direction

With the voltage in an ac circuit periodically changing from positive to negative, the current in the circuit also changes direction. Figure 15-25 examines the positive half cycle, the negative half cycle, and the combined sine wave.

During the positive half cycle, the sine wave generator is assigned a polarity of positive on the top and negative on the bottom. As with a dc voltage, current flows from

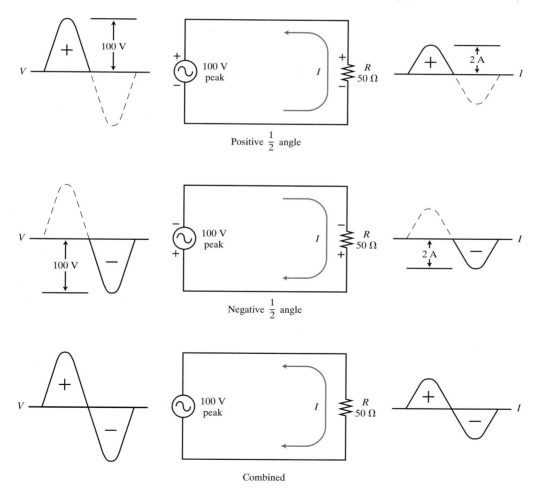

Figure 15-25. AC current flows in both directions.

negative to positive through the load resistor. The voltage applied to this circuit is 100 volts peak. Using Ohm's law ($i = v/R$) the current is calculated to be two amps peak. The current waveform is the same shape as the voltage, with its amplitude adjusted to the proper height.

During the negative half cycle, the sine wave generator is assigned a polarity of negative on the top and positive on the bottom. The negative half cycle performs the same as the positive half cycle. The only difference is that the current travels in the opposite direction.

The direction of the current has no effect on the performance of a purely resistive circuit. If the load is a light bulb, for example, the light produced is the same as a dc voltage equal to the rms value of the sine wave.

With the voltage from a ac waveform periodically hitting zero, you might expect a light bulb to flicker when plugged into an outlet. Yet, the light does not flicker on and off as a typical sine wave swings through its cycle from maximum positive, through zero, to maximum negative. This is because the light produced is a result of heat. The heat from the filament cannot change at the rate of a typical ac sine wave. Keep in mind, residential electricity has a frequency of 60 cycles per second. The period is 16.7 milliseconds. Each half cycle is on for 8.33 milliseconds. Changing this fast does not allow the filament to cool off.

AC Power in a Resistive Circuit

Power developed across a resistive load in an ac circuit is exactly in phase with the voltage and current. To calculate power, use the power formulas as with a dc circuit. *The amplitude values used for voltage and current must also be used to express the unit of measure for power.*

Figure 15-26 shows an ac circuit with a single resistive load. The voltage applied to this circuit is 50 volts peak. With a 200 ohm load, the current is calculated to be 0.25

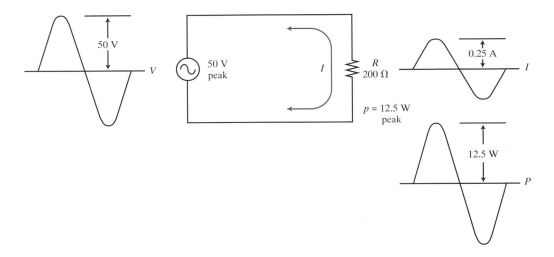

Figure 15-26. AC power in a resistive circuit is in phase with voltage and current.

amps peak. Power is calculated as 12.5 watts peak. The phase relationships of power, current, and voltage in this resistive circuit are all the same.

Series Resistance AC Circuits

Solving a series circuit uses the same circuit laws for dc and ac circuits. You can review series dc circuits in Chapter 6. Figure 15-27 is the series circuit used in sample problem 12. The waveforms of input voltage, current, and voltage drops are all in phase. The amplitudes of the waveforms are a result of the values of circuit components.

All values in this circuit are given in rms. To show the amplitude of the waveforms, rms is converted to peak.

Sample problem 12. _____

With the circuit shown in figure 15-27, calculate current and voltage drops. Convert each value to peak.

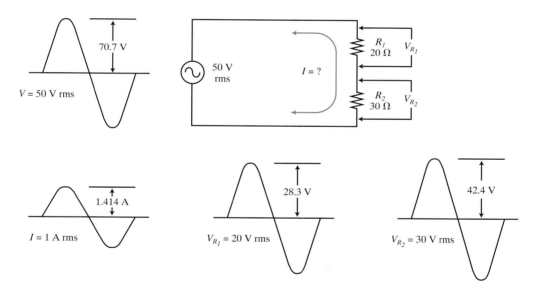

Figure 15-27. Basic circuit laws are used with a series resistive ac circuit just as in dc circuits. (Sample problem 12)

1. Total resistance, found using the same formula as with dc circuits:

    Formula: $R_T = R_1 + R_2$

    Substitution: $R_T = 20\ \Omega + 30\ \Omega$

    Answer: $R_T = 50\ \Omega$

2. Use Ohm's law to calculate current:

    Formula: $I = \dfrac{V}{R}$

Substitution: $I = \dfrac{50\text{ V rms}}{50\ \Omega}$

Answer: $I = 1$ amp rms

3. Voltage drop across R_1:

Formula: $V_{R_1} = I \times R_1$

Substitution: $V_{R_1} = 1\text{ A} \times 20\ \Omega$

Answer: $V_{R_1} = 20$ V rms

4. Voltage drop across R_2:

Formula: $V_{R_2} = I \times R_2$

Substitution: $V_{R_2} = 1\text{ A} \times 30\ \Omega$

Answer: $V_{R_2} = 30$ V rms

5. Convert rms to peak:

Formula: $\text{peak} = \dfrac{\text{rms}}{0.707}$

Substitutions and answers:

$$I\text{ peak} = \dfrac{1\text{ A rms}}{0.707}$$

$$I = 1.414\text{ A peak}$$

$$V_{R_1}\text{ peak} = \dfrac{20\text{ V rms}}{0.707}$$

$$V_{R_1} = 28.3\text{ V peak}$$

$$V_{R_2}\text{ peak} = \dfrac{30\text{ V rms}}{0.707}$$

$$V_{R_2} = 42.4\text{ V peak}$$

Parallel Resistance AC Circuits

Parallel circuits require calculating branch currents. Solving a parallel circuit, like solving a series circuit, uses the same circuit laws for dc and ac. Parallel dc circuits are covered in Chapter 7. Figure 15-28 is the circuit used for sample problem 13. The relationships of each current value can be seen by the amplitudes of the waveforms. Total current is the sum of the individual branch currents.

All values in this circuit are given in peak-to-peak. The calculations are performed in peak-to-peak and their respective waveforms are also shown in this form.

Sample problem 13. _____

Calculate the branch currents and total current of the circuit shown in figure 15-28.

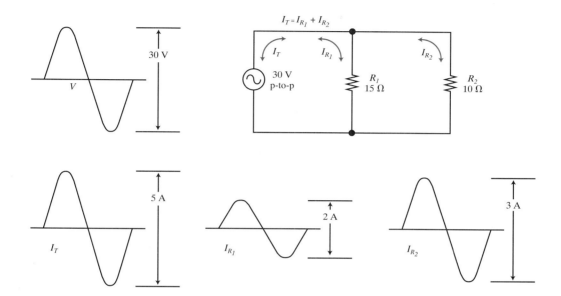

Figure 15-28. Basic circuit laws are used to solve parallel ac resistive circuits just as in dc circuits. (Sample problem 13)

1. Branch currents:

   Formula: $I = \dfrac{V}{R}$

   Substitutions and answers:

   $$I_{R_1} = \frac{30 \text{ V p-to-p}}{15 \ \Omega}$$

   $$I_{R_1} = 2 \text{ A p-to-p}$$

   $$I_{R_2} = \frac{30 \text{ V p-to-p}}{10 \ \Omega}$$

   $$I_{R_2} = 3 \text{ A p-to-p}$$

2. Total current:

   Formula: $I_T = I_{R_1} + I_{R_2}$

Substitution: $I_T = 2\,A + 3\,A$

Answer: $I_T = 5\,A$ p-to-p

15.11 OSCILLOSCOPE MEASUREMENTS OF AC WAVEFORMS

In Chapter 3, the oscilloscope was introduced as a tool to measure dc voltages. The procedures for reading the oscilloscope screen were also introduced. Whenever appropriate, the oscilloscope is also included in the lab activities. Continue to practice with this important instrument.

Review of the Oscilloscope Screen

A division on the oscilloscope is one major block. Each division has four small subdivision markings. These parts of a division can be labeled 0.2, 0.4, 0.6, and 0.8. This allows the screen to be read to one decimal place.

Voltage is read in the vertical direction. To find the voltage of a waveform, count the number of divisions from a reference point to the waveform. Multiply the number of divisions times the volts per division setting.

Time is measured from left to right. To find the time, count the number of divisions from a reference point to the next point of measurement. Multiply the number of divisions times the time per division setting.

Sample problem 14.

Find the peak-to-peak voltage and the period of the square wave in figure 15-29.

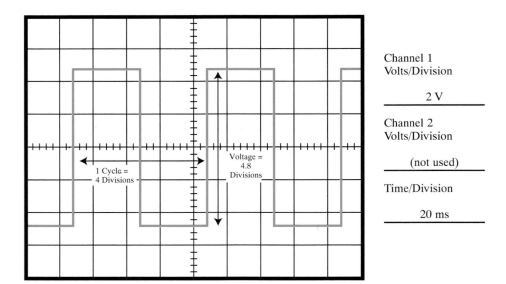

Figure 15-29. Find the peak-to-peak voltage and the period. (Sample problem 14.)

1. Voltage:

   Number of divisions: 4.8

   Volts/division: 2 V

   Peak-to-peak: 9.6 V

2. Period (time of one cycle):

   Number of divisions: 4

   Time/division: 20 ms

   Period: 80 ms

Sample problem 15. _____

Find the peak-to-peak voltage and the period of the square wave in figure 15-30.

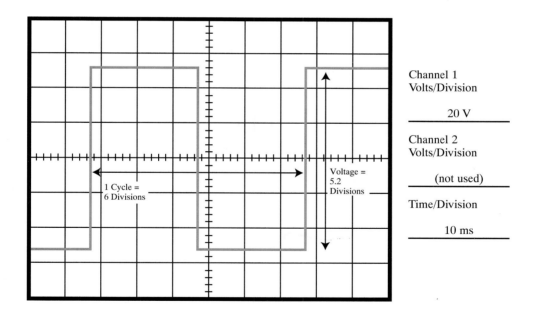

Figure 15-30. Find the peak-to-peak voltage and the period. (Sample problem 15)

1. Voltage:

   Number of divisions: 5.2

   Volts/division: 20 V

   Peak-to-peak: 104 V

2. Period (time of one cycle):

Number of divisions: 6

Time/division: 10 ms

Period: 60 ms

Sample problem 16. _____

Find the peak-to-peak voltage and the period of the ac waveform in figure 15-31.

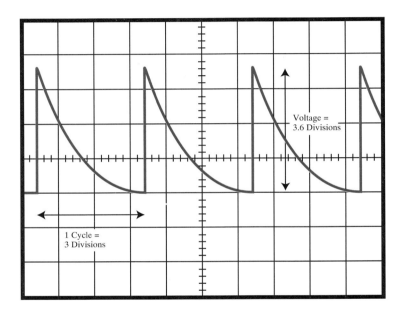

Channel 1
Volts/Division

_____ 10 V

Channel 2
Volts/Division

_____ (not used)

Time/Division

_____ 100 ms

**Figure 15-31. Find the peak-to-peak voltage and the period.
(Sample problem 16)**

1. Voltage:

Number of divisions: 3.6

Volts/division: 10 V

Peak-to-peak: 36 V

2. Period (time of one cycle):

Number of divisions: 3

Time/division: 100 ms

Period: 300 ms

Sample problem 17. _____

Find the peak-to-peak voltage and the period of the ac waveform in figure 15-32.

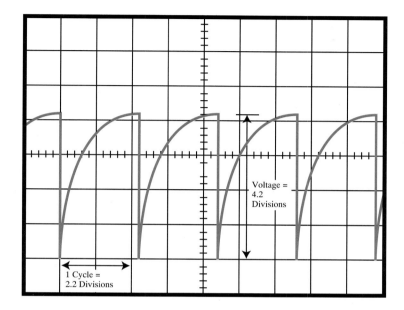

Channel 1
Volts/Division

50 mV
_____

Channel 2
Volts/Division

(not used)
_____

Time/Division

10 μs
_____

Voltage =
4.2
Divisions

1 Cycle =
2.2 Divisions

**Figure 15-32. Find the peak-to-peak voltage and the period.
(Sample problem 17)**

1. Voltage:

   Number of divisions: 4.2

   Volts/division: 50 mV

   Peak-to-peak: 210 mV

2. Period (time of one cycle):

   Number of divisions: 2.2

   Time/division: 10 μs

   Period: 22 μs

SUMMARY

- AC waveforms have a changing voltage with a pattern that repeats itself periodically.
- The frequency of a waveform is the number of cycles in one second.
- A square wave is an ac waveform with the effect of a dc voltage switching on and off or switching from a maximum to a minimum value.
- Square waves can be formed by the combined harmonic frequencies of a sine wave.
- The period, cycle, and frequency are all measured the same way regardless of the type of waveform.
- The sine wave is the waveform produced by generators and is used in standard household electricity.
- RMS and the average value are easily calculated for sine waves.
- When two sine waves with the same frequency are compared, they can have a phase shift ranging from 0° to 360°. The phase shift is the difference in the starting and ending points of the two waveforms.
- When using an oscilloscope to make measurements of voltage or time, count the number of divisions and multiply by the setting of the volts/division switch or the time/division switch respectively.

KEY WORDS AND TERMS GLOSSARY

ac voltage: A voltage that changes polarity.

amplitude: The strength of a signal, measured in either voltage, current, or power. When viewed on an oscilloscope, it is the height of the waveform and is measured as either peak or peak-to-peak.

average value: In a sine wave, the average is calculated by multiplying 0.636 times the peak. A half-wave rectified sine wave has a dc voltage equal to one half the average value of the sine wave. A full-wave rectifier produces a dc voltage equal to the average.

circumference: The distance around a circle.

cycle: On an ac waveform, from one point to the point where the waveform next repeats itself.

cycles per second (cps): An obsolete unit of measure for frequency.

dc voltage: A voltage that has a continuous polarity.

frequency: The number of cycles in one second. The unit of measure is hertz (Hz) or a multiple such as kilohertz (kHz), and megahertz (MHz). Cycles per second, cps, is an older unit of measure, generally considered obsolete.

harmonic frequency: A multiple of a sine wave's fundamental frequency. The second harmonic is twice the fundamental frequency and the third harmonic is three times the fundamental frequency.

hertz (Hz): The unit measure of frequency.

instantaneous value: Any point on a sine wave at an instant in time. The sine wave is made up of an infinite series of points, each having an instantaneous value.

lagging phase shift: A phase shift resulting from the waveform appearing after the reference.

leading phase shift: A phase shift resulting from the waveform appearing prior to the reference.

peak: A measurement of an ac waveform from zero to either the maximum positive value or the maximum negative value.

peak-to-peak: A measurement of an ac waveform from the maximum positive value to the maximum negative value.

period: The length of time required for an ac waveform to complete one cycle. Unit of measure is seconds or part of a second, such as a millisecond or microsecond.

phase shift: The difference in time between two waveforms. Phase shift is measured in electrical degrees ranging from 0° to 180°.

radian: A term from geometry referring to the distance from the center to the circumference of the circle. The circumference is the distance around the circle and can be calculated by the formula: $c = 2 \times \pi \times r$. Therefore, there are 2π radians forming the radius of a circle. A radian can be defined as the distance along the circumference of the circle equal to the radius of the circle. One radian equals 57.3°.

radius: Distance from the center to the circumference of a circle.

ramp waveform: A type of repetitive waveform. See figure 15-6. Also called a sawtooth wave.

rms value: Stands for *r*oot *m*ean *s*quare. The effective value of a sine wave. When a sine wave is applied to a resistive circuit, it produces the same amount of heat as a dc voltage equal to the rms. AC voltmeters read the rms value of a sine wave.

sawtooth wave: A type of repetitive waveform. See figure 15-6. Also called a ramp waveform.

sine wave: A periodic alternating waveform. It is the shape of the voltage used in residential electricity.

symmetrical waveform: A waveform with equal values of positive and negative voltages. With a symmetrical waveform, the peak-to-peak value is equal to twice the peak value.

triangular wave: A type of repetitive waveform. See figure 15-7.

waveform: The shape of the voltage over a period of time. Waveforms are usually thought of as the picture that would be obtained if the voltage were viewed with an oscilloscope.

KEY FORMULAS

Formula 15.A

$$f = \frac{1}{t}$$

Formula 15.B

$$t = \frac{1}{f}$$

Formula 15.C

$$v = V_{max} \times \sin \theta$$

Formula 15.D

$$rad = \frac{\pi \, rad}{180°} \times degrees$$

Formula 15.E

$$deg = \frac{180°}{\pi \, rad} \times radians$$

Formula 15.F

$$rms = peak \times 0.707$$

Formula 15.G

$$avg = peak \times 0.636$$

TEST YOUR KNOWLEDGE

Do not write in this text. Please use a separate sheet of paper.
1. Define a waveform. SHAPE OF VOLTAGE OVER AMOUNT OF TIME
2. What are two general characteristics of an ac waveform?
3. The amplitude of a waveform is measured from where to where?
4. One cycle of a waveform is measured from where to where?
5. What is the unit of measure of period?
6. For a symmetrical waveform, what is the relationship between peak and peak-to-peak? P TO P IS TWICE THAT OF PEAK
7. If a symmetrical waveform has a peak value of 20 volts, what is the peak-to-peak voltage? 40 v

8. What is the peak value for a symmetrical waveform having a peak-to-peak voltage of 100 volts?

9. What is the formula to calculate the period when frequency is known?

10. Calculate the period of a 1500 Hz ac signal.

11. What is the formula to calculate frequency when the period is known?

12. Calculate the frequency of a signal having a period of 4 ms.

13. What is the formula to calculate the instantaneous values of a sine wave?

14. Calculate the instantaneous value of a sine wave at 60 degrees with a peak value of 100 volts.

15. At how many electrical degrees will the instantaneous value of 50 volts occur on a sine wave with a peak value of 100 volts?

16. Convert $3\pi/2$ rad to degrees.

17. Convert 45 degrees to radians.

18. When an ac voltmeter is used to measure a sine wave voltage, will it measure peak, peak-to-peak, rms, or average?

19. For a sine wave with a peak value of 40 volts, calculate peak-to-peak, rms, and average.

20. For a sine wave with a peak-to-peak value of 240 volts, calculate peak, rms, and average.

21. For a sine wave with an rms value of 30 volts, calculate peak, peak-to-peak, and average.

22. For a sine wave with an average value of 50 volts, calculate peak, peak-to-peak, and rms.

23. Refer to the sine wave in figure 15-33. Use the volts per division to find: peak-to-peak, peak, rms, and average. Also, use the time per division to find period and frequency.

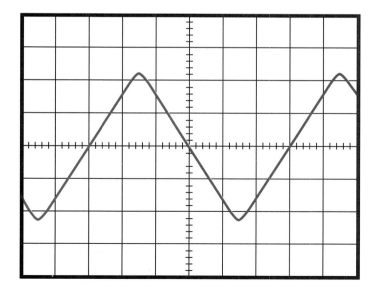
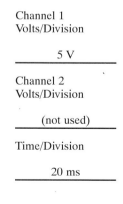

Channel 1
Volts/Division

5 V

Channel 2
Volts/Division

(not used)

Time/Division

20 ms

Figure 15-33. Problem #23.

24. Refer to the sine wave in figure 15-34 represented by channel 1. Use the volts per division to find: peak-to-peak, peak, rms, and average. Also, use the time per division to find period and frequency.

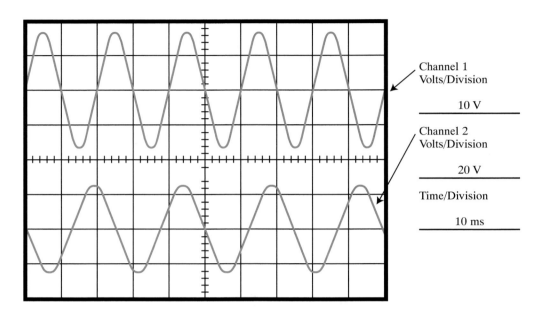

Figure 15-34. Problems #24 and #25.

25. Refer to the sine wave in figure 15-34 represented by channel 2. Use the volts per division to find: peak-to-peak, peak, rms, and average. Also, use the time per division to find period and frequency.

Chapter 16
Transformers

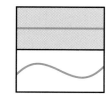

Upon completion of this chapter, you will be able to:
* Define technical terms related to transformers.
* Describe how a transformer is constructed.
* Describe how magnetism is used to pass voltage from one coil to another.
* Perform calculations using the turns ratio, voltage ratio, and current ratio.
* Perform calculations of transformer power and current.
* Describe transformer losses and ways to reduce these losses.
* Calculate transformer efficiency.
* Calculate the reflected resistance of a transformer and perform impedance matching.
* Recognize the effects of transformer loading.
* Use a multimeter to troubleshoot a defective transformer.

A **transformer** uses magnetism to link one coil of wire to another. The magnetic coupling of coils can produce several different results. It can produce an increase in voltage (with a corresponding decrease in current), a decrease in voltage (with a corresponding increase in current), or equal voltages from the input to the output.

Transformers have a wide range of applications. One high-power application is where the power companies use transformers to increase the voltage on transmission lines. This decreases the power lost during transmission. The power company then decreases the voltage for use by the consumer.

A low-power application of transformers is in an ac adaptor used to power small electronic devices. The transformer reduces the 120 volts coming from an outlet to a much lower voltage. Nine volts is common.

Transformers can also be found in computers, televisions, microwave ovens, and stereos. Almost all electrical appliances use a transformer to convert the 120 volts from a wall outlet to a lower voltage for use in the equipment.

Transformers are also used to match the resistance of a load to the internal resistance of the power source. This produces a maximum transfer of power.

16.1 MUTUAL INDUCTANCE

When current flows through a conductor, a magnetic field is developed. In addition, if a conductor is exposed to a changing magnetic field, current is induced. Other properties of inductance are discussed in Chapter 12.

The effects of mutual inductance are shown in figure 16-1. The coils are positioned to allow sharing of the same magnetic field. An iron core concentrates the magnetic flux and improves coupling between the primary and secondary coils. To convert the magnetic energy into electrical current, a continually changing magnetic field is required. A sine wave applied to the primary coil produces the best results for inducing a voltage in the secondary coil.

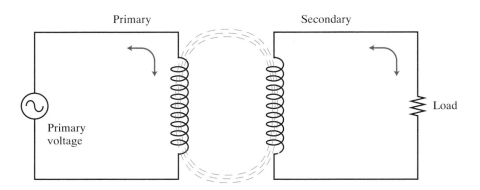

Figure 16-1. Mutual inductance results in transformer action.

16.2 TRANSFORMER CONSTRUCTION

The input of a transformer is called the **primary windings.** Voltage is applied to the primary windings, and current moves through the coil. This produces a magnetic field. The amount of current in the primary windings is determined by a combination of the resistance of the primary windings and the secondary current.

The **secondary windings** make up the output of the transformer. The magnetic field developed in the primary windings causes current to flow in the secondary windings. This current supplies the load.

Transformer Core

A core serves two purposes. First, a core holds the coils of wire in a firm position. Second, a core maximizes the magnetic coupling between the primary and secondary windings.

Figure 16-2 shows selected types of core styles. The open core is frequently made using a stiff cardboard. The windings are sometimes wound on top of each other. Different wiring patterns can be used. This type of transformer is used in radio frequency (RF) applications. These open core transformers are made to be tunable. The transformers are tuned by moving a ferrite core inside the cardboard cylinder. This changes the frequency response of the transformer.

The iron cores shown in figure 16-2 are used in applications such as power transformers, audio circuits, and isolation transformers. The iron core allows a large number

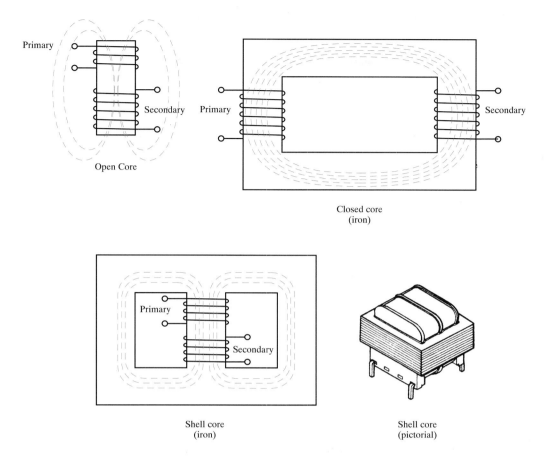

Figure 16-2. In transformer construction the primary and secondary coils are wound on the same core.

of windings to be wound on opposite sides of the core, rather than the windings being placed on top of each other.

Input/Output Phase Relationships

The primary windings on a transformer are the input. The secondary windings are the output. The phase relationship between the windings can be either 0°, exactly in phase, or 180°, the opposite phase relation. The input signal, applied to the primary windings, is the reference signal. The signal coming from the secondary windings is compared to it.

Figure 16-3 shows a transformer with the windings wrapped around the core in such a way as to cause a 0° phase shift. The drawing allows the use of the left hand rule (Chapter 11) to compare the direction of flux to the direction of the current. In this drawing, the polarity of the primary and secondary windings are both the same. This results in the 0° phase shift. The schematic symbol to the right shows the use of phasing dots to indicate this phase relationship.

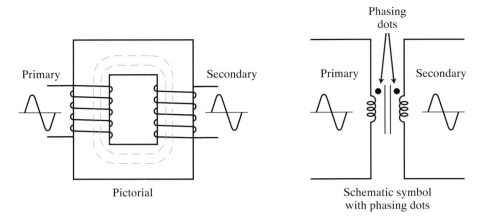

Figure 16-3. Transformer voltages are in phase when the windings have the same relationship to the magnetic field.

Figure 16-4 shows a transformer with the windings wound to produce a 180° phase shift. Compare direction of windings in this drawing. Notice that they are wound in opposite directions. The phasing dots in the schematic symbol show the opposite phase relation.

Do not confuse a 0° or 180° phase shift with series aiding and opposing fields. In the case of a transformer, there is only *one* magnetic field. Therefore, there is nothing to aid or oppose. Also, the phase relationship does not affect the amount of voltage produced. The need to understand the phase relationship comes in use when working with an oscilloscope, comparing input and output voltages.

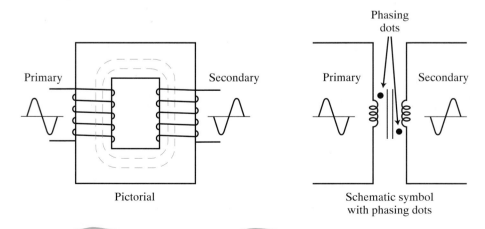

Figure 16-4. Transformer voltages are 180° out of phase when the windings are wound in opposite directions.

16.3 VARIATIONS OF TRANSFORMERS

All transformers change electrical energy into magnetism in the primary windings, where the input voltage is applied. This magnetism induces a voltage in the secondary coil, where a load can be connected.

Basic Schematic Symbols

The schematic symbols for transformers tell something about their electrical characteristics. Figure 16-5 shows three schematic symbols for transformers. These schematics represent the three basic transformers. All other transformers are a variation of these three.

Air core transformers are the type used in radio frequency applications. RF circuits cannot use iron cores because the core's response to high frequency sine waves is too slow.

Iron core transformers respond best to relatively low frequencies, typically under 20,000 hertz. In schematic symbols, the two parallel lines are used to indicate an iron core.

A **shielded transformer** has a magnetic shield around the outside of the windings. The purpose of the shield is to contain the magnetic flux, preventing it from interacting with other nearby circuits.

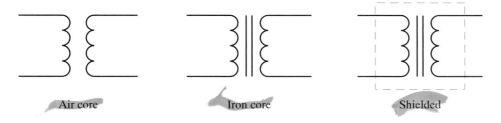

Air core   Iron core   Shielded

Figure 16-5. Schematic symbols for three types of transformers.

Center-Tapped Secondary

A **center-tapped secondary** produces two voltages, each equal to one half the total secondary voltage. Either half can be used separately, or they can be used together. Refer to figure 16-6.

The center tap can be used as a zero volt reference point. When compared to this point, the opposite sides of the secondary windings have opposite polarities. This means that the sine waves produced from the two sides of the secondary windings are 180 degrees out of phase. These dual sine waves are used in applications such as a full wave rectifier, in Chapter 26.

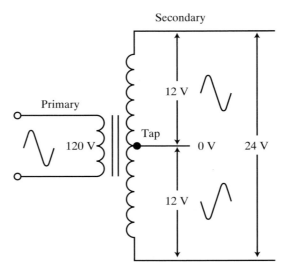

Figure 16-6. This transformer has a center-tapped secondary allowing for three separate outputs.

Multiple Secondary Windings

Multiple secondary windings are used in applications where it is necessary to produce several different voltages from the same power source. The sample in figure 16-7 could be used in an application such as a television power supply. The 600 volt windings would supply the high voltage section of the television. The windings producing 6.3 volts are used to power the filaments of vacuum tubes. The five volt windings are used for the electronic controls.

Other applications of multiple windings include such devices as computers. The logic circuitry in a computer uses +5 V in some sections. Other sections use +15 V and −15 V.

It is not necessary to use all of the windings that are provided. If the windings are left unconnected, as an open circuit, there is no current. Thus, those windings consume no power. The windings that are not used should be protected. There still will be voltage present in them.

Dual-Primary Transformer

A transformer with two primary windings is used where the electrical equipment can be connected to either 120 volts or 240 volts. This is a common situation with large commercial machinery. A higher voltage requires less current to develop the same amount of power.

Figure 16-8 shows a **dual-primary transformer** wired to a 120 volt supply. The two windings are connected in parallel. Each coil draws one half the total current. If one of the coils is disconnected, the other coil carries the full load current. A current this large might be more than the windings are designed to handle and could result in failure.

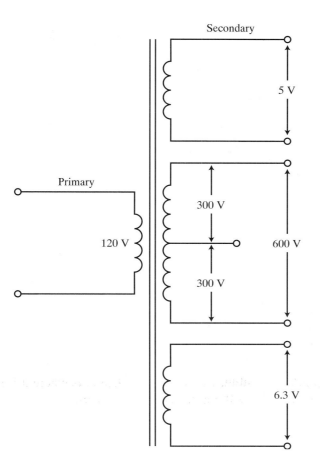

Figure 16-7. Typical power supply transformers have multiple secondary windings.

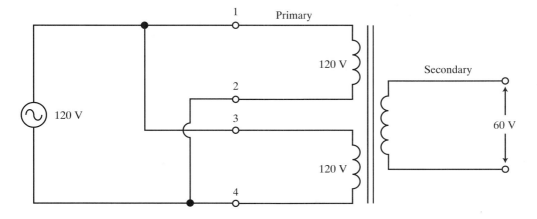

Figure 16-8. Dual-primary transformer wired in parallel for use with the lower primary voltage.

Figure 16-9 is a dual-primary transformer wired for 240 volt operation. In this configuration, the windings are in series. With twice the voltage, the current is one half the total of the 120 volt operation. With either the 120 volt or the 240 volt connection, the individual coils receive the same amount of current. In both circuits, the load voltage is the same.

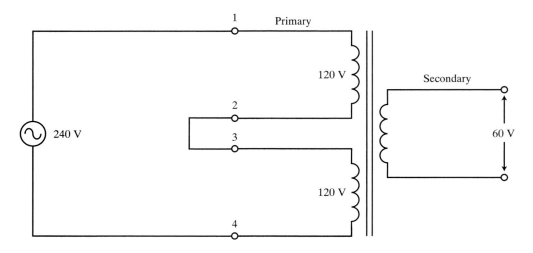

Figure 16-9. This dual-primary transformer is wired in series for use with the higher primary voltage.

Autotransformer

An **autotransformer** has only one coil, used for both the primary and secondary. A tap can slide along the coil to select a desired output voltage. See figure 16-10. One application of the autotransformer is on test benches. The autotransformer allows the technician to plug a circuit into an electrical source with a smaller fuse and a more convenient on/off switch. The technician also has the option of using a voltage different from the electrical wall outlet.

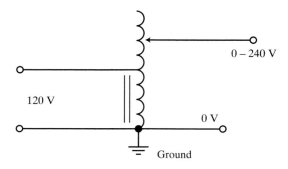

Figure 16-10. Autotransformers use a single coil. A variable sliding tap selects the output voltage.

Isolation Transformer

An **isolation transformer** has a turns ratio of 1:1. Its function is to isolate electrical equipment from earth ground.

Residential wiring uses earth ground as a third wire. When the third wire is properly connected, as in most situations, the ground connection provides protection from electrical shock.

Many electrical appliances, including televisions, stereos, and other consumer products, use only a two conductor cord. As seen in figure 16-11, one conductor is usually connected to the equipment chassis. It is inside and safely away from whoever is operating the appliance. A technician, however, must come in contact with the chassis. If the hot wire is connected to the chassis, there is a shock hazard.

The secondary of an isolation transformer does not use ground as a conductor. By plugging an appliance into the isolation transformer, the shock hazard is removed. See the bottom of figure 16-11.

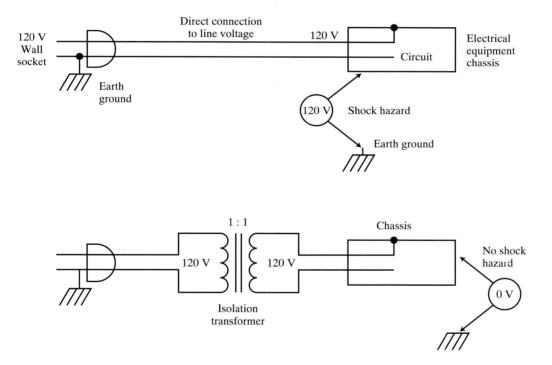

Figure 16-11. When an appliance has one conductor connected to the chassis, there is a significant shock hazard to the technician. An isolation transformer (bottom) eliminates this hazard.

Physical Construction

As with all electronic components, the physical construction is determined by the application. Generally speaking, a larger package is needed for a higher power duty. Smaller sizes, to the point of miniaturizing, are used on circuit boards.

Figure 16-12 shows some of the transformers used in electronic circuits. All the transformers shown here, except the variable power transformer, would be found inside the cabinet of electronic equipment. A variable power transformer is an autotransformer that is used to vary the voltage to a circuit and provide extra protection for a technician.

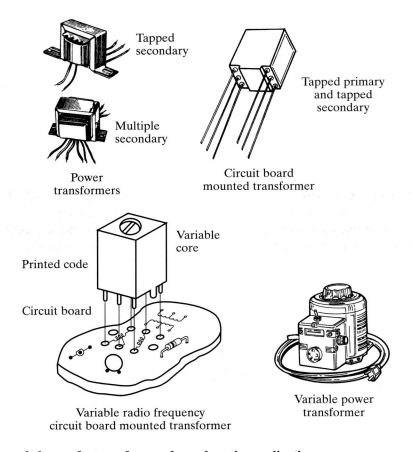

Figure 16-12. Size and shape of a transformer depend on the application.

16.4 TRANSFORMER TURNS RATIO

The voltage developed in the secondary windings of a transformer is determined by the ratio of the number of turns in the primary windings to the number of turns in the secondary windings. The turns ratio does not state the actual number of turns. It is a ratio, reduced to lowest terms. For example, figure 16-13 shows a transformer with 800 turns in the primary and 400 turns in the secondary. The turns ratio would be stated as 2:1.

Note in this figure that the relationship of the primary and secondary voltages is equal to the turns ratio. The turns ratio, combined with the voltage ratio can be used as a formula to calculate unknown values.

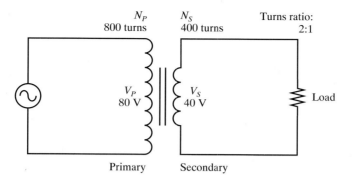

Figure 16-13. Step-down transformers have fewer turns in the secondary than in the primary. (Sample problems 1 and 2)

Transformer Voltage Ratio

A **step-up transformer** has a larger number of turns in the secondary than the primary. In a step-up transformer, the voltage is higher in the secondary.

A **step-down transformer** has a smaller number of turns in the secondary than the primary. In a step-down transformer, the voltage is lower in the secondary.

The formula used to calculate the turns ratio and voltage ratio is the same for both step-up and step-down transformers. The voltage ratio is calculated with formula 16.A.

Formula 16.A

$$\frac{N_P}{N_S} = \frac{V_P}{V_S}$$

N_P is the number of turns in the primary.

N_S is the number of turns in the secondary.

V_P is the primary voltage.

V_S is the secondary voltage.

Sample problem 1. _____

With the turns ratio given for the step-down transformer in figure 16-13, determine the secondary voltage with 120 V applied to the primary.

Formula: $\dfrac{N_P}{N_S} = \dfrac{V_P}{V_S}$

Substitution: $\dfrac{2}{1} = \dfrac{120\ V}{V_S}$

Rearranging: $V_S = \dfrac{120\ V \times 1}{2}$

Answer: $V_S = 60\ V$

Sample problem 2.

With the step-down transformer in figure 16-13, how much primary voltage is needed to produce 50 V across the load?

Formula: $\dfrac{N_P}{N_S} = \dfrac{V_P}{V_S}$

Substitution: $\dfrac{2}{1} = \dfrac{V_P}{50\ V}$

Rearranging: $V_P = \dfrac{2 \times 50\ V}{1}$

Answer: $V_P = 100\ V$

Sample problem 3.

With the step-up transformer shown in figure 16-14, what is the voltage available to the load when 50 volts is applied to the primary?

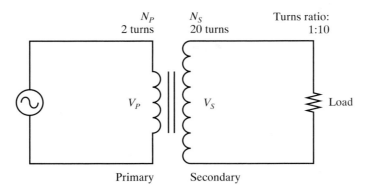

Figure 16-14. Step-up transformers have a greater number of turns in the secondary than in the primary. (Sample problem 3)

Formula: $\dfrac{N_P}{N_S} = \dfrac{V_P}{V_S}$

Substitution: $\dfrac{1}{10} = \dfrac{50\ V}{V_S}$

Rearranging: $V_S = \dfrac{50 \text{ V} \times 10}{1}$

Answer: $V_S = 500 \text{ V}$

Transformer Power

A transformer converts energy from one form to another, but it does not create energy. Power is electrical energy. Therefore, power of the primary windings is equal to power of the secondary windings, when losses are not considered. If the primary to secondary power was viewed as a ratio, it would be 1:1.

Formula 16.B

$$P_P = P_S$$

P_P is the primary power.

P_S is the secondary power.

An ideal transformer is one with no losses, such as the example shown in figure 16-15. This example is a step-up transformer with a turns ratio of 1:10. The primary voltage is 10 volts, producing a secondary voltage of 100 volts with current of 0.1 amps.

Sample problem 4. _____

Calculate the secondary and primary power of the circuit in figure 16-15.

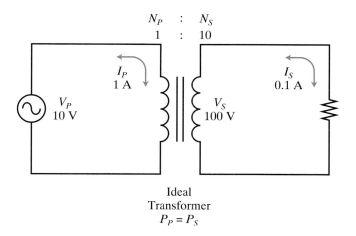

Figure 16-15. Power in the secondary equals power in the primary of an ideal transformer. (Sample problem 4)

Secondary power:

Formula: $P_S = I_S \times V_S$

Substitution: $P_S = 0.1 \text{ A} \times 100 \text{ V}$

Answer: $P_S = 10 \text{ W}$

Primary power (no losses):

Formula: $P_P = P_S$

Answer: $P_P = 10 \text{ W}$

Transformer Current Ratio

Current in a transformer is the inverse of the voltage ratio. *In a step-up transformer, the current is stepped down. In a step-down transformer, the current is stepped up.* This can be proven using the power formula in the voltage ratio, as follows:

Power formula:

$$P = I \times V$$

Power in a transformer:

$$P_P = P_S$$

Substitute into the power formula:

$$I_P \times V_P = I_S \times V_S$$

Transpose to form a ratio:

$$\frac{V_P}{V_S} = \frac{I_S}{I_P}$$

When combined with the voltage/turns ratio:

$$\frac{V_P}{V_S} = \frac{N_P}{N_S} = \frac{I_S}{I_P}$$

This formula relates the current ratio to the turns ratio.

Formula 16.C

$$\frac{N_P}{N_S} = \frac{I_S}{I_P}$$

N_P is turns of the primary.

N_S is turns of the secondary.

I_S is current of the secondary.

I_P is current of the primary.

Sample problem 5. _____

Using the circuit in figure 16-16, determine the primary current if the load current is increased to two amps.

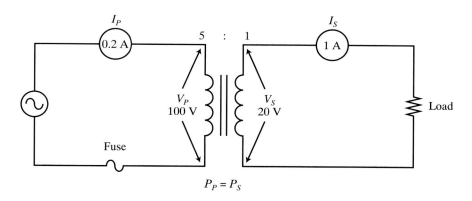

Figure 16-16. Example of an ideal step-down transformer. (Sample problem 16)

Formula: $\frac{V_P}{V_S} = \frac{I_S}{I_P}$

Substitution: $\frac{100\text{ V}}{20\text{ V}} = \frac{2\text{ A}}{I_P}$

Rearranging: $I_P = \frac{2\text{ A} \times 20\text{ V}}{100\text{ V}}$

Answer: $I_P = 0.4\text{ A}$

Power Company Application

An electric power company must supply voltage and current to a large number of customers. As discussed in Chapter 5, wire size is determined by the amount of current. In Chapter 4, Ohm's law showed that voltage drop is determined by resistance.

The power company must determine the most cost-effective means of delivering electricity to the consumer. Wire run for long distances has more resistance. Large wires could be used to reduce the resistance, but large wires are very heavy and expensive.

Figure 16-17 is an illustration of the use of transformers. This example covers only 10 residential customers to keep it simple. In this example, 120 kW of power is needed by the customers. The value of 120 kW is the maximum power needed throughout the process. At the power plant, a generator is capable of producing a certain amount of voltage and a maximum current. This generator produces a constant 2400 volts. The current varies to match the load demand.

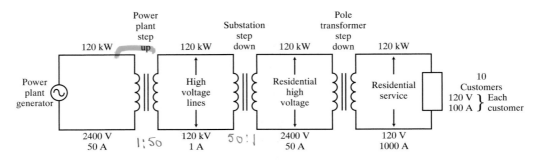

Figure 16-17. Simplified illustration of how transformers are used to deliver electricity to residential customers.

The generator feeds the electricity into a large step-up transformer. The voltage is stepped up 50 times. The current is stepped down by the same ratio. This very large voltage is carried over the long distance transmission lines. The advantage of using such large voltages is that the amount of current is reduced. Less current means a much smaller wire size can be used, resulting in less cost and a lighter wire. The total amount of power carried in the wire is still the same.

Substation transformers are located in the general area of the customer, within a few miles. High voltage lines enter the substation. In the substation, the voltage is reduced to much safer levels. In the example shown, the substation transformer uses a 50:1 turns ratio.

This medium level voltage is carried at the top of power poles, as seen in figure 16-18. From there the electricity moves to a transformer located a short distance from the customer. The figure shows a pole-mounted transformer. Transformers are also located on the ground in tamper proof boxes in areas where the electrical service is run underground.

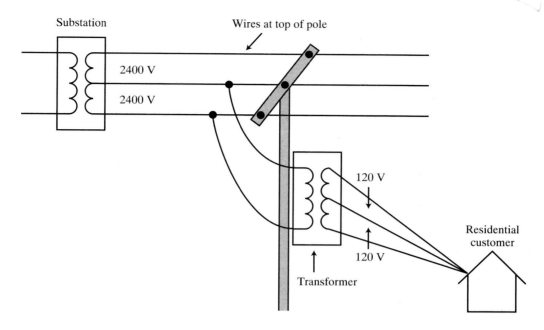

Figure 16-18. Voltage must be stepped down for customers to use.

16.5 TRANSFORMER CORE LOSSES

As with everything else, transformers are not perfect. When discussing theory of operation, it is easiest to start with an understanding of the ideal. However, it is also necessary to understand the performance of nonideal devices.

Losses produce heat. In electricity, heat comes from power. Generally speaking, heat is an indication power is being consumed. When heat is produced from a device that is not intended to produce heat, it is a clear indication of wasted power. Figure 16-19 shows that power lost in the transformer is not available to the load and must be subtracted from the input power. Transformers have three categories in which power is lost: copper losses, hysteresis loss, eddy currents. Each of these power losses can be represented as a series resistor, as shown in figure 16-20. Voltage is dropped across these equivalent series resistances that oppose current.

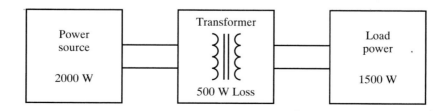

Figure 16-19. This block diagram represents the power loss that takes place in every transformer.

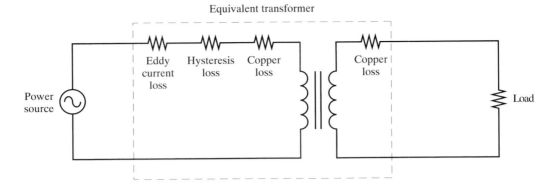

Figure 16-20. Schematic showing transformer losses as equivalent series resistances.

Copper Losses

The wire used to make the windings of a transformer has some resistance. Even though most wire is thought of as having zero resistance, the coil in a transformer may have several hundred to several thousand turns. This is a very long piece of wire. In addition, the wire that is used in many transformers is very small in diameter. This also gives it a higher resistance.

Losses due to the copper wire are called **I²R losses.** Remember, I²R is a formula used to calculate power. Therefore, copper losses are power losses.

Hysteresis Loss

Hysteresis is the amount of magnetization or flux density (*B*) that lags the magnetizing force (*H*) because of molecular friction, discussed in Chapter 11. In the case of transformers, the greatest cause of hysteresis loss is residual magnetism. The loss of energy comes from overcoming this residual magnetism.

Figure 16-21 is a hysteresis loop. A **hysteresis loop** represents the amount of energy needed to create a magnetic field in the core of the transformer. The curve starts at the center, the point labeled a, with no residual magnetism. As the voltage increases in the positive direction, the flux density (*B*) increases. To continue increasing the strength of the magnetic field, a stronger magnetizing force (*H*) must be applied.

At point b, the voltage has reached its peak positive and the magnetic field can increase no further. The sine wave decreases and so does the magnetic field, but at a somewhat slower pace. As the magnetizing force passes through zero on the H axis, an amount of flux remains. This is *residual magnetism.*

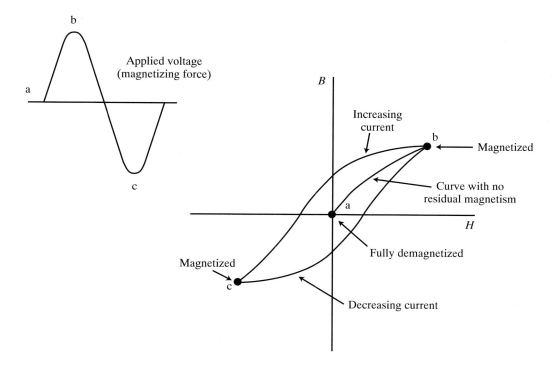

Figure 16-21. Hysteresis loops show flux density (*B*) vs magnetizing force (*H*).

The magnetizing force sine wave continues in the negative direction, causing the magnetic flux to flow in the opposite direction. At point c, the sine wave has reached its negative peak and begins to fall back to zero. The magnetic field, again, does not collapse at the same rate as the sine wave. This lagging effect results in a residual magnetism in both the positive and negative half cycles of the magnetizing force.

The lower the frequency of the applied voltage, the less significant the hysteresis loss is. Transformers are given a **frequency response rating** that indicates the best frequency applications for that transformer. Power transformers, with an iron core, used on the 60 hertz line frequency have very small hysteresis losses because of the low frequency.

Radio frequencies, however, are so high that an iron core is not practical. This is the reason air core transformers are used with radio frequencies.

Eddy Currents

Eddy currents result from a voltage induced in the iron core of transformers and other electromagnets. These eddy currents oppose the current producing the magnetic field. This results in a loss of power.

Figure 16-22 shows the difference between the eddy currents in a solid iron core and a core made with laminations. In the solid core, the large area allows a large eddy current to circulate. To prevent these large eddy currents, transformer cores are manufactured using thin slices of iron that are glued together. The slices are called **laminations.** The laminations break up the path for eddy currents. The result is much smaller losses. A laminated core can be made as large as any solid core.

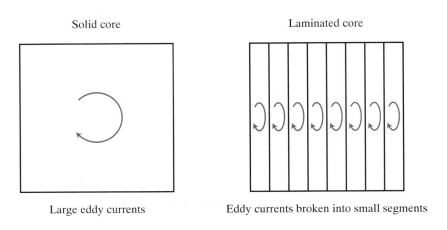

Solid core   Laminated core

Large eddy currents   Eddy currents broken into small segments

Figure 16-22. Laminated cores reduce the effects of eddy currents. This makes transformers more efficient.

DC Voltages in Transformers

Any current applied to a coil produces a magnetic field. However, if it is not a changing current, the magnetism produced will not induce a voltage in another coil.

When a dc voltage is applied to a transformer, it creates a magnetic field that must be overcome by the ac in order to produce voltage in the secondary. The dc voltage acts like residual magnetism and causes high losses.

It is possible for the dc voltage to be strong enough to stop the transformer from producing any further magnetism. When this happens, the transformer is saturated and does not produce secondary current.

16.6 CALCULATING EFFICIENCY

When working with transformers, a useful quantity to know is the transformers efficiency. **Efficiency** is the ratio of power output to power input. Efficiency is expressed in percentage.

Formula 16.D

$$\text{eff} = \frac{P_{out}}{P_{in}} \times 100\%$$

eff is efficiency expressed as a percentage.

P_{out} is power in the secondary circuit, in watts.

P_{in} is power applied to the primary circuit, in watts.

Formula 16.E

$$\text{eff} = \frac{V_s \times I_s}{V_P \times I_P} \times 100\%$$

eff is the efficiency expressed as a percentage.

V_S is the voltage in the secondary, measured in volts.

I_S is the current in the secondary, measured in amperes.

V_P is the current in the primary, measured in volts.

I_P is the current in the primary, measured in amperes.

Sample problem 6. _____

Calculate the efficiency of the transformer shown in figure 16-23.

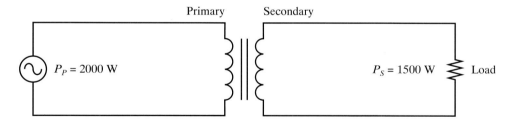

Figure 16-23. Calculate the efficiency. (Sample problem 6)

Formula: $\text{eff} = \dfrac{P_{out}}{P_{in}} \times 100\%$

Substitution: $\text{eff} = \dfrac{1500 \text{ W}}{2000 \text{ W}} \times 100\%$

Answer: $\text{eff} = 75\%$

Sample problem 7. _____

The power transformer in figure 16-24 is rated to have an efficiency of 80 percent. How much power must be applied to the primary to obtain 60 watts at the load?

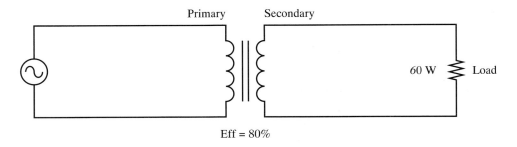

Eff = 80%

Figure 16-24. Find the power applied to the primary. (Sample problem 7)

Formula: $\text{eff} = \dfrac{P_{out}}{P_{in}} \times 100\%$

Substitution: $80\% = \dfrac{60\text{ W}}{P_{in}} \times 100\%$

Rearranging: $P_{in} = \dfrac{60\text{ W}}{80\%} \times 100\%$

Answer: $P_{in} = 75\text{ W}$

Sample problem 8. _____

Use the circuit measurements shown in figure 16-25 to determine the efficiency of the transformer.

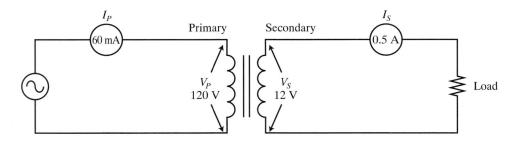

Figure 16-25. Calculate the efficiency. (Sample problem 8)

Formula: $\text{eff} = \dfrac{V_S \times I_S}{V_P \times I_P} \times 100\%$

Substitution: $\text{eff} = \dfrac{12\text{ V} \times 0.5\text{ A}}{120\text{ V} \times 60\text{ mA}} \times 100\%$

Intermediate step: $\text{eff} = \dfrac{6\text{ W}}{7.2\text{ W}} \times 100\%$

Answer: $\text{eff} = 83.3\%$

Sample problem 9. _____

Find the primary current in figure 16-26.

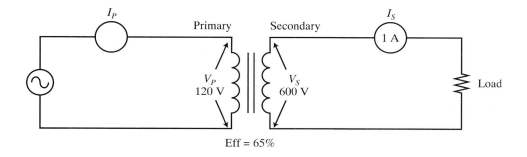

Figure 16-26. Find the current in the primary windings. (Sample problem 9)

Formula: $\text{eff} = \dfrac{V_S \times I_S}{V_P \times I_P} \times 100\%$

Substitution: $65\% = \dfrac{600\text{ V} \times 1\text{ A}}{120\text{ V} \times I_P} \times 100\%$

Rearranging: $I_P = \dfrac{600\text{ V} \times 1\text{ A}}{120\text{ V} \times 65\%} \times 100\%$

Answer: $I_P = 7.7\text{ A}$

Sample problem 10.

How much current is flowing through the load of the transformer circuit shown in figure 16-27?

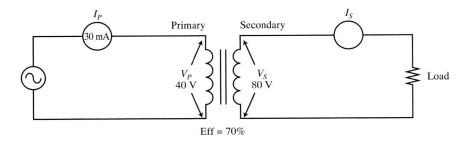

Eff = 70%

Figure 16-27. Find the current in the load. (Sample problem 10)

Formula: $$\text{eff} = \frac{V_S \times I_S}{V_P \times I_P} \times 100\%$$

Substitution: $$70\% = \frac{80 \text{ V} \times I_S}{40 \text{ V} \times 30 \text{ mA}} \times 100\%$$

Intermediate step: $$\frac{70\%}{100\%} = \frac{80 \text{ V} \times I_S}{1.2 \text{ W}}$$

Rearranging: $$I_S = \frac{70\%}{100\%} \times \frac{1.2 \text{ W}}{80 \text{ V}}$$

Answer: $$I_S = 10.5 \text{ mA}$$

Sample problem 11.

How much voltage will be measured across the secondary of the transformer in figure 16-28?

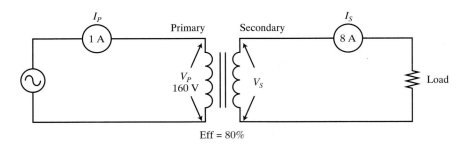

Eff = 80%

Figure 16-28. Find the voltage across the secondary. (Sample problem 11)

Formula: $$\text{eff} = \frac{V_S \times I_S}{V_P \times I_P} \times 100\%$$

Substitution:	$80\% = \dfrac{V_S \times 8\,A}{160\,V \times 1\,A} \times 100\%$	

Intermediate step: $\dfrac{80\%}{100\%} = \dfrac{V_S \times 8\,A}{160\,W}$

Rearranging: $V_S = \dfrac{80\%}{100\%} \times \dfrac{160\,W}{8\,A}$

Answer: $V_S = 16\,V$

16.7 LOAD RESISTANCE REFLECTED TO THE PRIMARY

When connecting a voltage source to a circuit, you need to know how much current is required. With a transformer circuit, the load resistance connected to the secondary determines the current drawn in the primary.

There are two ways to calculate primary resistance. First is Ohm's law. This is used if the primary current and voltage are known. The second way is with the **resistance ratio.** This method is used if the load resistance and turns ratio are known. The resistance ratio is also called the *impedance ratio*. Impedance is a term used to describe ac resistance. Impedance is discussed in detail in Chapters 18 and 19.

Calculating Reflected Resistance Using Ohm's Law

To calculate the resistance of the primary using Ohm's law, it is necessary to know the primary voltage and primary current. The only values given in many circuits are the primary voltage, turns ratio, and load resistance. The turns ratio allows calculating of the secondary voltage. Use Ohm's law to find the secondary current. Then use the turns ratio to find primary current. Finally, use Ohm's law to find resistance in the primary circuit. This resistance is a reflection of the load resistance.

Sample problem 12. ───────────────────────────────

Calculate the primary resistance of the circuit in figure 16-29 using Ohm's law and turns ratios.

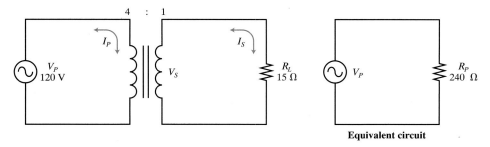

Equivalent circuit

Figure 16-29. Load resistance is reflected to the source voltage through the turns ratio of the transformer. (Sample problems 12 and 13)

1. Use turns/voltage ratio to calculate secondary voltage:

   Formula: $\dfrac{N_P}{N_S} = \dfrac{V_P}{V_S}$

   Substitution: $\dfrac{4}{1} = \dfrac{120 \text{ V}}{V_S}$

   Rearranging: $V_S = \dfrac{120 \text{ V} \times 1}{4}$

   Answer: $V_S = 30 \text{ V}$

2. Use Ohm's law to calculate secondary current:

   Formula: $I_S = \dfrac{V_S}{R_L}$

   Substitution: $I_S = \dfrac{30 \text{ V}}{15 \text{ }\Omega}$

   Answer: $I_S = 2 \text{ A}$

3. Use turns/current ratio to calculate primary current:

   Formula: $\dfrac{N_P}{N_S} = \dfrac{I_S}{I_P}$

   Substitution: $\dfrac{4}{1} = \dfrac{2 \text{ A}}{I_P}$

   Rearranging: $I_P = \dfrac{2 \text{ A} \times 1}{4}$

   Answer: $I_P = 0.5 \text{ A}$

4. Use Ohm's law to calculate primary resistance:

   Formula: $R_P = \dfrac{V_P}{I_P}$

   Substitution: $R_P = \dfrac{120 \text{ V}}{0.5 \text{ A}}$

   Answer: $R_P = 240 \text{ }\Omega$

Calculating Reflected Resistance Using the Resistance Ratio

The resistance ratio can be used to find loading of the voltage source. The resistance ratio formula requires the transformer turns ratio and the load resistance. The resistance ratio is derived by combining Ohm's law with the turns/voltage/current ratios.

Formula 16.F

$$\frac{R_P}{R_S} = \left(\frac{N_P}{N_S}\right)^2$$

R_P is the resistance of the primary, measured in ohms.

R_S is the resistance of the secondary, measured in ohms.

N_P is the number of turns in the primary.

N_S is the number of turns in the secondary.

Sample problem 13. _____

Determine the reflected resistance of the circuit in figure 16-29, using the resistance ratio.

Formula:	$\dfrac{R_P}{R_S} = \left(\dfrac{N_P}{N_S}\right)^2$
Substitution:	$\dfrac{R_P}{15\ \Omega} = \left(\dfrac{4}{1}\right)^2$
Rearranging:	$R_P = \left(\dfrac{4}{1}\right)^2 \times 15\ \Omega$
Intermediate step:	$R_P = 16 \times 15\ \Omega$
Answer:	$R_P = 240\ \Omega$

Impedance Matching Transformer

To achieve maximum transfer of power it is necessary for the load resistance to be equal to the resistance of the voltage source. In figure 16-30, a speaker is connected to an audio amplifier. An impedance matching transformer is used to reflect the eight ohm speaker resistance to the amplifier as 200 ohms, which is the amplifier's resistance.

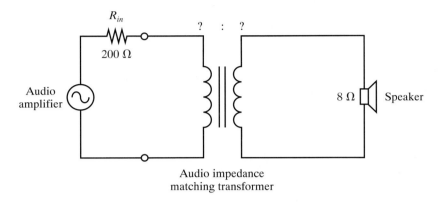

Figure 16-30. An impedance matching transformer is used in audio circuits to achieve maximum transfer of power. (Sample problem 14)

Sample problem 14. _____

Determine the turns ratio of the transformer needed in figure 16-30 to reflect the eight ohm speaker resistance as a load equal to the amplifier's internal resistance (200 ohms).

Formula:
$$\frac{R_P}{R_S} = \left(\frac{N_P}{N_S}\right)^2$$

Substitution:
$$\frac{200\ \Omega}{8\ \Omega} = \left(\frac{N_P}{N_S}\right)^2$$

Intermediate step:
$$25 = \left(\frac{N_P}{N_S}\right)^2$$

Intermediate step:
$$\frac{N_P}{N_S} = \sqrt{25} = 5$$

Answer:
$$\frac{N_P}{N_S} = \frac{5}{1}$$

Impedance matching transformers with resistance ratios of 200:8 are a standard catalog item for use with audio amplifiers.

16.8 LOADING A TRANSFORMER

When a sine wave flows through an inductor the current lags behind the applied voltage by 90°. This is represented by the phasors in figure 16-31. In phasor diagrams, a pure resistance has both current and voltage phasors drawn on the horizontal. An ideal transformer is purely inductive (has no resistance). It is drawn as a vertical phasor. Phasors are used to give a graphical representation of the relative size and angle of two quantities. Phasors will be discussed in detail in Chapter 20.

Figure 16-31. Ideal transformers have a 90° phase angle between the primary voltage and current.

Energizing Current

If a transformer were ideal, with no coil resistance, there would be no loss of power to energize the coil. With no load connected across the secondary, there would be an extremely small primary current. Figure 16-32 shows that an *actual* transformer requires a slight amount of **energizing current** to develop the magnetic field, even with no load connected. The phasor diagram shows a shift from the ideal 90° to 78°. The phase shift represents losses due to coil resistance.

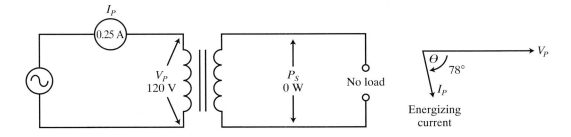

Figure 16-32. Transformers require an energizing current to develop a magnetic field due to the slight coil resistance.

Power Transformer with a Light Load

The transformer in figure 16-33 has a load of approximately 10 percent of the full load. The primary current is only slightly greater than the current needed to energize. Efficiency is increased to 33 percent with the increased secondary current.

A further shift in the phasor diagram at the right of figure 16-33 shows that the secondary current causes the transformer to act more like a resistor. The phasor diagram of current in pure resistance, as mentioned previously, is in phase with the voltage at 0°. By becoming more like a resistive current, the transformer becomes more efficient.

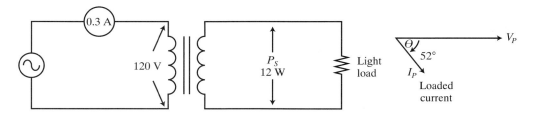

Figure 16-33. Power transformer with a light load. The primary current here is only slightly above the energizing current.

Medium and Heavy Loads

Figure 16-34 shows a transformer with a load equal to approximately 50 percent of full secondary power. The transformer's efficiency increases to 56 percent with the increased load. The phasor diagram shows the current becomes more resistive as it increases.

Figure 16-35 shows the transformer with its heaviest load. Efficiency is 87 percent. The phasor diagram shows the primary current and voltage have almost the same phase angle. When the current has a 0° phase angle, the power produced has no inductive losses. However, it is not possible to achieve 100 percent efficiency. There will always be some power losses.

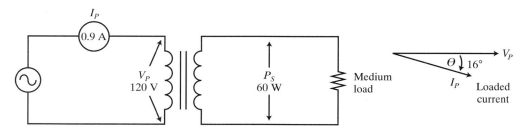

Figure 16-34. Power transformer with a medium load.

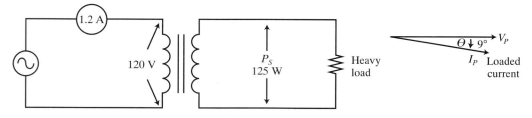

Figure 16-35. Power transformer with a heavy load. The transformers efficiency has improved to 87 percent.

16.9 TRANSFORMER RATINGS

All electronic components have ratings of maximum operating values. In addition, physical measurements are important. They are used to determine if the device fits into a certain size package or if it can be mounted on a circuit board. Transformers also have an additional consideration. They can be very heavy with their large number of turns of copper wire.

The turns ratio of a transformer is usually not given. Instead the primary voltage is given with the nominal secondary for a nominal secondary current. The power rating can be stated in either of two forms. The rating comes as a maximum wattage or as a maximum current at a specified voltage.

Resistance ratios, also called impedance ratios, are usually not given for power transformers. Impedance ratios, however, are listed for applications where load to circuit matching is necessary. Impedance, as you will recall, is ac resistance.

16.10 TRANSFORMER TROUBLESHOOTING

Keep in mind how a transformer is formed. A transformer is two coils of wire, or one coil if it is an autotransformer. A coil can have an open (a broken wire) or a short (where wires are touching together).

Measurements from a normal working circuit, figure 16-36, are used as a reference to examine a sample troubleshooting circuit. Four measurements are given: the current and voltage for both primary and secondary circuits. The circuit in figure 16-36 also contains a fuse, which can help with troubleshooting problems.

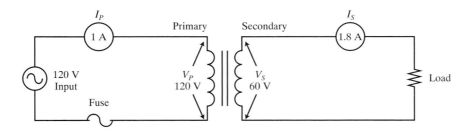

Figure 16-36. Measurements in a normal circuit are used to compare to a defective circuit.

Shorted Primary Windings

A short circuit results in an excessively high amount of current. A short can occur as a result of the transformer windings getting so hot that some of their insulation melts.

Symptoms, shown in figure 16-37:
• The fuse blows, removing voltage from the circuit.
• All measurements are zero with the blown fuse, except the applied voltage.
• The transformer is cold since there is no current.

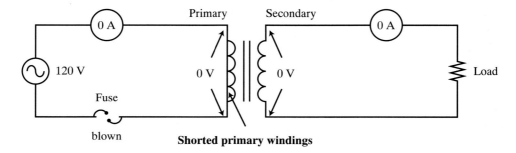

Figure 16-37. Shorted primary windings will blow the fuse.

Using an ohmmeter (applied voltage removed):
• Check the fuse with the meter to verify it is blown (infinite resistance).
• Placing the ohmmeter across the primary should show zero resistance. Be sure to use the lowest resistance range.
• Secondary should show a normal resistance measurement. The resistance is determined by the actual number of turns. A reading of 100 ohms to a few thousand is a good approximation.

Shorted Secondary Windings

Shorted secondary windings can be a result of a defect in the load or a defect in the transformer. It is possible for the load to be defective and draw so much current it damages the transformer. If a short is suspected in the secondary, it is best to remove the load to isolate the problem.

Symptoms, shown in figure 16-38:
• Excessive heat radiates from the transformer, to the point of melting insulation. Smoke is possible.
• The primary current is much larger than normal. However, it may not be enough to blow the fuse, especially if the fuse is oversized.

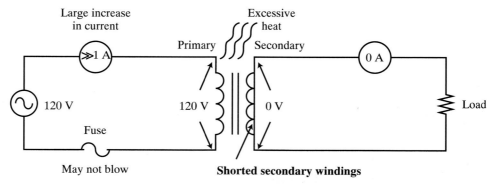

Figure 16-38. Shorted secondary windings do not always blow the fuse. However, it does cause excessive heat buildup in the transformer.

- The primary voltage equals the applied voltage.
- There are zero readings in the secondary.

Using an ohmmeter (applied voltage removed):
- The primary resistance is normal.
- The secondary resistance is zero. Be sure to use the lowest ohms scale possible.

Partially Shorted Secondary Windings

Excessive current through the load can cause the transformer to get too hot. Too much heat can result in failure of the insulation on the windings. It is possible for only a few of the windings, or a section of windings, to be shorted together.

Symptoms, shown in figure 16-39:
- There is a decrease in secondary voltage and current. Amount of decrease is dependent on how many turns are shorted.
- There is a decrease in primary current.
- Excessive heat is seen in the transformer.

Using the ohmmeter (applied voltage removed):
- The reading on the primary is normal.
- The reading on the secondary is lower. An accurate measurement must be taken to see a difference from the normal value.

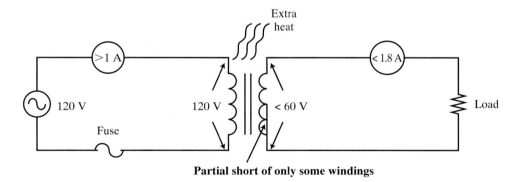

Figure 16-39. A partial short causes extra heat and a decrease in secondary readings.

Open Primary Windings

An open means the wire no longer provides a current path. An external open occurs at the point where wires are connected. Other opens can occur internally in the transformer, usually due to excessive heat.

Symptoms, shown in figure 16-40:
• There is no primary current.
• There is no secondary voltage or current.
• The transformer is cold. No heat whatsoever is produced.
• The applied voltage is measured across the primary.

Using an ohmmeter (applied voltage removed):
• An infinite resistance is found in the primary.
• A normal resistance is found in the secondary.

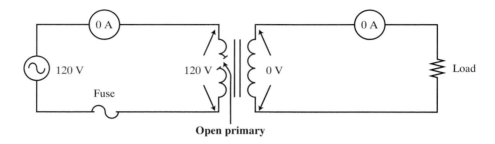

Figure 16-40. Open primary windings stop all current.

Open Secondary Windings

Opens frequently occur at the connections of the wires to the transformer leads. Check there first whenever an open is suspected. With an open in the secondary, the primary still develops a magnetic field. However, it does not have to supply secondary current, so only an energizing current will be present.

Symptoms, shown in figure 16-41:
• The primary current is lower than normal. How much lower depends on the size of the load when the normal measurements were taken.
• The primary voltage is equal to the applied voltage.

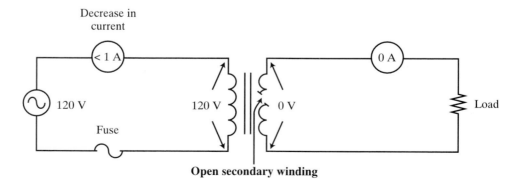

Figure 16-41. Open secondary windings result in a decrease in primary current and zero secondary readings.

- The secondary voltage and current are zero.
- The transformer is cold. Only a very slight amount of heat is developed from the energizing current.

Using an ohmmeter (applied voltage removed):
- Normal resistance is measured in the primary.
- Infinite resistance is measured in the secondary.

SUMMARY

- Transformers use magnetic coupling to link the primary and secondary windings.
- The turns ratio of the transformer determines the amount of voltage and current measured at the secondary as compared to the primary windings.
- A step-up transformer has a larger number of turns in the secondary windings than in the primary windings. It steps up the voltage but steps down the current.
- A step-down transformer has a smaller number of turns in the secondary windings than in the primary windings. It steps down the voltage but steps up the current.
- Power in an ideal transformer is the same in both the primary and secondary windings.
- Transformer losses result in a lower secondary power.
- Efficiency is the ratio of power out to power in.
- When troubleshooting a transformer, remember that a transformer is two coils of wire. An open circuit results in no current, and a short circuit results in too much current.

KEY WORDS AND TERMS GLOSSARY

autotransformer: A transformer that has only one coil used for both the primary and the secondary.

eddy current: Electrical current flowing within the core of an electromagnet resulting from voltage induced in the iron core. Eddy currents oppose the current producing the magnetic field resulting in a loss of power.

efficiency: The ratio of power output available to the load to the power input from the source.

energizing current: The slight amount of current needed to develop the magnetic field in a coil.

frequency response rating: A rating that indicates the best frequency applications for a transformer.

hysteresis: The amount of magnetization or flux density (B) that lags the magnetizing force (H) because of molecular friction.

hysteresis loop: Represents the amount of energy needed to create a magnetic field in the core of a transformer.

isolation transformer: A transformer with the function of isolating a portion of a circuit.

I²R losses: Losses due to the copper wire in a transformer.

primary windings: The input of a transformer. The current through the primary windings develops a magnetic field. The amount of current in the primary is determined by a combination of primary resistance and secondary current.

resistance ratio: One method of calculating the primary resistance.

secondary windings: The output coil of the transformer. The magnetic field developed in the primary induces voltage in the secondary through magnetic induction.

shielded transformer: A transformer with a magnetic shield on the outside of the windings. The shield prevents the magnetic flux from interfering with nearby circuits.

step-down transformer: A transformer with a smaller number of turns in the secondary than the primary. In a step-down transformer, secondary voltage is lower than primary voltage.

step-up transformer: A transformer with a larger number of turns in the secondary than the primary. In a step-up transformer, the secondary voltage is higher than primary voltage.

transformer: An electrical device which uses magnetism to link one coil of wire to another. Magnetic coupling of coils can result in an increase in voltage, decrease in voltage, or equal voltages from the input to the output.

KEY FORMULAS

Formula 16.A

$$\frac{N_P}{N_S} = \frac{V_P}{V_S}$$

Formula 16.B

$$P_P = P_S$$

Formula 16.C

$$\frac{N_P}{N_S} = \frac{I_S}{I_P}$$

Formula 16.D

$$\text{eff} = \frac{P_{out}}{P_{in}} \times 100\%$$

Formula 16.E

$$\text{eff} = \frac{V_s \times I_s}{V_p \times I_p} \times 100\%$$

Formula 16.F

$$\frac{R_P}{R_S} = \left(\frac{N_P}{N_S}\right)^2$$

TEST YOUR KNOWLEDGE

Do not write in this text. Please use a separate sheet of paper.
1. A transformer operates on the principle of _____ induction.
2. Describe how a transformer is constructed.
3. Describe how magnetism is used to pass voltage from one coil to another.
4. State two purposes for the transformer core.
5. What is the advantage of a transformer with multiple secondary windings?
6. Draw two wiring diagrams showing how to connect a dual-primary transformer:
    a. Low voltage.
    b. High voltage.
7. How is the voltage adjusted with an autotransformer?
8. What is the main function of an isolation transformer?
9. In a transformer with a turns ratio of 10:1, what is the secondary voltage when 50 volts are applied to the primary? Is this transformer a step-up or step-down?
10. A transformer with a turns ratio of 1:20 has 10 volts applied to the primary. What is the secondary voltage? Is this transformer a step-up or step-down?
11. What is the primary voltage applied to a transformer with a turns ratio of 25:1 if the secondary voltage is measured to be 10 volts? Is this transformer a step-up or step-down?
12. A secondary voltage of 50 volts is required from a transformer with a turns ratio of 1:50. What is the primary voltage? Is this transformer a step-up or step-down?
13. What is the turns ratio of a transformer with 600 volts on the secondary when 50 volts are applied to the primary? Is this transformer a step-up or step-down?
14. What is the turns ratio of a transformer with 120 volts applied to the primary and 12 volts on the secondary? Is this transformer a step-up or step-down?
15. What current can be expected in a transformer with 120 volts on the primary and 60 volts on the secondary if there is a secondary current of five amps? (Ignore losses.) Is this transformer a step-up or step-down?
16. A certain transformer has a two amp fuse in the primary circuit. This primary circuit is connected to 120 volts. If the secondary voltage is rated for 20 volts, what is the maximum secondary current? (Ignore losses.) Is this transformer a step-up or step-down?

17. Calculate the efficiency of a transformer with 800 watts measured at the secondary and a primary power of 1000 watts.

18. If a transformer is rated for 75 percent efficiency, what is the power in the primary with a secondary power of 4000 watts?

19. A transformer is measured to have a primary voltage of 120 volts, a secondary current of two amps, and a secondary voltage of 60 volts. Calculate the primary current if the efficiency is 80 percent.

20. In a transformer circuit, secondary current is measured at four amps through a five ohm load. If the input voltage is 120 volts, calculate the reflected resistance to the primary. Assume an efficiency of 100 percent.

21. Calculate the reflected resistance of a circuit with a 20 ohm load and a transformer with a turns ratio of 6:1.

22. State three types of transformer losses. Also, briefly describe how to reduce the losses.

23. Describe the effects of dc voltages applied to a transformer.

24. What happens to the secondary voltage of a transformer when the load current is increased from zero to the maximum rated current?

25. Given each of the following conditions, state if the transformer is defective with an open primary, shorted primary, open secondary, or shorted secondary.

    a. Fuse is blown, voltage and current are zero, secondary resistance is normal, primary resistance is zero.

    b. Primary voltage normal, primary current too low, secondary voltage and current zero, secondary resistance infinity, and primary resistance normal.

    c. High primary current, secondary current and voltage are too low, excessive heat produced, secondary resistance low, and primary resistance normal.

Chapter 17
Electric Generators and Motors

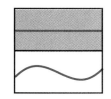

Upon completion of this chapter, you will be able to:
- Define technical terms related to generators and motors.
- Describe the basic generator action to produce electricity.
- State the functions of the armature and fields of a generator.
- Name the parts of a commutator and describe their function.
- List and explain the methods of exciting a generator field.
- Describe the basic action of an electric motor.
- Describe the operation of a dc motor.
- Describe the operation of an ac motor.
- State the ratings for motors.
- Perform basic horsepower calculations.
- Perform basic torque calculations.

Generators and motors are very similar in construction. There are two primary differences, the work produced and the energy used to produce the work. **Generators** use mechanical energy to produce electricity. **Motors** use electrical energy to produce mechanical force.

Generators and motors, in a sense, are opposites. Magnetism is used as a means of converting from one form of energy to the other. Transformers, studied in Chapter 16, also use magnetism to change energy from one form to another.

17.1 BASIC GENERATOR ACTION

The action of a basic generator can be represented by a coil of wire rotating in a magnetic field. As the conductor cuts through the magnetic lines of force, a voltage is induced in the wire. This concept was discussed in detail in Chapter 15, with the generator producing a sine wave. Refer to figure 15-11 for a drawing of the basic generator concept. Also, review figure 15-12 to see how a sine wave is developed as the generator rotates.

In this chapter, the emphasis is placed on how electricity is removed from the generator. When motors are studied, later in this chapter, the emphasis will be on using electricity to make the motor spin.

Armature and Fields

The **armature** is the coil of wire in a generator spinning past the magnetic fields. The **fields** of a generator contain the magnetic field. Fields may be made of permanent magnets or electromagnets.

In figure 17-1, a two-pole generator is shown with the armature spinning through the magnetic fields. It should be noticed that a generator produces a sine wave, even though some generators are classified as dc.

The spinning action of the armature is produced by some mechanical force. The mechanical force driving the armature could be: the blades of a windmill, the paddles of a waterwheel, the belts connected to the engine of an automobile, or steam spinning the turbines in a nuclear power plant.

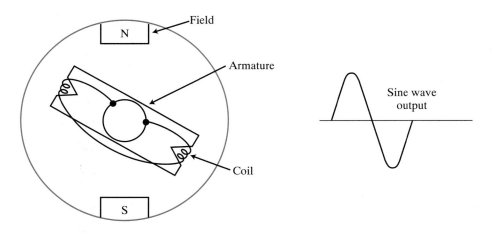

Figure 17-1. In a basic generator, the armature spins inside of a magnetic field. The generator's output is shown to the right.

AC Generator

The magnetic field induces a voltage in the armature windings as the windings pass through the lines of flux. A spinning armature presents a problem. How are the windings connected to the external load circuit? If the circuit wires were connected directly to the armature coil, the spinning action would twist the wires and break them.

Examine figure 17-2. Armature coil wires are attached directly to a pair of **slip rings.** The slip rings spin with the armature. **Brushes** make contact with the slip rings to bring the current to an outside load. The output of this ac generator, using slip rings and brushes, is a sine wave.

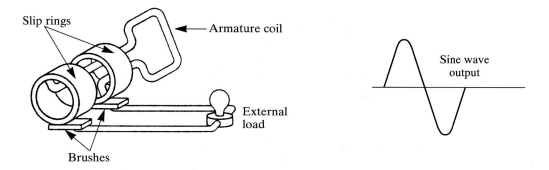

Figure 17-2. AC generators use slip rings and brushes to produce a sine wave output.

DC Generator

A dc generator operates in a similar manner to the ac generator. The armature holds a coil, which spins inside a magnetic field. The magnetic field induces voltage in the form of a sine wave. The difference in the dc generator is the manner in which the coil is attached to the external circuit.

As shown in figure 17-3, the dc generator uses a split-ring commutator and brushes. A **commutator** maintains the current in the same direction, *producing a rectified sine wave output.*

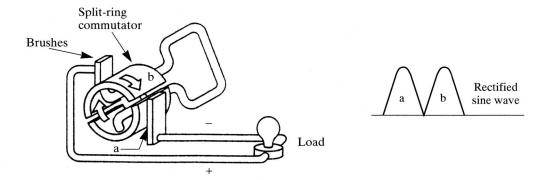

Figure 17-3. DC generators use a commutator and brushes to produce a DC voltage in the form of a rectified sine wave.

17.2 COMMUTATOR ACTION

A **split ring commutator** rotates with the coil as the coil produces the sine wave to pass the generated electricity to the load. When the sine wave is in the positive half cycle, one ring is positive and the other is negative. When the coil changes polarity as it rotates, the rings also change polarity. Thus, the output remains the same polarity since both the coil and rings change at the same time.

Refer to figure 17-4. In part A, the coil is in the lowest voltage position. It is at this point that the coil and commutator change polarity. In part B, the coil and commutator are passing through the peak voltage of one half cycle. In part C, there is, again, a polarity change. In part D, the voltage produced has an output of the same polarity as part B because both the coil and commutator changed polarity.

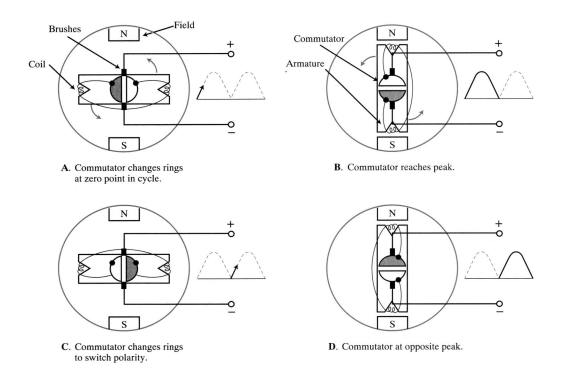

A. Commutator changes rings
at zero point in cycle.

B. Commutator reaches peak.

C. Commutator changes rings
to switch polarity.

D. Commutator at opposite peak.

Figure 17-4. Commutator action produces an output voltage with a constant polarity.

Brushes

An expanded view of a brush assembly is shown in figure 17-5. The brushes are spring-mounted to make contact with the slip rings in ac generators, and with the commutator in dc generators. The brushes connect the external load circuit to the armature.

The brush is used with both generators and motors and is quite efficient. However, there are some problems. Slip-rings and commutators are made with copper and have slots to recess the wires from the coils. The brushes are made with a carbon compound which is somewhat softer than the copper. The spinning action of the armature causes a constant rubbing on the brushes, wearing them down.

The wearing of the brushes is seldom perfectly uniform. Uneven wear can lead to sparking between the brushes and rings. Any sparking is a loss of power, and it also

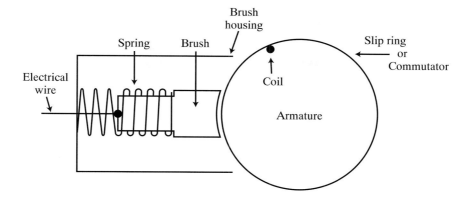

Figure 17-5. Brush assemblies are used in both ac and dc generators. This is an expanded view of a brush assembly.

leads to further wear. Eventually, the brushes wear down to where they do not make enough contact for operation. At this time, it is necessary to replace the brushes. The tension spring should also be replaced at this time.

Armature

To this point in the text, the armature has been represented by one coil spinning inside the magnetic field. In an actual generator, the armature is made with many coils. As shown in figure 17-6, slots are cut into the iron core armature that allow the coils to be recessed. The brushes and commutator assembly are attached to the end of the armature.

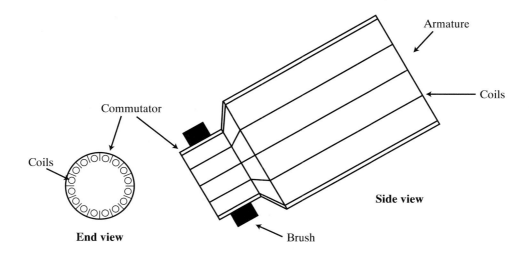

Figure 17-6. Armatures contain many coils.

17.3 EXCITING THE FIELDS

When a magnetic field is produced, it is **excited.** There are five basic methods of producing the magnetic fields in a generator.
- Permanent magnets.
- Independent dc voltage.
- Self-excited shunt connections.
- Self-excited series connections.
- Combinations of self-excited shunt and series connections.

Each of these methods has advantages and disadvantages.

Permanent Magnet Fields

Permanent magnets are used in very small generators. There are no windings and the field strength is uniform regardless of the load conditions. The advantages of using permanent magnets are a lower building cost and a much simpler construction. The disadvantages of using permanent magnets are the limited amount of output voltage and current that can be produced. There is also no speed or voltage regulation.

Figure 17-7 shows a four-pole permanent magnet generator. A typical application for this generator is to power the lights on a bicycle. The output voltage, applied to the load, is taken from the set of brushes. A sine wave can be produced if slip rings are used. Pulsating dc is developed when a commutator is used.

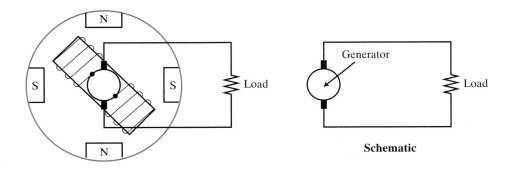

Figure 17-7. Four pole dc generator with permanent magnet fields.

Independently Excited Generator

The limited output of a permanent magnet generator can be overcome, to some degree, using electromagnetic poles. A separate dc voltage can be applied to the field windings, as shown in figure 17-8. This is used in an **independently excited generator.**

The advantage of this method is a constant, strong field strength. Field current, which determines strength, is controlled by a rheostat. Although this figure uses a variable resistor as the control, the rheostat can be replaced by electronic circuitry. This circuitry can be designed to accurately make adjustments to compensate for changing load demands.

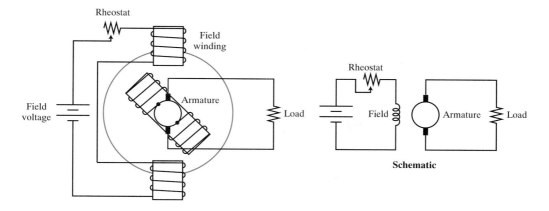

Figure 17-8. Independently excited generator.

The disadvantage of an independently excited generator is the need for a separate dc voltage source. The energy used to develop the magnetic field is not applied to the load. Therefore, the field energy is wasted power. This lowers the overall efficiency of the generator.

The schematic diagram for this generator appears to be two separate circuits. Electrically, they are separate. However, they are connected magnetically. In terms of total power consumption, which also includes the mechanical driving force, the field circuit and the armature circuit are one system.

Self-Excited Shunt DC Generator

The need for a separate dc power supply for the field windings can be overcome. A small amount of the voltage developed by the armature can be used to energize the field. The **self-excited shunt generator** connects the field winding in parallel with the armature winding, as shown in figure 17-9.

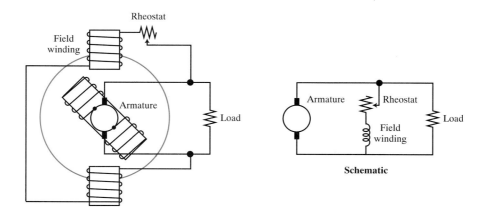

Figure 17-9. Self-excited shunt dc generator.

As with all parallel circuits, the current from the source is split to the parallel branches. The armature is the source of current in a generator. A small amount of current from the armature energizes the field. Field current is very small in comparison to load current.

A voltage is induced into the armature by the magnetism in the field windings. The initial voltage is developed by residual magnetism left in the cores of the poles. It may be necessary to apply a charge to the field when first starting the generator.

As the load draws more or less current, the field strength also changes. This self-changing field strength gives the shunt generator a constant output voltage. This is an advantage of this type of winding connection.

Self-Excited Series DC Generator

A **self-excited series generator** has field windings connected in series with the armature and load. The same current that goes through the load also passes through the field windings. The field windings have a small number of turns of a large size wire. This keeps the resistance low. The field windings must have a low resistance to reduce the voltage drop. The field windings must also have a high current capacity, a capacity equal to that of the armature. Figure 17-10 shows the connections for a series generator.

It is the current that creates a magnetic field. An increase in current also increases the output voltage, up to a certain point. The field windings will reach a saturation point where there is no further increase in magnetism or output voltage. If the generator is operated in this range, there can be a large change in the output resistance, yet the current will remain constant.

The series wound generator is used in applications where a constant current is required. An example is a series circuit arc lamp. Arc lamps produce a very intense light. They are used in the production of silk screens. A large amount of current is needed to maintain the arc in these devices.

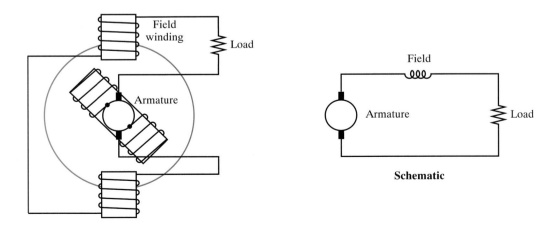

Figure 17-10. Self-excited series dc generator.

Compound Generator

A **compound generator** has both a series field winding and a shunt field winding. By using the combination of field windings, the advantages of each can be utilized. The shunt generator produces a constant voltage, from no load to full load. The series generator produces a constant current over a wide range of load resistances. The compound generator has the most constant output possible.

The four-pole generator shown in figure 17-11 is an example of one wiring configuration. The shunt winding can be connected in different configurations with respect to the load and armature. The connection of the shunt winding, in relation to the load and armature, varies the generator's operating characteristics.

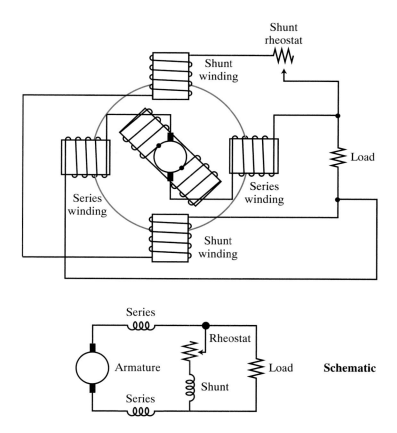

Figure 17-11. Compound dc generator.

Effects of Load Current

With no load applied to a generator, the armature speed increases until it reaches some maximum. The fields increase to a point of saturation and the output voltage is at its maximum value.

As the load current increases, the output begins to drop due to system losses. System voltage drops include the winding resistance, a decrease in the field current, and armature reaction.

Armature reaction is magnetic resistance. As the armature coil passes through the magnetic field, the induced voltage opposes the magnetic field. An increase in armature current also increases the armature's back emf.

17.4 BASIC MOTOR ACTION

Motor action is the attraction and repulsion of like and unlike magnetic fields. In figure 17-12, motor action is used to push the rotor away from like poles and toward unlike poles.

A motor is constructed in a manner very similar to a generator. A generator uses an external driving force to turn the armature, converting mechanical energy to electrical energy. In a motor, electricity is *applied*. The electricity turns the armature converting electrical energy to mechanical energy.

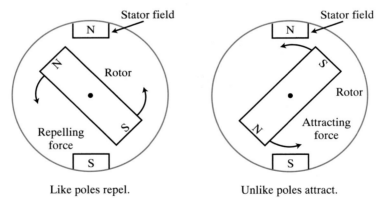

Figure 17-12. Motor action causes the rotor to spin.

Parts of a Motor

The armature is also called the **rotor** in an electric motor. The rotor is the part of a motor that spins. In a motor, the rotor drives the gears connected to the output shaft. The rotor is connected to the output shaft to perform mechanical work.

The **stator** contains the stationary magnetic fields of a motor. In almost all motors, the stator is an electromagnet. The stator serves the same function as the field windings in a generator. The stator produces a magnetic field.

17.5 DC MOTORS

A dc motor does *not* necessarily mean that the voltage applied to it is dc. AC voltages are used with dc motors. Brushes and a commutator convert the ac voltage to the dc voltage that drives the motor.

Electricity to drive the dc motor is applied to the brushes. The brushes press against the commutator, allowing current to flow. This creates a magnetic field in the rotor. The stator windings also receive current, either as a series connection, shunt connection, or combination of the two. The magnetic fields of the rotor and stator oppose and attract each other, depending on the position of the commutator. This creates the motor action.

A counter emf is developed in the rotor coil as it rotates through the stator fields. The counter emf is equal to the applied voltage, minus the voltage drop due to the windings.

Effect of Motor Load

Motor load is a mechanical force applied to the output shaft, which is connected to the rotor. An increase in the mechanical load draws more current from the power source. This can be explained as follows:

1.   An increasing load causes the motor speed to decrease.
2.   A decrease in speed causes a decrease in counter emf, which appears as a reduction in resistance to the electrical power source.
3.   The decrease in resistance allows more current to flow, which develops a stronger magnetic field, and increases the strength (torque) of the motor to drive the heavier load.

The end result is a constant rotor speed with increased current and torque.

Starting and Speed Control

Before the rotor begins spinning, there is no counter emf. Counter emf makes up the majority of electrical resistance. The resistance of the rotor coil is very low and starting currents can be as much as ten times higher than the running current. An example of the large current draw during start-up can be seen as the lights in a room dim when a large machine, such as a table saw or washing machine, is turned on.

Protection must be provided to prevent excessively high starting currents. Figure 17-13 is a combination starter circuit and speed control. When starting, the variable resistor applies full resistance to the rotor coil and minimum resistance to the stator. If a higher speed is desired, less resistance is applied to the rotor, and more resistance is applied to the stator.

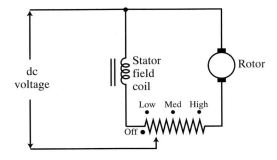

Figure 17-13. Starting and speed control of a dc motor using a variable resistor.

Direction of Rotation

With a dc motor, changing the direction of rotation is the simple matter of changing the polarity applied to either the rotor or the stator. In figure 17-14, a double-pole switch is used to change the polarity of the voltage applied to the rotor winding. Polarity is determined by a comparison of the magnetic fields.

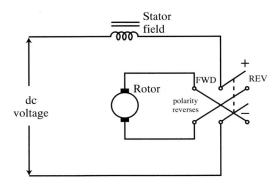

Figure 17-14. A switch changes the polarity of the rotor windings, which reverses the dc motor.

17.6 AC MOTORS

Motors operating on ac voltages fall into two general categories, induction and synchronous. **Induction motors** have a magnetically revolving stator field. Induction motors do not require stator windings. In **synchronous motors,** the rotor matches its speed to the line frequency. At **synchronous speed** the rotor turns at a speed equal to the speed of the rotating magnetic field.

Induction Motors

Induction motors are usually identified by the method that is used to start them. The basic types are split-phase, capacitor start, and shaded-pole.

One of the primary advantages of an induction motor is that it has no brushes, commutator, or slip-rings. This results in less parts to wear out. In addition, an induction motor can be installed directly to residential wiring with a minimum of controls. The induction motor is used in most home appliances. Some examples are: washing machines, clothes dryers, stationary power tools, refrigerators, and air conditioners.

The squirrel-cage rotor, shown in figure 17-15, is one type of rotor used with induction motors. It is made with heavy copper bars and shorting end rings, used to electrically connect the bars together. The bars act like armature coils, having the added advantage of handling very high currents.

The basic induction motor, shown in figure 17-16, has pulsating magnetic fields as the stator windings follow the applied sine wave voltage. At rest, rotor currents are set as shown in the first part of the figure. There is essentially no opposition to the passage of magnetic flux.

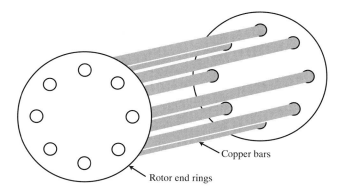

Figure 17-15. Drawing of a squirrel-cage rotor. This type of rotor is used in ac induction motors.

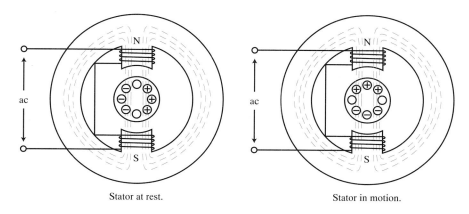

Stator at rest.   Stator in motion.

Figure 17-16. Current in a basic induction motor.

With rotation, the induced voltage in the rotor lags the stator voltage by almost 90 degrees. Currents are established, as shown in the second part of figure 17-16. The position of the poles appears to have magnetically rotated.

Induction motors require some form of automatic starting. The rotor can not start spinning by itself. Once it starts spinning it is able to continue. There are several different methods used to start induction motors. These methods include:
- Split-phase induction.
- Capacitor-start induction.
- Shaded-pole induction.

Split-Phase Induction Motors

The **split-phase induction motor** uses two types of stator coils. A starting coil is placed 90 electrical degrees from the running coil. The starting coil has high resistance with low inductance. The running coil is the opposite, with a low resistance and high inductance. The different characteristics of the coils induces a rotating field in the stator, starting the motor.

Figure 17-17 is the schematic diagram of the split-phase motor. Since the rotor does not draw power from the applied voltage it is shown not to have any electrical connection with the power source. The starting and running coils are in parallel. The starting winding has a **centrifugal switch** connected in series to remove the starting winding from the circuit. The switch is physically mounted on the rotor. Centrifugal force, created by the spinning motion of the rotor, opens the electrical contacts of the switch at approximately 75 percent of full speed. This removes the starting coil from the circuit.

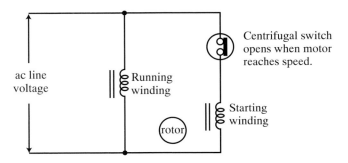

Figure 17-17. Split-phase induction motor.

Capacitor-Start Induction Motor

The **capacitor-start induction motor,** shown in figure 17-18 has a schematic similar to the split-phase motor, except it also has a capacitor in series with the starting winding. The capacitor is used to aid in the phase shift of the magnetic fields.

Some capacitor-type motors do not use a centrifugal switch. In these motors, the starting coil has resistance and inductance characteristics that are the same as the running coil. The capacitor creates the needed phase shift for starting. It is in the circuit at

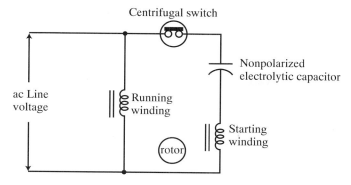

Figure 17-18. Capacitor-start motor.

all times. The advantage of this type of motor is that there is no centrifugal switch to malfunction.

Shaded-Pole Induction Motor

In a **shaded-pole induction motor,** a small portion of the stator field coil has a notch cut into it. Around the notch is a copper ring, called a **shading ring,** see figure 17-19.

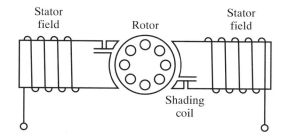

Figure 17-19. Note the notch cut into the shaded-pole induction motor.

Current is induced in the shading ring. This current develops an emf that *opposes changes* in the applied sine wave. Examine figure 17-20. When the sine wave is increasing, there is more flux developed on the side of the pole opposite the shading ring. Notice in the figure that the lines of flux are closer together on the left side. As the sine wave reaches maximum, the changes in the applied voltage are much less. The flux density is even throughout the core in center of figure 17-20. As the sine wave starts to decrease (right), the emf developed in the shading ring opposes the change. This adds to the field strength.

As a result of the shading ring, the field strength moves across the pole face with the changes in the applied sine wave. This moving field starts the rotor in motion.

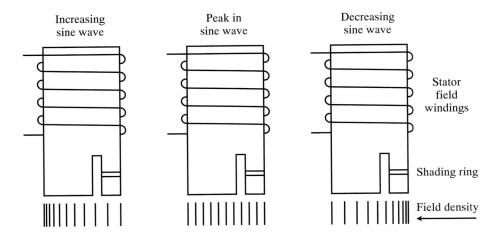

Figure 17-20. The sine wave causes the field strength to move in a shaded pole.

Rotor Slippage

With an induction motor, the rotor can not rotate at full synchronous speed. If the rotor did rotate at full synchronous speed, there would be no induced emf in the rotor bars. Rotor current would stop and rotation would also stop.

With no mechanical load applied to the output shaft, the rotor will rotate at very close to synchronous speed. Losses occur due to bearing friction and resistance of the bars. As the load applied to the output shaft increases, the mechanical pressure requires more rotor current to maintain the magnetic fields. The increase in current requires a shifting in the position of the induced voltage.

The difference between the speed of the rotating magnetic field of the stator and the running speed of the motor is called the **rotor slip.**

Synchronous Motors

Rotors for a synchronous motor match their speed to the frequency of the applied ac voltage. These rotors are made of a material with good residual magnetism. Synchronous motor rotors hold a magnetic field for a longer period of time than the core material of an induction motor.

In figure 17-21, the stator field coil is shown as a continuous coil. It is typical with synchronous motors to use a large number of poles. The rotor is also cut to have many individual poles. The higher the number of poles, the more stable the output speed as the load varies.

The residual magnetism of the rotor lags behind the changing sine wave. During the first half cycle, the magnetic relationship is with opposite poles on the stator and rotor, as shown in figure 17-22. When the sine wave cycle changes in the stator, the rotor presents

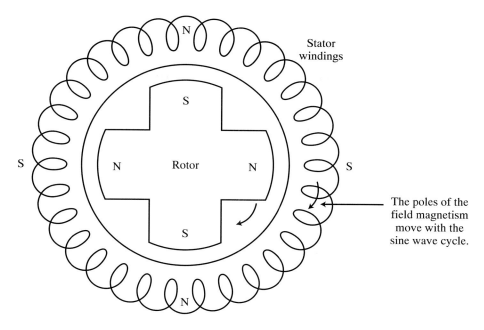

Figure 17-21. In a synchronous motor, the sine wave causes the field magnetism to rotate. The rotor locks on to follow the field.

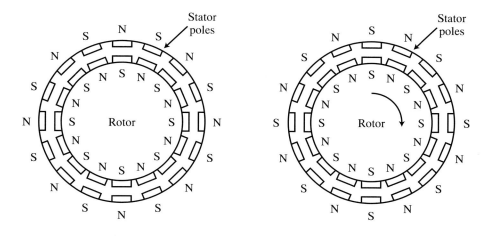

Figure 17-22. Poles rotate in the synchronous motor.

a like pole, which is repelled and forced to move to the next position. The rotor spins, always trying to have opposite poles lined up.

Synchronous motors are used in applications where the speed of the motor is critical. Typical applications include electric clocks and other timers. In many motors, the rotor is attached to a gear train. A gear train allows the rotor to spin at high revolutions per minute (rpm), while the speed of the output shaft is much slower. Remember, the

60 hertz line frequency that is used in homes and businesses is 60 cycles per second, or 3600 cycles per minute. The second hand on an electric clock moves at one revolution per minute.

17.7 MOTOR RATINGS

Electric motors have a very wide range of sizes. There are miniatures used to drive electric wrist watches. Larger sizes are used for household clocks, fans, power tools, the starter in an automobile, and windshield wipers. Even larger motors power major household appliances, winches, and large cranes. The motors shown in figure 17-23 are small motors used in electronic equipment.

Motors are given ratings in a series of areas.

- Horsepower.
- Torque.
- Voltage.
- Frequency.
- Output speed.
- Duty cycle.

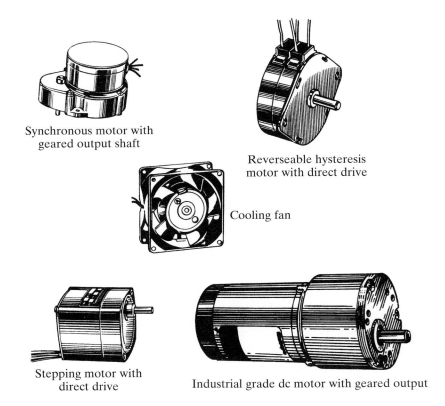

Synchronous motor with geared output shaft

Reverseable hysteresis motor with direct drive

Cooling fan

Stepping motor with direct drive

Industrial grade dc motor with geared output

Figure 17-23. Electric motors find many uses in small electronic equipment.

Ratings can be applied to all motors, regardless of size. However, some ratings are meaningless in certain applications. When examining a motor rating, be sure to consider the application for which you are using the motor.

Horsepower

The strength of a motor is the amount of mechanical force it can produce. There are three basic measurements used to measure the work capacity of a motor: horsepower, torque, and watts.

Horsepower (hp) is defined as moving 33,000 pounds one foot in one minute, or moving 550 pounds one foot in one second. The motor used in a household appliance, such as a washing machine, is typically 1/2 horsepower. One horsepower also equals 746 watts. The electrical rating for work is the watt. Generally speaking, when the wattage is given for a motor, it is to supply information of the electrical input requirements. The formula for conversion of horsepower to watts is:

Formula 17.A

$$\text{Power} = \frac{746 \text{ watts}}{1 \text{ horsepower}} \times \text{motor rating (hp)}$$

Sample problem 1.

What is the electric power, in watts, of a 1/2 horsepower motor, assuming an average load?

Formula: $\text{Power} = \dfrac{746 \text{ watts}}{1 \text{ horsepower}} \times \text{motor rating (hp)}$

Substitution: $\text{Power} = \dfrac{746 \text{ watts}}{1 \text{ horsepower}} \times 1/2 \text{ hp}$

Answer: $\text{Power} = 373 \text{ watts}$

Torque

The rotational force of a motor is the **torque.** Torque in a motor is related to the force produced on the conductors when rotating through the magnetic field. In terms of a rating, it is a measure of the force produced at the output shaft. If the rotor drives a gear train, the reduced speed from the gear train produces a higher torque than the rotor is capable of without gears. Torque is measured in foot-pounds (ft-lb) or ounce-inches (oz-in) in the cks system. In the mks system, torque is measured in newton meters (N·m). A typical small synchronous motor could have a rated torque of 100 oz-in (180 N·m).

Formula 17.B

$$\text{Torque} = \text{force} \times \text{distance}$$

Sample problem 2. _____

Calculate the torque of a small electric motor if it lifts a 30 ounce load six inches.

Formula: Torque = force × distance

Substitution: Torque = 30 ounces × 6 inches

Answer: Torque = 180 ounce-inches

Voltage

A voltage rating is given, as it is for all electrical devices. Motors can be found in any voltage rating desired. Both dc and ac voltages are available. Motor strength is not determined by the voltage rating. High torque motors are found in low voltages as well as high voltages. In the same way, low torque motors can be found in all voltages.

Frequency

AC motors have a rating for the frequency of the input voltage. This is especially important for synchronous motors, as the output speed varies with frequency. Typical motor frequencies are: 50 Hz (European), 60 Hz (U.S.), and 400 Hz (military). Other frequencies are available for special applications.

Output Speed

Small motors may be purchased with and without gear trains. The **speed rating** given is the speed at the output shaft of the motor. With certain applications, the rotor speed is also given. Output speed is measured in revolutions per minute (rpm). Motor speed without a gear train can be quite high, with very low torque. A gear train reduces the speed at the output shaft and increases the torque.

Duty Cycle

Motors have large values of rotor current, which can lead to high power consumption. The heat generated must be dissipated to prevent damage. In general, heat is dissipated easier with a larger part size or by attaching fan blades as part of the rotor.

In certain applications, a large surge of power is required for a short period of time. The motor is then given time to rest, allowing it to cool before the next surge of power. In applications such as this, the duty cycle affects the size and style of the motor. The **duty cycle** is the amount of time the motor is on as compared to the time the motor is off.

SUMMARY

- Generator action is the production of electricity by passing a conductor through a magnetic field.
- A generator is made of two basic parts, an armature and fields.
- To access the electricity produced: an ac generator uses slip rings and brushes, a dc generator uses split rings and brushes.
- Different methods of exciting the generator field have different advantages and disadvantages.
- Basic motor action is the producing of a mechanical force by the attraction and repulsion of magnetic fields.
- In a motor, the rotor is the part that rotates and the stator contains the stationary magnetic fields.
- AC motors fall into two general categories, induction and synchronous.
- Motors are rated for mechanical strength, horsepower, torque, and electrical characteristics.

KEY WORDS AND TERMS GLOSSARY

armature: The moving coil of wire in a generator spinning past the magnetic fields.

armature reaction: Magnetic resistance to the armature created when the armature passes through the magnetic field.

brush: Spring-mounted conductive material for making contact with the slip rings in ac generators or commutators of dc generators. The brushes connect the external load circuit to the armature.

capacitor-start induction motor: A split-phase motor with a capacitor in series with the starting winding.

centrifugal switch: A switch operated by the rotating motion of a motor. When the rotor reaches a certain speed, the switch contacts open.

commutator: A mechanical device used with generators to produce pulsating dc from a sine wave, similar to rectification. Commutators are also used with motors to change the applied sine wave to dc.

compound generator: A generator with both a series field winding and a shunt field winding.

duty cycle: The amount of time a circuit is on, as compared to the time off.

excited: The current producing a magnetic field. There are five basic methods of producing the magnetic fields in a generator: permanent magnets, independent dc voltage, self-excited shunt connections, self-excited series connections, combination of shunt and series connections.

field: The part of a generator containing the magnetic field. They may be made of permanent magnets or electromagnets.

generator: A device that uses mechanical energy to produce electrical energy.

horsepower (hp): The force required to move 33,000 pounds one foot in one minute, or 550 pounds one foot in one second. One horsepower also equals 746 watts.

independently excited generator: A generator in which the voltage to the field windings comes from a separate source.

induction motor: Motor having a magnetically revolving stator field. Induction motors do not require stator windings.

motor: A device that uses mechanical force to produce electrical energy.

motor action: The attraction and repulsion of like and unlike magnetic fields.

motor load: A mechanical force that is applied to the output shaft, which is connected to the rotor.

rotor: The part of an electric motor that spins. It is the same as the armature in a generator. In a motor, the rotor drives the gears connected to the output shaft. The rotor is the driving force of the mechanical energy produced. The rotor can be a series of laminated permanent magnets.

rotor slip: The difference between the speed of the rotating magnetic field of the stator and the running speed of the motor.

self-excited series generator: A generator that has the field winding connected in series with the armature and the load.

self-excited shunt generator: A generator that has the field winding connected in parallel with the armature winding.

shading ring: A small portion of the stator field coil has a notch cut into it in a shaded pole motor. Around the notch is a copper ring, called a *shading ring.* Current is induced in the shading ring.

slip rings: The armature coil wires are attached directly to a pair of *slip rings.* The slip rings spin with the armature.

speed rating: The speed at the output shaft of the motor.

split ring commutator: Rotates with the coil as the coil produces the sine wave to pass the generated electricity to the load.

split-phase induction motor: A motor with two types of stator coils. A starting coil is placed 90 electrical degrees apart from the running coil. The running coil has low resistance with high inductance. The starting coil has high resistance with low inductance. The different characteristics of the coils induce a rotating field in the stator, starting the motor.

stator: The stationary magnetic fields of a motor. In almost all motors, the stator is electromagnetic. It serves much the same function as the field windings in a generator.

synchronous motors: The rotor matches its speed to the line frequency.

synchronous speed: The speed required for the rotor to turn to equal the speed of the rotating magnetic field.

torque: 1. Rotational force. 2. Torque in a motor is related to the force produced on the conductors when rotating through the magnetic field. 3. Torque rating, is a measure of the force produced at the output shaft. 4. Torque is measured in force × distance: foot-pounds (ft-lb) or ounce-inches (oz-in).

KEY FORMULAS

Formula 17.A

$$\text{Power} = \frac{746 \text{ watts}}{1 \text{ horsepower}} \times \text{motor rating (hp)}$$

Formula 17.B

$$\text{Torque} = \text{force} \times \text{distance}$$

TEST YOUR KNOWLEDGE

Do not write in this text. Please use a separate sheet of paper.

1. How are generators and motors different?
2. Describe the basic generator action of producing electricity.
3. State the functions of the armature and fields of a generator.
4. Name the parts of a commutator and describe their function.
5. List and explain the methods of exciting a generator field.
6. Describe the basic action of an electric motor.
7. Describe the operation of a dc motor.
8. Describe the operation of an ac motor.
9. State the ratings for motors.
10. Calculate the wattage of a 3.5 horsepower electric motor.
11. Calculate the horsepower of a motor if it has an electric power rating of 1500 watts.
12. Calculate the torque exerted by an electric motor lifting two pounds a distance of three feet.

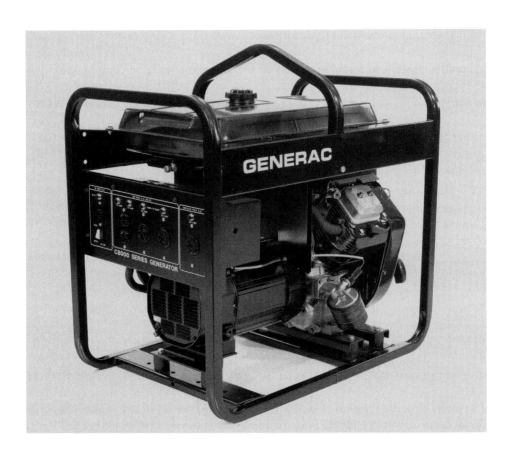

Generators convert mechanical energy into electrical energy. This generator produces electricity in 12 volt dc, 120 volt ac, and 240 volt ac levels. (Generac Corp.)

Chapter 18
Inductive Reactance and Impedance

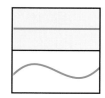

Upon completion of this chapter, you will be able to:
- Define reactance and impedance.
- Apply Ohm's law to inductive reactive circuits.
- Calculate inductive reactance.
- Examine the variables in the inductive reactance formula.
- Calculate the total reactance of series and parallel circuits.
- Examine the phase shift of current vs voltage in a purely inductive circuit.
- Construct phasor diagrams to show relationships in a series circuit with both reactance and resistance.
- Calculate impedance in a series circuit.
- Compare the effects of a larger and smaller X_L as compared to resistance in a series circuit.
- Construct phasor diagrams to show relationships in a parallel circuit with both reactance and resistance.
- Calculate impedance in a parallel circuit.
- Compare the effects of a larger and smaller X_L as compared to resistance in a parallel circuit.
- Measure phase angle with an oscilloscope.

   Reactance is the ac resistance of an inductor or a capacitor. This chapter deals with **inductive reactance,** which is the ac resistance of an inductor. Chapter 19 is capacitive reactance, the ac resistance of a capacitor. Reactance does not include the resistance of the component or any other internal losses.

   Impedance is the total ac resistance of a circuit containing both resistance and reactance. Impedance is discussed in detail both in this chapter, with inductive reactance, and Chapter 19, with capacitive reactance. The term impedance refers to the ac resistance of a wide variety of components used in electronics.

18.1 DC AND AC RESISTANCE OF AN INDUCTOR

An electrical current produces magnetism in a wire. The produced magnetism induces a counter emf (voltage) in the wire. As presented in Chapter 12, the counter emf is the property of an inductor that opposes any change in current. It is this opposition to a changing current that gives the inductor its reactance in an ac circuit.

With a dc circuit, the inductor opposes the current for a period of time equal to five time constants, as discussed in Chapter 14. After that period of time, the steady state current is opposed only by the circuit resistance.

In figure 18-1, the dc circuit is assumed to have reached steady state. A current of five amps is calculated by the resistance and applied voltage. After steady state is reached, the inductor has no effect on current. The inductor acts like a piece of wire to the steady-flowing dc current.

Figure 18-2 applies an ac voltage to the series RL circuit. The inductor has an inductive reactance of 1000 ohms. With a combined ac resistance of 1000 ohms, the current is measured at 0.05 amps (50 mA). The ac resistance, with the inductor, is much higher than with the resistor alone. Later in this chapter, calculation of the total ac resistance, the impedance, is discussed. The impedance of a circuit is not found by adding the resistance with the reactance. It is found using vectors, in a process called phasor addition.

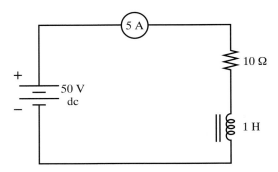

Figure 18-1. This dc circuit shows 10 ohms combined resistance.

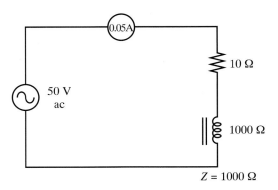

Figure 18-2. With an ac voltage applied, this circuit shows 1000 ohms combined ac resistance.

18.2 CALCULATING INDUCTIVE REACTANCE

Inductive reactance cannot be measured with an ohmmeter. If an ohmmeter is used with an inductor, only the dc resistance of the coil is measured.

Reactance can be measured indirectly. It is found by measuring current and the applied voltage. Ohm's law is used with the measured values to calculate the circuit's effective ac resistance. The letter symbol for inductive reactance is X_L. When using Ohm's law, X_L is used in place of R in the formulas. The unit of measure for reactance is **ohms.**

Sample problem 1. _____

Use Ohm's law to calculate the inductive reactance, X_L, of the circuit shown in figure 18-3.

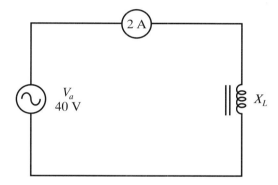

Figure 18-3. Calculate X_L given current and the voltage. (Sample problem 1)

    Formula: $X_L = \dfrac{V_a}{I}$

    Substitution: $X_L = \dfrac{40 \text{ V}}{2 \text{ A}}$

    Answer: $X_L = 20 \ \Omega$

Sample problem 2. _____

Use Ohm's law to calculate the current in the circuit shown in figure 18-4.

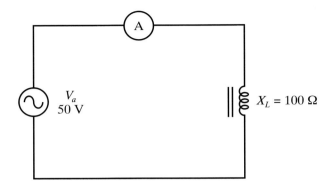

Figure 18-4. Calculate current given X_L and the voltage. (Sample problem 2)

Formula: $I = \dfrac{V_a}{X_L}$

Substitution: $I = \dfrac{50 \text{ V}}{100 \text{ }\Omega}$

Answer: $I = 0.5 \text{ A}$

The Inductive Reactance Formula

Inductive reactance can be calculated when the values of frequency and inductance are known. When X_L is known, the formula can be rearranged to find the values of either frequency or inductance.

Formula 18.A

$$X_L = 2\pi f L$$

X_L is inductive reactance, measured in ohms.

2π can be either 6.28 or 2 × π on a calculator.

f is frequency, measured in hertz.

L is inductance, measured in henrys.

Sample problem 3. _____

Use the values given in figure 18-5 to calculate the inductive reactance.

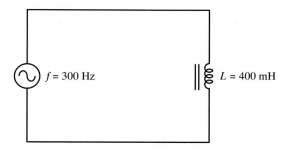

Figure 18-5. Calculate the inductive reactance. (Sample problem 3)

Formula: $X_L = 2\pi f L$

Substitution: $X_L = 2 \times \pi \times 300 \text{ Hz} \times 400 \text{ mH}$

Note: the value of π on an electronic calculator is used here.

Answer: $X_L = 754 \ \Omega$

Sample problem 4. _____

A circuit having an applied voltage with a frequency of 16 kHz has inductance of 45 mH. Calculate the inductive reactance.

Formula: $X_L = 2\pi f L$

Substitution: $X_L = 2 \times \pi \times 16 \text{ kHz} \times 45 \text{ mH}$

Answer: $X_L = 4524 \ \Omega$

Sample problem 5. _____

Using the values given in figure 18-6 to calculate the value of inductance.

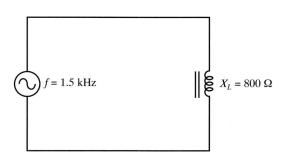

Figure 18-6. Calculate the inductive reactance. (Sample problem 5)

Formula: $X_L = 2\pi fL$

Rearranging: $L = \dfrac{X_L}{2\pi f}$

Substitution: $L = \dfrac{800 \ \Omega}{2 \times \pi \times 1.5 \ \text{kHz}}$

Answer: $L = 84.9 \ \text{mH}$

Sample problem 6.

Calculate the value for an inductor that will produce an inductive reactance of 250 ohms, with a frequency of 60 hertz.

Formula: $L = \dfrac{X_L}{2\pi f}$

Substitution: $L = \dfrac{250 \ \Omega}{2 \times \pi \times 60 \ \text{Hz}}$

Answer: $L = 0.663 \ \text{H} = 663 \ \text{mH}$

Sample problem 7.

Use the values given in the circuit of figure 18-7 to determine the frequency of the applied voltage.

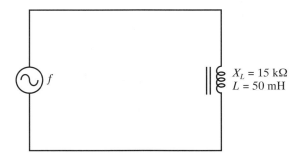

Figure 18-7. Calculate the frequency of the applied voltage. (Sample problem 7)

Formula: $X_L = 2\pi fL$

Rearranging: $f = \dfrac{X_L}{2\pi L}$

Substitution: $f = \dfrac{15 \ \Omega}{2 \times \pi \times 50 \ \text{mH}}$

Answer: $f = 47.7 \ \text{kHz}$

Sample problem 8. _____

At what frequency will a circuit containing a 50 mH inductor have an inductive reactance of 20 Ω?

Formula: $f = \dfrac{X_L}{2\pi L}$

Substitution: $f = \dfrac{20 \ \Omega}{2 \times \pi \times 50 \ \text{mH}}$

Answer: $f = 63.7 \ \text{Hz}$

Effect of Frequency and Inductance on Inductive Reactance

Both frequency and inductance are directly related to the value of inductive reactance. With an inductor of a specified value, if the frequency is increased, the reactance is also increased. In the same manner, if the frequency is set at a predetermined value, an increase in inductance increases the reactance. A decrease in frequency or a decrease in inductance causes a decrease in reactance.

Figure 18-8 contains two graphs of the effects on inductive reactance. One graph shows inductive reactance vs frequency. The other graph diagrams inductive reactance vs inductance.

18.3 REACTANCES IN SERIES AND PARALLEL

Reactances, when connected in series or parallel, combine to form a total reactance in the same manner as resistors in a dc circuit. If the inductors are in series, add the reactances. If the inductors are in parallel, use the reciprocal formula.

Formula 18.B (inductors in series)

$$X_{L_T} = X_{L_1} + X_{L_2} + X_{L_3} + \ldots X_{L_N}$$

X_{L_T} is the total inductive reactance, measured in ohms.

X_{L_1} through X_{L_N} are the individual inductive reactances, measured in ohms.

Formula 18.C (inductors in parallel)

$$\frac{1}{X_{L_T}} = \frac{1}{X_{L_1}} + \frac{1}{X_{L_2}} + \frac{1}{X_{L_3}} + \ldots \frac{1}{X_{L_N}}$$

X_{L_T} is the total inductive reactance, measured in ohms.

X_{L_1} through X_{L_N} are the individual inductive reactances, measured in ohms.

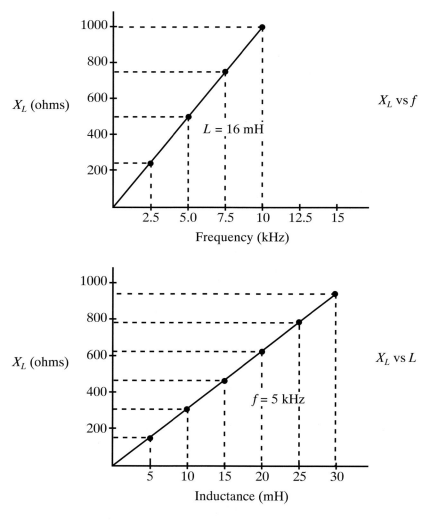

Figure 18-8. Diagraming the effects on inductive reactance.

Sample problem 9. (series)

Find the total inductive reactance of the circuit shown in figure 18-9.

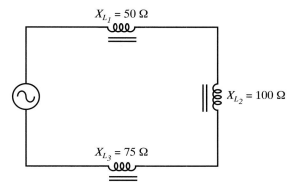

Figure 18-9. Series inductances add. (Sample problem 9)

Formula: $X_{L_T} = X_{L_1} + X_{L_2} + X_{L_3} + ... X_{L_N}$

Substitution: $X_{L_T} = 50\ \Omega + 100\ \Omega + 75\ \Omega$

Answer: $X_{L_T} = 225\ \Omega$

Sample problem 10. (series)

With the circuit shown in figure 18-10, find the total reactance by first finding the reactance of the individual inductors.

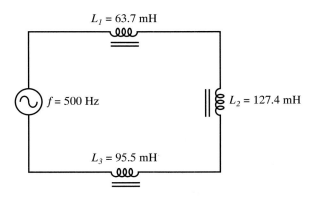

Figure 18-10. Find the total inductance. (Sample problems 10 and 11)

1. Individual reactances:

    Formula: $X_L = 2\pi f L$

    Substitution: $X_{L_1} = 2 \times \pi \times 500 \text{ Hz} \times 63.7 \text{ mH}$

    Answer: $X_{L_1} = 200 \ \Omega$

    Substitution: $X_{L_2} = 2 \times \pi \times 500 \text{ Hz} \times 127.4 \text{ mH}$

    Answer: $X_{L_2} = 400 \ \Omega$

    Substitution: $X_{L_3} = 2 \times \pi \times 500 \text{ Hz} \times 95.5 \text{ mH}$

    Answer: $X_{L_3} = 300 \ \Omega$

2. Total reactance:

    Formula: $X_{L_T} = X_{L_1} + X_{L_2} + X_{L_3} + \dots X_{L_N}$

    Substitution: $X_{L_T} = 200 \ \Omega + 400 \ \Omega + 300 \ \Omega$

    Answer: $X_{L_T} = 900 \ \Omega$

Sample problem 11. (series)

Once again, use the circuit shown in figure 18-10. This time, find the total reactance by first calculating total inductance using the series inductance formula found in Chapter 12. Then use total inductance to calculate total reactance.

1. Total inductance:

    Formula: $L_T = L_1 + L_2 + L_3 + \dots L_N$

    Substitution: $L_T = 63.7 \text{ mH} + 127.4 \text{ mH} + 95.5 \text{ mH}$

    Answer: $L_T = 286.6 \text{ mH}$

2. Total reactance:

    Formula: $X_{L_T} = 2\pi f L$

    Substitution: $X_{L_T} = 2 \times \pi \times 500 \text{ Hz} + 286.6 \text{ mH}$

    Answer: $X_{L_T} = 900 \ \Omega$

Sample problem 12. (parallel)

Find the total reactance of the circuit shown in figure 18-11.

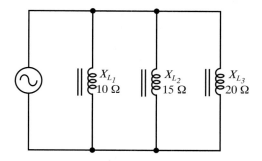

Figure 18-11. Parallel inductive reactances use the reciprocal formula. (Sample problem 12)

Formula:

$$\frac{1}{X_{L_T}} = \frac{1}{X_{L_1}} + \frac{1}{X_{L_2}} + \frac{1}{X_{L_3}} + \cdots \frac{1}{X_{L_N}}$$

Substitution:

$$\frac{1}{X_{L_T}} = \frac{1}{10 \ \Omega} + \frac{1}{15 \ \Omega} + \frac{1}{20 \ \Omega}$$

Decimal step:

$$\frac{1}{X_{L_T}} = 0.1 + 0.0667 + 0.05 = 0.2167$$

Answer:

$$X_{L_T} = 4.61 \ \Omega$$

Sample problem 13. (parallel)

Find the total reactance of the circuit in figure 18-12 by first finding the reactances of the individual inductors.

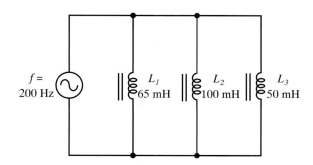

Figure 18-12. Find the total inductive reactance. (Sample problems 13 and 14)

1. Individual reactances:

    Formula: $X_L = 2\pi fL$

    Substitution: $X_{L_1} = 2 \times \pi \times 200 \text{ Hz} \times 65 \text{ mH}$

    $X_{L_1} = 81.7 \ \Omega$

    Substitution: $X_{L_2} = 2 \times \pi \times 200 \text{ Hz} \times 100 \text{ mH}$

    $X_{L_2} = 125.7 \ \Omega$

    Substitution: $X_{L_3} = 2 \times \pi \times 200 \text{ Hz} \times 50 \text{ mH}$

    $X_{L_3} = 62.8 \ \Omega$

2. Total reactance:

    Formula: $\dfrac{1}{X_{L_T}} = \dfrac{1}{X_{L_1}} + \dfrac{1}{X_{L_2}} + \dfrac{1}{X_{L_3}} + \dots \dfrac{1}{X_{L_N}}$

    Substitution: $\dfrac{1}{X_{L_T}} = \dfrac{1}{81.7 \ \Omega} + \dfrac{1}{125.7 \ \Omega} + \dfrac{1}{62.8 \ \Omega}$

    Decimal step: $\dfrac{1}{X_{L_T}} = 0.0122 + 0.00796 + 0.0159 = 0.0360$

    Answer: $X_{L_T} = 27.7 \ \Omega$

Sample problem 14. _____

Once again, use the circuit shown in figure 18-12. This time, find the total reactance by first calculating total inductance using the parallel inductance formula found in Chapter 13. Then, use total inductance to calculate total reactance.

1. Total inductance:

    Formula: $\dfrac{1}{L_T} = \dfrac{1}{L_1} + \dfrac{1}{L_2} + \dfrac{1}{L_3} + \dots \dfrac{1}{L_N}$

    Substitution: $\dfrac{1}{L_T} = \dfrac{1}{65 \text{ mH}} + \dfrac{1}{100 \text{ mH}} + \dfrac{1}{50 \text{ mH}}$

    Decimal step: $\dfrac{1}{L_T} = 15.4 + 10 + 20 = 45.4$

    Answer: $L_T = 22 \text{ mH}$

2. Total reactance:

Formula: $X_{L_T} = 2\pi fL$

Substitution: $X_{L_T} = 2 \times \pi \times 200 \text{ Hz} + 22 \text{ mH}$

Answer: $X_{L_T} = 27.7 \ \Omega$

18.4 CURRENT AND VOLTAGE IN PURE INDUCTIVE REACTANCE

A **pure inductive reactance** does not contain any resistance. It has only inductance. This is, of course, a theoretical condition. It is discussed for the purposes of understanding the relationships of voltage and current in an inductor.

When an ac voltage is applied to an inductor, the current is delayed due to the building up of the magnetic field. The current takes the shape of a sine wave (just like the voltage), except it will be delayed by 90°.

The amount of current is determined by the ac resistance in the circuit. If the circuit is a pure inductance, the ac resistance is the inductive reactance. Ohm's law can be used to calculate the current in the same manner as dc circuits were calculated in Chapters 6 and 7.

The circuit in figure 18-13 is a pure inductance connected to a sine wave voltage source. The current waveform is shown to start at zero. Notice that the voltage across the inductor reaches its peak 90° earlier than the current waveform.

All of the corresponding points of voltage lead the current by 90°. It can also be said that the current through an inductor lags the voltage across the inductor by 90°.

Also shown in figure 18-13 is a phasor diagram showing the relationship of voltage and current in a purely inductive circuit. A **phasor** is a vector, used to show the magnitude (size) and direction. The angle between the two phasors is the **phase angle.**

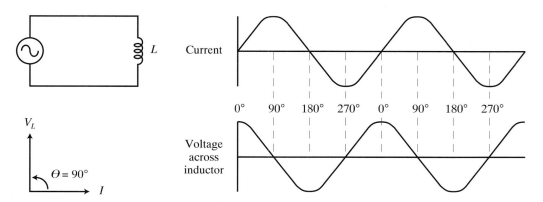

Figure 18-13. Voltage leads current by 90° in pure inductance.

18.5 SERIES RL AC CIRCUIT

Current through a resistor does not have a phase shift in relation to the voltage dropped across it. A very important point to remember is the current is the same at all points in a series circuit.

In a series circuit, the waveform for current is used as the reference. See figure 18-14. The resistor voltage drop waveform, V_R, is exactly in phase with the current. Also shown are the waveforms for inductor voltage, V_L, and the applied voltage, V_a. The inductor voltage leads the current, and resistor voltage, by 90°.

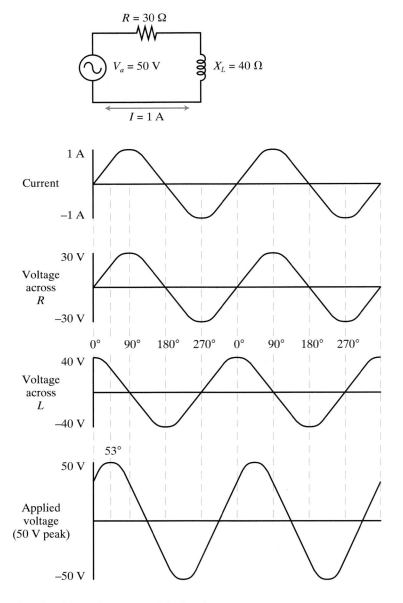

Figure 18-14. Series circuit with resistance and inductive reactance.

Phase Angle (θ) and Opposite Angle (ϕ)

The time difference between the applied voltage and the resistor voltage is called the circuit **phase shift,** or operating angle. The phase shift is represented by θ, the Greek letter theta. The amount of phase shift is determined by the relative sizes of the inductor and resistor. Calculations for this circuit are shown in sample problem 15. The terms "phase angle" and "phase shift" may be interchanged.

When comparing the shift of the inductor voltage with the applied voltage, V_L leads V_a by an amount equal to $90 - \theta$. This is referred to as the **opposite angle.** The opposite angle gets its name because in the phasor diagram, it is at the opposite side *from* θ. The opposite angle is represented by the Greek letter phi, ϕ.

Series Circuit Phasor Diagrams

There are two types of phasor diagrams used to solve a series circuit, voltage, and impedance. These are shown in figure 18-15. Each of these phasor diagrams can be drawn using a parallelogram or a triangle.

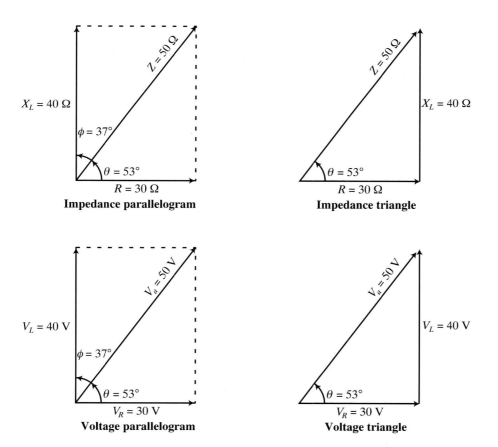

Figure 18-15. Phasor diagrams. (Sample problem 15)

The reference arrow is drawn as a horizontal line, 0°, starting at zero and pointing to the right. The reference is always the resistor phasor. It is either the resistor voltage or the resistor value. The current, if drawn, is also at 0°.

The inductor phasor is plotted straight up. If a parallelogram is to be used (figure 18-15, left side), the starting point for the inductor phasor is at zero (at the point where the resistor phasor begins). If a triangle is to be used (figure 18-15, right side), the inductor phasor is drawn straight up from the end of the resistor phasor.

Resultant Phasor

The resultant phasor, which connects the other two phasors, is the hypotenuse of the triangle, or the opposite corners of the parallelogram. This resultant phasor will be equal to either impedance or applied voltage. The resultant is found using the **Pythagorean theorem,** stated as:

$$a = \sqrt{b^2 + c^2}$$

This can be modified to correspond to the phasors.

a is the resultant phasor.

b and *c* are the initial two phasors.

Impedance in Series Circuit

Impedance is the total ac resistance, measured in ohms, of a circuit containing both resistance and reactance. The impedance has the same phase angle as the circuit voltage. The value of impedance is always larger than either the resistance or reactance. If either the resistance or reactance is very large in comparison to the other, the impedance will have a value very close to that of the larger quantity.

Impedance in series is calculated using the "square root of the sum of the squares formula." This is another way of stating the Pythagorean theorem, the formula for finding the hypotenuse of a right triangle.

There are several important formulas for calculating phasor diagrams in a series RL circuit. The impedance is calculated by:

Formula 18.D

$$Z = \sqrt{X_L^2 + R^2}$$

Z is the impedance, measured in ohms.

X_L is the inductive reactance, measured in ohms.

R is the resistance, measured in ohms.

The applied voltage is calculated by:

Formula 18.E

$$V_a = \sqrt{V_L^2 + V_R^2}$$

V_a is the applied voltage, measured in volts.

V_L is the voltage across the inductance, measured in volts.

V_R is the voltage across the resistance, measured in volts.

The phase angle is calculated by:

Formula 18.F

$$\theta = \tan^{-1} \frac{X_L}{R}$$

θ is the phase angle, measured in degrees.

X_L is the inductive reactance, measured in ohms.

R is the resistance, measured in ohms.

and

Formula 18.G

$$\theta = \tan^{-1} \frac{V_L}{V_R}$$

θ is the phase angle, measured in degrees.

V_L is the voltage across the inductance, measured in volts.

V_R is the voltage across the resistance, measured in volts.

The opposite angle is calculated by:

Formula 18.H

$$\phi = 90 - \theta$$

ϕ is the opposite angle, measured in degrees.

θ is the phase angle, measured in degrees.

The current is calculated by:

Formula 18.I

$$I = \frac{V}{Z}$$

I is the current, measured in amperes.

V is the voltage, measured in volts.

Z is the impedance, measured in ohms.

Sample problem 15. _____

In an ac circuit, a 30 ohm resistor is connected in series with a 40 ohm inductive reactance. Look back to figure 18-14. Calculate the information needed to plot the phasor diagrams shown in figure 18-15.

1. Calculate impedance and draw the impedance parallelogram and impedance triangle.

    Formula: $Z = \sqrt{X_L^2 + R^2}$

    Substitution: $Z = \sqrt{40^2 + 30^2}$

    Intermediate step: $Z = \sqrt{2500}$

    Answer: $Z = 50$ ohms

2. Calculate the operating angle, θ.

    Formula: $\theta = \tan^{-1} \dfrac{X_L}{R}$

    Substitution: $\theta = \tan^{-1} \dfrac{40\ \Omega}{30\ \Omega}$

    Intermediate step: $\theta = \tan^{-1} 1.333$

                          Note: to find the angle whose tangent is 1.333, enter the number into the calculator and use the inverse tan function.

    Answer: $\theta = 53.1°$

3. Calculate the opposite angle, ϕ. This angle is used when making certain measurements with the oscilloscope.

    Formula: $\phi = 90 - \theta$

    Substitution: $\phi = 90 - 53.1$

    Answer: $\phi = 36.9°$

4. Calculate circuit current using Ohm's law.

Formula: $I = \dfrac{V_A}{Z}$

Substitution: $I = \dfrac{50 \text{ V}}{50 \text{ }\Omega}$

Answer: $I = 1 \text{ A}$

5. Calculate voltage drops using Ohm's law.

a. Inductor voltage:

Formula: $V_L = I \times X_L$

Substitution: $V_L = 1 \text{ A} \times 40 \text{ }\Omega$

Answer: $V_L = 40 \text{ V}$

b. Resistor Voltage:

Formula: $V_R = I \times R$

Substitution: $V_R = 1 \text{ A} \times 30 \text{ }\Omega$

Answer: $V_R = 30 \text{ V}$

18.6 COMPARING X_L AND R IN A SERIES CIRCUIT

The relationship of the sizes of the inductive reactance and the resistance determine the shape of the phasor diagrams. If the reactance and resistance are equal, the triangle will have equal length sides and the phase angle will be 45°. If the reactance is smaller than the resistance, the triangle will be wider than it is tall. This produces a phase angle *less than* 45°. If the reactance is larger, the triangle will be taller than it is wide. This produces a phase angle *greater than* 45°.

Equal Resistance and Reactance

In a circuit with equal reactance and resistance, voltage drops across each component are equal and the phase angle is 45°, as shown in figure 18-16. In this figure, the value of inductance is given. The inductive reactance must be calculated using the reactance formula discussed earlier in this chapter.

Sample problem 16. _____

Calculate the inductive reactance, impedance, and phase angle of the circuit shown in figure 18-16.

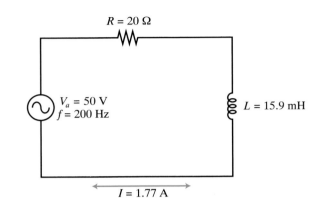

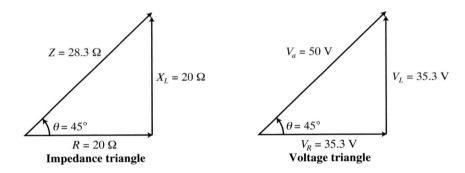

Figure 18-16. Diagram for when _R_ equals X_L in a series circuit. (Sample problem 16)

1. Calculate inductive reactance:

   Formula: $X_L = 2\pi fL$

   Substitution: $X_L = 2 \times \pi \times 200 \text{ Hz} \times 15.9 \text{ mH}$

   Answer: $X_L = 20 \ \Omega$

2. Calculate impedance:

   Formula: $Z = \sqrt{X_L^2 + R^2}$

   Substitution: $Z = \sqrt{20^2 + 20^2}$

   Answer: $Z = 28.3 \ \Omega$

3. Calculate the phase angle:

Formula: $\theta = \tan^{-1} \dfrac{X_L}{R}$

Substitution: $\theta = \tan^{-1} \dfrac{20\ \Omega}{20\ \Omega}$

Answer: $\theta = 45°$

Resistance Smaller than Reactance

When the resistance in a circuit is small compared to the reactance, the circuit impedance appears more inductive. The phase shift is closer to 90°, the characteristic of pure inductance.

Sample problem 17. _____

Calculate the impedance and phase angle of the circuit shown in figure 18-17.

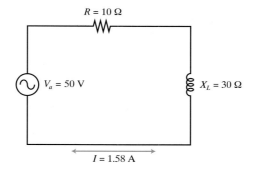

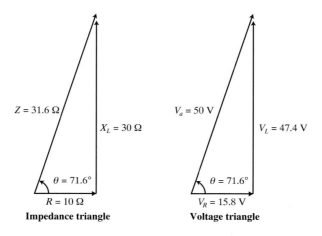

Figure 18-17. Diagram for when R is small in comparison to X_L in a series circuit. (Sample problem 17)

1. Impedance:

    Formula: $Z = \sqrt{X_L^2 + R^2}$

    Substitution: $Z = \sqrt{30^2 + 10^2}$

    Answer: $Z = 31.6\ \Omega$

2. Phase angle:

    Formula: $\theta = \tan^{-1}\dfrac{X_L}{R}$

    Substitution: $\theta = \tan^{-1}\dfrac{30\ \Omega}{10\ \Omega}$

    Answer: $\theta = 71.6°$

Resistance Larger than Reactance

When the resistance is large and the reactance is small, the circuit has characteristics more like the resistor. The impedance will be slightly larger than the resistor value. The phase angle will be closer to 0°.

Sample problem 18. _____

Using the circuit in figure 18-18, calculate the impedance and phase angle.

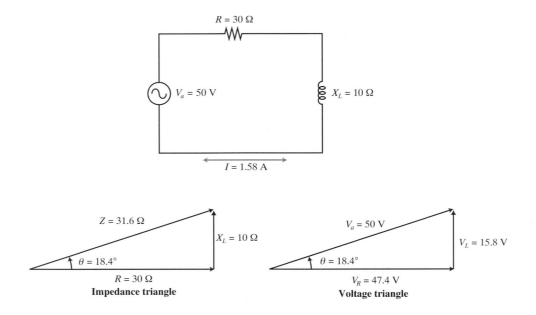

Figure 18-18. Diagram for when *R* is large in comparison to X_L in a series circuit. (Sample problem 18)

1. Impedance:

    Formula: $Z = \sqrt{X_L^2 + R^2}$

    Substitution: $Z = \sqrt{10^2 + 30^2}$

    Answer: $Z = 31.6 \ \Omega$

2. Phase angle:

    Formula: $\theta = \tan^{-1} \dfrac{X_L}{R}$

    Substitution: $\theta = \tan^{-1} \dfrac{10 \ \Omega}{30 \ \Omega}$

    Answer: $\theta = 18.4°$

18.7 PARALLEL RL AC CIRCUIT

In Chapter 7, you learned that voltage is the same throughout a parallel circuit. Current, in a parallel circuit, splits to each branch. The values of the components of the individual branches determines the amount of current.

The parallel circuit in figure 18-19 has several key points:

• Applied voltage is the reference at 0°.
• Resistive current is in phase with the applied voltage.
• Inductive current lags the voltage by 90°.
• Total current is found by using the "square root of the sum of the squares."

Parallel Circuit Phasor Diagram

The phasor diagrams for a parallel circuit in figure 18-19 are shown in figure 18-20. This figure shows both the parallelogram and the triangle methods. Resistive current, I_R, is plotted starting at zero and pointing to the right. Inductive current, I_L, is drawn down. This indicates an angle of −90°.

The negative angles are a reflection of the inductive current lagging the resistive current. In the series circuit, the inductive voltage and impedance have a positive angle.

Impedance in a Parallel Circuit

Impedance in a parallel circuit is found using Ohm's law. It does not require the use of phasors. The impedance has a phase angle equal to the angle for total current.

The total current is calculated by:

Formula 18.J

$$I_T = \sqrt{I_L^2 + I_R^2}$$

I_T is the total current, measured in amperes.

I_L is the current through the inductance, measured in amperes.

I_R is the current through the resistance, measured in amperes.

The phase angle is calculated by:

Formula 18.K

$$\theta = \tan^{-1} \frac{-I_L}{I_R}$$

θ is the phase angle, measured in degrees.

I_L is the current across the inductance, measured in amperes.

I_R is the current across the resistance, measured in amperes.

The opposite angle is calculated by:

Formula 18.L

$$\phi = -90 - \theta$$

ϕ is the opposite angle, measured in degrees.

θ is the phase angle, measured in degrees.

The impedance is calculated by:

Formula 18.M

$$Z = \frac{V_a}{I_T}$$

Z is the impedance, measured in ohms.

I_T is the total current, measured in amperes.

V_a is the applied voltage, measured in volts.

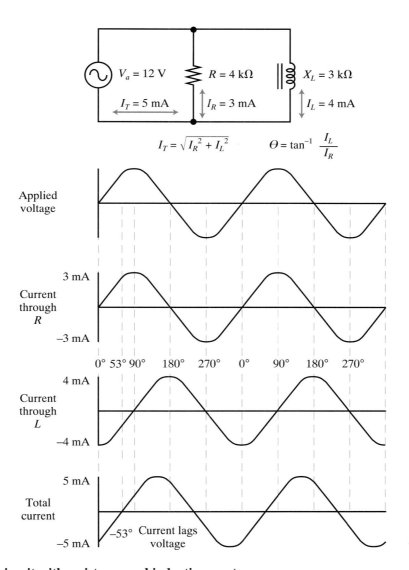

Figure 18-19. Parallel circuit with resistance and inductive reactance.

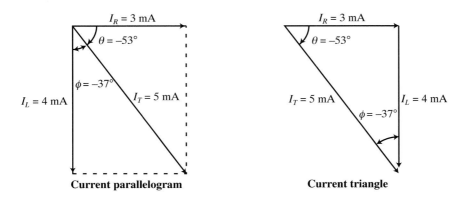

Figure 18-20. Phasor diagrams. (Sample problem 19)

Sample problem 19. _____

Using the values given in the circuit in figure 18-19, calculate branch currents, total current, phase angle, and impedance. Phasor drawings for this circuit are shown in figure 18-20.

1. Calculate branch currents using Ohm's law.

    a.   Resistive current:

    Formula: $I_R = \dfrac{V_a}{R}$

    Substitution: $I_R = \dfrac{12\ \text{V}}{4\ \text{k}\Omega}$

    Answer: $I_R = 3\ \text{mA}$

    b.   Inductive current:

    Formula: $I_L = \dfrac{V_a}{X_L}$

    Substitution: $I_L = \dfrac{12\ \text{V}}{3\ \text{k}\Omega}$

    Answer: $I_L = 4\ \text{mA}$

2. Calculate total current using the Pythagorean theorem.

    Formula: $I_T = \sqrt{I_L^2 + I_R^2}$

    Substitution: $I_T = \sqrt{(4\ \text{mA})^2 + (3\ \text{mA})^2}$

    Answer: $I_T = 5\ \text{mA}$

3. Calculate the phase angle as shown in the completed phasor diagrams.

    Formula: $\theta = \tan^{-1} \dfrac{-I_L}{I_R}$

    Substitution: $\theta = \tan^{-1} \dfrac{-4\ \text{mA}}{3\ \text{mA}}$

    Intermediate step: $\theta = \tan^{-1} -1.333$

    Answer: $\theta = -53.1°$

4. Calculate impedance using Ohm's law. The phase angle for the impedance is the same as for the total current.

Formula: $Z = \dfrac{V_a}{I_T}$

Substitution: $Z = \dfrac{12 \text{ V}}{5 \text{ mA}}$

Answer: $Z = 2.4 \text{ k}\Omega$

18.8 COMPARING X_L AND R IN A PARALLEL CIRCUIT

In a parallel circuit, it is the branch currents that determine the operating characteristics. Keep in mind, more current flows with less resistance. Since the parallel circuit uses a phasor *current* triangle, the length of the phasors are the inverse of the branch resistance.

Equal Resistance and Reactance

Equal R and X_L produce equal branch currents. As it did in the series circuit, the triangle has equal sides with a phase angle of 45°. The current triangle for a parallel RL circuit has a negative phase angle.

Sample problem 20. _____

Using the values of resistance and inductance given for the circuit shown in figure 18-21, calculate inductive reactance, branch currents, total current, phase angle, and impedance.

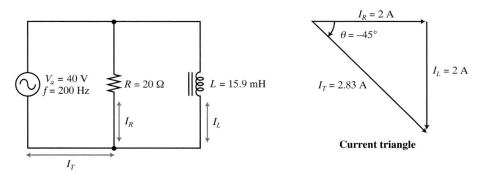

Figure 18-21. Diagram for when R equals X_L in a parallel circuit. (Sample problem 20)

1. Calculate inductive reactance.

    Formula: $X_L = 2\pi f L$

    Substitution: $X_L = 2 \times \pi \times 200 \text{ Hz} \times 15.9 \text{ mH}$

    Answer: $X_L = 20 \ \Omega$

2. Calculate branch currents using Ohm's law.
    a. Resistive current:

    Formula: $I_R = \dfrac{V_a}{R}$

    Substitution: $I_R = \dfrac{40 \text{ V}}{20 \text{ }\Omega}$

    Answer: $I_R = 2$ A

    b. Inductive current:

    Formula: $I_L = \dfrac{V_a}{X_L}$

    Substitution: $I_L = \dfrac{40 \text{ V}}{20 \text{ }\Omega}$

    Answer: $I_L = 2$ A

3. Calculate total current using the Pythagorean theorem.

    Formula: $I_T = \sqrt{I_L^2 + I_R^2}$

    Substitution: $I_T = \sqrt{2^2 + 2^2}$

    Answer: $I_T = 2.83$ A

4. Calculate the phase angle as shown in the completed phasor diagrams.

    Formula: $\theta = \tan^{-1} \dfrac{-I_L}{I_R}$

    Substitution: $\theta = \tan^{-1} \dfrac{-2 \text{ A}}{2 \text{ A}}$

    Intermediate step: $\theta = \tan^{-1} -1$

    Answer: $\theta = -45°$

5. Calculate impedance using Ohm's law. The phase angle for impedance is the same as for the total current.

    Formula: $Z = \dfrac{V_a}{I_T}$

    Substitution: $Z = \dfrac{40 \text{ V}}{2.83 \text{ A}}$

    Answer: $Z = 14.1$ Ω

Resistance Larger than Reactance

As shown in figure 18-22, a larger value of resistance in a parallel circuit has a lower branch current. The total current in the circuit is closer to the value of the inductive current. This results in a phase angle closer to –90°.

If the resistance were so large that the current I_R could be ignored, the circuit would behave like a purely inductive circuit, discussed earlier in this chapter. The inductor would appear to be in series with the applied voltage. A purely inductive circuit has a phasor diagram with the current at 0° and the applied voltage at 90°. The current lags the voltage by 90°.

With a parallel circuit, the phasor diagram is for branch currents. The resistance current is drawn at 0°. Also remember, applied voltage is in phase with resistive current. Therefore, current lags voltage by 90°.

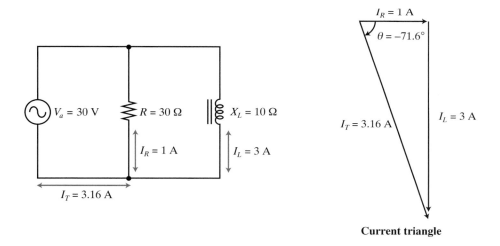

Current triangle

Figure 18-22. Diagram for when *R* is large in comparison to X_L in a parallel circuit.

Resistance Smaller than Reactance

If the reactance is large in a parallel circuit, less current travels through that branch. The total current then has the characteristics of the resistance branch. The total current has a phase angle closer to 0°, as shown in figure 18-23. If the inductive reactance is large enough, it will appear not to exist in the parallel circuit.

18.9 PHASE ANGLE ON AN OSCILLOSCOPE

An oscilloscope can be used to measure the phase angle of a circuit by comparing two voltages. It is used only with series circuits because an oscilloscope cannot measure current directly. When making measurements, the applied voltage is used as the reference. The resistive or inductive voltages are compared to the applied voltage.

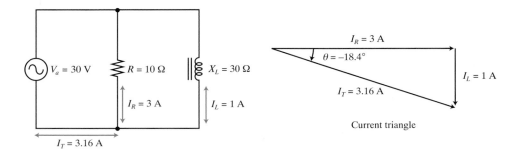

Figure 18-23. Diagram for when R is small in comparison to X_L in a parallel circuit.

Figures 18-24, 18-25, and 18-26 are all measuring a phase angle of 45°. The difference in the appearance of the three figures is caused by the difference in the number of divisions in the horizontal direction used for one cycle.

Steps in Measuring Phase Angle

1. Determine how many divisions are in one full cycle of the reference voltage, V_a.
2. Divide 360° (one cycle) by the number of divisions. This number is the **degrees per division.**
3. Count the divisions between the starting points of the two sine waves to one decimal place. This number is the **time difference.**
4. Multiply the time difference by the degrees per division. The result is the phase angle.

Sample problem 21. _____

Determine the phase angle on the oscilloscope shown in figure 18-24.

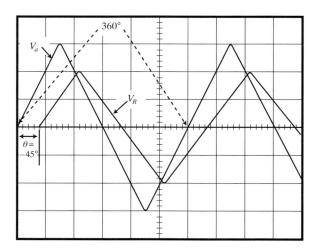

Figure 18-24. Oscilloscope adjusted for one cycle of V_a in six divisions. (Sample problem 21)

1. Divisions in one cycle:

$$360° = 6 \text{ divisions}$$

2. Degrees per division:

$$\frac{360°}{6} \text{ div} = 60° \text{ per division}$$

3. Time difference between sine waves:

$$-0.75 \text{ divisions}$$

4. Phase angle:

$$\theta = -0.75 \times 60°$$

$$\theta = -45°$$

Sample problem 22. _____

Determine the phase angle on the oscilloscope shown in figure 18-25.

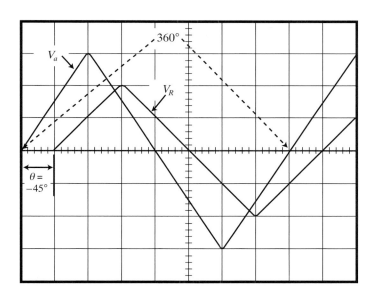

Figure 18-25. Oscilloscope adjusted for one cycle of V_a in eight divisions. (Sample problem 22)

1. Divisions in one cycle:

$$360° = 8 \text{ divisions}$$

2. Degrees per division:

$$\frac{360°}{8 \text{ div}} = 45° \text{ per division}$$

3. Time difference between sine waves:

$$-1.0 \text{ division}$$

4. Phase angle:

$$\theta = -1 \times 45°$$

$$\theta = -45°$$

Sample problem 23. _____

Determine the phase angle on the oscilloscope shown in figure 18-26.

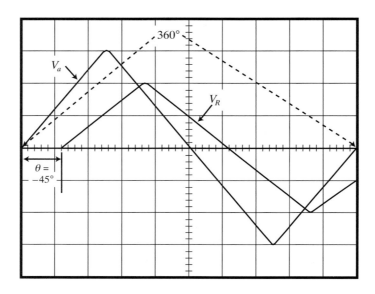

Figure 18-26. Oscilloscope adjusted for one cycle of V_a in ten divisions. (Sample problem 23)

1. Divisions in one cycle:

$$360° = 10 \text{ divisions}$$

2. Degrees per division:

$$\frac{360°}{10 \text{ div}} = 36° \text{ per division}$$

3. Time difference between sine waves:

$$-1.25 \text{ divisions}$$

4. Phase angle:

$$\theta = -1.25 \times 36°$$

$$\theta = -45°$$

Measuring Phase Angle in a Circuit

When making phase angle measurements in a circuit, the oscilloscope ground must be connected to the circuit ground. Two oscilloscope leads are needed. The reference channel is connected from ground to the applied voltage. The positive lead is connected to the point in the circuit that is being measured. In figures 18-27 and 18-28, the second channel is connected across the resistor.

A parallelogram of the circuit voltages is also shown. There are two angles shown in the parallelogram: θ, the actual phase angle, and ϕ, the opposite angle. The phase angle, θ, is measured on the oscilloscope by comparing V_a to V_R, as shown in figure 18-27. The opposite angle, ϕ, is measured when comparing V_a to V_L, as shown in figure 18-28.

Sample problem 24. _____

Determine the phase angle of the two sine waves on the oscilloscope shown in figure 18-27.

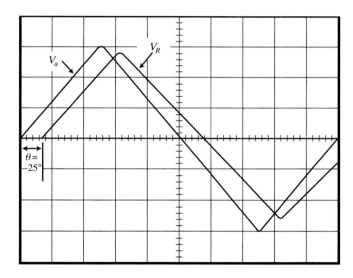

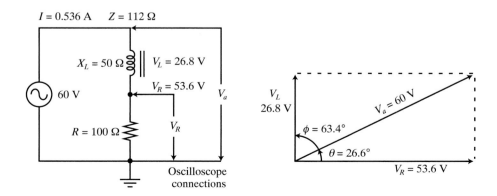

Figure 18-27. Angle θ is measured when comparing V_R to V_a on the oscilloscope. (Sample problem 24)

1. Divisions in one cycle:

$$360° = 10 \text{ divisions}$$

2. Degrees per division:

$$\frac{360°}{10 \text{ div}} = 36° \text{ per division}$$

3. Time difference between sine waves:

$$0.7 \text{ divisions}$$

4. Phase angle:

$$\theta = 0.7 \times 36°$$

$$\theta = -25°$$

The angle is negative because V_R lags V_a.

Measuring the Opposite Angle on the Oscilloscope

The opposite angle, ϕ, is measured with the oscilloscope when the reference measures V_a and the second channel measures V_L. As shown in figure 18-28, the circuit has been turned over in comparison to figure 18-27. The ground connection is made at the bottom of the inductor, rather than at the resistor.

The parallelogram in figure 18-28 shows the inductor voltage leads the applied voltage by an angle equal to $90° - \theta$. This is also seen on the oscilloscope. The reference is V_a, which has its starting point at zero. The waveform of V_L starts at a point prior to zero. The angle between the two waveforms is measured at the midpoint of the sine waves.

Sample problem 25. _____

Determine the angle between the two waveforms shown on the oscilloscope in figure 18-28. This is the opposite angle, ϕ.

1. Divisions in one cycle:

$$360° = 10 \text{ divisions}$$

2. Degrees per division:

$$\frac{360°}{10 \text{ div}} = 36° \text{ per division}$$

3. Time difference between sine waves:

$$1.7 \text{ divisions}$$

4. Opposite angle:

$$\phi = 1.7 \times 36°$$

$$\phi = 61.2°$$

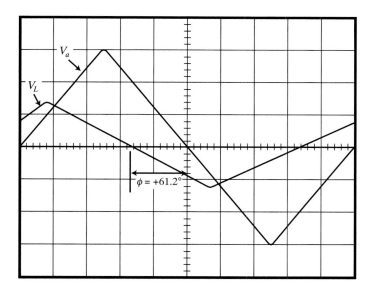

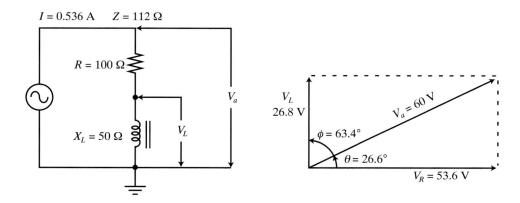

Figure 18-28. Angle ϕ is measured when comparing V_L to V_a on the oscilloscope. (Sample problem 25)

The oscilloscope measurement can vary from the calculations by several degrees. This is frequently the situation when making measurements. It is very difficult to accurately set up and read the oscilloscope screen. Calculations, in addition, are theoretical values. They are difficult to match in a circuit.

SUMMARY

- Reactance is the ac resistance of an inductor or capacitor.
- Impedance is the total ac resistance of a resistor in a circuit with a reactive component.

- Ohm's law can be used to calculate the reactance of a circuit if the current and voltage are known.
- Inductive reactance can be calculated using the formula, $X_L = 2\pi fL$.
- Both frequency and inductance are directly related to the value of inductive reactance.
- Reactances can be used in the same formulas as resistance to calculate the total reactance in series and parallel circuits.
- In a purely inductive circuit, voltage leads current by 90°.
- A circuit containing both resistance and inductance will have a phase shift between 0° and 90°. If the resistance is larger than the inductance, the phase shift is closer to 0°. If the resistance is smaller than the inductance, the phase shift is closer to 90°.
- In a series circuit, phasor diagrams are for voltage drops or ohmic values of the components.
- In a series circuit with R and X_L the phase shift is positive.
- In a parallel circuit, phasor diagrams are for current in each branch.
- In a parallel circuit with R and X_L the phase shift is negative.
- Phase shift is measured with an oscilloscope by measuring the difference in time between the two sine waves of V_R and V_L.

KEY WORDS AND TERMS GLOSSARY

impedance: The total ac resistance of a circuit containing both resistance and reactance. Unit of measure is ohms. Impedance has a phase angle equal to the phase angle of the circuit voltage.

inductive reactance (X_L): The ac resistance of an inductor. Unit of measure is ohms.

opposite angle(ϕ): When comparing the shift of the inductor voltage with the applied voltage, V_L will lead V_a by an amount equal to $90° - \theta$. This is referred to as the *opposite angle* because in the triangle, it is at the opposite side from θ.

phase angle (θ): The time difference between the applied voltage and the resistor voltage. Also called phase shift.

phase shift (θ): The time difference between the applied voltage and the resistor voltage. Also called phase angle.

phasor: A vector used to show the magnitude (size) and direction of an electrical quantity.

pure inductive reactance: A theoretical condition of an inductor in which the reactance contains no resistance.

Pythagorean theorem: A mathematical formula used to calculate the length of the hypotenuse of a right triangle, such as formed by the phasors of an ac circuit.

reactance: The ac resistance of an inductor or capacitor. The unit of measure is the ohm. Letter symbol is X_L for inductive reactance and X_C for capacitive reactance.

time difference: The difference between the starting points of two sine waves.

KEY FORMULAS

Formula 18.A

$$X_L = 2\pi fL$$

Formula 18.B

$$X_{L_T} = X_{L_1} + X_{L_2} + X_{L_3} + \ldots X_{L_N}$$

Formula 18.C

$$\frac{1}{X_{L_T}} = \frac{1}{X_{L_1}} + \frac{1}{X_{L_2}} + \frac{1}{X_{L_3}} + \ldots \frac{1}{X_{L_N}}$$

Formula 18.D

$$Z = \sqrt{X_L^2 + R^2}$$

Formula 18.E

$$V_a = \sqrt{V_L^2 + V_R^2}$$

Formula 18.F

$$\theta = \tan^{-1} \frac{X_L}{R}$$

Formula 18.G

$$\theta = \tan^{-1} \frac{V_L}{V_R}$$

Formula 18.H

$$\phi = 90 - \theta$$

Formula 18.I

$$I = \frac{V}{Z}$$

Formula 18.J

$$I_T = \sqrt{I_L^2 + I_R^2}$$

Formula 18.K

$$\theta = \tan^{-1}\frac{-I_L}{I_R}$$

Formula 18.L

$$\phi = -90 - \theta$$

Formula 18.M

$$Z = \frac{V_a}{I_T}$$

TEST YOUR KNOWLEDGE

Do not write in this text. Please use a separate sheet of paper.
1. How is reactance measured?
2. Use Ohm's law to calculate the inductive reactance of a circuit with an applied voltage of 60 volts and current through the inductor of three amps.
3. Use the inductive reactance formula to calculate X_L of a 500 mH inductor in a circuit with a frequency of 200 hertz.
4. Using the inductive reactance formula as a guide, what is the effect on X_L when there is an increase in frequency?
5. Using the inductive reactance formula as a guide, what is the effect on X_L when there is a decrease in the value of inductance?
6. Calculate the total reactance of three inductive reactances connected in series: X_{L_1} = 40 Ω, X_{L_2} = 80 Ω, X_{L_3} = 30 Ω.
7. Calculate the total reactance of three inductive reactances connected in parallel: X_{L_1} = 40 Ω, X_{L_2} = 80 Ω, X_{L_3} = 30 Ω.
8. Draw a phasor diagram showing the relationship between voltage and current in a purely inductive circuit.
9. In the ac circuit shown in figure 18-29:
   a. calculate X_L.
   b. calculate Z.
   c. plot the impedance triangle.
   d. calculate the operating angle.

   e. calculate the opposite angle.
   f. calculate the circuit current.
   g. calculate the voltage drops for R and X_L.
   h. draw the voltage triangle.

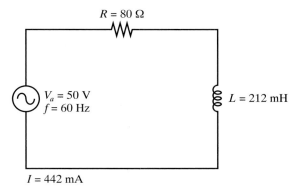

Figure 18-29. Problem #9.

10. In the ac circuit shown in figure 18-30:
   a. calculate Z.
   b. plot the impedance triangle.
   c. calculate the operating angle.
   d. calculate the opposite angle.
   e. calculate the circuit current.
   f. calculate the voltage drops for R and X_L.
   g. draw the voltage triangle.

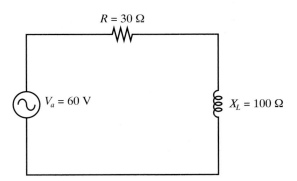

Figure 18-30. Problem #10.

11. In the ac circuit shown in figure 18-31:
    a.   calculate Z.
    b.   plot the impedance triangle.
    c.   calculate the operating angle.
    d.   calculate the opposite angle.
    e.   calculate the circuit current.
    f.   calculate the voltage drops for R and X_L.
    g.   draw the voltage triangle.

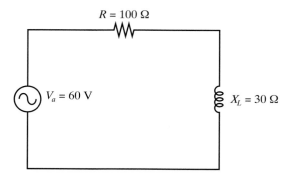

Figure 18-31. Problem #11.

12. In the ac circuit shown in figure 18-32:
    a.   calculate X_L.
    b.   calculate the resistance branch current.
    c.   calculate the reactive branch current.
    d.   calculate the total current.
    e.   plot the current triangle.
    f.   calculate the phase angle.
    g.   calculate the impedance.

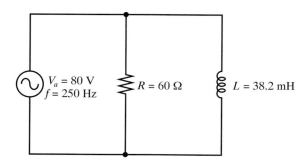

Figure 18-32. Problem #12.

13. In the ac circuit shown in figure 18-33:
    a. calculate the resistance branch current.
    b. calculate the reactive branch current.
    c. calculate the total current.
    d. plot the current triangle.
    e. calculate the phase angle.
    f. calculate the impedance.

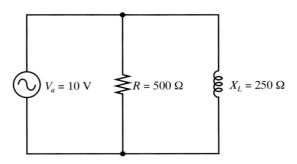

Figure 18-33. Problem #13.

14. In the ac circuit shown in figure 18-34:
    a. calculate the resistance branch current.
    b. calculate the reactive branch current.
    c. calculate the total current.
    d. plot the current triangle.
    e. calculate the phase angle.
    f. calculate the impedance.

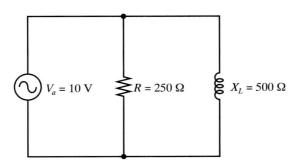

Figure 18-34. Problem #14.

15. **Determine** the phase shift of the RL circuit represented by the oscilloscope drawing in figure 18-35.

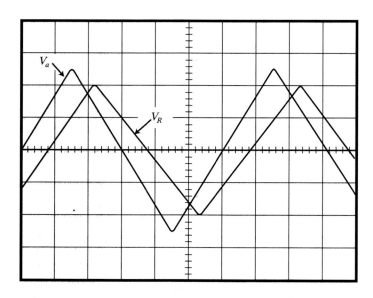

Figure 18-35. Problem #15.

Chapter 19
Capacitive Reactance
and Impedance

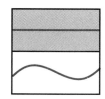

Upon completion of this chapter, you will be able to:
• Define capacitive reactance and describe its effects on an ac circuit.
• Use Ohm's law to calculate capacitive reactance with current and voltage measurements.
• Use the capacitive reactance formula with values of frequency and capacitance.
• Calculate total reactance in series and parallel capacitive reactance circuits.
• Identify the relationships of voltage and current in a purely capacitive circuit.
• Calculate voltage drops and impedance in a series RC ac circuit.
• Plot the phasor diagrams for voltage and impedance in a series RC ac circuit.
• Calculate phase shift and opposite angle for a series circuit.
• Describe the changes in circuit parameters for series circuits with a value of resistance larger and smaller than the capacitive reactance.
• Calculate branch currents and total current in a parallel RC ac circuit.
• Calculate impedance in a parallel circuit.
• Plot the current triangle for a parallel RC circuit.
• Compare the effects of a resistance larger and smaller than the value of capacitive reactance in a parallel circuit.
• Read an oscilloscope screen to measure phase shift and opposite angle in an RC circuit.

Capacitive reactance is the ac resistance of a capacitor. It is represented by X_C with the unit of measure of ohms. Capacitive reactance has phasor diagrams that point in the opposite direction of those for inductive reactance in Chapter 19.

Impedance is the total ac resistance, regardless of what type of reactance is present. Impedance is calculated using phasor addition in series circuits. In parallel circuits, impedance is calculated using Ohm's law.

19.1 CAPACITANCE IN DC AND AC CIRCUITS

In Chapter 14, capacitance was used in a dc circuit in relation to time constants. During the first five time constants, the capacitor is building up a voltage and current is flowing in the circuit. After a period of five time constants, a steady state condition is reached. The capacitor is fully charged and there is no current. Figure 19-1 shows a capacitor in a dc circuit after the steady state condition has been reached. Notice, the ammeter shows 0 A of current.

In an ac circuit, however, a capacitor allows current to pass, as shown in figure 19-2. The ammeter is registering 3.5 A of current. The amount of current is limited by the impedance of the circuit. Recall that the impedance is the total ac resistance of a circuit. It is a combination of resistance and reactance.

In a dc circuit, after the capacitor reaches full charge, the dielectric stops any further current in the circuit. In an ac circuit, the capacitor charges and discharges at a rate equal to the frequency of the applied voltage. This allows current to flow in the circuit.

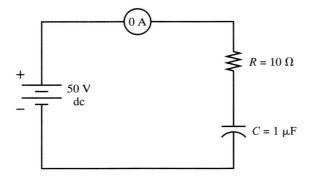

Figure 19-1. After reaching a steady-state condition, no current flows in a dc circuit with capacitors.

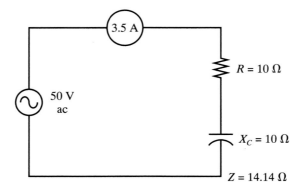

Figure 19-2. Current flows in an ac circuit with capacitors, limited by the combined ac resistance of the circuit.

A capacitor is constructed with two plates, separated by a dielectric. Capacitor construction can be reviewed in Chapter 13. The dielectric is an insulator, which does not allow current to pass from one plate to the other. However, it is not necessary for the electrons in the circuit to actually pass through the capacitor to have current in a circuit. Current is the movement of electrons. Therefore, the charge and discharge of a capacitor produces current in a dc circuit or in an ac circuit. In most ac circuits, a capacitor is constantly in the process of charging or discharging.

19.2 CALCULATING CAPACITIVE REACTANCE

Ohm's law can be used to calculate circuit values in a circuit with capacitive reactance the same way as with circuits containing inductance reactance or resistance. In a circuit, reactance is calculated by measuring voltage and resistance, then using Ohm's law.

Sample problem 1. _____

Use Ohm's law to calculate the value of capacitive reactance, X_C, for the circuit in figure 19-3.

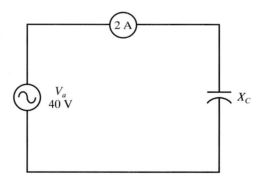

Figure 19-3. Calculate X_C given the voltage and current. (Sample problem 1)

Formula: $X_C = \dfrac{V_a}{I}$

Substitution: $X_C = \dfrac{40 \text{ V}}{2 \text{ A}}$

Answer: $X_C = 20 \text{ }\Omega$

Sample problem 2. _____

With the circuit shown in figure 19-4, determine the amount of current, using Ohm's law.

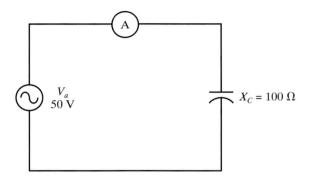

Figure 19-4. Calculate the current given the voltage and X_C.
(Sample problem 2)

Formula: $I = \dfrac{V_a}{X_C}$

Substitution: $I = \dfrac{50 \text{ V}}{100 \text{ }\Omega}$

Answer: $I = 0.5 \text{ A}$

Capacitive Reactance Formula

Capacitive reactance can be calculated using a formula when the values of frequency and capacitance are given. The formula can also be rearranged to solve for either of the other two variables, frequency and capacitance. The formula for the capacitive reactance is written:

Formula 19.A

$$X_C = \frac{1}{2\pi fC}$$

X_C is capacitive reactance, measured in ohms.

2π can be either 6.28 or $2 \times \pi$ on a calculator.

f is frequency, measured in hertz.

C is capacitance, measured in farads.

Sample problem 3. _____

Calculate the capacitive reactance for the circuit in figure 19-5.

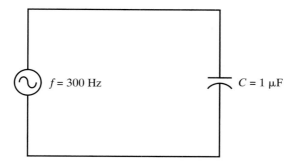

$f = 300$ Hz $C = 1\ \mu\text{F}$

Figure 19-5. Calculate the capacitive reactance. (Sample problem 3)

Formula: $X_C = \dfrac{1}{2\pi f C}$

Substitution: $X_C = \dfrac{1}{2 \times \pi \times 300\ \text{Hz} \times 1\ \mu\text{F}}$

Answer: $X_C = 531\ \Omega$

Sample problem 4. _____

Calculate the capacitive reactance in a circuit with an input frequency of 60 Hz and capacitance of 0.05 μF.

Formula: $X_C = \dfrac{1}{2\pi f C}$

Substitution: $X_C = \dfrac{1}{2 \times \pi \times 60\ \text{Hz} \times 0.05\ \mu\text{F}}$

Answer: $X_C = 53.1\ \text{k}\Omega$

Sample problem 5. _____

What is the capacitance in the circuit of figure 19-6?

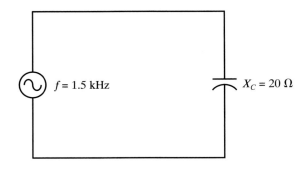

Figure 19-6. Calculate the capacitive reactance. (Sample problem 5)

Formula: $X_C = \dfrac{1}{2\pi f C}$

Rearranging: $C = \dfrac{1}{2\pi f X_C}$

Substitution: $C = \dfrac{1}{2 \times \pi \times 1.5 \text{ kHz} \times 20 \text{ }\Omega}$

Answer: $C = 5.31 \text{ }\mu\text{F}$

Sample problem 6. _____

Determine the value of capacitance needed at a frequency of 400 Hz to produce a reactance of 1000 Ω.

Formula: $X_C = \dfrac{1}{2\pi f C}$

Rearranging: $C = \dfrac{1}{2\pi f X_C}$

Substitution: $C = \dfrac{1}{2 \times \pi \times 400 \text{ Hz} \times 1000 \text{ }\Omega}$

Answer: $C = 0.4 \text{ }\mu\text{F}$

Sample problem 7.

With the values given for capacitance and reactance in figure 19-7, calculate the input frequency.

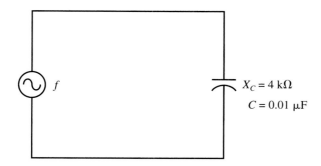

Figure 19-7. Calculate the capacitance and the reactance. (Sample problem 7)

Formula: $X_C = \dfrac{1}{2\pi f C}$

Rearranging: $f = \dfrac{1}{2\pi C X_C}$

Substitution: $f = \dfrac{1}{2 \times \pi \times 0.01\ \mu F \times 4\ k\Omega}$

Answer: $f = 3980$ Hz

Sample problem 8.

At what frequency will a five μF capacitor produce one kΩ of reactance?

Formula: $X_C = \dfrac{1}{2\pi f C}$

Rearranging: $f = \dfrac{1}{2\pi C X_C}$

Substitution: $f = \dfrac{1}{2 \times \pi \times 5\ \mu F \times 1\ k\Omega}$

Answer: $f = 31.8$ Hz

19.3 EFFECT OF FREQUENCY ON CAPACITIVE REACTANCE

Capacitive reactance is inversely related to frequency and capacitance. An increase in either frequency or capacitance causes a decrease in reactance. This is not a linear relationship, however. As the frequency approaches zero, the reactance changes rapidly toward infinity. At high frequencies, the reactance approaches zero.

Figure 19-8 lists the results for calculations of X_C using a 10 μF capacitor while varying the frequency. These results are graphed in figure 19-9. The frequency varies in factors of 10.

The shape of this curve results from the formula having the variables in the denominator. If any variable in the denominator approaches zero, the result approaches infinity. If a variable in the denominator approaches infinity, the result approaches zero.

Figure 19-8. Table of results to plot the curve for X_C using a 10 μF capacitor.

$$C = 10 \text{ μF}$$

$$X_C = \frac{1}{2\pi f C}$$

f	X_C
1 Hz	16 kΩ
10 Hz	1.6 kΩ
100 Hz	160 Ω
1 kHz	16 Ω
10 kHz	1.6 Ω
100 kHz	0.16 Ω
1 MHz	0.016 Ω

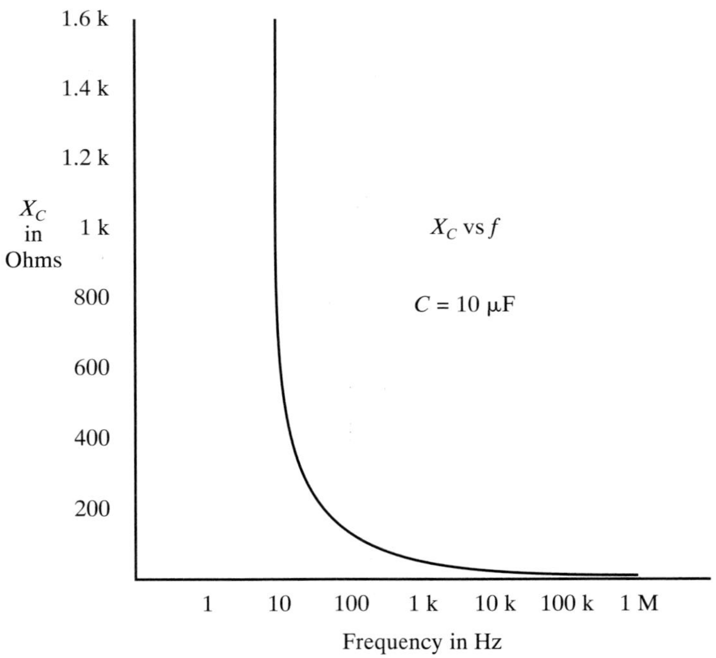

Figure 19-9. Curve of X_C vs frequency. This shows the effects of frequency on reactance.

19.4 SERIES AND PARALLEL REACTANCES

Reactances connected in series and parallel can be calculated using the same formulas as resistance. The type of reactance, inductive or capacitive, does not affect the calculations. Series reactances add. Parallel reactances use the reciprocal formula.

Formula 19.B

$$X_{C_T} = X_{C_1} + X_{C_2} + X_{C_3} + \ldots X_{C_N}$$

X_{C_T} is the total capacitive reactance, measured in ohms.

X_{C_1} through X_{C_N} are the individual capacitive reactances, measured in ohms.

Formula 19.C

$$\frac{1}{X_{C_T}} = \frac{1}{X_{C_1}} + \frac{1}{X_{C_2}} + \frac{1}{X_{C_3}} + \ldots \frac{1}{X_{C_N}}$$

X_{C_T} is the total capacitive reactance, measured in ohms.

X_{C_1} through X_{C_N} are the individual capacitive reactances, measured in ohms.

Sample problem 9. _____

Find the total capacitive reactance of the circuit shown in figure 19-10.

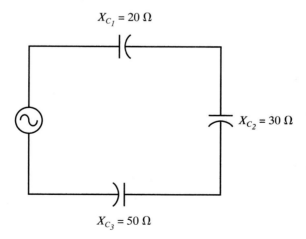

Figure 19-10. Total reactances in series add. (Sample problem 9)

Formula: $X_{C_T} = X_{C_1} + X_{C_2} + X_{C_3} + ...X_{C_N}$

Substitution: $X_{C_T} = 20\ \Omega + 30\ \Omega + 50\ \Omega$

Answer: $X_{C_T} = 100\ \Omega$

Sample problem 10. _____

With the circuit shown in figure 19-11, find the total reactance by first finding the individual capacitive reactances using the reactance formula.

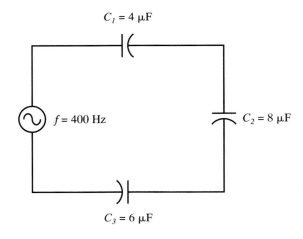

Figure 19-11. Total capacitances in series use the reciprocal formula. (Sample problems 10 and 11)

1. Individual reactances.

    Formula: $X_C = \dfrac{1}{2\pi fC}$

    a. Substitution: $X_{C_1} = \dfrac{1}{2 \times \pi \times 400\ \text{Hz} \times 4\ \mu\text{F}}$

    Answer: $X_{C_1} = 99.5\ \Omega$

    b. Substitution: $X_{C_2} = \dfrac{1}{2 \times \pi \times 400\ \text{Hz} \times 8\ \mu\text{F}}$

    Answer: $X_{C_2} = 49.7\ \Omega$

c. Substitution: $X_{C_3} = \dfrac{1}{2 \times \pi \times 400 \text{ Hz} \times 6 \text{ μF}}$

Answer: $X_{C_3} = 66.3 \text{ Ω}$

2. Total reactance.

Formula: $X_{C_T} = X_{C_1} + X_{C_2} + X_{C_3} + \dots X_{C_N}$

Substitution: $X_{C_T} = 99.5 \text{ Ω} + 49.7 \text{ Ω} + 66.3 \text{ Ω}$

Answer: $X_{C_T} = 215.5 \text{ Ω}$

Sample problem 11. _____

Using the circuit shown in figure 19-11, find the total reactance by first calculating total capacitance. Use the series capacitance formula found in Chapter 13. Then, use the total capacitance to calculate total reactance.

1. Total capacitance.

Formula: $\dfrac{1}{C_T} = \dfrac{1}{C_1} + \dfrac{1}{C_2} + \dfrac{1}{C_3} + \dots \dfrac{1}{C_N}$

Substitution: $\dfrac{1}{C_T} = \dfrac{1}{4 \text{ μF}} + \dfrac{1}{8 \text{ μF}} + \dfrac{1}{6 \text{ μF}}$

Decimal step: $\dfrac{1}{C_T} = 250{,}000 + 125{,}000 + 167{,}000$

Answer: $C_T = 1.85 \text{ μF}$

2. Total reactance.

Formula: $X_{C_T} = \dfrac{1}{2\pi f C}$

Substitution: $X_{C_T} = \dfrac{1}{2 \times \pi \times 400 \text{ Hz} \times 1.85 \text{ μF}}$

Answer: $X_{C_T} = 215.1 \text{ Ω}$

Sample problem 12. _____

Calculate total reactance using the values given in the circuit of figure 19-12.

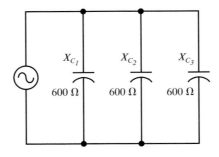

Figure 19-12. Total reactances in parallel use the reciprocal formula. (Sample problem 12)

Formula: $\dfrac{1}{X_{C_T}} = \dfrac{1}{X_{C_1}} + \dfrac{1}{X_{C_2}} + \dfrac{1}{X_{C_3}} + \dots \dfrac{1}{X_{C_N}}$

Substitution: $\dfrac{1}{X_{C_T}} = \dfrac{1}{600\ \Omega} + \dfrac{1}{600\ \Omega} + \dfrac{1}{600\ \Omega}$

Decimal step: $\dfrac{1}{X_{C_T}} = 1.67\ \text{E--03} + 1.67\ \text{E--03} + 1.67\ \text{E--03}$

Answer: $X_{C_T} = 200\ \Omega$

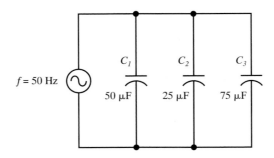

Figure 19-13. Capacitors in parallel add. (Sample problems 13 and 14)

Sample problem 13. _____

Find the total reactance of the circuit in figure 19-13 by first finding the individual capacitive reactances.

1. Individual reactances.

Formula: $X_C = \dfrac{1}{2\pi fC}$

a. Substitution: $X_{C_1} = \dfrac{1}{2 \times \pi \times 50 \text{ Hz} \times 50 \text{ }\mu\text{F}}$

Answer: $X_{C_1} = 63.7 \text{ }\Omega$

b. Substitution: $X_{C_2} = \dfrac{1}{2 \times \pi \times 50 \text{ Hz} \times 25 \text{ }\mu\text{F}}$

Answer: $X_{C_2} = 127.3 \text{ }\Omega$

c. Substitution: $X_{C_3} = \dfrac{1}{2 \times \pi \times 50 \text{ Hz} \times 75 \text{ }\mu\text{F}}$

Answer: $X_{C_3} = 42.4 \text{ }\Omega$

2. Total reactance.

Formula: $\dfrac{1}{X_{C_T}} = \dfrac{1}{X_{C_1}} + \dfrac{1}{X_{C_2}} + \dfrac{1}{X_{C_3}} + \dots \dfrac{1}{X_{C_N}}$

Substitution: $\dfrac{1}{X_{C_T}} = \dfrac{1}{63.7 \text{ }\Omega} + \dfrac{1}{127.3 \text{ }\Omega} + \dfrac{1}{42.4 \text{ }\Omega}$

Decimal step: $\dfrac{1}{X_{C_T}} = 15.7 \text{ E–}03 + 7.86 \text{ E–}03 + 23.6 \text{ E–}03$

Answer: $X_{C_T} = 21.2 \text{ }\Omega$

_____

Sample problem 14. _____

Calculate total reactance of the circuit in figure 19-13 by first finding total capacitance. Use the formula for parallel capacitance found in Chapter 13. Then, calculate total reactance.

1. Total capacitance.

Formula: $C_T = C_1 + C_2 + C_3 + \dots C_N$

Substitution: $C_T = 50 \text{ }\mu\text{F} + 25 \text{ }\mu\text{F} + 75 \text{ }\mu\text{F}$

Answer: $C_T = 150 \text{ }\mu\text{F}$

2. Total reactance.

   Formula: $X_{C_T} = \dfrac{1}{2\pi fC}$

   Substitution: $X_{C_T} = \dfrac{1}{2 \times \pi \times 50 \text{ Hz} \times 150 \text{ }\mu\text{F}}$

   Answer: $X_{C_T} = 21.2 \text{ }\Omega$

19.5 CURRENT AND VOLTAGE IN PURE CAPACITIVE REACTANCE

A **pure capacitive reactance** does not contain any resistance, it has only capacitance. This is a theoretical condition for the purposes of understanding the current and voltage in an ac circuit with capacitance.

The charge of a capacitor is the voltage between the two plates. This charge takes time to build up and to discharge. The current supplies electrons to one plate, while an equal number of electrons leave the opposite plate. The voltage across the two plates charges as high as the applied voltage will allow. Keep in mind, a sine wave changes from zero to maximum positive, through zero to maximum negative. The amount of time it takes for this change to take place determines what percentage of full charge the capacitor reaches.

Voltage Lags Current by 90 Degrees

In a purely capacitive ac circuit, the current waveform is the reference. It starts at zero degrees. As shown in figure 19-14, the voltage across the capacitor lags the current by 90°. The phasor diagram, also shown in figure 19-14, shows the current phasor plotted at 0° and the voltage phasor is plotted down at −90°.

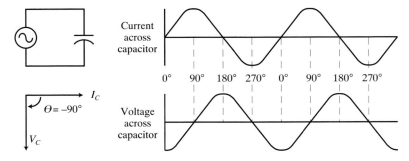

Figure 19-14. Voltage lags current by 90° in a purely capacitive circuit.

19.6 SERIES RC AC CIRCUIT

Current in the series circuit is the reference waveform, starting at 0°. The voltage dropped across the resistor has the same phase as the current, 0°. Voltage dropped across the capacitor lags by 90°. The applied voltage has a negative phase angle. The phase angle will be somewhere between 0° and 90°, as shown in figure 19-15.

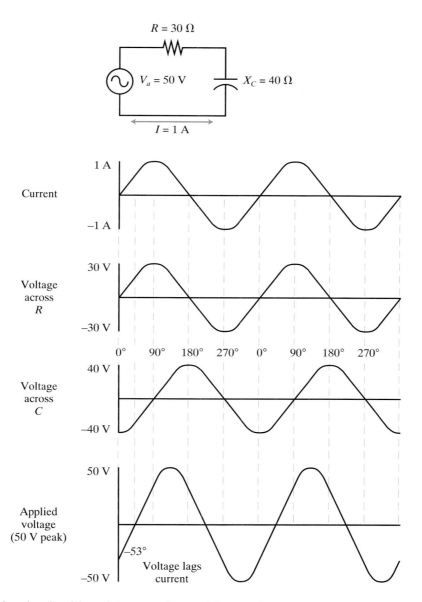

Figure 19-15. Series circuit with resistance and capacitive reactance.

Calculations for this circuit are in sample problem 15. The circuits phasor diagrams are in figure 19-16.

Phase Shift (θ) and Opposite Angle (ϕ)

The phase shift, θ, is the time difference between the applied voltage and the resistive voltage. In a series capacitive circuit, the phase angle is negative. The angle between impedance and resistance is also equal to the phase angle.

The time difference between the applied voltage and the capacitive voltage is equal to the opposite angle, ϕ. The opposite angle is calculated by subtracting θ from $-90°$.

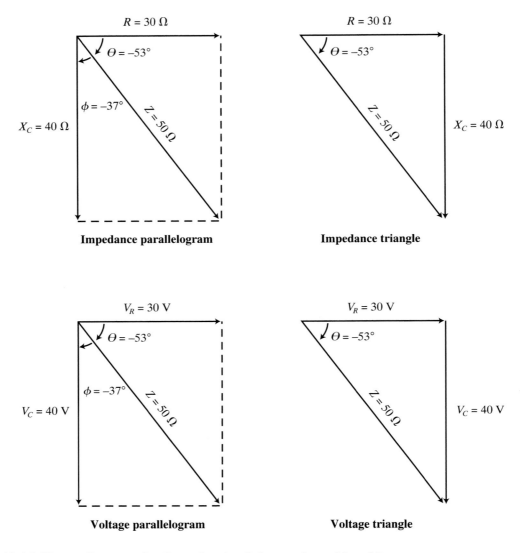

Figure 19-16. Phasor diagrams for the series circuit in sample problem 15.

This angle is important when measuring circuits using an oscilloscope. To make this calculation, the negative sign used with θ must be included.

To calculate phasor diagrams in series RC circuits, use the following formulas.

Formula 19.D

$$\text{Impedance: } Z = \sqrt{X_C^2 + R^2}$$

Z is the impedance, measured in ohms.

X_C is the capacitive reactance, measured in ohms.

R is the resistance, measured in ohms.

Formula 19.E

$$\text{Applied voltage: } V_a = \sqrt{V_C^2 + V_R^2}$$

V_a is the applied voltage, measured in volts.

V_C is the voltage across the capacitance, measured in volts.

V_R is the voltage across the resistance, measured in volts.

Formula 19.F

$$\text{Phase angle: } \theta = \tan^{-1} \frac{-X_C}{R} \text{ or } \theta = \tan^{-1} \frac{-V_C}{V_R}$$

θ is the phase angle, measured in degrees.

X_C is the capacitive reactance, measured in ohms.

R is the resistance, measured in ohms.

Formula 19.G

$$\text{Opposite angle: } \phi = -90 - \theta$$

ϕ is the opposite angle, measured in degrees.

θ is the phase angle, measured in degrees.

Sample problem 15.

With the circuit shown in figure 19-15, calculate the information needed to plot the phasor diagrams shown in figure 19-16.

1. Calculate the impedance.

    Formula: $Z = \sqrt{X_C^2 + R^2}$

    Substitution: $Z = \sqrt{40^2 + 30^2}$

    Answer: $Z = 50 \ \Omega$

2. Calculate the phase angle.

    Formula: $\theta = \tan^{-1} \frac{-X_C}{R}$

    Substitution: $\theta = \tan^{-1} \frac{-40 \ \Omega}{30 \ \Omega}$

Intermediate step: $\theta = \tan^{-1} -1.333$

Answer: $\theta = -53°$

3. Calculate the opposite angle, ϕ, used in the parallelogram.

   Formula: $\phi = -90 - \theta$

   Substitution: $\phi = -90 - -53°$

   Answer: $\phi = -37°$

4. Calculate current using Ohm's law.

   Formula: $I = \dfrac{V}{Z}$

   Substitution: $I = \dfrac{50\ V}{50\ \Omega}$

   Answer: $I = 1\ A$

5. Calculate voltage drops, using Ohm's law.

   a. Capacitor voltage:

      Formula: $V_C = I \times X_C$

      Substitution: $V_C = 1\ A \times 40\ \Omega$

      Answer: $V_C = 40\ V$

   b. Resistor voltage:

      Formula: $V_R = I \times R$

      Substitution: $V_R = 1\ A \times 30\ \Omega$

      Answer: $V_R = 30\ V$

19.7 COMPARING X_C AND R IN A SERIES CIRCUIT

The relative difference between the sizes of resistance and reactance determines the shape of the circuit triangles and the resultant phase shift. If X_C is larger than R, the triangle and phase shift will be closer to –90°. If the resistor is larger, the result will be closer to 0°. If reactance and resistance are equal, the resultant will be halfway between 0° and –90°, or a phase angle of –45°.

Equal Resistance and Reactance

With equal values of circuit components, the voltage drop across each is equal and the operating phase angle is –45°, as shown in figure 19-17. The circuit shown in this figure gives the value of the capacitor as opposed to the capacitive reactance. The capacitive reactance must be calculated.

Sample problem 16. _____

With the circuit in figure 19-17, calculate capacitive reactance, impedance, and phase angle.

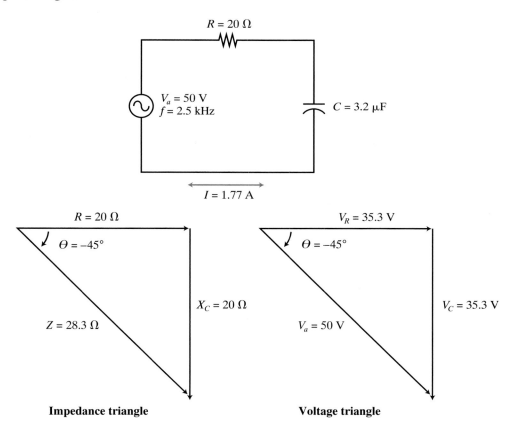

Figure 19-17. Diagram for when R is equal to X_C in a series circuit. (Sample problem 16)

1. Calculate capacitive reactance.

    Formula: $X_C = \dfrac{1}{2\pi f C}$

    Substitution: $X_C = \dfrac{1}{2 \times \pi \times 2.5 \text{ kHz} \times 3.2 \text{ μF}}$

    Answer: $X_C = 20\ \Omega$

2. Calculate impedance.

    Formula: $Z = \sqrt{X_C^2 + R^2}$

    Substitution: $Z = \sqrt{20^2 + 20^2}$

    Answer: $Z = 28.3\ \Omega$

3. Calculate the phase angle.

    Formula: $\theta = \tan^{-1} \dfrac{-X_C}{R}$

    Substitution: $\theta = \tan^{-1} \dfrac{-20\ \Omega}{20\ \Omega}$

    Intermediate step: $\theta = \tan^{-1} -1.0$

    Answer: $\theta = -45°$

Resistance Smaller than Reactance

When the resistance in a circuit is smaller than the capacitive reactance, the circuit appears more like the reactive component. The phase angle is larger than −45°. The impedance is closer in value to the reactance. If the resistance in the circuit is very small, the phase angle becomes very close to 90° and the impedance will be very close to the value of reactance.

Sample problem 17. _____

Using the circuit values given in figure 19-18, calculate impedance and phase angle.

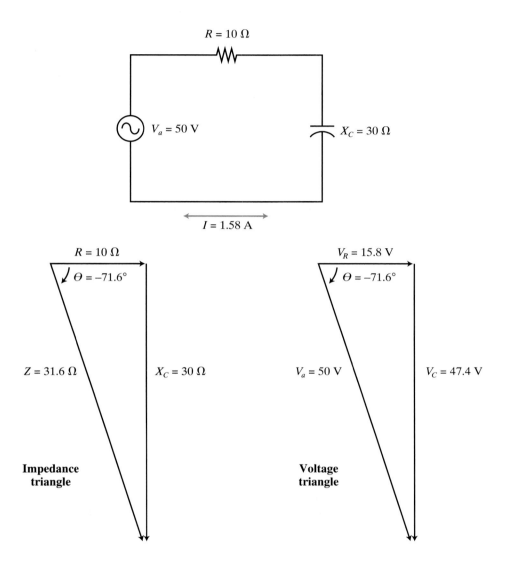

Figure 19-18. Diagram for when R is small in comparison to X_C in a series circuit. (Sample problem 17)

1. Calculate impedance.

   Formula: $Z = \sqrt{X_C^2 + R^2}$

   Substitution: $Z = \sqrt{30^2 + 10^2}$

   Answer: $Z = 31.6 \ \Omega$

2. Calculate the phase angle.

Formula: $\theta = \tan^{-1} \dfrac{-X_C}{R}$

Substitution: $\theta = \tan^{-1} \dfrac{-30 \; \Omega}{10 \; \Omega}$

Intermediate step: $\theta = \tan^{-1} -3.0$

Answer: $\theta = -71.6°$

Resistance Larger than Reactance

When the resistance of the circuit is larger than the reactance, the circuit appears to have the characteristics of the resistor. Phase angle is less than $-45°$. The impedance is closer to the value of the resistor. If the resistance is very large, the angle becomes very close to $0°$ and impedance will be close to the value of resistance.

Sample problem 18.

Use the circuit values given in figure 19-19 to find impedance and phase angle.

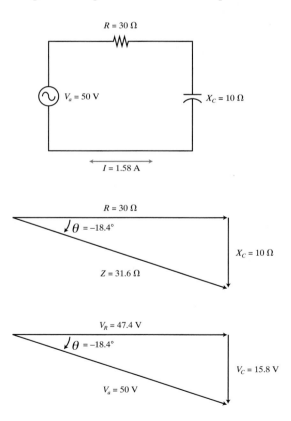

Figure 19-19. Diagram for when R is large in comparison to X_C in a series circuit. (Sample problem 18)

1. Calculate impedance.

   Formula: $Z = \sqrt{X_C^2 + R^2}$

   Substitution: $Z = \sqrt{10^2 + 30^2}$

   Answer: $Z = 31.6 \ \Omega$

2. Calculate the phase angle.

   Formula: $\theta = \tan^{-1} \dfrac{-X_C}{R}$

   Substitution: $\theta = \tan^{-1} \dfrac{-10 \ \Omega}{30 \ \Omega}$

   Intermediate step: $\theta = \tan^{-1} -0.333$

   Answer: $\theta = -18.4°$

19.8 PARALLEL RC AC CIRCUIT

In a parallel circuit, the voltage is the same at all points, while the current splits into individual branches. The applied voltage is the reference at 0°. A phasor diagram is required for the currents in a parallel circuit.

Examine figure 19-20. Current in the resistor branch has the same phase angle as the applied voltage. Current in the capacitor branch leads the applied voltage by 90°. As you will recall from earlier in this chapter, voltage lags current in a capacitive circuit. With the applied voltage as the reference, current must have a positive angle. Total current has a phase shift between 0° and 90°.

Figure 19-20 shows a parallel RC circuit. The currents represented by sine waves and their relationships to each other are shown. Calculations for this circuit are in sample problem 19 and phasor diagrams are drawn in figure 19-21.

Impedance in a Parallel RC Circuit

As shown in figure 19-21, total current is calculated using phasor diagrams. The impedance is found using Ohm's law with the applied voltage and total current. The phase angle for the impedance is equal to the phase angle of the total current.

To calculate phasor diagrams in parallel RC circuits, use the following formulas.

Formula 19.H

$$\text{Total current: } I_T = \sqrt{I_C^2 + I_R^2}$$

I_T is the total current, measured in amperes.

I_C is the current through the capacitance, measured in amperes.

I_R is the current through the resistance, measured in amperes.

Formula 19.I

$$\text{Phase angle: } \theta = \tan^{-1} \frac{I_C}{I_R}$$

θ is the phase angle, measured in degrees.

I_C is the current across the capacitance, measured in amperes.

I_R is the current across the resistance, measured in amperes.

Formula 19.J

$$\text{Opposite angle: } \phi = 90 - \theta$$

ϕ is the opposite angle, measured in degrees.

θ is the phase angle, measured in degrees.

Formula 19.K

$$\text{Impedance: } Z = \frac{V_a}{I_T}$$

Z is the impedance, measured in ohms.

I_T is the total current, measured in amperes.

V_a is the applied voltage, measured in volts.

Sample problem 19. _____

Using the circuit values given in figure 19-20, calculate the information needed to plot the parallelogram and triangle in figure 19-21.

1. Calculate branch currents.

    a. Resistive current:

   Formula: $I_R = \dfrac{V_a}{R}$

   Substitution: $I_R = \dfrac{12 \text{ V}}{4 \text{ k}\Omega}$

   Answer: $I_R = 3 \text{ mA}$

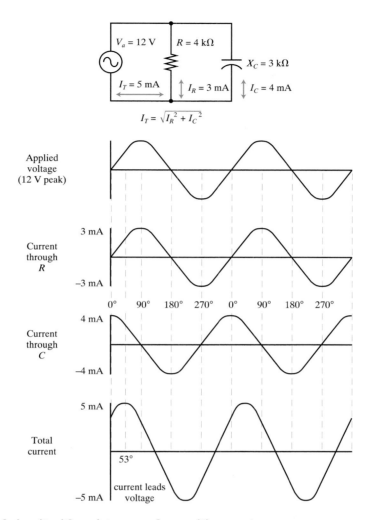

Figure 19-20. Parallel circuit with resistance and capacitive reactance.

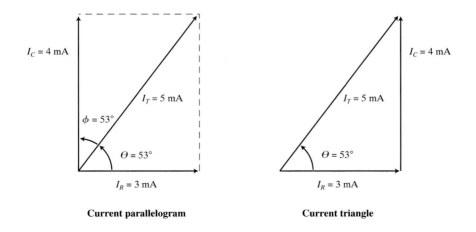

Current parallelogram   Current triangle

Figure 19-21. Phasor diagrams for the parallel circuit in sample problem 19.

b. Capacitive current:

Formula: $I_C = \dfrac{V_a}{X_C}$

Substitution: $I_C = \dfrac{12\text{ V}}{3\text{ k}\Omega}$

Answer: $I_C = 4\text{ mA}$

2. Calculate total current.

Formula: $I_T = \sqrt{I_L^2 + I_R^2}$

Substitution: $I_T = \sqrt{(4\text{ mA})^2 + (3\text{ mA})^2}$

Answer: $I_T = 5\text{ mA}$

3. Calculate phase angle.

Formula: $\theta = \tan^{-1} \dfrac{I_L}{I_R}$

Substitution: $\theta = \tan^{-1} \dfrac{4\text{ mA}}{3\text{ mA}}$

Intermediate step: $\theta = \tan^{-1} 1.333$

Answer: $\theta = 53.1°$

19.9 COMPARING X_C AND R IN A PARALLEL CIRCUIT

In a parallel circuit, branch current determines the circuit performance. The amount of current in each branch is determined by the size of the resistance (or reactance) in the branch. A smaller resistance (or reactance) allows for a larger amount of current.

The branch in a parallel circuit with the largest current has the longest phasor in the phasor diagram. If the larger current is capacitive, the phase angle will be closer to +90°. If the larger current is resistive, the phase angle will be closer to 0°.

Equal Resistance and Reactance

In a parallel capacitive circuit, the current triangle is plotted with a positive phase angle. Capacitive current is plotted up and resistive current is plotted to the right. Any circuit with R equal to X, either X_L or X_C, produces a phase angle of 45°. In a series RC circuit, θ was −45°. In a parallel RC circuit, θ is +45°.

With equal values of resistance and reactance, a little geometry can be used to make the calculation of the resultant phasor easier. The resistance and reactance phasor produce a right isosceles triangle. This is a right triangle that has equal sides and opposite angles of 45°. The resultant phasor, either Z or V_a in the series RC circuit, or I_T in the parallel RC circuit will have a value equal to 1.414 ($\sqrt{2}$) times the value resistance or the reactance phasor.

Sample problem 20.

With the values given in figure 19-22, find capacitive reactance, branch currents, total current, and impedance.

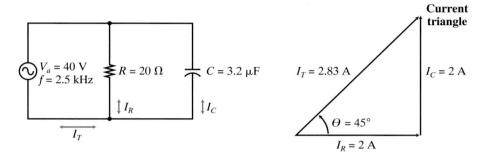

Figure 19-22. Diagram for when R is equal to X_C in a parallel circuit. (Sample problem 20)

1. Calculate capacitive reactance.

    Formula: $X_C = \dfrac{1}{2\pi fC}$

    Substitution: $X_C = \dfrac{1}{2 \times \pi \times 2.5 \text{ kHz} \times 3.2 \text{ }\mu\text{F}}$

    Answer: $X_C = 20 \text{ }\Omega$

2. Calculate branch currents.

    a. Resistive branch current:

        Formula: $I = \dfrac{V}{R}$

        Substitution: $I = \dfrac{40 \text{ V}}{20 \text{ }\Omega}$

        Answer: $I = 2 \text{ A}$

    b. Capacitive branch current:

        Formula: $I = \dfrac{V}{X_C}$

        Substitution: $I = \dfrac{40 \text{ V}}{20 \text{ }\Omega}$

        Answer: $I = 2 \text{ A}$

3. Calculate total current.

Formula: $I_T = \sqrt{I_R^2 + I_C^2}$

Substitution: $I_T = \sqrt{2^2 + 2^2}$

Answer: $I_T = 2.83$ A

4. Calculate phase angle.

Formula: $\theta = \tan^{-1} \dfrac{I_C}{I_R}$

Substitution: $\theta = \tan^{-1} \dfrac{2\text{ A}}{2\text{ A}}$

Answer: $\theta = 45°$

5. Calculate impedance.

Formula: $Z = \dfrac{V}{I_T}$

Substitution: $Z = \dfrac{40\text{ V}}{2.83\text{ A}}$

Answer: $Z = 14.1$ Ω

Resistance Larger than Reactance

With a resistance larger than reactance in a parallel circuit, as shown in figure 19-23, the circuit has a larger amount of capacitive current. The triangle has an angle greater than 45°. With a much larger resistor, the phase angle is very close to 90°. With a very large resistor, the circuit appears not to have a resistive branch and performs like a pure capacitive circuit.

Resistance Smaller than Reactance

In a parallel circuit with the resistor branch smaller than the reactive branch, the resistor has a larger amount of current. The phasor triangle has an angle closer to 0°, as shown in figure 19-24.

In a circuit with a very small resistance, or one with an exceptionally large reactance, the capacitor branch has little effect on the circuit. Its performance is like that of a purely resistive circuit. A resistor causes no phase shift in an ac circuit.

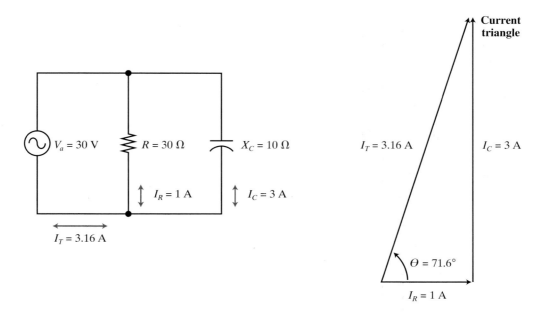

Figure 19-23. Diagram for when R is large in comparison to X_C in a parallel circuit.

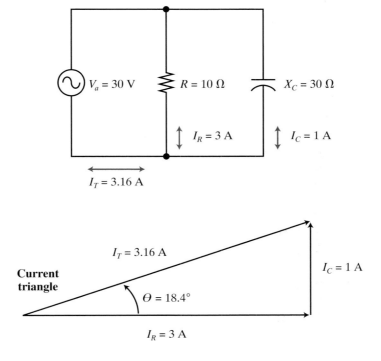

Figure 19-24. Diagram for when R is small in comparison to X_C in a parallel circuit.

19.10 PHASE SHIFTER NETWORKS

An application of the RC circuit with ac voltages is as a **phase shifter.** In figure 19-25, the phasor parallelogram is rotated to place the V_a phasor at 0°. This makes V_a the reference voltage with a phase of 0°. By rotating the parallelogram, the angle between the resistor voltage and applied voltage becomes positive. Earlier in this chapter, with series circuits, the parallelogram was drawn with the resistor voltage at 0°, in which case the angles are negative.

An output voltage can be taken across the capacitor for a negative (lagging) phase shift. The output also can be taken across the resistor for a positive (leading) phase shift.

In these circuits, a variable resistor is used. The variable resistor provides an adjustment to set the phase angle as needed. This type of circuit is effective for angles up to approximately ±60°. Beyond 60° the voltage divider causes too great a loss.

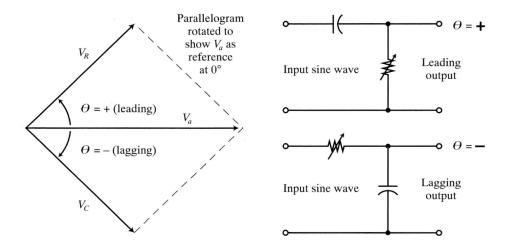

Figure 19-25. Shown are schematics for basic RC phase shifter circuits. These circuits can be used for phase shifts of up to 60°.

Cascaded Phase Shift Networks

In figure 19-26, RC phase shift circuits are connected together to create a larger phase angle. In this manner, phase shifts of greater than 60° are easily achieved. Each RC section produces its own phase shift. These shifts are added to the other sections.

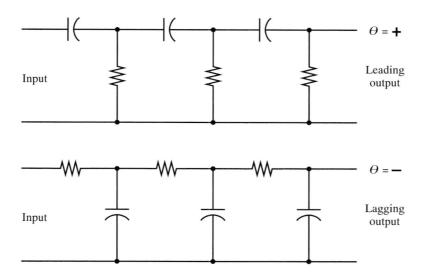

Figure 19-26. Phase shifter networks can be cascaded. This setup allows for phase shifts greater than 60°.

For example, if each RC segment were adjusted for 45°, three sections would produce a 135° phase shift.

If the resistor is the parallel component, the output is taken across the resistor to produce a positive (leading) phase angle. With the capacitor as the parallel component, a negative (lagging) phase angle is produced.

19.11 OSCILLOSCOPE MEASUREMENTS OF PHASE AND OPPOSITE ANGLES

Figures 19-27 and 19-28 use the phase shifter circuit to demonstrate the procedures for measuring phase with an oscilloscope. The parallelogram is rotated to show V_a as the reference at 0°.

Phase angle θ is measured with the reference channel connected across the applied voltage, V_a, and the other channel connected across the resistor, V_R. The phase shifter circuit is a series circuit. Therefore, the phase shift is a positive, leading angle.

Opposite angle ϕ is measured with the reference channel connected across the applied voltage, V_a and the other channel connected across the capacitor, V_C. The opposite angle is a negative, lagging angle.

Steps in Measuring Angles

1. Determine how many divisions are in one full cycle of the reference voltage, V_a.
2. Divide 360° (one cycle) by the number of divisions. This is the degrees per division.

3. Count the divisions, to one decimal place, between the starting points of the two sine waves. This is the time difference.
4. Multiply the time difference by the degrees per division. The result is the phase angle (or opposite angle).

Sample problem 21. _____

Determine the phase angle on the oscilloscope in figure 19-27.

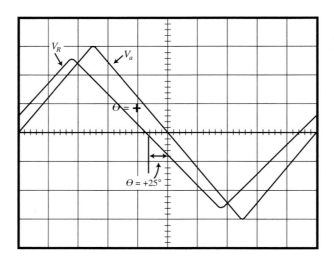

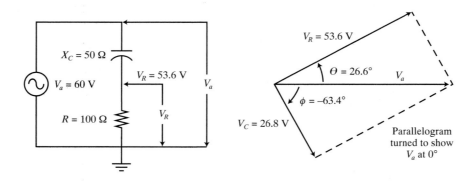

Figure 19-27. Determine the phase angle shown on the oscilloscope. (Sample problem 21)

1. Divisions in one cycle:

$$360° = 10 \text{ divisions}$$

2. Degrees per division:

$$\frac{360°}{10 \text{ div}} = 36° \text{ per division}$$

3. Time difference between sine waves:

Leading by 0.7 divisions

4. Phase angle:

$$\theta = 0.7 \times 36°$$

$$\theta = 25°$$

Note, frequently the calculated angle is slightly different from the measured value due to inaccuracies when reading the oscilloscope.

Sample problem 22. _____

Determine the opposite angle on the oscilloscope in figure 19-28.

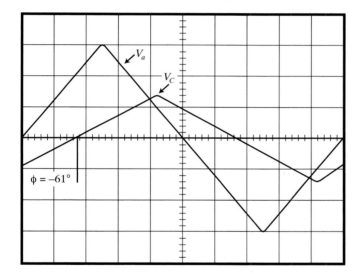

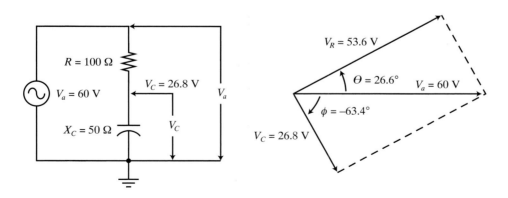

Figure 19-28. Determine the opposite angle shown on the oscilloscope. (Sample problem 22)

1. Divisions in one cycle:

$$360° = 10 \text{ divisions}$$

2. Degrees per division:

$$\frac{360°}{10 \text{ div}} = 36° \text{ per division}$$

3. Time difference between sine waves:

Lagging by 1.7 divisions

4. Opposite angle:

$$\theta = -1.7 \times 36°$$

$$\theta = -61°$$

SUMMARY

- Capacitive reactance is the ac resistance of a capacitor.
- Current flows in an ac circuit containing capacitance.
- Capacitive reactance can be calculated using Ohm's law when the values of circuit voltages and current are given in a series circuit.
- Capacitive reactance can be calculated using a formula based on frequency and capacitance.
- Frequency and capacitance are both inversely related to capacitive reactance. An increase in either f or C decreases X_C. A decrease in f or C causes an increase in X_C.
- Total reactance, whether connected in series or parallel, is calculated using the same formulas as pure resistive circuits.
- In the theoretical condition of a circuit with a pure capacitive reactance, voltage lags current by 90°.
- Two phasor diagrams are plotted for a series circuit, impedance and voltage.
- In a series RC circuit the phase angle is negative.
- Impedance in a series circuit is calculated using the Pythagorean theorem.
- In a series circuit, when resistance is smaller than reactance, the circuit characteristics appear more like a pure capacitance.
- In a series circuit, when resistance is larger than the reactance, the circuit characteristics appear more like a pure resistance.
- In a parallel circuit, branch currents and total current are calculated. The phasor diagram uses the values of the currents.
- In a parallel RC circuit the phase angle is positive.
- Impedance in a parallel circuit is calculated using Ohm's law with the values of voltage and total current.
- In a parallel circuit, the branch with the larger current has the greatest effect on the circuit's characteristics.

- Phase shift and the opposite angle are measured on an oscilloscope by measuring the time difference between the voltage of the component and the applied voltage.

KEY WORDS AND TERMS GLOSSARY

capacitive reactance (X_C): The resistance a capacitor offers to an ac signal. The unit of measure for capacitive reactance is ohms.

phase shifter: A circuit that shifts the phase of an electrical signal.

pure capacitive reactance: A theoretical condition of a capacitor in which the reactance contains no resistance.

KEY FORMULAS

Formula 19.A

$$X_C = \frac{1}{2\pi f C}$$

Formula 19.B

$$X_{C_T} = X_{C_1} + X_{C_2} + X_{C_3} + \ldots X_{C_N}$$

Formula 19.C

$$\frac{1}{X_{C_T}} = \frac{1}{X_{C_1}} + \frac{1}{X_{C_2}} + \frac{1}{X_{C_3}} + \ldots \frac{1}{X_{C_N}}$$

Formula 19.D

$$Z = \sqrt{X_C^2 + R^2}$$

Formula 19.E

$$V_a = \sqrt{V_C^2 + V_R^2}$$

Formula 19.F

$$\theta = \tan^{-1} \frac{-X_C}{R} \text{ or } \theta = \tan^{-1} \frac{-V_C}{V_R}$$

Formula 19.G

$$\phi = -90 - \theta$$

Formula 19.H

$$I_T = \sqrt{I_L^2 + I_R^2}$$

Formula 19.I

$$\theta = \tan^{-1} \frac{I_L}{I_R}$$

Formula 19.J

$$\phi = 90 - \theta$$

Formula 19.K

$$Z = \frac{V_a}{I_T}$$

TEST YOUR KNOWLEDGE

Do not write in this text. Please use a separate sheet of paper.

1. Define capacitive reactance and describe its effects on an ac circuit.
2. Use Ohm's law to calculate capacitive reactance in a series circuit with a two amp current and applied voltage of 100 volts.
3. Use the capacitive reactance formula to calculate X_C in a circuit with a frequency of one kHz and 2.50 µF capacitance.
4. Calculate total reactance in a circuit with $X_{C_1} = 50\ \Omega$ in series with $X_{C_2} = 200\ \Omega$.
5. Calculate total reactance in a circuit with $X_{C_1} = 50\ \Omega$ in parallel with $X_{C_2} = 200\ \Omega$.
6. In a purely capacitive circuit, what is the relationship between voltage and current?

7. Calculate capacitive reactance, impedance, current, and voltage drops using the series ac circuit shown in figure 19-29.

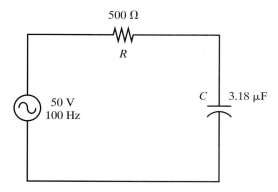

Figure 19-29. Problems #7, #8, #9, and #10.

8. Plot the impedance phasor diagram for the circuit in problem 7.
9. Plot the voltage phasor diagram for the circuit in problem 7.
10. Calculate the phase shift and opposite angle for the circuit in problem 7.
11. Using the circuit shown in figure 19-30, calculate branch currents, total current, and impedance.

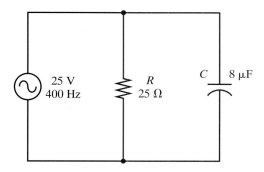

Figure 19-30. Problems #11, #12, and #13.

12. Plot the current phasor diagram for the circuit in problem 11.
13. Calculate the phase shift and opposite angle for the circuit in problem 11.
14. Draw three impedance phasor diagrams for a series circuit:
    a. $R = 100\ \Omega$ and $X_C = 100\ \Omega$.
    b. $R = 200\ \Omega$ and $X_C = 100\ \Omega$.
    c. $R = 100\ \Omega$ and $X_C = 200\ \Omega$.

15. For the diagrams from problem 14, describe the difference in phase shift.
16. For the second and third diagrams from problem 14, describe the effect on impedance.
17. Draw three current phasor diagrams for a parallel circuit:
   a. $R = 100 \ \Omega$ and $X_C = 100 \ \Omega$.
   b. $R = 200 \ \Omega$ and $X_C = 100 \ \Omega$.
   c. $R = 100 \ \Omega$ and $X_C = 200 \ \Omega$.
18. For the diagrams from problem 17, describe the difference in phase shift.
19. For the second and third diagrams from problem 17, describe the effect on impedance.
20. Read the oscilloscope screen in figure 19-31. Measure the phase shift and opposite angle for an RC circuit.

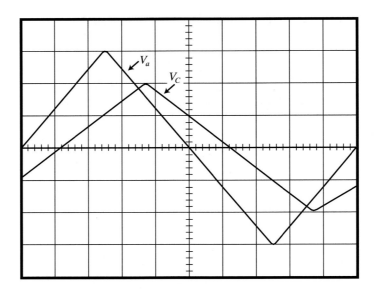

Figure 19-31. Problem #20.

Chapter 20
Phasors and Complex Numbers

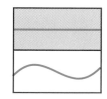

Upon completion of this chapter, you will be able to:
- Use phasor addition to solve series and parallel circuits.
- Use phasor addition to determine the effective reactance in a circuit with both inductance and capacitance.
- Mathematically represent the impedance of a circuit as resistance +/− j-operator for the value of reactance (rectangular form).
- Mathematically represent the impedance of a circuit as a quantity at its phase angle (polar form).
- Convert complex numbers from rectangular to polar and polar to rectangular.
- Write complex numbers using the values given in a circuit.
- Use complex numbers to solve series and parallel circuits.

Phasors were used in Chapters 18 and 19 to solve series and parallel circuits containing resistance and either inductive reactance or capacitive reactance. This chapter combines both types of reactance in the same series or parallel circuit.

Phasors demonstrate the opposite characteristics of X_L and X_C. The two phasors are plotted in opposite directions. This results in one phasor canceling, or partially canceling, the other phasor. In circuits containing more than one component of the same type, the phasors add.

20.1 PHASOR ADDITION

Circuits often contain many components: resistors, capacitors, and inductors. If two of the components are the same, their effects are combined together by adding their phasors. If the components have opposite reactances (X_L and X_C), the component's phasors are subtracted from one another. **RLC circuits** have a least one resistor, one capacitor, and one inductor.

Series circuits have an impedance triangle and a voltage triangle, as shown in figure 20-1. The impedance triangle has phasors for resistance and reactance. The voltage triangle has phasors representing the voltage drops across the series components.

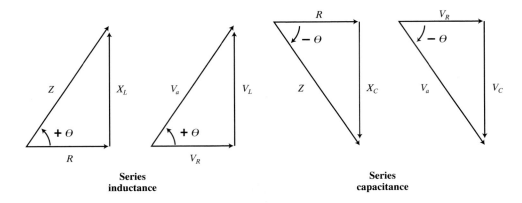

Figure 20-1. Phasors used in series circuits.

The voltage drops in an RLC series circuit *appear* to add to a larger value than the applied voltage. A series circuit containing both X_L and X_C has some cancellation due to the opposite phasors created by the inductors and capacitors.

A parallel circuit has a current triangle, shown in figure 20-2. The phasors represent the current in each branch of the parallel circuit. Notice that the direction of the current phasor in a parallel circuit is opposite the direction of its corresponding phasor in a series circuit. In a series circuit, inductive reactance is positive, and in a parallel circuit, inductive current is negative.

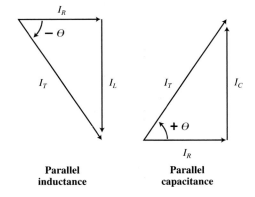

Figure 20-2. Phasors used in parallel circuits.

Like Phasors Add

Components of the same type have similar phasors. These phasors are classified as **like phasors**. They are plotted end-to-end. The phasors are added together to find the resultant phasor.

Figure 20-3 shows the phasors for three resistors connected in series. The phasors are all drawn at 0° and connected end-to-end. Addition is used to find the resultant

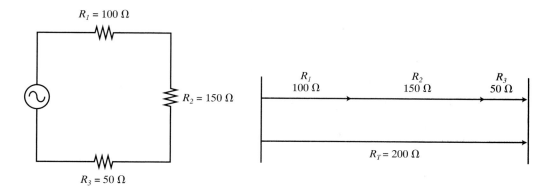

Figure 20-3. Resistor phasors are connected end-to-end. The resultant phasor can be found by simple addition.

phasor. Although this drawing does not show any other components, there can be reactance mixed in the circuit without changing the manner in which the three resistance phasors are plotted.

Figure 20-4 has a resistor and two inductors. The inductive reactances are added by connecting the phasors end-to-end. They are plotted at +90°. The one resistance phasor is plotted at 0°. Although this circuit has two inductors, the impedance triangle is produced the same as those in Chapter 18.

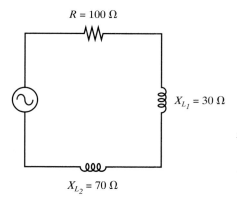

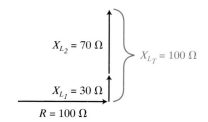

Figure 20-4. The phasors for the inductors are added end-to-end.

Phasors for capacitive reactance in series are plotted at −90°. They are added by connecting the phasors end-to-end, as shown in figure 20-5.

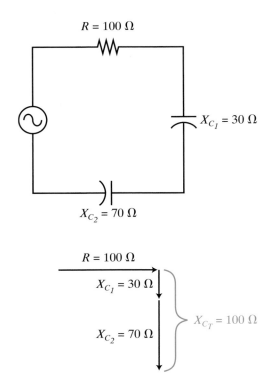

Figure 20-5. The phasors for the capacitors are added end-to-end.

Opposite Phasors Subtract

Capacitors and inductors have **opposite phasors.** When these components are contained in the same circuit, they work to cancel each other. Figure 20-6 shows the results of a circuit that has one reactance larger than the other. In these diagrams, resistance is plotted at 0°. It is used as a reference. The reactance phasors have been plotted at the end of the resistor phasor. X_L is positive and X_C is negative. The smaller phasor (X_C in the figure on the left and X_L in the figure on the right) is redrawn as a dotted line starting from the end of the larger phasor. You can see that the opposite phasors subtract, leaving a resultant phasor equal to the difference.

An alternate method is to add the phasors, using positive and negative numbers to represent the direction of the phasors. When X_L is larger than X_C, the resultant is positive. When X_L is smaller than X_C, the resultant is negative. The cancellation of reactance phasors has no effect on the resistance phasor.

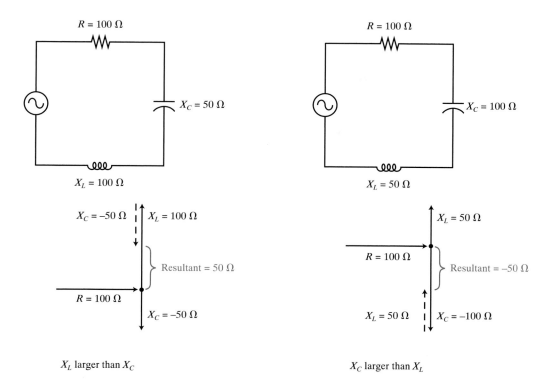

Figure 20-6. Opposite phasors, capacitive and inductive, subtract.

20.2 NET REACTANCE IN SERIES CIRCUITS

Net reactance is the effective reactance in a circuit containing both X_L and X_C. As shown in figure 20-7, an equivalent circuit can be drawn to show the net reactance in series with the resistance.

A series circuit requires two phasor diagrams, an impedance diagram and a voltage drop diagram. The impedance diagram is drawn first. Determine the net reactance by subtracting the smaller reactance from the larger. Calculate the circuit impedance using X_{net} and R in an impedance triangle.

The impedance is used to calculate the current via Ohm's law. The individual voltage drops are also calculated using Ohm's law. Examine the voltage drops shown in figure 20-7. Notice, if simple addition is used to compare the voltage drops to the applied voltage, the result is higher than the applied voltage. Voltage drops *must* be added using phasor arithmetic, as shown in the voltage triangle.

Sample problem 1. _____

Using the circuit shown in figure 20-7, find the net reactance, impedance, phase angle, current, and voltage drops.

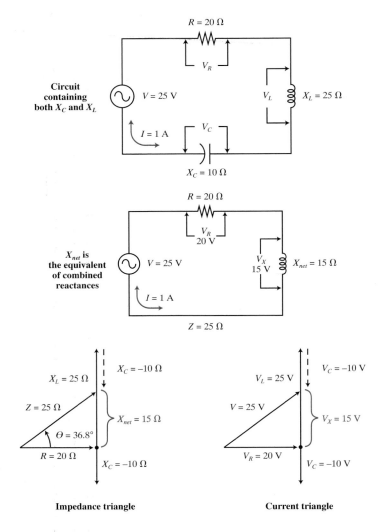

Figure 20-7. Net reactance in a series circuit with X_L larger than X_C. (Sample problem 1)

1. Find the net reactance by subtracting the smaller phasor (X_L or X_C) from the larger.

   Formula: $X_{net} = X_L - X_C$

   Substitution: $X_{net} = 25 \; \Omega - 10 \; \Omega$

   Answer: $X_{net} = 15 \; \Omega$ (inductive)

2. Calculate impedance using the Pythagorean theorem.

Formula: $Z = \sqrt{X_{net}^2 + R^2}$

Substitution: $Z = \sqrt{(15\ \Omega)^2 + (20\ \Omega)^2}$

Intermediate step: $Z = \sqrt{625}$

Answer: $Z = 25\ \Omega$

3. Calculate phase angle using the tangent.

Formula: $\theta = \tan^{-1} \dfrac{X_{net}}{R}$

Substitution: $\theta = \tan^{-1} \dfrac{15\ \Omega}{20\ \Omega}$

Intermediate step: $\theta = \tan^{-1} 0.75$

Answer: $\theta = 36.8°$

4. Calculate current using Ohm's law.

Formula: $I = \dfrac{V}{Z}$

Substitution: $I = \dfrac{25\ V}{25\ \Omega}$

Answer: $I = 1\ A$

5. Calculate voltage drops using Ohm's law.

Voltage across R:

Formula: $V_R = I \times R$

Substitution: $V_R = 1\ A \times 20\ \Omega$

Answer: $V_R = 20\ V$

Voltage across X_L:

Answer: $V_{X_L} = 25\ \Omega$

Voltage across X_C:

Answer: $V_{X_C} = 10\ V$

Voltage across X_{net}:

Answer: $V_{X_{net}} = 15\ V$

Sample problem 2. _____

With the circuit shown in figure 20-8, find the net reactance, impedance, phase angle, current, and voltage drops.

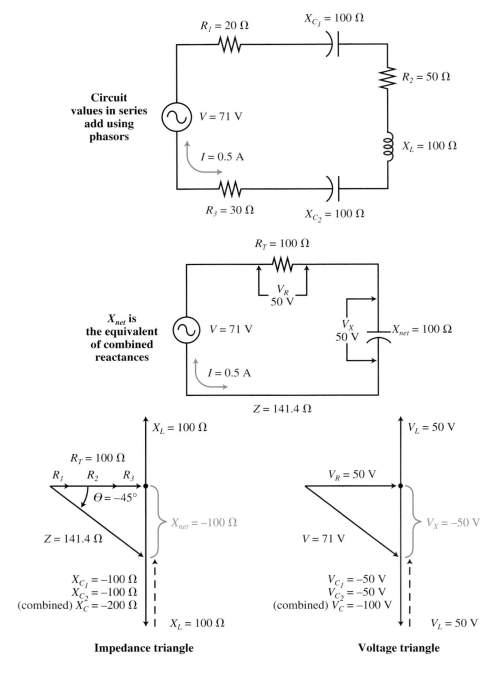

Figure 20-8. Net reactance in a series circuit with X_C larger than X_L. (Sample problem 2)

1. Find the net reactance by subtracting the smaller phasor (X_L or X_C) from the larger.

    Formula: $X_{net} = X_C - X_L$

    Substitution: $X_{net} = 200\ \Omega - 100\ \Omega$

    Answer: $X_{net} = 100\ \Omega$ (capacitive)

2. Calculate impedance using the Pythagorean theorem.

    Formula: $Z = \sqrt{X_{net}^2 + R^2}$

    Substitution: $Z = \sqrt{(100\ \Omega)^2 + (100\ \Omega)^2}$

    Answer: $Z = 141.4\ \Omega$

3. Calculate phase angle using the tangent. Include the negative sign with X_{net}.

    Formula: $\theta = \tan^{-1} \dfrac{X_{net}}{R}$

    Substitution: $\theta = \tan^{-1} \dfrac{-100\ \Omega}{100\ \Omega}$

    Intermediate step: $\theta = \tan^{-1} -1$

    Answer: $\theta = -45°$

4. Calculate current using Ohm's law.

    Formula: $I = \dfrac{V}{Z}$

    Substitution: $I = \dfrac{71\ V}{141.4\ \Omega}$

    Answer: $I = 0.5\ A$

5. Calculate voltage drops using Ohm's law.

    Voltage across R:

    Formula: $V_R = I \times R$

    Substitution: $V_R = 0.5\ A \times 100\ \Omega$

    Answer: $V_R = 50\ V$

    Voltage across X_{net}:

    Answer: $V_{X_{net}} = 50\ V$

20.3 NET REACTANCE IN PARALLEL CIRCUITS

In a parallel circuit, net reactance is not calculated directly, as it is with a series circuit. In a parallel circuit, phasors for branch currents are used to find the net current. The net current is used to find net reactance. Figure 20-9 shows a parallel circuit containing X_L and X_C, with no resistance.

The process of charging and discharging a capacitor and inductor in parallel develops a *circulating current* within the parallel branches. Inductive current has a –90° phase and the capacitive current has a phase of +90°. Combining the two currents results in 180° phase shift, having a cancellation effect. Circulating current is equal to the value of the smaller current.

Sample problem 3. _____

Using the circuit shown in figure 20-9, calculate the circulating current, net current, and net reactance.

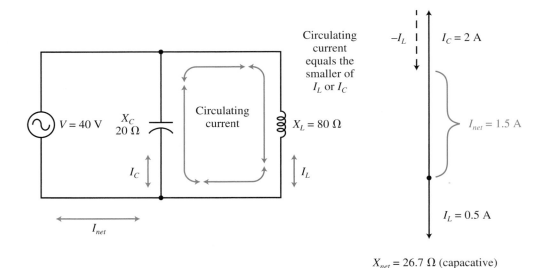

Figure 20-9. Reactive currents in parallel have a cancellation effect. (Sample problem 3)

1. Circulating current.

    I_L is the smaller value = 0.5 A

2. Calculate net current by subtracting phasors.

    Formula: $I_{net} = I_C - I_L$

    Substitution: $I_{net} = 2\,A - 0.5\,A$

    Answer: $I_{net} = 1.5\,A$ (capacitive)

3. Calculate net reactance using Ohm's law.

Formula: $X_{net} = \dfrac{V}{I_{net}}$

Substitution: $X_{net} = \dfrac{40\ V}{1.5\ A}$

Answer: $X_{net} = 26.7\ \Omega$ (capacitive)

Parallel Circuits Containing Resistance and Reactance

When a parallel circuit contains both resistance and reactance, as shown in figure 20-10, the total current is the combination of both the resistive and the net reactive currents. It does not matter where the resistor is located in the parallel circuit. Circulating current is not affected by the resistance.

The net reactance is the circuit's effective reactance. The type of net reactance, either capacitive or inductive, is determined by the larger branch current. The circuit's impedance contains both the resistance and the net reactance. In a parallel circuit, impedance is calculated using Ohm's law with the applied voltage and the total current.

Sample problem 4. _____

Using the circuit shown in figure 20-10, find net current (I_C and I_L), net reactance, total current, and impedance.

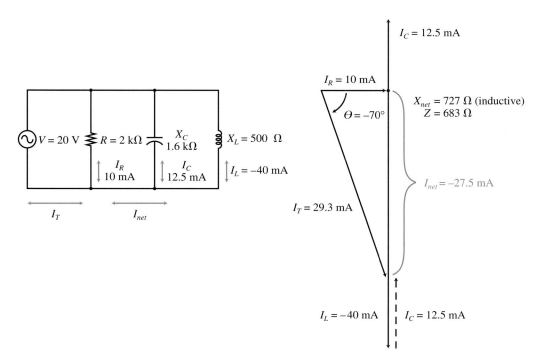

Figure 20-10. Parallel circuit containing resistance and reactance. (Sample problem 4)

1. Calculate net current by subtracting phasors.

    Formula: $I_{net} = I_L - I_C$

    Substitution: $I_{net} = 40 \text{ mA} - 12.5 \text{ mA}$

    Answer: $I_{net} = 27.5 \text{ mA (inductive)}$

2. Calculate net reactance using Ohm's law.

    Formula: $X_{net} = \dfrac{V}{I_{net}}$

    Substitution: $X_{net} = \dfrac{20 \text{ V}}{27.5 \text{ mA}}$

    Answer: $X_{net} = 727 \ \Omega \text{ (inductive)}$

3. Calculate total current using the current triangle.

    Formula: $I_T = \sqrt{I_R^{\ 2} + I_{net}^{\ 2}}$

    Substitution: $I_T = \sqrt{(10 \text{ mA})^2 + (27.5 \text{ mA})^2}$

    Answer: $I_T = 29.3 \text{ mA}$

4. Calculate impedance using Ohm's law.

    Formula: $Z = \dfrac{V}{I_T}$

    Substitution: $Z = \dfrac{20 \text{ V}}{29.3 \text{ mA}}$

    Answer: $Z = 683 \ \Omega$

Adding Branches to the Parallel Circuit

When solving dc parallel circuits, the total resistance can be found using one of two methods. The first method is the reciprocal formula. The second method is by first finding the branch currents, then using Ohm's law.

Parallel circuits containing reactance must be solved using branch currents with phasor arithmetic. If the reactances are like, their currents add. If the reactances are unlike, their currents subtract. If the circuit contains both resistance and reactance, the Pythagorean theorem solves for the impedance.

Sample problem 5. _____

Using the circuit shown in figure 20-11, find currents of like reactances, net current, net reactance, total current, and impedance.

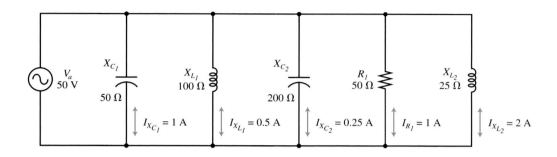

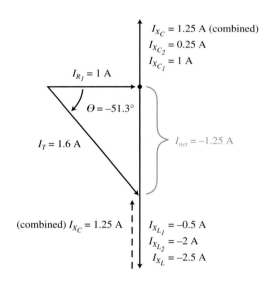

Figure 20-11. The phasors for branch currents in a parallel circuit add and subtract. (Sample problem 5)

1. Calculate currents of like reactances.

    a. Capacitive current:

    Formula: $I_C = I_{X_{C_1}} + I_{X_{C_2}}$

    Substitution: $I_C = 1\text{ A} + 0.25\text{ A}$

    Answer: $I_C = 1.25\text{ A}$

b. Inductive current:

   Formula: $I_L = I_{X_{L_1}} + I_{X_{L_2}}$

   Substitution: $I_L = 0.5 \text{ A} + 2 \text{ A}$

   Answer: $I_L = 2.5 \text{ A}$

2. Calculate net current by subtracting unlike phasors.

   Formula: $I_{net} = I_L - I_C$

   Substitution: $I_{net} = 2.5 \text{ A} - 1.25 \text{ A}$

   Answer: $I_{net} = 1.25 \text{ A}$ (inductive)

3. Calculate net reactance using Ohm's law.

   Formula: $X_{net} = \dfrac{V}{I_{net}}$

   Substitution: $X_{net} = \dfrac{50 \text{ V}}{1.25 \text{ A}}$

   Answer: $X_{net} = 40 \ \Omega$ (inductive)

4. Calculate total current using the current triangle.

   Formula: $I_T = \sqrt{I_R^2 + I_{net}^2}$

   Substitution: $I_T = \sqrt{1^2 + 1.25^2}$

   Answer: $I_T = 1.6 \text{ A}$

5. Calculate impedance using Ohm's law.

   Formula: $Z = \dfrac{V}{I_T}$

   Substitution: $Z = \dfrac{50 \text{ V}}{1.6 \text{ A}}$

   Answer: $Z = 31.25 \ \Omega$

20.4 COMPLEX NUMBERS IN RECTANGULAR FORM

The **j-operator** is a mathematical tool used to represent reactances. The **j-term** is also called an **imaginary number. Complex numbers** are numbers that contain a real number *and* an imaginary number. For example, 10 + j5 is a complex number. The 10 is the *real* part of the complex number. The j5 is the *imaginary* part, or the j-term, of the complex number. The j-term represents the reactive component.

Using phasors, resistance is plotted at 0°. See figure 20-12, 0° is labeled the real axis. The real axis contains positive numbers at 0° and negative numbers at 180°. The negative direction of the real axis, 180°, also represents j^2, which is equal to –1. The imaginary axis runs vertically, with +j at +90° and –j at –90°.

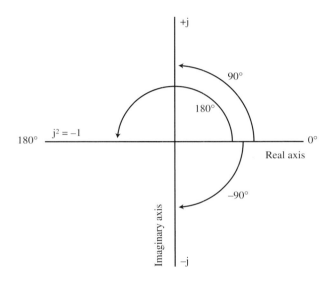

Figure 20-12. The j-operator is a tool used to identify the reactive components.

Rectangular Coordinates

Rectangular coordinates locate a point on a graph that is divided into four quadrants by the real axis and imaginary axis, figure 20-13. Each quadrant has a relationship to positive and negative real and imaginary values.
- First quadrant: + real and + j-term.
- Second quadrant: – real and + j-term.
- Third quadrant: – real and – j-term.
- Fourth quadrant: + real and – j-term.

In figure 20-13, four points are located, one in each quadrant. The points are located with paired numbers. These numbers make up an ordered pair. An **ordered pair** contains a real term, written first, and an imaginary term, written second.

In the first quadrant, the point (3,+j4) is located by moving 3 places in the positive direction along the real axis and 4 places in the positive direction along the imaginary axis. Dotted lines are drawn to the point where they intersect. This is the location of the point. You used this method when drawing the parallelogram for phasors.

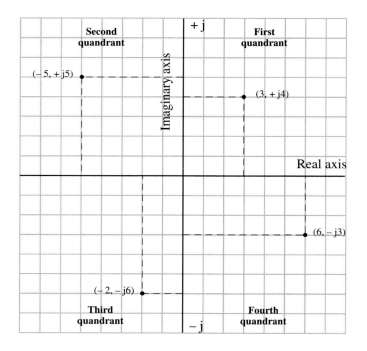

Figure 20-13. Note the relationships between the four ordered pairs and the two axes.

Other points in figure 20-13 have been selected in the remaining three quadrants. In the second quadrant is the point (–5,+j5). The third quadrant has the point (–2,–j6). The fourth quadrant contains the point (6,–j3).

Impedance in Rectangular Form

Writing the rectangular coordinates for the X component and the R component gives the ordered pair equal to the impedance. When written as impedance, the ordered pair is written as an arithmetic expression.

Figure 20-14 shows two examples of impedance written in rectangular form. When $R = 5 \ \Omega$ and $X_L = +j5 \ \Omega$, $Z = 5 + j5 \ \Omega$. When $R = 3 \ \Omega$ and $X_C = -j4 \ \Omega$, $Z = 3 - j4 \ \Omega$.

Keep in mind when using the rectangular form of a complex number, the coordinates are the lengths of the sides of the impedance triangle. The impedance, in rectangular form, is the hypotenuse expressed as the lengths of the sides, rather than as the length of the hypotenuse.

20.5 COMPLEX NUMBERS IN POLAR FORM

A number expressed in **polar form** has a length and a direction. Polar numbers are drawn using phasors that radiate from the center of concentric circles with increasingly larger radii. See figure 20-15. The polar form is equivalent to the hypotenuse of a phasor triangle. Numbers written in polar form show a value followed be an angle. Some examples are $50 \angle 45°$, $2 \angle 67°$, and $20 \angle \text{-}30°$.

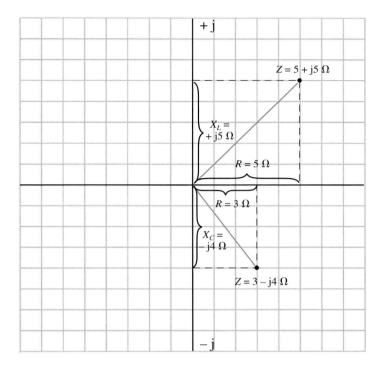

Figure 20-14. Impedance expressed in rectangular form.

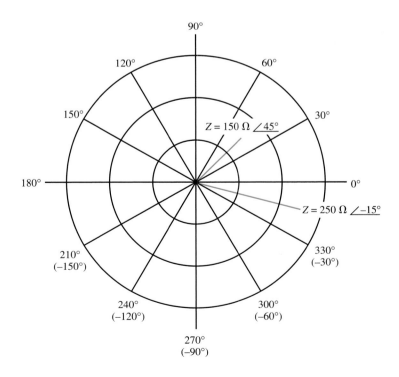

Figure 20-15. Impedance expressed in polar form.

The circumference of the outer circle is labeled in degrees. Directly to the right is 0°. Straight up is labeled +90°. Opposite 0°, to the left, is 180°. Straight down is 270°. The degrees of a circle move positive when traveling in the counterclockwise direction. The circle can be labeled in negative degrees by moving in the clockwise direction.

Two sample impedance phasors are shown in figure 20-15. In this figure, the distance between each of the circles is 100 Ω. The sample phasors are drawn to show their respective length and angle.

20.6 CONVERTING NUMBERS IN COMPLEX FORM

The purpose of writing circuit values as complex numbers is to allow easier manipulation of the arithmetic expressions. Converting between rectangular form and polar form uses three basic functions from trigonometry. These functions are sine (sin), cosine (cos), and tangent (tan).

Examine figure 20-16. The three sides of a right triangle are identified. One side is opposite (opp) θ. One side is adjacent (adj) θ. One side is the hypotenuse (hyp) of the triangle. Notice that angle theta, θ, dictates the identification of the sides.

In relation to electrical values, the opposite side is reactive, the adjacent side is resistive, and the hypotenuse is the total. These values can be measured in ohms using a series circuit impedance triangle, measured in volts using a series circuit voltage triangle, or they can be measured in amps using a parallel circuit current triangle.

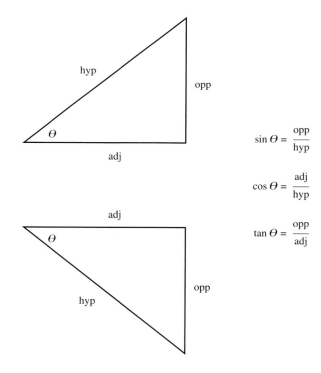

$$\sin \theta = \frac{\text{opp}}{\text{hyp}}$$

$$\cos \theta = \frac{\text{adj}}{\text{hyp}}$$

$$\tan \theta = \frac{\text{opp}}{\text{adj}}$$

Figure 20-16. These triangles show the relationship between the three basic trigonometric functions.

Converting Rectangular Form to Polar Form

The rectangular form of a complex number is the sides of the triangle. This form is expressed in reactance or resistance. These numbers can be given as the circuit values of reactance and resistance or calculated as voltage drops or branch currents. The hypotenuse of the triangle is the polar form. It is written as a length at the angle theta.

To convert rectangular form into polar form involves two steps. In the first step, the Pythagorean theorem (hyp = $\sqrt{\text{adj}^2 + \text{opp}^2}$) is used.

Formula 20.A

$$Z = \sqrt{R^2 + X^2}$$

Z is the impedance, measured in ohms.

R is the resistance, measured in ohms.

X is the reactance, measured in ohms.

In the second step, the angle theta is calculated using the tangent function. The tangent function is rearranged to solve for the angle:

Formula 20.B

$$\theta = \tan^{-1} \frac{\text{opp } (X)}{\text{adj } (R)}$$

θ is the angle, measured in degrees.

R is the resistance, measured in ohms.

X is the reactance, measured in ohms.

Write the polar form as a number at an angle: $Z \angle \theta$.

Sample problem 6. _____

Convert the following circuit values from rectangular form into polar, $R = 87\ \Omega$, $X_L = +j50\ \Omega$. The triangle is shown in figure 20-17.

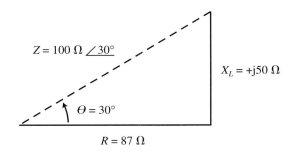

Figure 20-17. Given *R* and X_L, write *Z* in polar form. (Sample problem 6)

1. Find the length of the hypotenuse using the Pythagorean theorem.

    Formula: $Z = \sqrt{R^2 + X_L^2}$

    Substitution: $Z = \sqrt{87^2 + 50^2}$

    Answer: $Z = 100 \ \Omega$

2. Find the operating angle.

    Formula: $\theta = \tan^{-1} \dfrac{X_L}{R}$

    Substitution: $\theta = \tan^{-1} \dfrac{50 \ \Omega}{87 \ \Omega}$

    Answer: $\theta = 30°$

3. Express impedance in polar form.

    $$Z = 100 \ \Omega \ \angle 30°$$

Sample problem 7. _____

The branch currents of a circuit are given in rectangular form, $I_R = 6$ A and $I_L = -j6$ A. Find the value of total current in polar form. See figure 20-18.

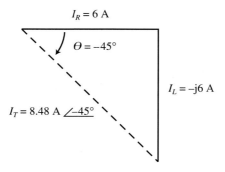

Figure 20-18. Given branch currents for I_R and I_L, write the total current in polar form. (Sample problem 7)

1. Find the length of the hypotenuse using the Pythagorean theorem.

    Formula: $I_T = \sqrt{I_R^2 + I_L^2}$

    Substitution: $I_T = \sqrt{6^2 + 6^2}$

    Answer: $I_T = 8.48$ A

2. Find the operating angle.

      Formula: $\theta = \tan^{-1} \dfrac{I_L}{I_R}$

      Substitution: $\theta = \tan^{-1} \dfrac{-j6 \text{ A}}{6 \text{ A}}$

      Answer: $\theta = -45°$

3. Express total current in polar form.

$$I_T = 8.48 \text{ A } \angle -45°$$

Converting Polar Form to Rectangular Form

    A number expressed in polar form is converted to its rectangular equivalent by using the sine and cosine trigonometric functions. The reactive value is the side of the triangle opposite theta. The resistive value is adjacent to theta. The reactive value is found using the sine formula.

Formula 20.C

$$\cos \theta = \frac{\text{adj}}{\text{hyp}}$$

θ is the angle, measured in degrees.

The resistive value is found using the cosine formula.

Formula 20.D

$$\sin \theta = \frac{\text{opp}}{\text{hyp}}$$

θ is the angle, measured in degrees.

The j-operator, for a reactive component, is positive for a positive theta and negative for a negative theta.

Sample problem 8. _____

Find the reactive and resistive branch currents when given the total current in polar form, $I_T = 150$ mA $\angle 40°$. The triangle is drawn in figure 20-19.

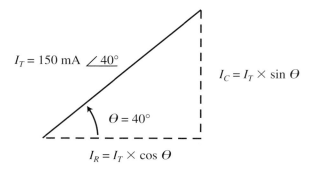

Figure 20-19. Given the polar form of total current, find the rectangular form. (Sample problem 8)

1. Resistive branch current.

   Formula: $\cos \theta = \dfrac{\text{adj}}{\text{hyp}}$

   Rearranging: $\text{adj} = \text{hyp} \times \cos \theta$

   Substitution: $I_R = 150$ mA $\times \cos 40°$

   Answer: $I_R = 115$ mA

2. Reactive branch current.

   Formula: $\sin \theta = \dfrac{\text{opp}}{\text{hyp}}$

   Rearranging: $\text{opp} = \text{hyp} \times \sin \theta$

   Substitution: $I_C = 150$ mA $\times \sin 40°$

   Answer: $I_C = +\text{j}96.4$ mA

3. Write total current in rectangular form.

   $$I_T = 115 + \text{j}96.4 \text{ mA}$$

Sample problem 9. _____

Find the voltage drops across the series components with a total voltage, expressed in polar form, of 25 V $\angle{-53°}$. Refer to figure 20-20.

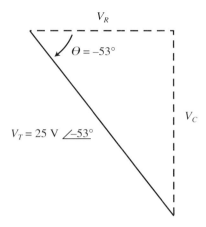

Figure 20-20. Find the voltage drops across the series components. (Sample problem 9)

1. Resistive voltage drop.

    Formula: $\cos \theta = \dfrac{\text{adj}}{\text{hyp}}$

    Rearranging: $\text{adj} = \text{hyp} \times \cos \theta$

    Substitution: $V_R = 25 \text{ V} \times \cos -53°$

    Answer: $V_R = 15 \text{ V}$

2. Reactive voltage drop.

    Formula: $\sin \theta = \dfrac{\text{opp}}{\text{hyp}}$

    Rearranging: $\text{opp} = \text{hyp} \times \sin \theta$

    Substitution: $V_C = 25 \text{ V} \times \sin -53°$

    Answer: $V_C = -j20 \text{ V}$

3. Write total current in rectangular form.

    $$V_T = 15 - j20 \text{ V}$$

Writing a Complex Number from Circuit Components

All circuit component values can be written in rectangular form, then converted to polar form. Figure 20-21 lists examples of circuit components and their equivalent rectangular and polar expressions.

	Rectangular	Polar
A $R = 10\ \Omega$	$Z = 10 + j0\ \Omega$	$Z = 10\ \Omega\ \angle 0°$
B $X_L = +j20\ \Omega$	$Z = 0 + j20\ \Omega$	$Z = 20\ \Omega\ \angle 90°$
C $X_C = -j30\ \Omega$	$Z = 0 - j30\ \Omega$	$Z = 30\ \Omega\ \angle{-90°}$
D $R = 15\ \Omega$ $X_C = -j20\ \Omega$	$Z = 15 - j20\ \Omega$	$Z = 25\ \Omega$ $\angle{-53.1°}$
E $R = 25\ \Omega$ $X_L = +j30\ \Omega$	$Z = 25 + j30\ \Omega$	$Z = 39\ \Omega\ \angle 50.2°$
F $R = 50\ \Omega$ $X_C = -j100\ \Omega$ $X_L = +j40\ \Omega$	$Z = 50 - j100 + j40\ \Omega$ $Z = 50 - j60\ \Omega$	$Z = 78.1\ \Omega$ $\angle{-50.2°}$
G $X_L = +j250$ $R = 100\ \Omega$ $X_C = -j150\ \Omega$	$Z = 100 + j250 - j150\ \Omega$ $Z = 100 + j100\ \Omega$	$Z = 141.4\ \Omega$ $\angle 45°$

Figure 20-21. Writing impedances as complex numbers.

Figure A is a resistor with no reactive component. The rectangular form of its impedance is written with the j-term having a value of zero. In polar form, this is expressed as the impedance having an angle of zero.

Figures B and C are reactive components with no resistance. The rectangular form shows a value of zero for resistance. The polar form expresses the impedance at an angle of 90°, either + or –.

Figures D and E show an impedance containing both resistance and reactance. The rectangular form states the values of each component. The polar form is the combined impedance, expressed as a quantity at an angle.

Figures F and G contain resistance with both inductive and capacitive reactances. The impedance can be written as a rectangular expression showing the value of each component. However, the expression should be simplified to show only the net reactance. The polar form is the combined values of the three components.

20.7 SOLVING A SERIES CIRCUIT USING COMPLEX NUMBERS

The basic procedures used to solve a dc series circuit are used when solving ac circuits. Total resistance (impedance) is the sum of the series components. Current and voltage drops are found using Ohm's law. The voltage drops in a series circuit should add to equal the applied voltage. *This addition must be performed in rectangular form.*

Because the current has a phase angle, the voltage drops across all series circuit components have a phase shift. This includes resistors. The voltage drop across a resistor has a reactive portion and the voltage drop across the reactive components has a resistive portion.

Arithmetic with complex numbers is slightly more difficult than arithmetic with rational numbers. There are several important rules to follow.

1. Add and subtract complex numbers in rectangular form.
    a. Add or subtract real values (resistive values) together.
    b. Add or subtract the j-terms (reactive values) together.
2. Multiply and divide complex numbers in polar form.
    a. Multiply the magnitudes, then *add* the angles.
    b. Divide the magnitudes, then *subtract* the angles (the numerator minus the denominator).
3. Reciprocals are found in polar form.
    a. The magnitude is inverted.
    b. The angle gets the opposite sign.

Sample problem 10. _____

Using the circuit values given in part A of figure 20-22, find: the impedance, current, and voltage drops. Express voltage drops in both polar and rectangular form to compare the values to the total voltage. Results are shown in part B of the figure.

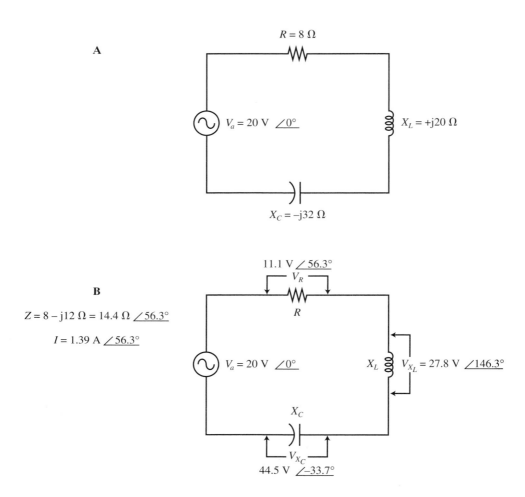

Figure 20-22. Solving a series circuit using complex numbers. (Sample problem 10)

1. Impedance in rectangular form.

Formula: $Z = R + X_L + X_C$

Substitution: $Z = 8\ \Omega + \mathrm{j}20\ \Omega - \mathrm{j}32\ \Omega$

Answer: $Z = 8 - \mathrm{j}12\ \Omega$ (rectangular)

2. Impedance in polar form.

   a. Magnitude:

   Formula: $Z = \sqrt{R^2 + X^2}$

   Substitution: $Z = \sqrt{8^2 + 12^2}$

   Answer: $Z = 14.4\ \Omega$

   b. Angle:

   Formula: $\theta = \tan^{-1}\dfrac{X}{R}$

   Substitution: $\theta = \tan^{-1}\dfrac{-12}{8}$

   Answer: $\theta = -56.3°$

   c. Impedance in polar form:

   $$Z = 14.4\ \Omega\ \angle{-56.3°}\text{(polar)}$$

3. Calculate current using numbers in polar form.

   Formula: $I = \dfrac{V}{Z}$

   Substitution: $I = \dfrac{20\ \text{V}\ \angle{0°}}{14.4\ \Omega\ \angle{-56.3°}}$

   Answer: $I = 1.39\ \text{A}\ \angle{56.3°}$

4. Resistive voltage drop, in polar form.

   Formula: $V_R = I \times R$

   Substitution: $V_R = 1.39\ \text{A}\ \angle{56.3°} \times 8\ \Omega\ \angle{0°}$

   Answer: $V_R = 11.1\ \text{V}\ \angle{56.3°}$ (polar)

5. Resistive voltage drop in rectangular form.

   a. Resistive portion of voltage drop:

   Formula: $R_V = V_R \times \cos\theta$

   Substitution: $R_V = 11.1\ \text{V} \times \cos 56.3°$

   Answer: $R_V = 6.16\ \text{V}$

b. Reactive portion of voltage drop:

Formula: $X_V = V_R \times \sin \theta$

Substitution: $X_V = 11.1 \text{ V} \times \sin 56.3°$

Answer: $X_V = 9.23 \text{ V}$

c. Rectangular form of resistive voltage drop:

$$V_R = 6.16 + j9.23 \text{ V (rectangular)}$$

6. Inductive voltage drop.

Formula: $V_L = I \times X_L$

Substitution: $V_L = 1.39 \text{ A} \underline{/56.3°} \times 20 \text{ }\Omega \underline{/90°}$

Answer: $V_L = 27.8 \text{ V} \underline{/146.3°}$ (polar)

$V_L = -23.1 + j15.4$ (rectangular)

7. Capacitive voltage drop.

Formula: $V_C = I \times X_C$

Substitution: $V_C = 1.39 \text{ A} \underline{/56.3°} \times 32 \text{ }\Omega \underline{/-90°}$

Answer: $V_C = 44.5 \underline{/-33.7°}$ (polar)

$V_C = 37.0 - j24.7$ (rectangular)

8. Compare voltage drops to applied voltage by adding in rectangular form. Add like terms.

$$
\begin{array}{r}
6.16 + j9.23 \\
-23.1 \ \ + j15.4 \\
37.0 \ \ - j24.7 \\
\hline
\end{array}
$$

$V_a = 20.06 - j0.07 \text{ V}$ (rectangular)

$V_a = 20 \text{ V} \underline{/0°}$ (polar) (checks with given circuit value)

20.8   SOLVING A PARALLEL CIRCUIT USING COMPLEX NUMBERS

As with dc circuits, the reciprocal formula is used to find the total impedance in an ac parallel circuit. To solve these ac circuits, first write the impedances of the individual branches in rectangular form. Then, change the impedances to polar form so they can be used in the reciprocal formula. The shortcut formulas used to solve dc circuits can also be applied to parallel circuits containing complex impedances. The reciprocal of impedance is **admittance (Y).** Admittance is measured in the same units as conductance, the siemen (S).

To use the reciprocal formula to find the total impedance:

1. Write down the impedances of all of the branches in rectangular form. Convert each branch impedance into polar form.
2. While the branch impedances are in polar form, calculate the admittance (reciprocal) of each one. These admittances are then converted back into rectangular form.
3. Add the branch admittances to find the total admittance.
4. Convert the total admittance back into polar form, and calculate the reciprocal one final time. The result is the total impedance in polar form.

Sample problem 11. _____

Find the total impedance of the circuit in figure 20-23, using the admittance method.

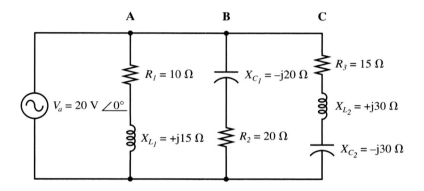

Figure 20-23. To solve a parallel circuit with complex branches find the impedance of each branch, then find the branch currents. (Sample problems 11 and 12)

1. Write each branch impedance in rectangular form and convert it to polar.

    Branch A: $Z_A = 10 + j15 \ \Omega = 18 \ \Omega \ \angle 56.3°$

    Branch B: $Z_B = 20 - j20 \ \Omega = 28.3 \ \Omega \ \angle -45°$

    Branch C: $Z_C = 15 + j0 \ \Omega = 15 \ \Omega \ \angle 0°$

2. Find the admittance of each by taking the reciprocal in polar form. Then convert each to rectangular form.

    Branch A: $Y_A = \dfrac{1}{Z_A} = \dfrac{1}{18 \ \Omega \ \angle 56.3°} = 0.0556 \ S \ \angle -56.3°$

    Branch A: $Y_A = 0.031 - j0.046 \ S$ (rectangular)

Branch B: $Y_B = \dfrac{1}{Z_B} = \dfrac{1}{28.3\ \Omega\ \angle{-45°}} = 0.0353\ S\ \angle{45°}$

Branch B: $Y_B = 0.025 + j0.025\ S$ (rectangular)

Branch C: $Y_C = \dfrac{1}{Z_C} = \dfrac{1}{15\ \Omega\ \angle{0°}} = 0.0667\ S\ \angle{0°}$

Branch C: $Y_C = 0.0667 + j0\ S$ (rectangular)

3. Add the admittances in rectangular form. Add like terms.

$$Y_A = 0.031\ -\ j0.046$$
$$Y_B = 0.025\ +\ j0.025$$
$$Y_C = 0.0667 + j0$$
$$Y_T = 0.1227 - j0.021\ S\ \text{(rectangular)}$$
$$Y_T = 0.1245\ S\ \angle{-9.7°}\ \text{(polar)}$$

4. Find total impedance by taking the reciprocal of admittance.

Formula: $Z_T = \dfrac{1}{Y_T}$

Substitution: $Z_T = \dfrac{1}{.1245\ S\ \angle{-9.7°}}$

Answer: $Z_T = 8.03\ \Omega\ \angle{9.7°}$ (polar)

Sample problem 12. _____

Use the circuit shown in figure 20-23 and the branch impedances calculated in sample problem 11 to find branch currents and total current. Write each current in both rectangular and polar form. Also, use the total current to calculate total impedance.

1. Summary of branch impedances in polar form.

   Branch A: $Z_A = 18\ \Omega\ \angle{56.3°}$

   Branch B: $Z_B = 28.3\ \Omega\ \angle{45°}$

   Branch C: $Z_C = 15\ \Omega\ \angle{0°}$

2. Use Ohm's law to calculate branch currents.

   a. Branch A:

   Formula: $I_A = \dfrac{V_a}{Z_A}$

Substitution: $I_A = \dfrac{20 \text{ V } \angle 0°}{18 \text{ } \Omega \text{ } \angle 56.3°}$

Answer: $I_A = 1.11 \text{ A } \angle -56.3°$ (polar)

$I_A = 0.616 - j0.924 \text{ A}$ (rectangular)

b. Branch B:

Formula: $I_B = \dfrac{V_a}{Z_B}$

Substitution: $I_B = \dfrac{20 \text{ V } \angle 0°}{28.3 \text{ } \Omega \text{ } \angle -45°}$

Answer: $I_B = .707 \text{ A } \angle 45°$ (polar)

$I_B = 0.5 + j0.5 \text{ A}$ (rectangular)

c. Branch C:

Formula: $I_C = \dfrac{V_a}{Z_C}$

Substitution: $I_C = \dfrac{20 \text{ V } \angle 0°}{15 \text{ } \Omega \text{ } \angle 0°}$

Answer: $I_C = 1.33 \text{ A } \angle 0°$ (polar)

$I_C = 1.33 + j0 \text{ A}$ (rectangular)

3. Find total current by adding branch currents in rectangular form. Also, change to polar form.

$$I_A = 0.616 \quad - j0.924 \text{ A}$$
$$I_B = 0.5 \quad + j0.5 \text{ A}$$
$$\underline{I_C = 1.33 \quad + j0 \text{ A}}$$
$$I_T = 2.446 \quad - j0.424 \text{ A (rectangular)}$$
$$I_T = 2.48 \text{ A } \angle -9.7° \text{ (polar)}$$

4. Use total current to calculate total impedance.

Formula: $Z = \dfrac{V}{I_T}$

Substitution: $Z = \dfrac{20 \text{ V } \angle 0°}{2.48 \text{ A } \angle -9.7°}$

Answer: $Z = 8.06 \text{ } \Omega \text{ } \angle 9.7°$ (polar)

20.9 SOLVING COMBINATION CIRCUITS USING COMPLEX NUMBERS

To solve ac series-parallel combination circuits, follow the same procedures that were used in solving dc series-parallel combination circuits. Combine the parallel branches to produce an equivalent series circuit. Then, solve the equivalent circuit to find total resistance (impedance).

Sample problem 13. _____

Find the total impedance of the circuit shown in figure 20-24, part 1.

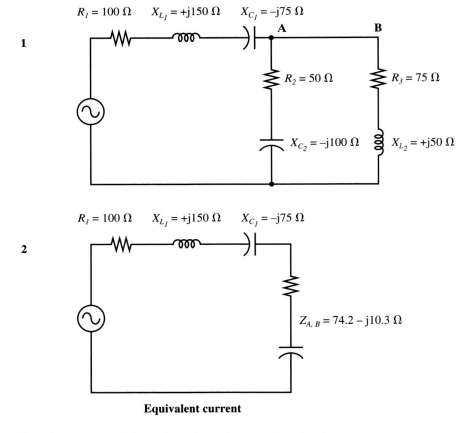

Equivalent current

Figure 20-24. Combine the parallel portion of a series-parallel circuit to make an equivalent series circuit. (Sample problem 13)

1. Write the impedance for the parallel branch A in rectangular form and polar form.

    Branch A: $Z_A = 50 - j100 \ \Omega$ (rectangular)

a. Convert to polar magnitude:

Formula: $Z_A = \sqrt{R^2 + X^2}$

Substitution: $Z_A = \sqrt{50^2 + 100^2}$

Answer: $Z_A = 112\ \Omega$

b. Convert to polar angle:

Formula: $\theta = \tan^{-1} \dfrac{X}{R}$

Substitution: $\theta = \tan^{-1} \dfrac{-100}{50}$

Answer: $\theta = -63.4°$

c. Impedance of branch A in polar form:

$$Z_A = 112\ \Omega \ \angle{-63.4°}\ \text{(polar)}$$

2. Write the impedance for the parallel branch B in rectangular and polar forms.

Branch B: $Z_B = 75 + j50$ W (rectangular)

a. Convert to polar magnitude:

Formula: $Z_A = \sqrt{R^2 + X^2}$

Substitution: $Z_A = \sqrt{75^2 + 50^2}$

Answer: $Z_A = 90.1\ \Omega$

b. Convert to polar angle:

Formula: $\theta = \tan^{-1} \dfrac{X}{R}$

Substitution: $\theta = \tan^{-1} \dfrac{50}{75}$

Answer: $\theta = 33.7°$

c. Impedance of branch A in polar form:

$$Z_B = 90.1\ \Omega \ \angle{33.7°}\ \text{(polar)}$$

3. Combine parallel branches A and B using the shortcut formula.

Formula: $$Z_{A,B} = \frac{Z_A \times Z_B}{Z_A + Z_B}$$

Substitution: $$Z_{A,B} = \frac{112 \angle{-63.4°} \times 90.1 \angle{33.7°}}{(50 - j100) + (75 + j50)}$$

Intermediate step: $$Z_{A,B} = \frac{10{,}091 \angle{-29.7°}}{125 - j50}$$

Intermediate step: $$Z_{A,B} = \frac{10{,}091 \angle{-29.7°}}{135 \angle{-21.8}}$$

Answer:
$$Z_{A,B} = 74.7 \ \Omega \ \angle{-7.9°} \text{ (polar)}$$
$$Z_{A,B} = 74 - j10.3 \ \Omega \text{ (rectangular)}$$

4. Refer to the equivalent circuit shown in part 2 of figure 20-24. Add like terms to find the total impedance.

$$
\begin{aligned}
R_1 &= 100 + j0 \ \Omega \\
X_{L_1} &= 0 + j150 \ \Omega \\
X_{C_1} &= 0 - j75 \ \Omega \\
Z_{A,B} &= 74 - j10.3 \ \Omega \\
\hline
Z_T &= 174 + j64.7 \ \Omega \text{ (rectangular)} \\
Z_T &= 186 \ \Omega \ \angle{20.4°} \text{ (polar)}
\end{aligned}
$$

SUMMARY

- Phasors are used to add values having both magnitude and direction.
- In a series circuit, total voltage is the phasor sum of the voltages across each component.
- In a parallel circuit, total current is the phasor sum of the current through each branch.
- Phasors contain positive and negative angles. Like phasors, phasors with the same positive or negative sign, add. Unlike phasors, phasors with opposite signs, subtract.
- Net reactance is the effective reactance in a circuit containing both X_C and X_L.
- In a series circuit, net reactance is calculated by adding or subtracting the phasors of the reactive components.
- In a parallel circuit, net reactance is calculated by adding or subtracting the phasors of the reactive branch currents.
- A complex number is a mathematical tool used to represent the combination of resistance and reactance in a circuit.
- The rectangular form of a complex number shows impedance of the circuit expressed using the values of resistance and reactance. These are the values used to form the sides of the impedance triangle.

- The polar form of a complex number shows the impedance of a circuit expressed using the effective combination of resistance and reactance. The polar form is the hypotenuse of the impedance triangle.
- Complex numbers are converted between rectangular form and polar form using the trigonometric functions of sine, cosine, and tangent as well as the Pythagorean theorem.

KEY WORDS AND TERMS GLOSSARY

admittance (*Y*): The ease with which alternating current flows in a circuit. It is the reciprocal of impedance. The unit of measure is siemens (S) or mhos (℧).

complex number: A number used to represent the quantities of an electronic circuit containing both resistance and reactance.

imaginary number: The j-term of a complex number, used to represent the reactive component.

j-operator: A mathematical tool used to represent reactances as complex numbers.

j-term: The imaginary part of a complex number.

like phasors: Phasors in circuits containing more than one of the same type of component. They are plotted end-to-end, with addition used to find the resultant.

net reactance: The effective reactance in a circuit containing both X_L and X_C.

opposite phasors: Phasors in circuits containing both capacitors and inductors. They are plotted end-to-end, with subtraction used to find the resultant.

ordered pair: Points located on a four-quadrant graph. The points are located with paired numbers, called ordered pairs, which contain a real term written first and an imaginary term written second.

polar form: A number expressed as a length and angle. It is used to describe the magnitude and direction of a phasor.

RLC circuit: Circuits that contain at least one resistor, one inductor, and one capacitor.

rectangular coordinates: The values to locate a point on a graph divided into four quadrants by the real axis and imaginary axis.

KEY FORMULAS

Formula 20.A

$$Z = \sqrt{R^2 + X^2}$$

Formula 20.B

$$\theta = \tan^{-1} \frac{\text{opp } (X)}{\text{adj } (R)}$$

Formula 20.C

$$\cos \theta = \frac{\text{adj}}{\text{hyp}}$$

Formula 20.D

$$\sin \theta = \frac{\text{opp}}{\text{hyp}}$$

TEST YOUR KNOWLEDGE

Do not write in this text. Please use a separate sheet of paper.

1. What do the two phasors represent in the impedance triangle?
2. On one graph, draw the phasors for the following components: $R_1 = 250\ \Omega$, $X_{C_1} = 100\ \Omega$, $X_{L_1} = 500\ \Omega$, $R_2 = 100\ \Omega$, $X_{C_2} = 450\ \Omega$, $X_{L_2} = 150\ \Omega$.
3. Use phasor addition to find the net reactance in a series circuit with $X_C = 40\ \Omega$ and $X_L = 60\ \Omega$. Is the result capacitive or inductive?
4. Draw an impedance triangle and solve for the impedance in a series circuit with $R = 400\ \Omega$, $X_C = 550\ \Omega$, and $X_L = 250\ \Omega$.
5. Find net reactance, impedance, current, phase angle, and voltage drops across each component in a series circuit with 50 volts applied to the following components: $R = 25\ \Omega$, $X_C = 100\ \Omega$, $X_L = 90\ \Omega$. Also, draw the impedance triangle and the voltage triangle.
6. A series circuit has 100 volts applied to the following components: $R_1 = 35\ \Omega$, $X_{C_1} = 75\ \Omega$, $R_2 = 65\ \Omega$, $X_{L_1} = 40\ \Omega$, $X_{C_2} = 25\ \Omega$, $X_{L_2} = 160\ \Omega$. Find the net reactance, impedance, current, and phase angle. Also, draw the impedance triangle and the voltage triangle.
7. Find the branch currents, circulating current, net current, and net reactance in a circuit with $X_C = 40\ \Omega$ in parallel with $X_L = 60\ \Omega$. The applied voltage is 10 volts. In addition, draw the phasor diagram.
8. A circuit has three parallel branches: $R = 100\ \Omega$, $X_C = 50\ \Omega$, and $X_L = 200\ \Omega$. If the applied voltage is 25 volts, determine the branch currents, net reactive current, total current, and the impedance. Also, draw the current triangle.
9. For each of the resistance and reactance combinations find the equivalent impedance in rectangular form and polar form. Draw a phasor diagram to represent the impedance in each form.
    a. $R = 40\ \Omega$ and $X_L = 50\ \Omega$
    b. $R = 5\ \text{k}\Omega$ and $X_C = 3\ \text{k}\Omega$
    c. $R = 0\ \Omega$ and $X_L = 35\ \Omega$
    d. $R = 100\ \Omega$ and $X_C = 0\ \Omega$

   e. $R = 80\ \Omega$ and $X_L = 60\ \Omega$

   f. $R = 20\ \Omega$ and $X_C = 10\ \Omega$

   g. $R = 300\ \Omega$ and $X_L = 400\ \Omega$

   h. $R = 25\ \Omega$ and $X_C = 75\ \Omega$

   i. $R = 4\ k\Omega$ and $X_L = 4\ k\Omega$

   j. $R = 10\ \Omega$ and $X_C = 10\ \Omega$

10. Use the complex number given to find the impedance in the other form.

	Rectangular	Polar
a.	. . . .	$10\ \Omega\ \angle 45°$
b.	$(35 + j25)\ \Omega$	. . . .
c.	. . . .	$15\ \Omega\ \angle{-30°}$
d.	$(50 - j100)\ \Omega$	. . . .
e.	. . . .	$40\ \Omega\ \angle 0°$
f.	$(40 + j40)\ \Omega$	. . . .
g.	. . . .	$100\ \Omega\ \angle{-37°}$
h.	$(0 - j150)\ \Omega$	. . . .
i.	. . . .	$50\ \Omega\ \angle 90°$
j.	$(10 + j0)\ \Omega$	. . . .

11. Use complex numbers to find the impedance, current, and voltage drops in a series circuit with an applied voltage of 25 volts, $R = 50\ \Omega$, and $X_C = 35\ \Omega$.

12. Use complex numbers to find the branch currents, total current, and impedance in a parallel circuit with an applied voltage of 100 volts, $R = 100\ \Omega$, and $X_L = 200\ \Omega$.

13. Refer to the circuit shown in figure 20-25.

   a. Find the impedance of each branch.

   b. Draw an equivalent circuit to show branch impedances.

   c. Calculate the branch currents.

   d. Calculate the total current.

   e. Calculate the total impedance.

14. Refer to the circuit shown in figure 20-26.

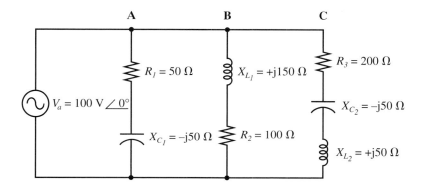

Figure 20-25. Problem #13.

a. Find the impedances for branches A and B.
b. Draw an equivalent series circuit.
c. Find the total impedance.

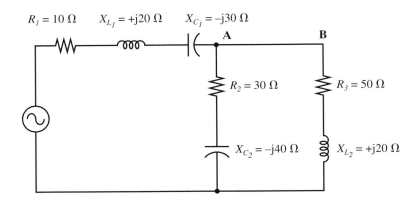

$R_1 = 10\ \Omega$ $X_{L_1} = +j20\ \Omega$ $X_{C_1} = -j30\ \Omega$

A   **B**

$R_2 = 30\ \Omega$ $R_3 = 50\ \Omega$

$X_{C_2} = -j40\ \Omega$ $X_{L_2} = +j20\ \Omega$

Figure 20-26. Problem #14.

Chapter 21
Circuit Theorems
Applied to AC

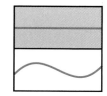

Upon completion of this chapter, you will be able to:
- Solve series and parallel circuits by treating each resistance and reactance combination as an impedance network.
- Use Kirchhoff's voltage law to solve complex ac circuits.
- Use Kirchhoff's current law to solve complex ac circuits.
- Use the superposition theorem to solve ac circuits with two voltage sources.
- Use Thevenin's theorem to simplify a complex ac circuit.
- Represent delta and wye transformations as three-terminal impedance networks.

The circuit theorems used to solve dc circuits are also used to solve ac circuits. This chapter discusses solving ac circuits with Kirchhoff's voltage and current laws, the superposition theorem, Thevenin's theorem, and delta-wye network transformations. For a review of the dc theorems, refer to Chapter 9.

21.1 REPRESENTING AN IMPEDANCE NETWORK

An impedance network can be formed from many different components such as resistors, inductors, and capacitors. Impedance is also associated with transformers, transistors, and sometimes an entire circuit.

On a schematic diagram, an impedance network is represented by a box, as shown in figure 21-1. The value of impedance assigned to the network is the *net value* of all of the components combined. Usually, an impedance network is shown in its polar form, although the rectangular form is needed to perform addition.

Adding Impedance Networks in Series

The total impedances of networks connected in series are added in the same manner as when adding individual components. Remember, the addition of impedances must be in rectangular form.

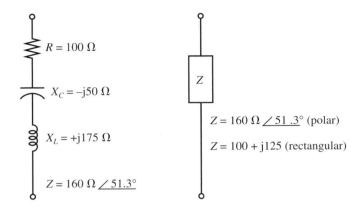

Figure 21-1. The complex impedance network on the left can be represented by a single symbol, the box to the right.

Sample problem 1. _____

    Find the total impedance of the circuit in figure 21-2.

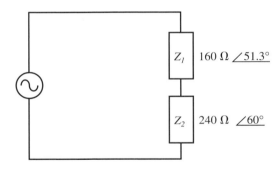

Figure 21-2. Find the total impedance by adding in rectangular form. (Sample problem 1)

1.   Convert Z_1 to rectangular form.

$$Z_1 = 160 \text{ } \Omega \text{ } \angle 51.3° \text{ (polar)}$$

    a. Resistance value:

        Formula: $R = Z \times \cos \theta$

        Substitution: $R = 160 \times \cos 51.3$

        Answer: $R = 100 \text{ } \Omega$

b. Reactance value:

Formula: $X = Z \times \sin \theta$

Substitution: $X = 160 \times \sin 51.3$

Answer: $X = +j125 \ \Omega$

$Z_1 = 100 + j125 \ \Omega$ (rectangular)

2.   Convert Z_2 to rectangular form.

$Z_2 = 240 \ W \ \underline{\angle 60°}$ (polar)

a. Resistance value:

Formula: $R = Z \times \cos \theta$

Substitution: $R = 240 \times \cos 60$

Answer: $R = 120 \ \Omega$

b. Reactance value:

Formula: $X = Z \times \sin \theta$

Substitution: $X = 240 \times \sin 60$

Answer: $X = +j208 \ \Omega$

$Z_2 = 120 + j208 \ \Omega$ (rectangular)

3.   Add in rectangular form.

Formula: $Z_T = Z_1 + Z_2$

Substitution: $Z_T = (100 + j125 \ \Omega) + (120 + j208 \ \Omega)$

Answer: $Z_T = 220 + j333 \ \Omega$ (rectangular)

Combining Impedances in Parallel

The total impedance for parallel connections is found using the same formulas that were used to solve for resistors connected in parallel. The shortcut formulas also apply.

The reciprocal of impedance is admittance. Therefore, using the reciprocal formula is the same as adding the admittances. Reciprocals are calculated in polar form. As discussed in Chapter 20, the admittance is found by taking the reciprocal of the impedance's magnitude and then changing the sign of the angle. The total admittance is the reciprocal of the total impedance.

Sample problem 2. _____

Find the total impedance of the circuit shown in figure 21-3 using the reciprocal formula.

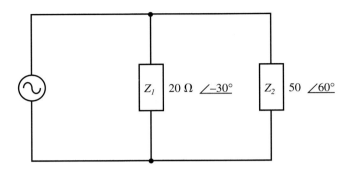

Figure 21-3. Total impedance in parallel is found by adding the admittances. (Sample problem 2)

Formula:

$$\frac{1}{Z_T} = \frac{1}{Z_1} + \frac{1}{Z_2}$$

Substitution:

$$\frac{1}{Z_T} = \frac{1}{20 \angle{-30°}} + \frac{1}{50 \angle{60°}}$$

Admittances:

$$\frac{1}{Z_T} = 0.05 \angle{30°} + 0.02 \angle{-60°}$$

Rectangular:

$$\frac{1}{Z_T} = (0.0433 + j0.025 \text{ S}) + (0.01 - j0.0173 \text{ S})$$

Intermediate step:

$$\frac{1}{Z_T} = 0.0533 + j0.0077 \text{ S} = 0.0539 \text{ S} \angle{8.22°}$$

Answer:

$$Z_T = 18.55 \ \Omega \ \angle{-8.22°} \text{ (polar)}$$

21.2 KIRCHHOFF'S CIRCUIT LAWS

Kirchhoff's laws were introduced in Chapter 9 along with the other dc circuit theorems. Although not specifically stated, these laws were also used in Chapters 6, 7, and 8 to verify circuit calculations.

In this chapter, Kirchhoff's laws are used in three ways.
- To verify that the voltage drops around a loop equal the applied voltage.
- To find an unknown voltage around a loop.
- To verify that the current into a junction equals the current out of the junction.

Verifying Voltage Drops around a Loop

Finding the voltage drops across impedance networks connected in series uses a procedure similar to the procedure used while solving dc circuits. The difference between solving ac and dc networks is in the use of complex numbers in the formulas. The Ohm's law formulas produce voltage drops in polar form. These voltage drops must be changed to rectangular form before they can be added.

Sample problem 3.
_____

Using the circuit in figure 21-4, find the total impedance, total current, and voltage drops. Then use Kirchhoff's voltage law to verify that the voltage drops add to equal the applied voltage.

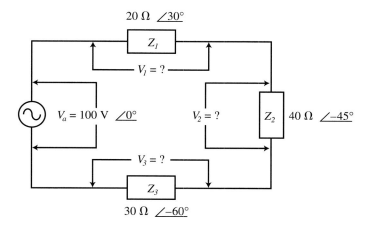

Figure 21-4. From Kirchhoff's voltage law, the vector sum of voltage drops around a closed loop equals the applied voltage. (Sample problem 3)

1. Convert each impedance network to rectangular form.

   Formulas: $R = Z \times \cos \theta$

   $X = Z \times \sin \theta$

   $Z_1 = 20\ \Omega\ \angle 30°$ (polar)
   $Z_1 = 17.3 + j10\ \Omega$ (rectangular)

   $Z_2 = 40\ \Omega\ \angle{-45°}$ (polar)
   $Z_2 = 28.3 - j28.3\ \Omega$ (rectangular)

   $Z_3 = 30\ \Omega\ \angle{-60°}$ (polar)
   $Z_3 = 15 - j26\ \Omega$ (rectangular)

2. Find the total impedance in rectangular form.

    Formula: $Z_T = Z_1 + Z_2 + Z_3$

    Adding:

    $$17.3 + j10$$
    $$28.3 - j28.3$$
    $$\underline{15 - j26}$$

    Answer: $Z_T = 60.6 - j44.3 \ \Omega$ (rectangular)

3. Convert to polar form.

    a. Magnitude of impedance:

        Formula: $Z = \sqrt{R^2 + X^2}$

        Substitution: $Z = \sqrt{60.6^2 + 44.3^2}$

        Answer: $Z = 75.1 \ \Omega$

    b. Phase angle theta:

        Formula: $\theta = \tan^{-1} \dfrac{X}{R}$

        Substitution: $\theta = \tan^{-1} \dfrac{-44.3}{60.6}$

        Answer: $\theta = -36.2°$

        $Z_T = 75.1 \ \Omega \ \angle{-36.2°}$

4. Calculate current with Ohm's law in polar form.

    Formula: $I = \dfrac{V}{Z}$

    Substitution: $I = \dfrac{100 \ V \ \angle{0°}}{75.1 \ \Omega \ \angle{-36.2°}}$

    Answer: $I = 1.33 \ A \ \angle{36.2°}$

5. Find voltage drops across each impedance network.

   Formula: $V = I \times Z$ (polar form)

   a. Voltage across Z_1:

      Substitution: $V_1 = 1.33 \text{ A } \underline{/36.2°} \times 20 \text{ } \Omega \text{ } \underline{/30°}$

      Answer: $V_1 = 26.6 \text{ V } \underline{/66.2°}$

   b. Voltage across Z_2:

      Substitution: $V_2 = 1.33 \text{ A } \underline{/36.2°} \times 40 \text{ } \Omega \text{ } \underline{/-45°}$

      Answer: $V_2 = 53.2 \text{ V } \underline{/-8.8°}$

   c. Voltage across Z_3:

      Substitution: $V_3 = 1.33 \text{ A } \underline{/36.2°} \times 30 \text{ } \Omega \text{ } \underline{/-60°}$

      Answer: $V_3 = 39.9 \text{ V } \underline{/-23.8°}$

6. Change each voltage drop to rectangular form.

   Formulas: $R = Z \times \cos \theta$ and $X = Z \times \sin \theta$

   $$V_1 = 26.6 \text{ V } \underline{/66.2°} \text{ (polar)}$$
   $$V_1 = 10.7 + j24.3 \text{ V (rectangular)}$$

   $$V_2 = 53.2 \text{ } \underline{/-8.8°} \text{ (polar)}$$
   $$V_2 = 52.6 - j8.1 \text{ V (rectangular)}$$

   $$V_3 = 39.9 \text{ } \underline{/-23.8°} \text{ (polar)}$$
   $$V_3 = 36.5 - j16.1 \text{ V (rectangular)}$$

7. Add the voltage drops in rectangular form. The result should equal the applied voltage of $100 + j0$.

   Formula: $V_a = V_1 + V_2 + V_3$

   Adding:

   $$
   \begin{array}{l}
   10.7 + j24.3 \\
   52.6 - j8.1 \\
   \underline{36.5 - j16.1}
   \end{array}
   $$

   Answer: $V_a = 99.8 + j0.1 \cong 100 + j0$

   There is a slight error because of rounding.

Finding Unknown Voltage Drops in a Loop

Another use for Kirchhoff's voltage law is to find an unknown voltage drop in a loop when the applied voltage and the remaining voltages are known. In this case, the known voltage drops are added together. Then they are subtracted from the applied voltage.

Sample problem 4. _____

Find the missing voltage in figure 21-5, using Kirchhoff's voltage law.

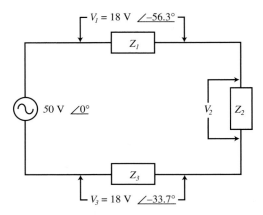

Figure 21-5. Find the voltage across Z_2. (Sample problem 4)

1. Change the given voltages to rectangular form.

    $V_a = 50$ V $\angle 0°$ (polar)
    $V_a = 50 + j0$ (rectangular)

    $V_1 = 18$ V $\angle{-56.3°}$ (polar)
    $V_1 = 10 - j15$ V (rectangular)

    $V_3 = 18$ V $\angle{-33.7°}$ (polar)
    $V_3 = 15 - j10$ V (rectangular)

2. Add the voltage drops together, and then subtract the result from the applied voltage to find the unknown.

    Formula: $\qquad V_2 = V_a - (V_1 + V_3)$

    Substitution: $\qquad V_2 = (50 + j0) - [(10 - j15) + (15 - j10)]$

    Intermediate step: $\quad V_2 = (50 + j0) - (25 - j25)$

    Answer: $\qquad V_2 = 25 + j25$ V (rectangular)
    $\qquad\qquad\quad V_2 = 35.4$ V $\angle 45°$ (polar)

Junction Currents

The current entering a junction equals the current leaving a junction. In a parallel circuit, the total current splits in the individual branches, as shown in figure 21-6. The total current, therefore, is the sum of the branch currents. When the currents are given in polar form, be sure to change the currents to their rectangular form first.

Sample problem 5. _____

Find the total current in the circuit shown in figure 21-6.

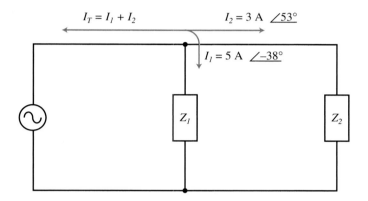

Figure 21-6. Kirchhoff's current law states that the vector sum of currents entering a junction equals the current leaving the junction. (Sample problem 5)

1. Change the branch currents to rectangular form.

$$I_1 = 5 \text{ A } \angle{-38°} \text{ (polar)}$$
$$I_1 = 3.9 - j3.1 \text{ A (rectangular)}$$

$$I_2 = 3 \text{ A } \angle{53°} \text{ (polar)}$$
$$I_2 = 1.8 + j2.4 \text{ A (rectangular)}$$

2. Add the currents in rectangular form.

Formula: $I_T = I_1 + I_2$

Substitution: $I_T = (3.9 - j3.1) + (1.8 + j2.4)$

Answer: $I_T = 5.7 - j0.7 \text{ A (rectangular)}$
$$I_T = 5.74 \text{ A } \angle{-7°} \text{ (polar)}$$

21.3   SUPERPOSITION THEOREM

The superposition theorem finds the current through the load by solving each voltage source independently. Then, the results from each voltage source are combined.

The load impedance in the circuit shown in figure 21-7 is supplied by two different voltage sources. When using the superposition theorem, one voltage source is replaced with a short circuit, as shown in figure 21-8. The total impedance as seen from the remaining voltage source can now be found. The total impedance, along with the applied voltage, is used to calculate the total current and the voltage drop across the series impedance. Finally, the voltage drop and current through the load impedance are calculated. This process must be repeated with the other voltage source in the circuit. Once this is accomplished the results from both voltage sources can be combined.

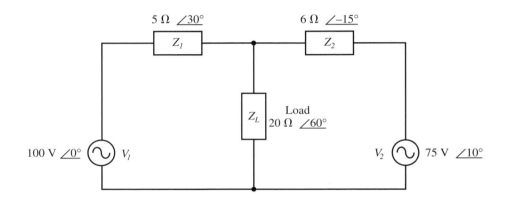

Figure 21-7. Superposition theorem is used to find the current through a load with two voltage sources. (Sample problem 6)

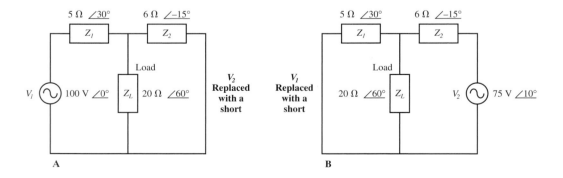

Figure 21-8. Superposition theorem replaces each voltage source with a short circuit. (Sample problem 6)

Sample problem 6. _____

Using the circuit shown in figure 21-7, find the current and voltage across the load impedance using the superposition theorem. Figure 21-8 shows the individual voltage sources replaced with a short circuit.

1. Find the total impedance as seen from V_1 with V_2 replaced with a short circuit. See circuit A of figure 21-8.

    a. Find the parallel combination of Z_2 and Z_L. Note, the reciprocal formula is used in this example.

    Formula:

    $$\frac{1}{Z_{P_1}} = \frac{1}{Z_2} + \frac{1}{Z_L}$$

    Substitution:

    $$\frac{1}{Z_{P_1}} = \frac{1}{6\ \Omega\ \angle{-15°}} + \frac{1}{20\ \Omega\ \angle{60°}}$$

    Intermediate step:

    $$\frac{1}{Z_{P_1}} = 0.167\ \text{S}\ \angle{15°} + 0.05\ \text{S}\ \angle{-60°}$$

    In rectangular form:

    $$\frac{1}{Z_{P_1}} = (0.161 + j.043\ \text{S}) + (0.025 - j.043\ \text{S})$$

    Intermediate step:

    $$\frac{1}{Z_{P_1}} = 0.186 + j0\ \text{S} = 0.186\ \text{S}\ \angle{0°}$$

    Answer:

    $$Z_{P_1} = 5.38\ \Omega\ \angle{0°} = 5.38 + j0\ \Omega$$

    b. Add the parallel equivalent to the series impedance.

    $$Z_1 = 5\ \Omega\ \angle{30°} = 4.33 + j2.5\ \Omega$$

    Formula: $Z_{T_1} = Z_1 + Z_{P_1}$

    Substitution: $Z_{T_1} = (4.33 + j2.5\ \Omega) + (5.38 + j0\ \Omega)$

    Answer: $Z_{T_1} = 9.71 + j2.5\ \Omega = 10\ \Omega\ \angle{14.4°}$

2. Find the circuit current supplied by V_1.

    Formula: $I_{T_1} = \dfrac{V_1}{Z_{T_1}}$

    Substitution: $I_{T_1} = \dfrac{100\ \text{V}\ \angle{0°}}{10\ \Omega\ \angle{14.4°}}$

    Answer: $I_{T_1} = 10\ \text{A}\ \angle{-14.4°}$

3. Find the voltage drop across the series impedance.

    Formula: $V_{Z_1} = I_{T_1} \times Z_1$

    Substitution: $V_{Z_1} = 10\ \text{A}\ \angle{-14.4°} \times 5\ \Omega\ \angle{30°}$

    Answer: $V_{Z_1} = 50\ \text{V}\ \angle{15.6°}$

4. Find the voltage drop across the equivalent parallel impedance (Z_{P_1}). V_{P_1} is equal to the voltage across the load (V_{L_1}) as supplied by V_1.

    Formula: $V_{L_1} = I_{T_1} \times Z_{P_1}$

    Substitution: $V_{L_1} = 10\ \text{A}\ \angle{-14.4°} \times 5.38\ \Omega\ \angle{0°}$

    Answer: $V_{L_1} = 53.8\ \text{V}\ \angle{-14.4°}$

5. Find the current through the load supplied by V_1.

    Formula: $I_1 = \dfrac{V_{L_1}}{Z_L}$

    Substitution: $I_1 = \dfrac{53.8\ \text{V}\ \angle{-14.4°}}{20\ \Omega\ \angle{60°}}$

    Answer: $I_1 = 2.69\ \text{A}\ \angle{-74.4°} = 0.72 - j2.59\ \text{A}$

6. Find the total impedance as seen from V_2 with V_1 replaced with a short circuit. See circuit B of figure 21-8.

    a. Find the parallel combination of Z_1 and Z_L. Note, the reciprocal formula is used in this example.

    Formula: $\dfrac{1}{Z_{P_2}} = \dfrac{1}{Z_1} + \dfrac{1}{Z_L}$

    Substitution: $\dfrac{1}{Z_{P_2}} = \dfrac{1}{5\ \Omega\ \angle{30°}} + \dfrac{1}{20\ \Omega\ \angle{60°}}$

    Intermediate step: $\dfrac{1}{Z_{P_2}} = 0.2\ \text{S}\ \angle{-30°} + 0.05\ \text{S}\ \angle{-60°}$

    In rectangular form: $\dfrac{1}{Z_{P_2}} = (0.173 - j0.1\ \text{S}) + (0.025 - j.043\ \text{S})$

    Intermediate step: $\dfrac{1}{Z_{P_2}} = 0.198 - j0.143\ \text{S} = 0.244\ \text{S}\ \angle{-35.8°}$

    Answer: $Z_{P_2} = 4.1\ \Omega\ \angle{35.8°} = 3.33 + j2.4\ \Omega$

b. Add the parallel equivalent to the series impedance.
$$Z_2 = 6 \ \Omega \ \angle{-15°} = 5.8 - j1.55 \ \Omega$$

Formula: $Z_{T_2} = Z_2 + Z_{P_2}$

Substitution: $Z_{T_2} = (5.8 - j1.55 \ \Omega) + (3.33 + j2.4 \ \Omega)$

Answer: $Z_{T_2} = 9.13 + j0.85 \ \Omega = 9.17 \ \Omega \ \angle{5.3°}$

7. Find the circuit current supplied by V_2.

Formula: $I_{T_2} = \dfrac{V_2}{Z_{T_2}}$

Substitution: $I_{T_2} = \dfrac{75 \text{ V} \ \angle{10°}}{9.17 \ \Omega \ \angle{5.3°}}$

Answer: $I_{T_2} = 8.18 \text{ A} \ \angle{4.7°}$

8. Find the voltage drop across the series impedance.

Formula: $V_{Z_2} = I_{T_2} \times Z_2$

Substitution: $V_{Z_2} = 8.18 \text{ A} \ \angle{4.7°} \times 6 \ \Omega \ \angle{-15°}$

Answer: $V_{Z_2} = 49.08 \text{ V} \ \angle{-10.3°}$

9. Find the voltage drop across the equivalent parallel impedance (Z_{P_2}), V_{P_2} is equal to the voltage across the load (V_{L_2}) as supplied by V_2.

Formula: $V_{L_2} = I_{T_2} \times Z_{P_2}$

Substitution: $V_{L_2} = 8.18 \text{ A} \ \angle{4.7°} \times 4.1 \ \Omega \ \angle{35.8°}$

Answer: $V_{L_2} = 33.5 \text{ V} \ \angle{40.5°}$

10. Find the current through the load supplied by V_2.

Formula: $I_2 = \dfrac{V_{L_2}}{Z_L}$

Substitution: $I_2 = \dfrac{33.5 \text{ V} \ \angle{40.5°}}{20 \ \Omega \ \angle{60°}}$

Answer: $I_2 = 1.675 \text{ A} \ \angle{-19.5°} = 1.58 - j0.56 \text{ A}$

11. Find the combined current through the load.

Formula: $I_L = I_1 + I_2$

Substitution: $I_L = (0.72 - j2.59\text{ A}) + (1.58 - j0.56\text{ A})$

Answer: $I_L = 2.3 - j3.15\text{ A} = 3.9\text{ A} \angle -53.9°$

12. Find the voltage across the load.

Formula: $V_L = I_L \times Z_L$

Substitution: $V_L = 3.9\text{ A} \angle -53.9° \times 20\ \Omega \angle 60°$

Answer: $V_L = 78\text{ V} \angle 6.1°$

21.4 THEVENIN'S THEOREM

Thevenin's theorem is especially useful when simplifying a circuit to determine the best load to use. This is very convenient in circuits such as amplifiers. To achieve the maximum transfer of power to a load, the impedance of the load must be equal to the impedance of the source voltage.

The **Thevenin equivalent impedance** is found by first removing the load. Then, the circuit impedance is calculated looking back from the load terminals. The Thevenin equivalent voltage is found by removing the load and determining the open circuit voltage.

Sample problem 7.

Find the Thevenin equivalent of the circuit shown in figure 21-9. Find the Thevenin equivalent impedance looking from the load terminals with the load removed. Z_1 is in parallel with Z_3. Z_2 is in series with their parallel equivalent.

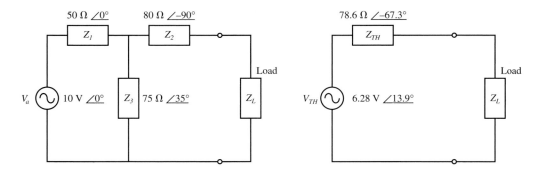

Figure 21-9. Thevenin's theorem is used to find the voltage and its internal resistance at the load terminals. (Sample problem 7)

1. Find the parallel equivalent of Z_1 and Z_3. Combine Z_1 with Z_3:

   Formula:

   $$\frac{1}{Z_P} = \frac{1}{Z_1} + \frac{1}{Z_3}$$

   Substitution:

   $$\frac{1}{Z_P} = \frac{1}{50 \ \Omega \ \angle 0°} + \frac{1}{75 \ \Omega \ \angle 35°}$$

   Intermediate step:

   $$\frac{1}{Z_P} = 0.02 \ \text{S} \ \angle 0° + 0.0133 \ \text{S} \ \angle{-35°}$$

   In rectangular form:

   $$\frac{1}{Z_P} = (0.02 + j0 \ \text{S}) + (0.011 - j0.0076 \ \text{S})$$

   Adding:

   $$\frac{1}{Z_P} = 0.031 - j0.0076 \ \text{S}$$

   To polar form:

   $$\frac{1}{Z_P} = 0.032 \ \text{S} \ \angle{-13.8°}$$

   Answer:

   $$Z_P = 31.25 \ \Omega \ \angle 13.8°$$

2. Find the Thevenin equivalent impedance by adding the parallel equivalent, Z_P, to Z_2. First change both impedances to rectangular form.

   $$Z_P = 31.25 \ \Omega \ \angle 13.8° = 30.3 + j7.5 \ \Omega$$
   $$Z_2 = 80 \ \Omega \ \angle{-90°} = 0 - j80 \ \Omega$$

   Formula: $Z_{TH} = Z_P + Z_2$

   Substitution: $Z_{TH} = (30.3 + j7.5 \ \Omega) + (0 - j80 \ \Omega)$

   Answer: $Z_{TH} = 30.3 - j72.5 \ \Omega = 78.6 \ \Omega \ \angle{-67.3°}$

3. To find the Thevenin equivalent voltage, you can ignore the series impedance, Z_2, because V_{TH} is an open circuit voltage. This example uses the voltage divider formula to solve for V_{TH}, however current and voltage drops also could be used.

   Formula: $$V_{TH} = V_a \times \frac{Z_3}{Z_1 + Z_3}$$

   Substitution: $$V_{TH} = 10 \ \text{V} \ \angle 0° \times \frac{75 \ \Omega \ \angle 35°}{(50 \ j0 \ \Omega) + (61.4 + j43 \ \Omega)}$$

Change denominator to rectangular form:

$$V_{TH} = 10 \text{ V } \angle 0° \text{ } \Omega \frac{75 \text{ } \Omega \text{ } \angle 35°}{(50 + j0 \text{ } \Omega) + (61.4 + j43 \text{ } \Omega)}$$

Add denominator, then change to polar form:

$$V_{TH} = 10 \text{ V } \angle 0° \times \frac{75 \text{ } \Omega \text{ } \angle 35°}{111.4 + j43 \text{ } \Omega}$$

$$V_{TH} = 10 \text{ V } \angle 0° \times \frac{75 \text{ } \Omega \text{ } \angle 35°}{119.4 \text{ } \Omega \text{ } \angle 21.1°}$$

Simplifying fraction: $V_{TH} = 10 \text{ V } \angle 0° \times 0.628 \text{ } \Omega \text{ } \angle 13.9°$

Answer: $V_{TH} = 6.28 \text{ V } \angle 13.9°$

21.5 THREE-TERMINAL IMPEDANCE NETWORKS

The impedance networks shown in figure 21-10 are examples of **three-terminal networks.** The top two networks are equivalent. They can be called **wye networks** or **tee networks** depending on how they are drawn. The bottom two networks are also equivalent. They are **delta networks** or **pi networks.**

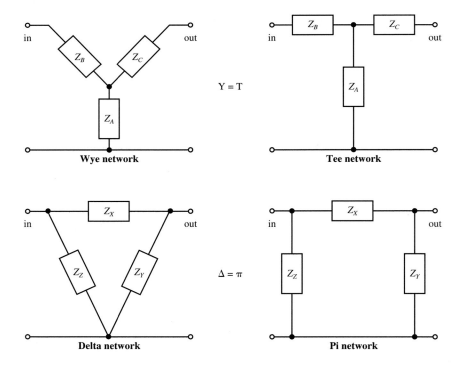

Figure 21-10. Three-terminal impedance networks are named by their shape.

The bottom line is common to both the input and the output in these networks. Filters are made up of different combinations of impedance networks. It is necessary to be able to calculate the relationships between the different formats to make design considerations flexible.

The following conversion formulas transform (change) one type of network to the other.

(Refer to figure 21-10 for letter designations.)

Delta to Wye:

Formula 21.A

$$Z_A = \frac{Z_Y Z_Z}{Z_X + Z_Y + Z_Z}$$

Formula 21.B

$$Z_B = \frac{Z_X Z_Z}{Z_X + Z_Y + Z_Z}$$

Formula 21.C

$$Z_C = \frac{Z_X Z_Y}{Z_X + Z_Y + Z_Z}$$

Wye to Delta:

Formula 21.D

$$Z_X = \frac{Z_A Z_B + Z_B Z_C + Z_A Z_C}{Z_A}$$

Formula 21.E

$$Z_Y = \frac{Z_A Z_B + Z_B Z_C + Z_A Z_C}{Z_B}$$

Formula 21.F

$$Z_Z = \frac{Z_A Z_B + Z_B Z_C + Z_A Z_C}{Z_C}$$

Sample problem 8. _____

Convert the tee network shown in figure 21-11 to a pi network. Use the list of transformation formulas for wye to delta:

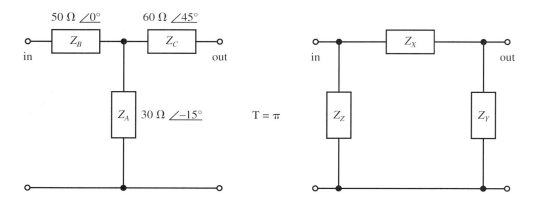

Figure 21-11. Transform a tee network into a pi network. (Sample problem 8)

1. Note that the numerators in the wye to delta formulas are all the same. Solve the numerator first, then use the result in each of the formulas. All values are in ohms.

   Formula for numerator: $Z_{NUM} = Z_A Z_B + Z_B Z_C + Z_A Z_C$

   Substitution: $Z_{NUM} = (30 \angle{-15°})(50 \angle{0°}) + (50 \angle{0°})(60 \angle{45°})$
   $+ (30 \angle{-15°})(60 \angle{45°})$

   Multiply: $Z_{NUM} = 1500 \angle{-15°} + 3000 \angle{45°} + 1800 \angle{30°}$

   In rectangular form: $Z_{NUM} = (1449 - j388) + (2121 + j2121) + (1559 + j900)$

   Adding: $Z_{NUM} = 5129 + j2633\ \Omega^2 = 5765\ \Omega^2\ \angle{27.2°}$

2. Substitute the polar value of the numerator into the formula and solve for Z_X.

   Formula: $Z_X = \dfrac{Z_{NUM}}{Z_A}$

   Substitute: $Z_X = \dfrac{5765\ \Omega^2\ \angle{27.2°}}{30\ \Omega\ \angle{-15°}}$

   Answer: $Z_X = 192\ \Omega\ \angle{42.2°}$

3. Substitute the polar value of the numerator into the formula and solve for Z_Y.

Formula: $Z_Y = \dfrac{Z_{NUM}}{Z_B}$

Substitute: $Z_Y = \dfrac{5765\ \Omega^2\ \angle 27.2°}{50\ \Omega\ \angle 0°}$

Answer: $Z_Y = 115\ \Omega\ \angle 27.2°$

4. Substitute the polar value of the numerator into the formula and solve for Z_Z.

Formula: $Z_Z = \dfrac{Z_{NUM}}{Z_C}$

Substitute: $Z_Z = \dfrac{5765\ \Omega^2\ \angle 27.2°}{60\ \Omega\ \angle 45°}$

Answer: $Z_Z = 96\ \Omega\ \angle -17.8°$

Sample problem 9. _____

Convert the delta network shown in figure 21-12 to a wye network. Use the list of transformation formulas for delta to wye.

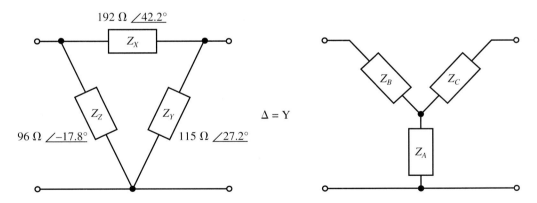

Figure 21-12. Transform a delta network into a wye network.
(Sample problem 9)

1. Note that the denominator in each of these formulas is the same. First solve the denominator, use the result in each of the formulas.

Denominator: $Z_{DEN} = Z_X + Z_Y + Z_Z$

Substitution: $Z_{DEN} = 192\ \Omega\ \angle42.2° + 115\ \Omega\ \angle27.2° + 96\ \Omega\ \angle-17.8°$

In rectangular form: $Z_{DEN} = (142 + j129\ \Omega) + (102 + j53\ \Omega) + (91 - j29\ \Omega)$

Answer: $Z_{DEN} = 335 + j153\ \Omega = 368\ \Omega\ \angle24.5°$

2. Use the formula to solve for Z_A.

Formula: $Z_A = \dfrac{Z_Y Z_Z}{Z_{DEN}}$

Substitution: $Z_A = \dfrac{(115\ \Omega\ \angle27.2°)\ (96\ \Omega\ \angle-17.8°)}{368\ \Omega\ \angle24.5°}$

Answer: $Z_A = 30\ \Omega\ \angle-15°$

3. Use the formula to solve for Z_B.

Formula: $Z_B = \dfrac{Z_X Z_Z}{Z_{DEN}}$

Substitution: $Z_B = \dfrac{(192\ \Omega\ \angle42.2°)\ (96\ \Omega\ \angle-17.8°)}{368\ \Omega\ \angle24.5°}$

Answer: $Z_B = 50\ \Omega\ \angle0°$

4. Use the formula to solve for Z_C.

Formula: $Z_C = \dfrac{Z_X Z_Y}{Z_{DEN}}$

Substitution: $Z_C = \dfrac{(192\ \Omega\ \angle42.2°)\ (115\ \Omega\ \angle27.2°)}{368\ \Omega\ \angle24.5°}$

Answer: $Z_C = 60\ \Omega\ \angle45°$

SUMMARY

- Resistor/inductance/capacitance combinations can be represented by an equivalent impedance network.
- Kirchhoff's voltage law applies to ac circuits. The voltage drops around the loop add to equal zero when using complex numbers to perform the arithmetic.
- Kirchhoff's current law applies to ac circuits. Current entering a junction equals the current leaving the junction.
- The superposition theorem can be used for ac circuits containing more than one voltage source.

- Thevenin's theorem is used to simplify complex ac circuits. The simplified circuit contains a Thevenin voltage and a Thevenin impedance connected to the load.
- Delta and wye transformations can be represented by three-terminal impedance networks.

KEY WORDS AND TERMS GLOSSARY

delta network: A three-terminal network that is connected in series in a closed loop. Also called a pi network.

pi network: A three-terminal network that is connected in series in a closed loop. Also called a delta network.

tee network: A three-terminal network where the components have one end connected in common and the other ends are connected to the power source and load. Also called a wye network.

Thevenin equivalent impedance: Impedance that would be measured at the load terminals if the load was removed, and the voltage source was removed and replaced with a short.

wye network: A three-terminal network where the components have one end connected in common and the other ends are connected to the power source and load. Also called a tee network.

KEY FORMULAS

Formula 21.A

$$Z_A = \frac{Z_Y Z_Z}{Z_X + Z_Y + Z_Z}$$

Formula 21.B

$$Z_B = \frac{Z_X Z_Z}{Z_X + Z_Y + Z_Z}$$

Formula 21.C

$$Z_C = \frac{Z_X Z_Y}{Z_X + Z_Y + Z_Z}$$

Formula 21.D

$$Z_X = \frac{Z_A Z_B + Z_B Z_C + Z_A Z_C}{Z_A}$$

Formula 21.E

$$Z_Y = \frac{Z_A Z_B + Z_B Z_C + Z_A Z_C}{Z_B}$$

Formula 21.F

$$Z_Z = \frac{Z_A Z_B + Z_B Z_C + Z_A Z_C}{Z_C}$$

TEST YOUR KNOWLEDGE

Do not write in this text. Please use a separate sheet of paper.

1. Find the total impedance of two networks connected in series: $Z_1 = 100\ \Omega\ \angle\!-45°$ and $Z_2 = 150\ \Omega\ \angle 60°$.
2. Find the total impedance of two networks connected in parallel:
    $Z_1 = 100\ \Omega\ \angle\!-45°$ and $Z_2 = 150\ \Omega\ \angle 60°$.
3. Use the circuit shown in figure 21-13:

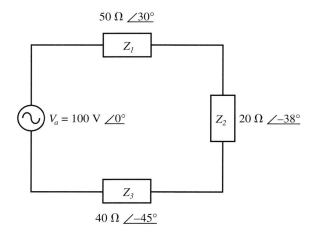

Figure 21-13. Problem #3.

    a.   Find the total impedance.
    b.   Find the circuit current.
    c.   Find the voltage across Z_1.
    d.   Find the voltage across Z_2.
    e.   Find the voltage across Z_3.
    f.   Use Kirchhoff's voltage law to show that the voltage drops add to equal the applied voltage.

4. Use Kirchhoff's voltage law to find the missing voltage drop in figure 21-14.

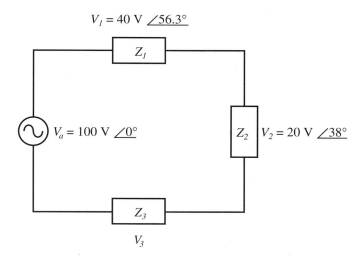

Figure 21-14. Problem #4.

5. Use Kirchhoff's current law to find current I_3 in figure 21-15.

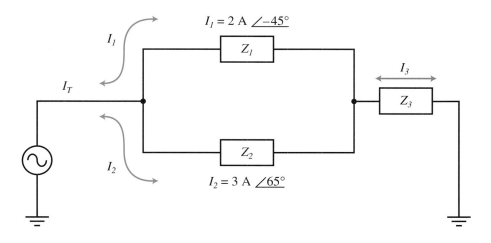

Figure 21-15. Problem #5.

6. Use the superposition theorem to solve the ac circuit with two voltage sources shown in figure 21-16. Find the current and voltage across the load.

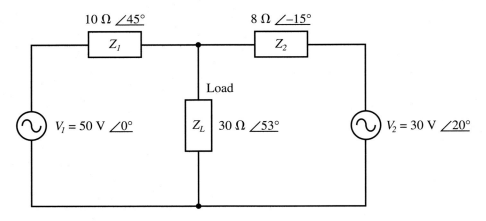

Figure 21-16. Problem #6.

7. Use Thevenin's theorem to simplify the complex ac circuit shown in figure 21-17. Also, draw the equivalent Thevenin circuit.

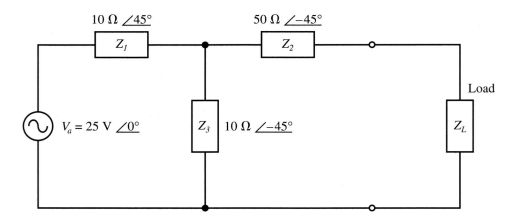

Figure 21-17. Problem #7.

8.   Convert the tee network in figure 21-18 to a pi network.

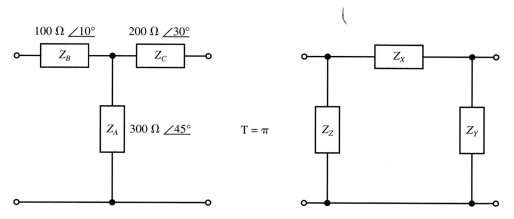

Figure 21-18. Problem #8.

9.   Convert the delta network in figure 21-19 to a wye network.

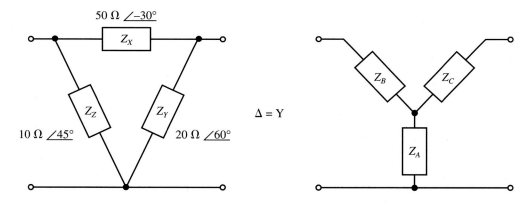

Figure 21-19. Problem #9.

Circuit theorems help break complex problems down to a manageable level. (Tektronix)

Chapter 22
AC Power

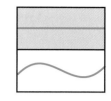

Upon completion of this chapter, you will be able to:
- Describe the power curve in an ac circuit.
- State the difference between absorbed and reflected ac power.
- Describe ac power in a purely resistive circuit.
- Describe ac power in a purely inductive circuit.
- Describe ac power in a purely capacitive circuit.
- Describe ac power in a circuit with resistance and reactance.
- Draw and label the phasor triangle for ac power.
- Define real power, reactive power, and apparent power.
- Define power factor.
- In an ac circuit, calculate the values for: impedance, phase angle, current, individual voltage drops, and power.
- Describe the need for power factor correction.
- Calculate the value of a capacitor needed for power factor correction in an inductive circuit.

Power is a measure of energy consumption in an electrical circuit. In dc circuits, power is dissipated across resistance. It is expressed as wattage. The power formulas multiply current times voltage to obtain the value of power.

In ac circuits, components other than resistors are also drawing current and developing a voltage drop. An ac circuit often has a phase shift between the current and voltage. When I and V are multiplied together, the result *does not* equal the power dissipated in the ac circuit. Power in ac circuits is represented in terms other than watts.

22.1 POWER CURVE IN AN AC CIRCUIT

In a circuit containing only one component, the phase shift between the current and voltage has one of three possible values: 0°, +90°, or –90°. Note, however, this assumes that the component is ideal. For example, an ideal resistor would contain only resistance. There would be no inductance or capacitance. An ideal inductor would contain only inductance. It would have zero resistance.

Absorbed Power and Reflected Power

The *power curve* of a single-component ac circuit is in the form of a sine wave. On a power curve, positive values of power are absorbed by the load. Negative values on the power curve are reflected back to the power source.

A pure resistive circuit produces an all-positive curve. All of the power is **absorbed power.** This means that all of the power is absorbed by the load. Pure reactance circuits, whether inductive or capacitive, produce curves with equal amounts of power on the positive and negative side of the curve. This results in a cancellation effect. Circuits that contain a combination of component characteristics, produce power curves containing phase shift, with a portion of the power having a cancellation effect.

The power absorbed by the load is power actually recorded on a power meter. The **reflected power,** the power reflected back to the power source, is not recorded by the meter. As a result, work is performed but there appears to have been no power consumption. An example of this is the electric motor, which is primarily inductive. The motor runs but a **wattmeter** cannot record the electricity used.

Power in an AC Circuit with Pure Resistance

An ac circuit containing only resistance has the current and voltage in phase, as shown in figure 22-1. The instantaneous power also has a sine wave pattern. It is calculated by multiplying the instantaneous current times the instantaneous voltage.

Examine the first half cycle, from 0° to 180°. The current and voltage have positive values and their product is also positive. As the *i* and *v* curves approach 180°, their values drop to zero, as does the power curve. Between 180° and 360° the *i* and *v* curves are both negative. Two negatives multiplied together produce a positive. Thus, the power curve is also positive during the second half cycle.

Later in this chapter, power factor is introduced. The power factor is a ratio of resistive power to total power. Power factor correction makes the electrical load have a power characteristic curve of a pure resistance. The power dissipated in a pure resistance is referred to as **real power** or **true power** and is measured in watts.

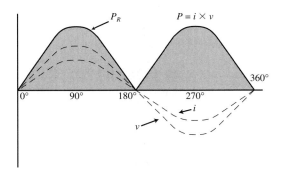

Figure 22-1. Instantaneous power in a pure resistive circuit. Voltage and current are in phase.

Power in an AC Circuit with Pure Inductance

In a circuit containing pure inductance, the current lags the voltage by 90°. For a review of inductive circuits, refer to Chapter 18. Figure 22-2 shows one complete cycle of the current and its relationship to voltage and power. Power is calculated by multiplying $i \times v$.

Refer to figure 22-2. At 0° the current is zero resulting in zero power. At 45° i and v have the same value resulting in peak power. At 90° the voltage crosses zero resulting in zero power.

Between 90° and 180° the voltage is negative while the current is positive. Therefore, between 90° and 180° the power curve is negative.

Between 180° and 270° both i and v are negative. Therefore, between 180° and 270° the power curve is positive. Finally, between 270° and 360° voltage is positive while the current is negative, resulting in a negative power curve.

An examination of this curve shows equal amounts of positive and negative power. The average of this curve is zero, therefore no true power can be measured. Power in a pure reactance is measured in **VAR,** which stands for **volt-ampere-reactive.**

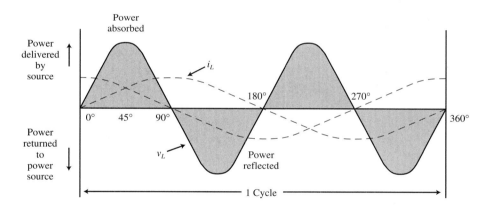

Figure 22-2. A pure inductance reflects an equal amount of power as it absorbs.

Power in an AC Circuit with Pure Capacitance

The capacitive power curve, shown in figure 22-3, is the opposite of the inductive curve. In a purely capacitive circuit, current leads voltage by 90°. Refer to Chapter 19 to review capacitive circuits.

Between 0° and 90° the voltage is negative while the current is positive, resulting in a negative power curve. Between 90° and 180° both i and v are positive, resulting in an all-positive power. Between 180° and 270° i is negative and v is positive. This produces a negative power curve. Finally, between 270° and 360° both current and voltage are negative, resulting in a positive curve.

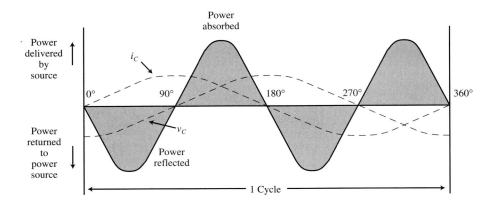

Figure 22-3. A pure capacitance reflects an equal amount of power as it absorbs.

A Circuit Containing Both Resistance and Reactance

With both resistance and reactance in a circuit, the power curve is a combination of the positive effects of the resistance and the positive and negative effects of the reactance.

The circuit that produced the curve shown in figure 22-4 had equal amounts of resistance and reactance. Notice that the power curve shows a much larger positive power than might have been expected from a circuit with a 45° phase angle. Consider, however, that a reactive circuit produces some positive power. This positive power is added to the positive power coming from the resistor.

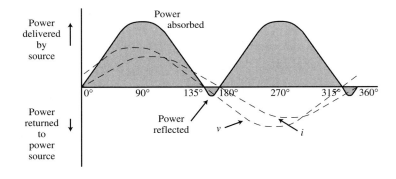

Figure 22-4. Power in a circuit containing equal amounts of resistance and reactance has a 45° phase shift.

22.2 THE POWER TRIANGLE

To graphically represent the effects of resistive power and reactive power, a vector triangle is used. See figure 22-5. The inductive triangle has a positive angle, while the capacitive triangle has a negative angle.

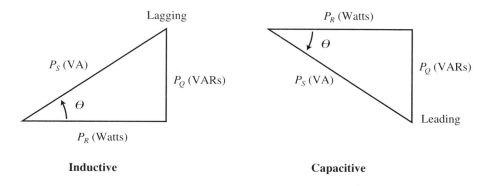

Figure 22-5. Power triangles.

Real power, the power dissipated in pure resistance, is shown as a vector plotted on the horizontal axis. Real power, indicated as P_R, is measured in watts.

Reactive power, the power dissipated in pure reactance, either inductive or capacitive, is shown as a vertical vector. Reactive power is labeled as P_Q and is measured in VAR. The reactive component determines if the triangle is positive or negative. An inductive load produces a positive angle and is classified as lagging, because it has a lagging current. A capacitive load produces a negative angle and is classified as leading, because it has a leading current.

The hypotenuse of the triangle represents the total or effective circuit power, and is given the name **apparent power.** Apparent power is represented by P_S and is measured in VA. The unit of measure, **VA,** stands for **volt-ampere.**

Power Factor

The **power factor (PF)** is the ratio of real power to apparent power. The power factor can be found using either of two formulas. One formula uses the ratio of the powers. The second formula uses the cosine of the phase angle.

The power factor is a ratio, so it has no units of measure. The value of power factor ranges from zero to one. A PF of zero is purely reactive. A PF of one is purely resistive. The power factor is frequently expressed as a percentage. For example: PF = 0.85 = 85 percent.

Formula 22.A

$$PF = \frac{P_R}{P_S}$$

PF is the power factor.

P_R is real power, measured in watts.

P_S is apparent power, measured in volt-amperes.

Formula 22.B

$$PF = \cos \theta$$

PF is the power factor.

θ is the phase angle.

Remember that PF has no units and has a value between zero and one.

Calculating Power in a Circuit

To find the power in an inductive circuit requires going through the calculations for finding the total impedance, current, and voltage drops. Then, the power can be found. Power can be found for each individual component or for the total circuit. If the circuit contains many components, the apparent power is the power of the combined circuit as seen from the power source.

Sample problem 1. _____

Using figure 22-6, calculate the following parameters: Z, θ, I, V_R, V_L, P_R, P_Q, P_S, and PF.

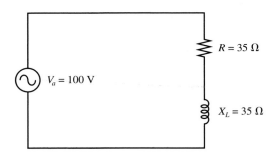

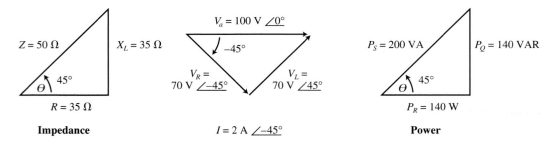

Figure 22-6. An inductive circuit with its three triangles. (Sample problem 1)

1. Calculate the impedance and plot the triangle.

   Formula: $Z = \sqrt{R^2 + X_L{}^2}$

   Substitution: $Z = \sqrt{35^2 + 35^2}$

   Answer: $Z = 50 \ \Omega$

2. Calculate the phase angle using the impedance triangle.

   Formula: $\theta = \tan^{-1} \dfrac{X_L}{R}$

   Substitution: $\theta = \tan^{-1} \dfrac{35 \ \Omega}{35 \ \Omega}$

   Answer: $\theta = 45°$

3. Current is found using Ohm's law.

   Formula: $I = \dfrac{V_a}{Z}$

   Substitution: $I = \dfrac{100 \text{ V} \angle 0°}{50 \ \Omega \ \angle 45°}$

   Answer: $I = 2 \text{ A} \ \angle -45°$

4. Voltage drop across the resistor.

   Formula: $V_R = I \times R$

   Substitution: $V_R = 2 \text{ A} \ \angle -45° \times 35 \ \Omega \ \angle 0°$

   Answer: $V_R = 70 \text{ V} \ \angle -45°$

5. Voltage drop across the inductor.

   Formula: $V_L = I \times X_L$

   Substitution: $V_L = 2 \text{ A} \ \angle -45° \times 35 \ \Omega \ \angle 90°$

   Answer: $V_L = 70 \text{ V} \ \angle +45°$

6. Calculate applied voltage, using the triangle, to compare to the actual circuit voltage.

   Formula: $V_a = \sqrt{V_R{}^2 + V_L{}^2}$

   Substitution: $V_a = \sqrt{70 \text{ V} \ \angle -45° \times 70 \text{ V} \ \angle +45°}$

   Answer: $V_a = 99 \ \angle 0° \cong 100 \ \angle 0°$

Notice that the voltage triangle has been rotated from its normal position due to the inclusion of the angle of the current.

7. Power dissipated in the resistor. Notice that angles are not included with power calculations. Real power is plotted on the horizontal axis.

    Formula: $P_R = I \times V_R$

    Substitution: $P_R = 2\,A \times 70\,V$

    Answer: $P_R = 140\,W$

8. Power dissipated in the inductor. Do not include angles with power calculations. Reactive power is plotted vertically (positive for inductance).

    Formula: $P_Q = I \times V_L$

    Substitution: $P_Q = 2\,A \times 70\,V$

    Answer: $P_Q = 140\,VAR$

9. Apparent power can be calculated two ways, either as a vector sum (the hypotenuse) or with the power formulas. The vector sum is used here.

    Formula: $P_S = \sqrt{P_R{}^2 + P_Q{}^2}$

    Substitution: $P_S = \sqrt{140^2 + 140^2}$

    Answer: $P_S = 198\,VA$

10. The power factor can be calculated two ways, either as the ratio of real power to apparent power or as the cosine of the phase angle. The cosine of the phase angle is used here.

    Formula: $PF = \cos\theta$

    Substitution: $PF = \cos 45$

    Answer: $PF = 0.707 = 70.7\%$

Sample problem 2.

Calculate the power and the power factor of the circuit shown in figure 22-7. All the other circuit parameters have been given.

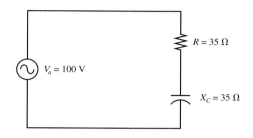

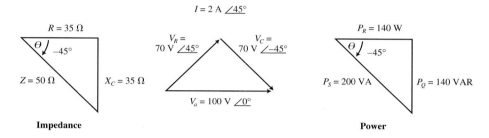

Figure 22-7. A capacitive circuit with its three triangles. (Sample problem 2)

1. Power dissipated in the resistor. Angles are not included with power calculations. Real power is plotted on the horizontal axis.

    Formula: $P_R = I \times V_R$

    Substitution: $P_R = 2 \text{ A} \times 70 \text{ V}$

    Answer: $P_R = 140 \text{ W}$

2. Power dissipated in the inductor. Do not include angles with power calculations. Reactive power is plotted vertically (negative for capacitance).

    Formula: $P_Q = I \times X_L$

    Substitution: $P_Q = 2 \text{ A} \times 70 \text{ V}$

    Answer: $P_Q = 140 \text{ VAR}$

3. Apparent power can be calculated two ways, either as a vector sum (the hypotenuse) or with the power formulas. The power formula is used here.

    Formula: $P_S = I \times V_a$

    Substitution: $P_S = 2 \text{ A} \times 100 \text{ V}$

    Answer: $P_S = 200 \text{ VA}$

4. Power factor can be calculated two ways, either as the ratio of real power to apparent power or as the cosine of the phase angle. The ratio of real power to apparent power is used here.

Formula: $PF = \dfrac{P_R}{P_S}$

Substitution: $PF = \dfrac{140 \text{ W}}{200 \text{ VA}}$

Answer: $PF = 0.7 = 70\%$

Power in Parallel Circuits

A circuit connected in series or parallel makes no difference to the power source. The source supplies the necessary power to the circuit, depending only on circuit values.

Figure 22-8 shows a parallel circuit converted to its series equivalent. To convert from a parallel circuit to a series circuit equivalent, use the reciprocal formula to find the equivalent impedance. Then convert the equivalent impedance into its component parts. The power can now be found using the series equivalent.

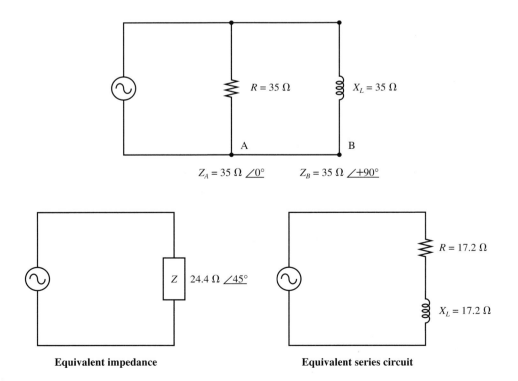

Figure 22-8. Shown is a parallel circuit with its series equivalent.

Equivalent Circuit when Given Power and Power Factor

In many applications a load is expressed by the amount of power it draws and its power factor. To best understand the circuit conditions, an equivalent circuit can be drawn and the circuit parameters studied.

Sample problem 3. _____

Calculate the current needed to supply an 840 watt motor with a 70 percent lagging power factor when it is connected to a 120 volt/60 hertz supply. Draw the equivalent circuit and related triangles.

1. A lagging power factor is an inductive load, as would be expected from a motor. The circuit drawn in figure 22-9 shows a motor connected in series with its internal resistance.

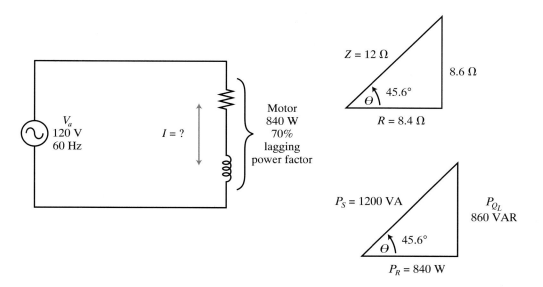

Figure 22-9. A motor is represented by its resistance and inductance. (Sample problem 3)

2. Using the cosine power factor formula, the phase angle can be found.

   Formula: $PF = \cos \theta$

   Substitution: $0.7 = \cos \theta$

   Rearranging: $\theta = \cos^{-1} 0.7$

   Answer: $\theta = 45.6°$

3. Real power is given in the problem as 840 watts. Calculate reactive power using the tangent of the angle.

    Formula: $\tan \theta = \dfrac{\text{Opposite}}{\text{Adjacent}} = \dfrac{P_Q}{P_R}$

    Rearranging: $P_Q = P_R \times \tan \theta$

    Substitution: $P_Q = 840 \text{ W} \times \tan 45.6°$

    Answer: $P_Q = 858 \text{ VAR}$

4. Complete the power triangle by finding apparent power.

    Formula: $P_S = \sqrt{P_R{}^2 + P_Q{}^2}$

    Substitution: $P_S = \sqrt{840^2 + 858^2}$

    Answer: $P_S = 1200 \text{ VA}$

5. Current is found using the power formulas with the applied voltage and apparent power.

    Formula: $I = \dfrac{P_S}{V_a}$

    Substitution: $I = \dfrac{1200 \text{ VA}}{120 \text{ V}}$

    Answer: $I = 10 \text{ A}$

22.3 POWER FACTOR CORRECTION

For maximum efficiency, it is necessary to supply power to a load with a **unity power factor.** A unity power factor has a value of one, and appears as a purely resistive load. A unity power factor appears not to have any reactive power. If a load is highly reactive, as would be the case with a large electric motor, the power factor is low. The cancellation effect of capacitive reactance and inductive reactance is used to improve the power factor.

   When a capacitor is placed in parallel with an inductor, the two reactive components supply current to each other. The effect is to reduce the line current, while maintaining the same voltage. A reduction in line current makes the system more efficient.

   The circuit shown in figure 22-9 is an example of a circuit in need of power factor correction. The inductive load needs a capacitor in parallel to make the correction. The capacitor must have the same reactive power as the reactive power of the motor. Notice, after solving sample problem 4, the current supplied by the power source has been reduced. The apparent power is also reduced. The motor still has the same current as before the capacitor was introduced. It also has the same voltage.

Sample problem 4. _____

Calculate the value of capacitance needed to correct the circuit in figure 22-9 to a unity power factor. Calculate the apparent power in the corrected circuit and draw the new schematic.

1. Reactive power must be equal in both inductance and capacitance. P_Q = 860 VAR. Calculate capacitive current using the power formulas.

    Formula: $I_C = \dfrac{P_Q}{V_a}$

    Substitution: $I_C = \dfrac{860 \text{ VAR}}{120 \text{ V}}$

    Answer: $I_C = 7.17 \text{ A}$

2. Find the value of capacitive reactance using Ohm's law.

    Formula: $X_C = \dfrac{V_a}{I_C}$

    Substitution: $X_C = \dfrac{120 \text{ V}}{7.17 \text{ A}}$

    Answer: $X_C = 16.7 \text{ } \Omega$

3. Calculate the value of capacitance using the reactance and the supply frequency. Formula 19-1 is used.

    Formula: $X_C = \dfrac{1}{2\pi f C}$

    Rearranging: $C = \dfrac{1}{2\pi f X_C}$

    Substitution: $C = \dfrac{1}{2 \times \pi \times 60 \text{ Hz} \times 16.7 \text{ } \Omega}$

    Intermediate step: $C = \dfrac{1}{6295.8}$

    Answer: $C = 159 \text{ } \mu F$

4. Schematic and phasor diagrams are shown in figure 22-10. Find the supply current of the corrected circuit. Apparent power is 840 VA, which is equal to the wattage of the motor because the reactive powers canceled.

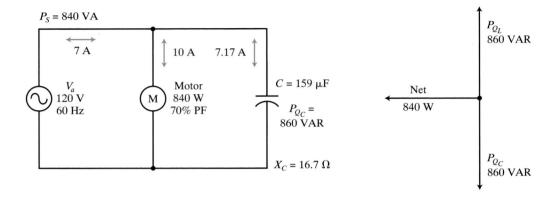

Figure 22-10. A capacitor is added in parallel with the motor to correct for power factor. (Sample problem 4)

Formula: $I = \dfrac{P_S}{V_a}$

Substitution: $I = \dfrac{840 \text{ VA}}{120 \text{ V}}$

Answer: $I = 7$ A

SUMMARY

- Resistive power is absorbed by the load, while reactive power is reflected back to the power source.
- Real power is the power dissipated in resistance and is measured in watts.
- Reactive power is the power dissipated in a reactive component, either inductive or capacitive, and is measured in VAR.
- Apparent power is the total power in an ac circuit containing both resistance and reactance. It is measured in VA.
- Power factor is the ratio of real power to apparent power. PF has no units and will range from zero to one.
- A purely resistive circuit has a power factor of unity and is the most efficient.
- Power factor correction is made by connecting a reactive component in parallel with the load to cancel some of the effects.

KEY WORDS AND TERMS GLOSSARY

absorbed power: Power that is absorbed by the load.

apparent power (P_S): The total power in an ac circuit containing both resistance and reactance. The hypotenuse of the power triangle represents the apparent power. Apparent power is measured in volt-amperes (VA).

power factor (PF): The ratio of real power to apparent power. Power factor can be found using either of two formulas: the ratio of the powers or the cosine of the phase angle. Power factor is a ratio so it has no units of measure. The value of power factor is always between zero and one. A PF of zero is purely reactive, and a PF of one is purely resistive.

reactive power (P_Q): Power dissipated by pure reactance.

real power (P_R): The power dissipated in a pure resistance. Usually the term is applied to ac circuits, but it also applies to dc circuits. Unit of measure is the watt.

reflected power: Power that is reflected back to the power source.

true power: The power dissipated in a pure resistance. Also called real power. Unit of measure is the watt.

unity power factor: A power factor of one. This provides maximum power efficiency to a load.

volt-ampere (VA): Unit of measure of apparent power.

volt-ampere-reactive (VAR): Unit of measure for power in the pure reactance portion of an ac circuit.

wattmeter: Meter that records how many watts of power are used.

KEY FORMULAS

Formula 22.A

$$PF = \frac{P_R}{P_S}$$

Formula 22.B

$$PF = \cos \theta$$

TEST YOUR KNOWLEDGE

1. Describe the power curve in an ac circuit containing a purely resistive load.
2. Describe the power curve in an ac circuit containing a purely reactive load.
3. In an ac circuit, what type of load absorbs ac power?
4. In an ac circuit, what type of load reflects ac power?

5. In a purely resistive circuit, the ac power is called _____ and is measured in _____.

6. In a purely inductive circuit, the ac power is called _____ and is measured in _____.

7. In a purely capacitive circuit, the ac power is called _____ and is measured in _____.

8. The total ac power in a circuit with resistance and reactance is called _____ and is measured in _____.

9. Draw and label the phasor triangle for ac power in a circuit with resistance and capacitance.

10. Draw and label the phasor triangle for ac power in a circuit with resistance and inductance.

11. Define power factor. What is the unit of measure? What is the range of values?

12. Describe the need for power factor correction.

13. In a series circuit with a resistance of 50 ohms and an inductor of 75 ohms connected to a 100 volt 0° source, calculate the following:
    a. Impedance.
    b. Phase angle.
    c. Current.
    d. Voltage drops.
    e. Real power.
    f. Reactive power.
    g. Apparent power.
    h. Power factor.

14. If the power factor in a circuit is leading, what type of reactive component has the largest value?

15. If the power factor in a circuit is lagging, what type of reactive component has the largest value?

16. Calculate the current needed to supply a 1000 watt motor with a 75 percent lagging power factor when connected to a 120 V/60 Hz supply. Also draw the equivalent circuit and related triangles.

17. Calculate the value of capacitor needed to correct to unity power factor the motor circuit described in Problem #16.

Chapter 23
Resonance

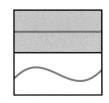

Upon completion of this chapter, you will be able to:
- Define resonance as it applies to ac circuits.
- Calculate the resonant frequency.
- List the characteristics of a series resonant circuit.
- Identify the typical curves for a series resonant circuit.
- Calculate values in a series resonant circuit.
- List the characteristics of a parallel resonant circuit.
- Identify the typical curves for a parallel resonant circuit.
- Calculate circuit values in a parallel resonant circuit.
- Define the bandwidth of resonant circuit.
- Define the Q of a resonant circuit.
- Calculate the bandwidth and circuit Q.

Resonance occurs in an ac circuit when the capacitive reactance is equal to the inductive reactance. The reactances cancel. The circuit displays purely resistive characteristics.

A resonant circuit is commonly used to select or reject certain frequencies. Radios use resonant circuits. A radio antenna is in contact with every possible radio frequency that is strong enough to induce a voltage. However, it is not desirable for the antenna to receive all signals at once. Using a resonant circuit, a selected band of frequencies are allowed to pass, while the others are rejected.

23.1   CALCULATING THE RESONANT FREQUENCY

The curves shown in figure 23-1 compare X_C to X_L. Capacitive reactance is high at low frequencies, then decreases as the frequency increases. Inductive reactance is low at low frequencies, then increases with an increase in frequency. Resonance occurs at the frequency where the curves cross.

The **resonant frequency** (f_r) is dependent on the values of the components. For every value of inductance, there is a value of capacitance that will produce resonance.

To arrive at the formula for finding the resonant frequency, start with the reactance formulas. Resonance is defined as when the reactances are equal, so the two formulas are equal to each other at the resonant frequency ($X_C = X_L$ at f_r). Combine the formulas and solve for the resonant frequency.

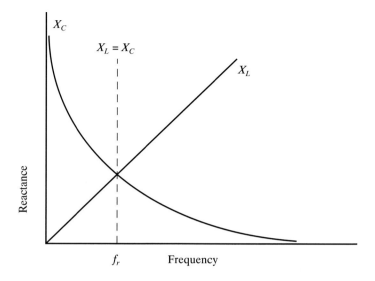

Figure 23-1. Reactance vs frequency graph. Resonance occurs when X_L equals X_C.

Formula 23.A (capacitive reactance)

$$X_C = \frac{1}{2\pi f C}$$

X_C is the capacitive reactance, measured in ohms.

f is the frequency, measured in hertz.

C is the capacitance, measured in farads.

Formula 23.B (inductive reactance)

$$X_L = 2\pi f L$$

X_L is the inductive reactance, measured in ohms.

f is the frequency, measured in hertz.

L is the inductance, measured in henrys.

Combining:

$$X_C = X_L$$

$$2\pi fL = \frac{1}{2\pi fC}$$

Rearranging:

$$f^2 = \frac{1}{2^2\pi^2 LC}$$

Take the square root of both sides to arrive at the formula to calculate the resonant frequency.

Formula 23.C

$$f_r = \frac{1}{2\pi\sqrt{LC}} \quad \text{or} \quad f_r = \frac{0.159}{\sqrt{LC}}$$

f_r is resonant frequency, measured in hertz.

L is inductance, measured in henrys.

C is capacitance, measured in farads.

Sample problem 1. _____

Calculate the resonant frequency of a circuit containing 0.1 H inductance and 25 μF capacitance.

Formula: $f_r = \dfrac{1}{2\pi\sqrt{LC}}$

Substitution: $f_r = \dfrac{1}{2\pi\sqrt{0.1\text{H} \times 25\ \mu\text{F}}}$

Remember to complete the square root first, and take the reciprocal last.

Answer: $f_r = 100\ \text{Hz}$

Sample problem 2. _____

Find the resonant frequency of a circuit containing a capacitor of 25 μF and an inductor of four mH.

Formula: $f_r = \dfrac{0.159}{\sqrt{LC}}$

Substitution: $f_r = \dfrac{0.159}{\sqrt{4\text{mH} \times 25\ \mu\text{F}}}$

Answer: $f_r = 503\ \text{Hz}$

Sample problem 3. _____

What value of capacitance is needed in a circuit containing a 20 mH inductor if the resonant frequency is to be 35 kHz?

Formula: $$f_r = \frac{1}{2\pi\sqrt{LC}}$$

Square both sides: $$f_r^2 = \frac{1}{2^2\pi^2 LC}$$

Rearranging: $$C = \frac{1}{2^2\pi^2 L f_r^2}$$

Substitution: $$C = \frac{1}{4 \times \pi^2 \times 20 \text{ mH} \times (35 \text{ kHz})^2}$$

Answer: $C = 0.001 \ \mu F$

Sample problem 4. _____

Find the value of inductance needed to produce a resonant frequency of 2.5 kHz in a circuit containing a 4 μF capacitor.

Formula: $$f_r = \frac{1}{2\pi\sqrt{LC}}$$

Square both sides: $$f_r^2 = \frac{1}{2^2\pi^2 LC}$$

Rearranging: $$L = \frac{1}{2^2\pi^2 C f_r^2}$$

Substitution: $$L = \frac{1}{4 \times \pi^2 \times 4 \ \mu F \times (2.5 \text{ kHz})^2}$$

Answer: $L = 1 \text{ mH}$

23.2 SERIES RESONANT CIRCUITS

A **series resonant circuit** has a capacitor and inductor in series. Any practical series circuit also has a resistance due to the inductor, resistance due to capacitive losses, and a resistance equivalent to other circuit losses. The circuit shown in figure 23-2 is a series resonant circuit. Resistor r_s represents all circuit resistance.

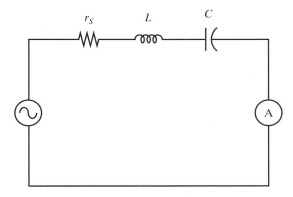

Figure 23-2. Practical series resonant circuits have some resistance due to component losses.

Characteristics of Series Resonance

There are a number of characteristics that describe a series resonant circuit. These traits are as follows:

- Net reactance $(X_L - X_C)$ equals zero due to cancellation.
- The impedance is at a minimum. Impedance at resonance is equal to the value of the series resistance.
- The current is at a maximum. Current at resonance is limited only by the resistance.
- The current is in phase with the applied voltage.
- There is a rise in voltage across the reactive components at resonance. L and C have reactance and voltage drops even though their net reactance is zero due to their opposite phase angles.
- There is a unity power factor. As described in Chapter 22, the power factor is the ratio of circuit reactance to resistance. When the reactance is zero, the power factor is one.
- When the applied frequency is higher than the resonant frequency, the net reactance is inductive.
- When the applied frequency is lower than the resonant frequency, the net reactance is capacitive.

Series Resonant Curves

Impedance and current curves are used to describe resonant characteristics. These curves are called **bell curves** because of their bell shape. In a series circuit, the impedance curve, figure 23-3, is high when the frequency is far from resonance. As resonance is neared, the impedance sharply decreases to a point equal to the series resistance.

The curve of the current, figure 23-4, is minimal at frequencies far from resonance. As the frequency nears f_r there is a sharp increase in current. Compare figures 23-3 and

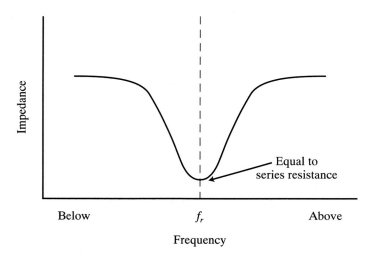

Figure 23-3. Impedance decreases sharply when approaching the resonant frequency in a series circuit.

23-4. These two curves are opposites, as would be expected in a series circuit. As the impedance falls in the circuit, the current rises. Figure 23-5 shows values used to plot typical *Z* and *I* curves.

Figure 23-6 shows an output voltage taken across the capacitor. This is a typical application of a series resonant circuit, notably in radio circuits. At resonance, the voltage developed is at its maximum value. This large voltage drop is caused by the large resonant current through the capacitive reactance.

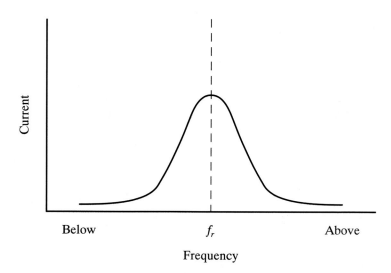

Figure 23-4. Current increases dramatically when approaching the resonant frequency in a series circuit.

Figure 23-5. Table demonstrating the relationships in a series circuit. These values can be used to plot Z and I curves.

$V = 100$ volts. $R = 100$ ohms. $L = 16$ mH. $C = 25$ μF

Frequency (hertz)	X_L (ohms)	X_C (ohms)	Net X $(X_L - X_C)$	Impedance (ohms)	Current (mA)
50	5	12800	$-j12795$	12800	7
500	50	1250	$-j1200$	1204	83
1000	100	625	$-j525$	535	187
1500	150	420	$-j270$	288	347
2000	200	320	$-j120$	156	641
2500	250	250	0	100	1000
3180	320	200	$+j120$	156	641
4240	420	150	$+j270$	288	347
6360	625	100	$+j525$	535	187
12700	1250	50	$+j1200$	1204	83
127000	12800	5	$+j12795$	12800	7

Figure 23-6. The high voltage developed across the capacitor at resonance can be used to select a radio frequency.

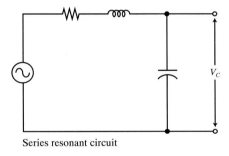
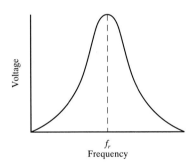

Series resonant circuit

Sample problem 5. _____

Using the circuit values given in figure 23-7, calculate:

1. The resonant frequency.

Also calculate the following quantities at resonance:

2. Inductive reactance.

3. Capacitive reactance.

4. Net reactance.

5. Impedance.

6. Current.

7. Voltage developed across the capacitor.

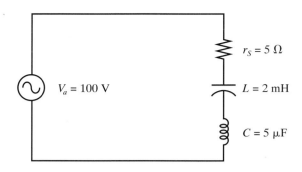

Figure 23-7. Find the resonant frequency and the circuit values at resonance. (Sample problem 5)

1. Resonant frequency.

    Formula: $f_r = \dfrac{1}{2\pi\sqrt{LC}}$

    Substitution: $f_r = \dfrac{1}{2\pi \sqrt{2 \text{ mH} \times 5 \text{ } \mu\text{F}}}$

    Answer: $f_r = 1590$ Hz

2. Inductive reactance at resonance.

    Formula: $X_L = 2\pi f L$

    Substitution: $X_L = 2 \times \pi \times 1590 \text{ Hz} \times 2 \text{ mH}$

    Answer: $X_L = 20 \text{ } \Omega$

3. Capacitive reactance at resonance.

    Formula: $X_C = \dfrac{1}{2\pi f C}$

    Substitution: $X_C = \dfrac{1}{2 \times \pi \times 1590 \text{ Hz} \times 5 \text{ } \mu\text{F}}$

    Answer: $X_C = 20 \text{ } \Omega$

4. Net reactance at resonance.

    Formula: $X_{net} = X_L - X_C$

    Substitution: $X_{net} = 20 \text{ } \Omega - 20 \text{ } \Omega$

    Answer: $X_{net} = 0$

5. Impedance at resonance.

    The circuit impedance equals series resistance at resonance.

    Answer: $Z = r_s = 5 \ \Omega$

6. Current at resonance.

    Formula: $I = \dfrac{V}{Z}$

    Substitution: $I = \dfrac{100 \ V}{5 \ \Omega}$

    Answer: $I = 20 \ A$

7. Voltage developed across the capacitor at resonance.

    Formula: $V_C = I \times X_C$

    Substitution: $V_C = 20 \ A \times 20 \ \Omega$

    Answer: $V_C = 400 \ V$

23.3 PARALLEL RESONANT CIRCUITS

A **parallel resonant circuit** has an inductor in parallel with a capacitor, as shown in figure 23-8. The circuit also includes a resistor in series with the inductor. This resistor represents the resistance of the coil in addition with other circuit losses. A parallel resonant circuit is also called a **tank circuit.**

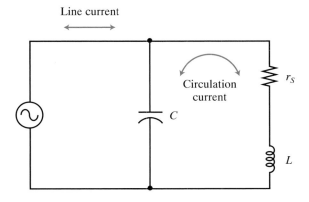

Figure 23-8. Parallel resonant circuits are often called tank circuits.

The charging and discharging of the inductor and capacitor creates a circulation current within a tank circuit. This produces a sine wave. With each cycle, the developed sine wave loses some of its amplitude across the series resistor. At the resonant frequency, these losses are a minimum. The voltage source need supply only enough current to overcome the resistor loss to sustain the sine wave.

Characteristics of Parallel Resonant Circuits

There are a number of characteristics that describe a parallel resonant circuit. These traits are as follows:

- Net reactance $(X_L - X_C)$ equals zero due to cancellation.
- The line impedance is at a maximum. Impedance at resonance is calculated with line current.
- The line current is at a minimum. Current at resonance is equal to the current through the series resistor, as if it were alone.
- The line current is in phase with the applied voltage.
- There is a unity power factor. When the reactance cancels, the power factor is one.
- When the applied frequency is higher than the resonant frequency, the net reactance is capacitive.
- When the applied frequency is lower than the resonant frequency, the net reactance is inductive.

Parallel Resonant Curves

The curves for impedance vs frequency and current vs frequency have the same characteristic bell shape as the curves for series resonant circuits. However, the curves for the parallel resonant circuit are the opposite of those for the series circuit.

The line current, in figure 23-9, sharply decreases at the resonant frequency. This indicates that the resonant frequency cannot pass the parallel circuit.

The line impedance, figure 23-10, has a sharp increase at the resonant frequency. This impedance curve is the reciprocal of the line current curve.

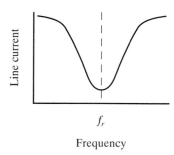

Figure 23-9. Line current decreases dramatically when approaching the resonant frequency in a parallel circuit.

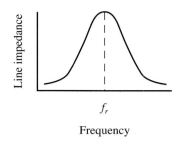

Figure 23-10. Line impedance increases sharply when approaching the resonant frequency in a parallel circuit.

Sample problem 6. _____

Using the circuit shown in figure 23-11, calculate:

1. The resonant frequency.

   Also calculate the following quantities at resonance:

2. The inductive reactance.

3. The capacitive reactance.

4. The branch impedances.

5. The branch currents.

6. The line current.

7. The equivalent impedance.

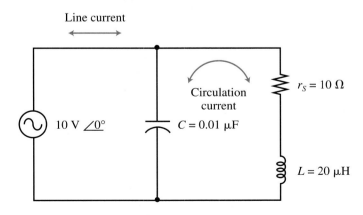

Figure 23-11. Find the resonant frequency and the circuit values at resonance. (Sample problem 6)

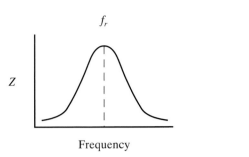

Impedance vs frequency

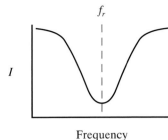

Current vs frequency

Figure 23-11. Continued

1. Resonant frequency. The same formula is used in calculating the resonant frequency in both series and parallel circuits.

    Formula: $f_r = \dfrac{1}{2\pi\sqrt{LC}}$

    Substitution: $f_r = \dfrac{1}{2 \times \pi \times \sqrt{20\ \mu H \times 0.01\ \mu F}}$

    Answer: $f_r = 356\ kHz$

2. Inductive reactance.

    Formula: $X_L = 2\pi f L$

    Substitution: $X_L = 2 \times \pi \times 356\ kHz \times 20\ \mu H$

    Answer: $X_L = 44.7\ \Omega$

3. Capacitive reactance.

    Formula: $X_C = \dfrac{1}{2\pi f C}$

    Substitution: $X_C = \dfrac{1}{2 \times \pi \times 356\ kHz \times 0.01\ \mu F}$

    Answer: $X_C = 44.7\ \Omega$

4. Branch impedances.

    a. Inductive branch impedance.

       Formula: $Z_L = r_s + jX_L$

       Answer: $Z_L = 10 + j44.7\ \Omega$ (rectangular)

       $Z_L = 45.8\ \Omega\ \underline{/77.4°}$ (polar)

b. Capacitive branch impedance.

> **Formula:** $Z_C = R - jX_C$
>
> **Answer:** $Z_C = 0 - j44.7 \ \Omega$ (rectangular)
>
> $Z_C = 44.7 \ \Omega \ \angle{-90°}$ (polar)

5. Branch currents.

a. Inductive branch current.

> **Formula:** $I_L = \dfrac{V}{Z_L}$
>
> **Substitution:** $I_L = \dfrac{10 \text{ V} \ \angle{0°}}{45.8 \ \Omega \ \angle{77.4°}}$
>
> **Answer:** $I_L = 0.218 \text{ A} \ \angle{-77.4°}$ (polar)
>
> $I_L = 0.0476 - j0.2127 \text{ A}$ (rectangular)

b. Capacitive branch current.

> **Formula:** $I_C = \dfrac{V}{Z_C}$
>
> **Substitution:** $I_C = \dfrac{10 \text{ V} \ \angle{0°}}{44.7 \ \Omega \ \angle{-90°}}$
>
> **Answer:** $I_C = 0.224 \text{ A} \ \angle{+90°}$ (polar)
>
> $I_C = 0 + j0.224 \text{ A}$ (rectangular)

6. Line current.

> **Formula:** $I_T = I_L + I_C$
>
> **Substitution:** $I_T = (0.0476 - j0.2127) + (0 + j0.224)$
>
> **Answer:** $I_T = 0.0476 + j0.0113 \text{ A}$ (rectangular)
>
> $I_T = 0.0489 \text{ A} \ \angle{13.3°}$ (polar)

7. Equivalent impedance.

> **Formula:** $Z_{EQ} = \dfrac{V}{I_T}$
>
> **Substitution:** $Z_{EQ} = \dfrac{10 \text{ V} \ \angle{0°}}{0.0489 \text{ A} \ \angle{13.3°}}$
>
> **Answer:** $Z_{EQ} = 204 \ \Omega \ \angle{-13.3°}$

23.4 BANDWIDTH AND CIRCUIT Q

Bandwidth is a range of frequencies close to the resonant frequency. The **bandwidth** is the width between two points on the response curve. These points occur where the current is equal to 70.7 percent of its maximum value. The upper and lower frequencies, labeled as f_2 and f_1 respectively in figure 23-12, identify the frequencies at the 70.7 percent point. Bandwidth is calculated as the difference between the upper and lower frequencies. The 70.7 percent point on a response curve is also equal to the **half-power point.** Recall in the power formula that $P = I^2R$. Squaring .707 gives .5, or one half of the maximum power.

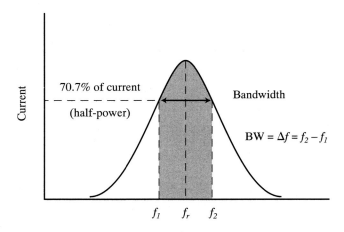

Figure 23-12. Bandwidth is the band of frequencies above and below resonance at the half-power point.

Formula 23.D

$$BW = \Delta f = f_2 - f_1$$

BW is the bandwidth, measured in hertz.

Δf is the difference between the upper and lower frequencies.

f_1 is the upper half-power point.

f_2 is the lower half-power point.

Selectivity is another term used to describe the bandwidth of a resonant circuit. **Selectivity** is usually described as the strength of the frequency, measured in volts, or microvolts. This measurement is used to determine the ability of the resonant circuit to select a specified band of frequencies, rejecting those frequencies that are higher and lower.

Circuit *Q*

   Q or **circuit *Q*,** also called the **quality factor,** is the ratio of the inductive reactance to the circuit resistance. A smaller resistance produces a larger *Q*. A larger *Q* produces a narrower bandwidth and a response curve with a higher maximum value. This can be seen in the response curves of figure 23-13.

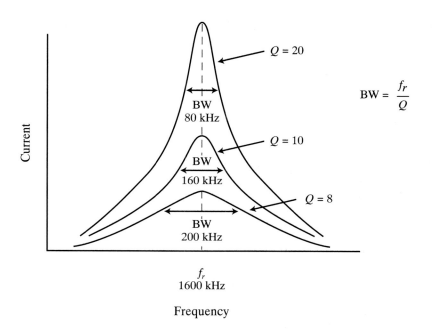

Figure 23-13. A larger circuit *Q* produces a narrower bandwidth.

Formula 23.E

$$Q = \frac{X_L}{r_s} \text{ (series or parallel circuits)}$$

Q is quality factor, a number with no units.

X_L is the inductive reactance, measured in ohms.

r_s is the series resistance at resonance, measured in ohms.

Formula 23.F

$$Q = \frac{Z_{EQ}}{X_L} \text{ (parallel circuits)}$$

Q is quality factor, a number with no units.

Z_{EQ} is the equivalent impedance, measured in ohms.

X_L is the inductive reactance, measured in ohms.

The bandwidth can also be described in terms of Q.

Formula 23.G

$$\text{BW} = \frac{f_r}{Q}$$

BW is the bandwidth, measured in hertz.

f_r is the frequency at resonance, measured in hertz.

Q is quality factor, a number with no units.

LC Product

Formula 23.C shows that there is a capacitance to match any value of inductance to produce some resonant frequency. Either the inductor or the capacitor value can be pre-selected. The desired resonant frequency can be obtained by choosing the proper value for the opposite reactive component.

When selecting the combination of reactive components, the desired circuit Q must be taken into consideration. A larger value of inductance produces a larger inductive reactance. This results in a higher circuit Q. A higher Q has a steeper response curve with a smaller bandwidth. This creates a more selective circuit. The combination of inductance and capacitance values in a resonant circuit is called the **LC product.**

Sample problem 7. _____

Calculate the Q and bandwidth of the circuit used for sample problem 5, figure 23-7. ($X_L = 20 \ \Omega$, $r_s = 5 \ \Omega$, $f_r = 1590$ Hz).

1. Circuit Q.

   Formula: $Q = \dfrac{X_L}{r_s}$

   Substitution: $Q = \dfrac{20 \ \Omega}{5 \ \Omega}$

   Answer: $Q = 4$ (no units)

2. Bandwidth.

   Formula: $BW = \dfrac{f_r}{Q}$

   Substitution: $BW = \dfrac{1590 \text{ Hz}}{4}$

   Answer: BW = 398 Hz

Sample problem 8. _____

Calculate the Q and bandwidth of the circuit used for sample problem 6, figure 23-11. Use the Q to calculate the equivalent parallel impedance. (X_L = 44.7 Ω, r_s = 10 Ω, f_r = 356 kHz)

1. Circuit Q.

   Formula: $Q = \dfrac{X_L}{r_s}$

   Substitution: $Q = \dfrac{44.7 \ \Omega}{10 \ \Omega}$

   Answer: Q = 4.47

2. Bandwidth.

   Formula: $BW = \dfrac{f_r}{Q}$

   Substitution: $BW = \dfrac{356 \text{ kHz}}{4.47}$

   Answer: BW = 79.6 kHz = 79,600 Hz

3. Equivalent impedance.

   Formula: $Z_{EQ} = Q \times X_L$

   Substitution: $Z_{EQ} = 4.47 \times 44.7 \ \Omega$

   Answer: Z_{EQ} = 200 Ω

Damping a Circuit

Damping is the widening of the bandwidth. The more resistance a circuit has, the wider the bandwidth. To damp a series circuit, place a resistor in series. A parallel circuit is damped by adding a resistor in parallel, as shown in figure 23-14.

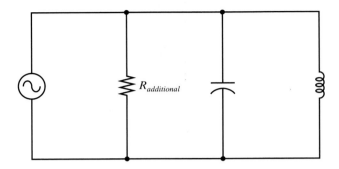

Figure 23-14. To widen the bandwidth of a parallel resonant circuit, addition resistance is added.

23.5 APPLICATIONS OF RESONANT CIRCUITS

Resonant circuits are used in radio circuits to accept or reject certain frequencies. Figure 23-15 shows two possible connections to the antenna. Variable capacitors are used to provide tuning.

In some applications, an antenna covers a broad band of frequencies, such as those of the AM broadcast band. In two-way communications, radios operate on a fixed frequency. These radio antennas are tuned to accept only that one frequency.

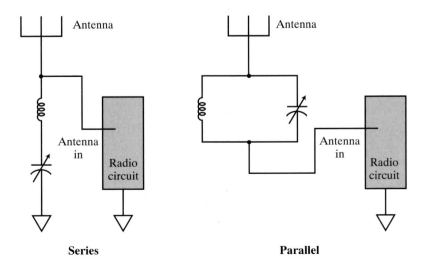

Figure 23-15. Resonant circuits are used to connect antennas to receiver circuits.

There are many other applications of resonant circuits. Some of these will be introduced in Chapter 24. The decision as to which type of circuit to use, series or parallel, is made based on the characteristic of the circuit. Figure 23-16 gives a summary of the resonant circuit characteristics.

Figure 23-16. Characteristics of series and parallel circuits at resonance.

Quantity	Series	Parallel
X_{net}	0	0
Resonant frequency	$f_r = \dfrac{1}{2\pi\sqrt{LC}}$	$f_r = \dfrac{1}{2\pi\sqrt{LC}}$
Impedance	minimum $Z = r_S$	maximum $Z_{EQ} = Q \times X_L$
Line current	maximum	minimum
Current in reactance	$I = I_{line}$	$I = Q \times I_{line}$
Voltage drop	$V_L = V_C = Q \times V_a$	V_a
X_{net} when $f > f_r$	inductive	capacitive
X_{net} when $f < f_r$	capacitive	inductive

SUMMARY

- Resonance is the characteristic of an ac circuit when X_C and X_L are equal.
- The resonant frequency is calculated using the formula:

$$f_r = \frac{1}{2\pi\sqrt{LC}}$$

- The two primary characteristics of a series resonant circuit are minimum impedance and maximum current at the resonant frequency.
- Two curves are used to describe a series resonant circuit. The impedance curve dips to near zero at resonance and the current curve reaches a maximum at resonance.
- The two primary characteristics of a parallel resonant circuit are maximum impedance and minimum current at the resonant frequency.
- Two curves are used to describe a parallel resonant circuit. The impedance curve rises to maximum at resonance and the current curve dips to near zero at resonance.
- The bandwidth of resonant circuit is the width of the range of frequencies near resonance having values equal to or greater than 70.7 percent of the maximum.
- The Q of a resonant circuit is the ratio of inductive reactance to circuit resistance.

KEY WORDS AND TERMS GLOSSARY

bandwidth: The range of frequencies close to the resonant frequency. Bandwidth is measured at the points on the response curve equal to 70.7 percent of maximum current, which is also equal to the –3 dB points.

bell curve: The impedance and current curves used to describe characteristics of a resonant circuit. The curves are given the name because of their bell shape.

circuit *Q*: The ratio of inductive reactance to the circuit resistance.

damping: The reduction of energy. In resonant circuits, resistance is added to increase the bandwidth.

half-power point: The point on the response curve of a resonant circuit equal to 70.7 percent of the maximum voltage. The half-power point is also equal to the –3 dB point, and it is the point used to measure the bandwidth.

LC product: The combination of inductance and capacitance values in a resonant circuit.

parallel resonant circuit: A circuit that has an inductor in parallel with a capacitor.

***Q*:** The ratio of inductive reactance to the circuit resistance.

quality factor (*Q*): The ratio of inductive reactance to the circuit resistance.

resonance: A condition that occurs in an ac circuit when capacitive reactance is equal to inductive reactance. The reactances cancel, leaving the circuit to display purely resistive characteristics.

resonant frequency (*f_r*): Frequency at which resonance occurs in a given circuit.

selectivity: A term used to describe the bandwidth of a resonant circuit. Selectivity is usually described as the strength of the frequency, measured in volts or microvolts. This measurement is used to determine the ability of the resonant circuit to select a specified band of frequencies.

series resonant circuit: A circuit that has an inductor in series with a capacitor.

tank circuit: A parallel resonant circuit.

KEY FORMULAS

Formula 23.A

$$X_C = \frac{1}{2\pi f C}$$

Formula 23.B

$$X_L = 2\pi f L$$

Formula 23.C

$$f_r = \frac{1}{2\pi\sqrt{LC}} \ \text{ or } \ f_r = \frac{0.159}{\sqrt{LC}}$$

Formula 23.D

$$BW = \Delta f = f_2 - f_1$$

Formula 23.E

$$Q = \frac{X_L}{r_s}$$

Formula 23.F

$$Q = \frac{Z_{EQ}}{X_L}$$

Formula 23.G

$$BW = \frac{f_r}{Q}$$

TEST YOUR KNOWLEDGE

Do not write in this text. Please use a separate sheet of paper.

1. Calculate the resonant frequency of a circuit containing 50 mH inductance and 400 µF capacitance.
2. Calculate the resonant frequency of a circuit containing 0.2 H inductance and 50 µF capacitance.
3. Calculate the resonant frequency of a circuit with 100 µF capacitance and 20 mH inductance.
4. What value of capacitance is needed to produce a resonant frequency at 50 kHz in a circuit with 30 mH inductance?
5. What value of inductance is needed to produce a resonant frequency at three kHz in a circuit with 10 µF capacitance?
6. State the typical characteristics of a series resonant circuit for each of the following:
   a. Net reactance.
   b. Impedance.
   c. Current.
   d. Phase angle of current.
   e. Relative amount of voltage drops.
   f. Power factor.
7. In a series circuit, when the applied frequency is higher than the resonant frequency, is the net reactance inductive or capacitive?
8. In a series circuit, when the applied frequency is lower than the resonant frequency, is the net reactance inductive or capacitive?

9. Does the characteristic curve for impedance in a series circuit have a maximum or minimum value at resonance?
10. Does the characteristic curve for current in a series circuit have a maximum or minimum value at resonance?
11. A six mH inductor is connected in series with a 10 μF capacitor and a 50 volt generator. The effective r_s in the circuit is eight ohms. Calculate the following:

    a. Resonant frequency.

    Calculate b through g at resonance.

    b. Inductive reactance.
    c. Capacitive reactance.
    d. Net reactance.
    e. Impedance.
    f. Current.
    g. Voltage across the capacitor.

12. State the typical characteristics of a parallel resonant circuit for each of the following:

    a. Net reactance.
    b. Impedance.
    c. Current.
    d. Phase angle of line current.
    e. Power factor.

13. In a parallel circuit, when the applied frequency is higher than the resonant frequency, is the net reactance inductive or capacitive?
14. In a parallel circuit, when the applied frequency is lower than the resonant frequency, is the net reactance inductive or capacitive?
15. Does the characteristic curve for impedance in a parallel circuit have a maximum or minimum value at resonance?
16. Does the characteristic curve for current in a parallel circuit have a maximum or minimum value at resonance?
17. A signal generator with 25 volts is applied to the parallel combination of a 0.05 μF capacitor and 40 μH inductor. The effective r_s of the inductor is 10 ohms. Calculate the following:

    a. Resonant frequency.

    Calculate b through g at resonance.

    b. Inductive reactance.
    c. Capacitive reactance.
    d. Branch impedance.
    e. Branch currents.
    f. Line current.
    g. Equivalent impedance of the tank circuit.

18. Calculate the bandwidth and circuit Q for the circuit of problem 11.
19. Calculate the bandwidth and circuit Q for the circuit of problem 17.
20. The bandwidth of the circuit in problem 11 needs to be doubled. What value of series resistance should be added?

Chapter 24
Filter Circuits

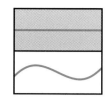

Upon completion of this chapter, you will be able to:
- Describe how to remove a dc voltage from an ac signal.
- Describe how to separate ac and dc voltages.
- Describe how a filter circuit separates different ac frequencies.
- Describe how a low pass filter removes high frequencies.
- Calculate the cutoff frequency and output voltage when a low pass filter is used as a voltage divider.
- Describe how a high pass filter removes low frequencies.
- Calculate the cutoff frequency and output voltage when a high pass filter is used as a voltage divider.
- Describe how a resonant circuit is used as a band pass filter.
- Describe how a resonant circuit is used as a band stop filter.

Filter circuits use the characteristics of inductive and capacitive reactance over the full range of frequencies, from dc (0 Hz) to radio frequencies. The properties of resonance also have an important part in filtering selected frequencies.

Filters can be used to remove a dc voltage from an ac signal, remove an ac ripple from a dc voltage, or separate ac and dc for use in the circuit. Filters are also used to pass selected frequencies while rejecting others.

24.1 COMBINED DC AND AC

Many electronic circuits contain both dc and ac voltages. Both voltages have their own function. In some situations, it is desirable to have the two voltages combined. In other applications, the dc and ac voltages must be separate. In some cases, one voltage must be removed.

Figure 24-1 shows the effects of combining dc and ac in the same circuit. The dc produces a voltage level that raises the center of the ac signal above ground. Without a dc voltage present, the sine wave would swing from five volts positive to five volts negative with its center at zero. By including the 25 volt dc supply, the shape of the sine

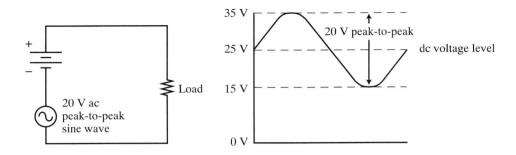

Figure 24-1. Both dc and ac power sources are combined to produce the signal on the right.

wave is not changed. It is still 20 volt peak-to-peak. However, the wave is lifted above the ground level. The center of the sine wave takes the value of the dc voltage.

A typical application of the circuit shown in figure 24-1 is a transistor amplifier. In a transistor amplifier, a dc voltage is used to bias the transistor for proper operation. The ac signal is the audio signal (music or voice) being amplified.

If an ac signal, with a dc voltage level, is to be amplified by *another* transistor stage, the dc voltage must be removed. If the dc is not removed it will adversely affect the next amplification stage.

Using a Capacitor to Remove DC Voltages

A capacitor offers a fairly low opposition to ac. The capacitor, however, offers infinite opposition to dc. A capacitor acts like an open circuit to dc voltages.

Figure 24-2 has a **blocking capacitor,** also called a **coupling capacitor,** in series with the dc and ac mixed signal. The capacitor charges to the dc voltage, then it stops any further dc. The ac signal is continually changing and passes through the capacitor. The end result is the sine wave has a zero volt center line. The dc voltage level is removed.

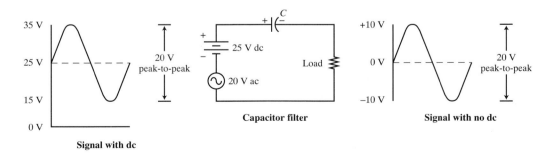

Figure 24-2. A blocking, or coupling, capacitor is used to remove the dc voltage from the ac signal.

Using a Transformer to Remove DC Voltages

A transformer only passes a changing voltage. A circuit, such as the one in figure 24-3, containing both ac and dc, will pass only the ac signal to the secondary windings of the transformer. The dc voltage produces a magnetic field. But since the magnetic field does not change, it does not induce a voltage in the secondary.

One problem with applying dc to a transformer is the dc voltage could saturate the transformer. If a transformer is **saturated,** it is producing the largest magnetic field that it possibly can. In this condition, the transformer will not pass any of the ac signal. The transformer's magnetic field has no room to fluctuate.

In some applications, 1:1 transformers are used. They are used both as a means of separating the dc from the ac and as part of a resonant circuit. In other applications, step-up transformers are used to do the separating and boost the signal voltage at the same time.

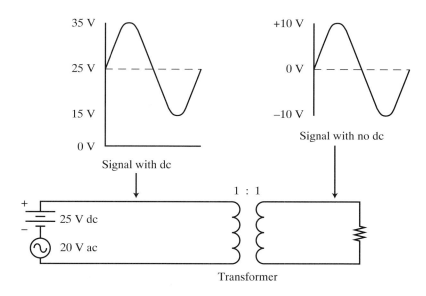

Figure 24-3. Transformers will remove the dc component from an ac circuit.

Using a Capacitor to Provide a Separate Path

In certain circuits, such as transistor amplifiers, it is necessary to provide a separate path for both the dc voltage level and ac signal. Figure 23-4 shows a **bypass capacitor,** which gives the ac signal a parallel path around the load resistor. Notice, the blocking capacitor shown in figure 24-2 is in series with the load. The bypass capacitor shown in figure 24-4 is in parallel with the load.

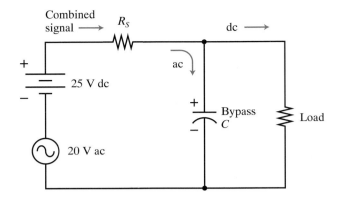

Figure 24-4. Bypass capacitors provide a path to ground for the ac part of a signal. Only the dc component passes to load.

24.2 SEPARATION OF SIGNALS BY FREQUENCY

Inductors and capacitors have opposite responses to many signals. This feature can be used to select the type of signal and how to deal with it. Figure 24-5 shows how capacitors and inductors respond to a wide range of frequencies.

The inductor passes both dc and low frequency ac voltages. The lower the frequency, the less opposition the inductor offers the signal. The capacitor will pass only ac voltages. The higher the frequency, the less opposition a capacitor offers the signal.

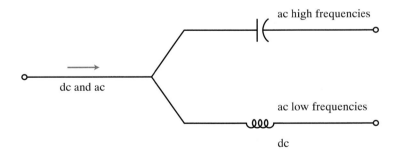

Figure 24-5. Separation of dc and ac can be accomplished using a capacitor and an inductor.

Response of an Inductor

An inductor offers the least amount of opposition to low frequencies. An inductor also poses less resistance when its inductance value is low. Figure 24-6 shows the effects in several situations. In part A, the low inductance value has a little effect on the

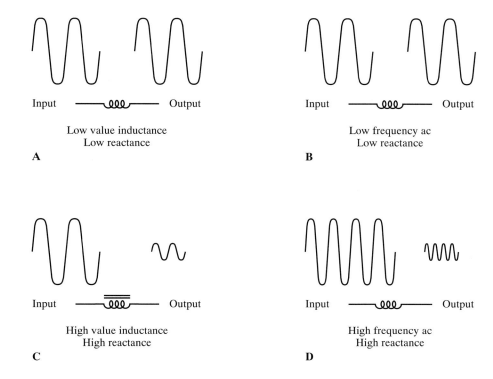

Figure 24-6. **Inductive reactance is lowest with low values of inductance and low frequencies.**

ac signal. In part B, a low frequency is passed through the inductor with little opposition. In part C, the large inductance value offers a high opposition to the signal and will drop a significant amount of voltage. In part D, the high frequency signal is met with much opposition.

Figure 24-6 demonstrates how reactance in an inductor varies. Examine the inductive reactance formula, $X_L = 2\pi fL$, to gain further understanding. If either the signal frequency or the value of the inductor is made larger, the reactance is also made larger. The reactance is directly related to the values of frequency and inductance.

Response of a Capacitor

A capacitor has its least opposition at high frequencies or with high values of capacitance. Figure 24-7 makes a comparison of different circuit characteristics and their effect on the ac signal. Part A shows that high values of capacitance offer little opposition. Part B shows that there is only slight opposition to high frequencies. In part C, low values of capacitance offer high opposition to the signal. In part D, high opposition is offered to low frequencies.

An examination of the capacitive reactance formula, $X_C = \dfrac{1}{2\pi fC}$ shows that reactance is inversely related to frequency and capacitance. This inverse relationship means that if either frequency or capacitance is increased, the reactance decreases. Also, if either f or C decrease, reactance increases.

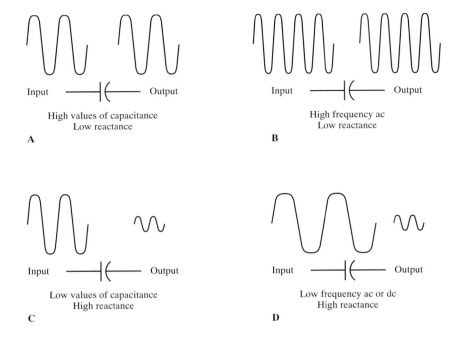

Figure 24-7. Capacitive reactance is lowest with high capacitance or high frequencies.

24.3 LOW PASS FILTERS

The property of inductors and capacitors to change reactance with a change in frequency is very useful. This property creates circuits that accept or reject certain frequencies. These circuits are simple filters. The position of the reactive component in the circuit determines if the filter is low pass or high pass.

Figure 24-8 shows three simple low pass filters. If the filter circuit contains an inductor, the inductor is in series with the ac signals. When a capacitor is used in a low

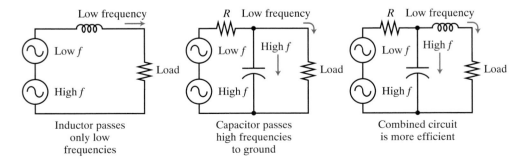

Figure 24-8. Three simple low pass filters.

pass filter, the capacitor is placed in parallel with the load. A high pass filter will have the inductor in parallel with the load or the capacitor in series with the signal.

Low Pass Filter Response Curve

The response curve for a filter shows which frequencies will be passed and which frequencies will be rejected. Figure 24-9 shows a response curve for one filter. This curve shows that all frequencies from zero Hz (dc) to some frequency where the curve rolls off (the 70.7 percent point) are passed equally. The curve continues to drop until it completely rejects any higher frequencies. The 70.7 percent point in filter circuit is called the **cutoff frequency,** and labeled f_c.

This curve is nearly ideal. In an actual circuit, the response curve does not have such a steep slope.

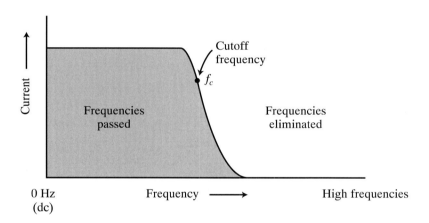

Figure 24-9. Typical response curve for a simple low pass filter.

Low pass filters can be used to remove a high frequency that is mixed with a low frequency, see figure 24-10. This input signal has a 10 kHz audio signal mixed with a 60 hertz signal. It is desired to have only the 60 hertz. The series inductor in the filter offers a low reactance to the 60 Hz signal while offering a high reactance to the 10 kHz signal. Also, the capacitor in parallel has a high reactance to the 60 Hz and a low reactance to the 10 kHz. The inductor should filter out most of the 10 kHz signal while passing the 60 Hz signal. Any of the 10 kHz signal that makes it past the inductor should be shorted to ground by the low impedance capacitor. The calculations in sample problem 1 demonstrate the response of both components.

Sample problem 1. _____

Using the circuit in figure 24-10, calculate the reactance at 60 Hz and 10 kHz for both the inductor and capacitor.

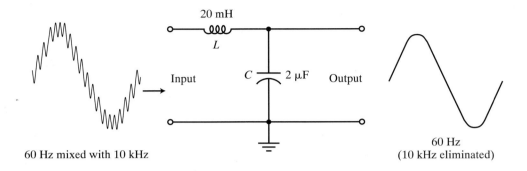

60 Hz mixed with 10 kHz

Input C ⎓ 2 μF   Output

60 Hz
(10 kHz eliminated)

Figure 24-10. A low pass filter removes an audio signal mixed with 60 Hz. (Sample problem 1)

1. Inductive reactance at 60 Hz.

    Formula: $X_L = 2\pi f L$

    Substitution: $X_L = 2 \times \pi \times 60 \text{ Hz} \times 20 \text{ mH}$

    Answer: $X_L = 7.5 \ \Omega$

2. Inductive reactance at 10 kHz.

    Formula: $X_L = 2\pi f L$

    Substitution: $X_L = 2 \times \pi \times 10 \text{ kHz} \times 20 \text{ mH}$

    Answer: $X_L = 1.3 \ k\Omega$

3. Capacitive reactance at 60 Hz.

    Formula: $X_C = \dfrac{1}{2\pi f C}$

    Substitution: $X_C = \dfrac{1}{2 \times \pi \times 60 \text{ Hz} \times 2 \text{ μF}}$

    Answer: $X_C = 1.3 \ k\Omega$

4. Capacitive reactance at 10 kHz.

    Formula: $X_C = \dfrac{1}{2\pi f C}$

Substitution: $X_C = \dfrac{1}{2 \times \pi \times 10 \text{ kHz} \times 2 \text{ μF}}$

Answer: $X_C = 8 \text{ Ω}$

24.4 DECIBELS AND THE –3 dB POINT

The **decibel (dB)** is a unit of measure related to the logarithmic response of the human ear. The decibel is also a useful measure for the frequency response of filters. It is useful for two reasons. First, response curves are usually plotted with frequency on a logarithmic scale (factors of 10). Second, decibels give a comparison of the input to the output.

Decibels can be calculated with either of two formulas. The use of decibels is shown here to help develop further concepts. The cutoff frequency on the response curve, labeled as 70.7 percent voltage or the half-power point, is also called the **–3 dB point.** Notice the negative sign in front of the 3 dB. The negative sign indicates a loss from the input to the output.

Formula 24.A

$$dB = 20 \log \frac{V_{out}}{V_{in}}$$

V_{out} is the output voltage, measured in volts.

V_{in} is the input voltage, measured in volts.

Formula 24.B

$$dB = 10 \log \frac{P_{out}}{P_{in}}$$

P_{out} is the output power, measured in watts.

P_{in} is the input power, measured in watts.

Sample problem 2. _____

Calculate the decibel equivalent on a response curve where the output voltage is equal to 70.7 percent of the input voltage.

Formula: $dB = 20 \log \dfrac{V_{out}}{V_{in}}$

Substitution: dB = 20 log 0.707

Use a calculator to find the log of 0.707, then multiply by 20.

Answer: −3 dB

___

Sample problem 3. _____

Calculate the decibel equivalent in a circuit in which the output power is one-half the input power.

Formula: $dB = 10 \log \dfrac{P_{out}}{P_{in}}$

Substitution: $dB = 10 \log 0.5$

Answer: −3 dB

___

Low Pass Filter Using a Reactive Voltage Divider

The low pass filters shown in figures 24-11 and 24-12 use a reactive component in combination with a resistor to form a voltage divider. When the reactance is equal to the resistance, in either of these circuits, the output is −3 dB as compared to the input.

Sample problem 4. _____

Calculate the output voltage of the circuit in figure 24-11. Determine if it is 70.7 percent of the input voltage.

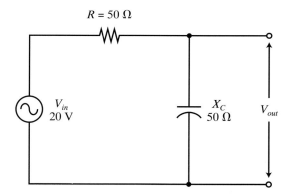

Figure 24-11. RC voltage dividers can be used as low pass filters. V_{out} **is taken across the capacitor. (Sample problem 4)**

1. Use the voltage divider formula to calculate the output voltage.

Formula: $V_{out} = V_{in} \times \dfrac{R_{out}}{R_T}$

Note: $R_T = Z = \sqrt{R^2 + X_C^{\,2}}$

Modify formula: $V_{out} = V_{in} \times \dfrac{X_C}{\sqrt{R^2 + X_C^{\,2}}}$

Substitution: $V_{out} = 20\ \text{V} \times \dfrac{50\ \Omega}{\sqrt{50^2\ \Omega + 50^2\ \Omega}}$

Intermediate step: $V_{out} = 20\ \text{V} \times \dfrac{50\ \Omega}{70.7\ \Omega}$

Answer: $V_{out} = 14.14\ \text{V}$

2. Find percentage of input voltage.

Formula: $\% = \dfrac{V_{out}}{V_{in}} \times 100$

Substitution: $\% = \dfrac{14.14\ \text{V}}{20\ \text{V}} \times 100$

Answer: $\% = 70.7\%$

Sample problem 5. _____

Calculate the output voltage of the circuit in figure 24-12. Determine what percentage the output voltage is of the input voltage.

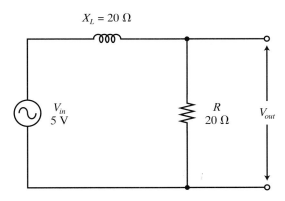

Figure 24-12. RL voltage dividers can also be used as low pass filters. V_{out} **is taken across the resistor. (Sample problem 5)**

1. Use the voltage divider formula to calculate the output voltage.

Modified formula: $V_{out} = V_{in} \times \dfrac{R}{\sqrt{R^2 + X_L^{\,2}}}$

Substitution: $V_{out} = 5\ \text{V} \times \dfrac{20\ \Omega}{\sqrt{20^2\ \Omega + 20^2\ \Omega}}$

Intermediate step: $V_{out} = 5\ \text{V} \times \dfrac{20\ \Omega}{28.28\ \Omega}$

Answer: $V_{out} = 3.54\ \text{V}$

2. Find percentage of input voltage.

Formula: $\% = \dfrac{V_{out}}{V_{in}} \times 100$

Substitution: $\% = \dfrac{3.54\ \text{V}}{5\ \text{V}} \times 100$

Answer: $\% = 70.8\% \cong 70.7\%$

Calculating the Cutoff Frequency

The cutoff frequency in a reactive voltage divider filter is the frequency at which the –3 dB point is reached. The cutoff frequency can be calculated if you know the value of the component and its reactance. The reactance formula is rearranged to solve for the frequency. The voltage does not have to be known because the frequency is not affected by voltage.

Sample problem 6. _____

Find the cutoff frequency for the circuit in figure 24-13. Assume reactance is equal to resistance.

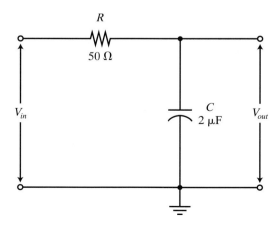

Figure 24-13. With a low pass RC filter, the cutoff frequency is obtained when X_C equals R. (Sample problem 6)

Formula: $X_C = \dfrac{1}{2\pi fC}$

Rearranging: $f = \dfrac{1}{2\pi X_C C}$

Substitution: $f = \dfrac{1}{2 \times \pi \times 50\ \Omega \times 2\ \mu F}$

Answer: $f = 1.59\ \text{kHz}$

Sample problem 7. _____

Find the cutoff frequency for the circuit in figure 24-14. Assume reactance is equal to resistance.

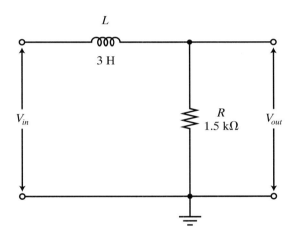

Figure 24-14. With a low pass RL filter, the cutoff frequency is obtained when X_L equals R. (Sample problem 7)

Formula: $X_L = 2\pi fL$

Rearranging: $f = \dfrac{X_L}{2\pi L}$

Substitution: $f = \dfrac{1.5\ \text{k}\Omega}{2 \times \pi \times 3\ \text{Hz}}$

Answer: $f = 79.6\ \text{Hz}$

Other Low Pass Filters

Inductors and capacitors can be combined in any number of stages to provide filtering of the signal. Generally speaking, the more stages involved, the better the filtering process. However, one problem with having many filtering stages is that every series component drops some of the voltage. Too many stages seriously weakens a signal.

Figure 24-15 shows two types of cascading filter stages. The **T-type filter** offers two inductors and the **pi-type filter** offers two capacitors. Low pass filter circuits have the inductor in series and the capacitor in parallel.

A resistor *can* be used in place of the inductor to provide a voltage drop for the capacitor. The resistor is less expensive than an inductor, but it provides no filtering action.

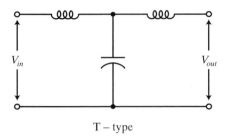
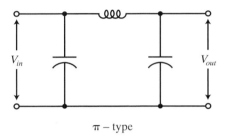

Figure 24-15. Other types of low pass filters.

24.5 HIGH PASS FILTERS

The major difference between low pass and high pass filters is placement of the components. Inductors in high pass filters are placed in parallel with the load. Capacitors are placed in series with the load.

Filters are classified according to the frequencies passed or rejected. Low pass filters accept the frequencies below cutoff. High pass filters accept the frequencies above cutoff. A typical response curve for a high pass filter is shown in figure 24-16.

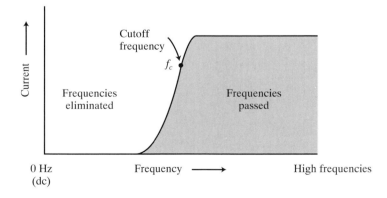

Figure 24-16. Typical response curve of a high pass filter.

Removing Unwanted Low Frequencies

A very common problem in audio circuits is a 60 cycle hum. The hum can be caused by nearby electrical power lines or from within the equipment's own power supply. If the 60 hertz is not filtered out, it is heard as a low hum.

Coupling capacitors in series with the signal, such as the one shown in figure 24-2, can be used to stop the 60 cycle hum. However, a better filter will include an inductor in parallel. See figure 24-17.

Each reactive component offers a different opposition to each different frequency contained in the circuit. In sample problem 8, notice how the change in reactance performs the desired filtering process. This same signal was passed through a low pass filter in figure 24-10. Compare the results of the two filters.

Sample problem 8. _____

Determine the reactance for each of the components shown in the circuit in figure 24-17, for both 60 Hz and 10 kHz.

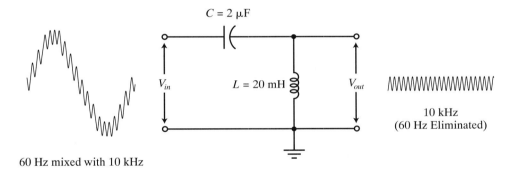

Figure 24-17. A high pass filter removes a 60 Hz signal from an audio signal. (Sample problem 8)

1. Inductive reactance at 60 Hz.

    Formula: $X_L = 2\pi f L$

    Substitution: $X_L = 2 \times \pi \times 60 \text{ Hz} \times 20 \text{ mH}$

    Answer: $X_L = 7.5 \ \Omega$

2. Inductive reactance at 10 kHz.

    Formula: $X_L = 2\pi f L$

    Substitution: $X_L = 2 \times \pi \times 10 \text{ kHz} \times 20 \text{ mH}$

    Answer: $X_L = 1.3 \text{ k}\Omega$

3. Capacitive reactance at 60 Hz.

   Formula: $X_C = \dfrac{1}{2\pi fC}$

   Substitution: $X_C = \dfrac{1}{2 \times \pi \times 60\ \text{Hz} \times 2\ \mu\text{F}}$

   Answer: $X_C = 1.3\ \text{k}\Omega$

4. Capacitive reactance at 10 Hz.

   Formula: $X_C = \dfrac{1}{2\pi fC}$

   Substitution: $X_C = \dfrac{1}{2 \times \pi \times 10\ \text{kHz} \times 2\ \mu\text{F}}$

   Answer: $X_C = 8\ \Omega$

High Pass Filter Using a Reactive Voltage Divider

The reactive voltage divider can be used for either a low pass or a high pass filter. The difference is the manner in which the components are connected. In the high pass filter the capacitor is connected in series and the resistor is in parallel with the signal. In the inductive high pass filter, the resistor is in series while the inductor is connected in parallel. In either case, the reactance is equal to the resistance at the cutoff frequency.

Sample problem 9. _____

Determine the cutoff frequency for the circuit shown in figure 24-18.

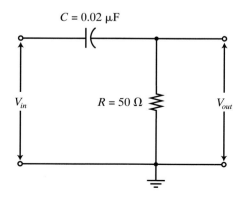

Figure 24-18. With a high pass RC filter, the cutoff frequency is obtained when X_C equals R. V_{out} is taken across the resistor. (Sample problem 9)

Formula: $X_C = \dfrac{1}{2\pi f C}$

Rearranging: $f = \dfrac{1}{2\pi X_C C}$

$X_C = 50\ \Omega$ at the cutoff frequency.

Substitution: $f = \dfrac{1}{2 \times \pi \times 50\ \Omega \times 0.02\ \mu F}$

Answer: $f = 159$ kHz

Sample problem 10. _____

Find the cutoff frequency for the circuit in figure 24-19. Assume reactance is equal to the resistance.

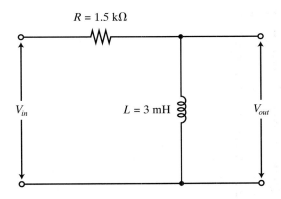

Figure 24-19. With a high pass RL filter, the cutoff frequency is obtained when X_L equals R. V_{out} is taken across the inductor. (Sample problem 10)

Formula: $X_L = 2\pi f L$

Rearranging: $f = \dfrac{X_L}{2\pi L}$

Substitution: $f = \dfrac{1.5\ k\Omega}{2 \times \pi \times 3\ mH}$

Answer: $f = 79.6$ kHz

Other Types of High Pass Filters

Like the low pass circuits, the high pass filters can have any number of stages added to improve the filtering. High pass filters have the capacitor in series and the inductor in parallel. Figure 24-20 shows a high pass T-type filter and a high pass pi-type filter.

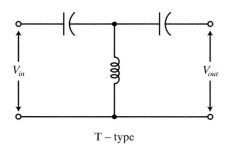

T – type

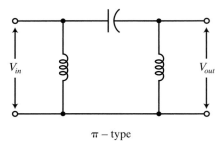

π – type

Figure 24-20. Other types of high pass filters.

24.6 RESONANT FILTERS

Series and parallel resonant circuits are ideal to pass or stop a specific band of frequencies. The filters that stop all but one specific band of frequencies are called **band pass filters.** Filters that do the reverse, stop only one specific band of frequencies, are called **band stop filters.** The resonant frequency is the **center frequency (f_0)** for the band that is passed or stopped by the filters. The bandwidth, as calculated from the circuit Q, determines the width of the band.

The manner in which the resonant circuit is connected determines if the filter is band pass or band stop. Consider the characteristics of the series and parallel resonant circuits. The series circuit offers minimum impedance at resonance, while the parallel circuit is at a maximum.

Series Resonant Band Pass

The series resonant circuit offers minimum impedance at resonance, equal to the series resistance of the resistor. By placing the resonant circuit in series with the output, the frequencies near resonance pass with the least opposition. Figure 24-21 is an example of a series resonant band pass filter.

The center frequency, labeled f_0 in filter circuits, is calculated using the frequency of resonance formula. Upper and lower cutoff frequencies are calculated using the circuit Q. Circuit Q is the inductive reactance divided by the total resistance. The total resistance is the sum of r_s and the parallel resistor R. The bandwidth is the resonant frequency divided by the Q. For review of these formulas and concepts, refer to Chapter 23.

Sample problem 11. _____

Using the values of the circuit shown in figure 24-21, determine: center frequency, inductive reactance, circuit Q, bandwidth, and the upper and lower cutoff frequencies.

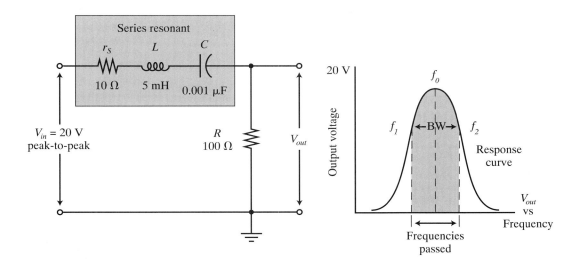

Figure 24-21. Series resonant band pass filter and its response curve. (Sample problem 11)

1. Center frequency.

    Formula: $f_0 = \dfrac{1}{2\pi\sqrt{LC}}$

    Substitution: $f_0 = \dfrac{1}{2 \times \pi \times \sqrt{5 \text{ mH} \times 0.001 \text{ μF}}}$

    Answer: $f_0 = 71.2$ kHz

2. Inductive reactance.

    Formula: $X_L = 2\,\pi f L$

    Substitution: $X_L = 2 \times \pi \times 71.2 \text{ kHz} \times 5 \text{ mH}$

    Answer: $X_L = 2.24$ kΩ

3. Circuit Q.

Formula: $Q = \dfrac{X_L}{r_s + R}$

Substitution: $Q = \dfrac{2.24 \text{ k}\Omega}{10 \ \Omega + 100 \ \Omega}$

Answer: $Q = 20.4$

4. Bandwidth.

Formula: $\text{BW} = \dfrac{f_0}{Q}$

Substitution: $\text{BW} = \dfrac{71.2 \text{ kHz}}{20.4}$

Answer: $\text{BW} = 3490 \text{ Hz}$

5. Upper cutoff frequency.

Formula: $f_2 = f_0 + \dfrac{\text{BW}}{2}$

Substitution: $f_2 = 71.2 \text{ kHz} + \dfrac{3490 \text{ Hz}}{2}$

Answer: $f_2 = 72,945 \text{ Hz}$

6. Lower cutoff frequency.

Formula: $f_1 = f_0 - \dfrac{\text{BW}}{2}$

Substitution: $f_1 = 71.2 \text{ kHz} - \dfrac{3490 \text{ Hz}}{2}$

Answer: $f_2 = 69,455 \text{ Hz}$

Parallel Resonant Band Pass

A parallel resonant circuit has maximum impedance at the resonant frequency. Other frequencies that are off resonance see a much lower impedance. Consequently, if a band pass circuit is desired, the resonant circuit is placed in parallel with the load. Frequencies near resonance pass by the circuit to the output. Off resonance frequencies are shorted to ground.

Sample problem 12. _____

Using the circuit in figure 24-22, determine: center frequency, inductive reactance, circuit Q, bandwidth, and the upper and lower cutoff frequencies.

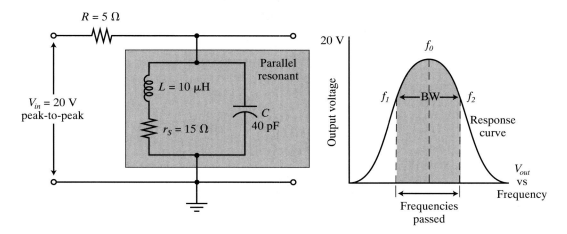

Figure 24-22. Parallel resonant band pass filter and its response curve. (Sample problem 12)

1. Center frequency.

    Formula: $f_0 = \dfrac{1}{2\pi\sqrt{LC}}$

    Substitution: $f_0 = \dfrac{1}{2 \times \pi \times \sqrt{10\ \mu H \times 40\ pF}}$

    Answer: $f_0 = 7.96$ MHz

2. Inductive reactance.

    Formula: $X_L = 2\pi f L$

    Substitution: $X_L = 2 \times \pi \times 7.96$ MHz $\times 10\ \mu H$

    Answer: $X_L = 500\ \Omega$

3. Circuit Q.

    Formula: $Q = \dfrac{X_L}{r_s}$

    Note, only r_s is considered in a parallel circuit.

Substitution: $Q = \dfrac{500\ \Omega}{15\ \Omega}$

Answer: $Q = 33.3$

4. Bandwidth.

Formula: $BW = \dfrac{f_0}{Q}$

Substitution: $BW = \dfrac{7.96\ \text{MHz}}{33.3}$

Answer: $BW = 239\ \text{kHz}$

5. Upper cutoff frequency.

Formula: $f_2 = f_0 + \dfrac{BW}{2}$

Substitution: $f_2 = 7.96\ \text{MHz} + \dfrac{239\ \text{kHz}}{2}$

Answer: $f_2 = 8.08\ \text{MHz}$

6. Lower cutoff frequency.

Formula: $f_1 = f_0 - \dfrac{BW}{2}$

Substitution: $f_2 = 7.96\ \text{MHz} - \dfrac{239\ \text{kHz}}{2}$

Answer: $f_2 = 7.84\ \text{MHz}$

Series Resonant Band Stop Filter

At the resonant frequency, the minimum impedance of the series circuit can be used to short the signal to ground. Figure 24-23 shows the series resonant circuit is connected in parallel with the output.

Off resonant frequencies pass through the series resistor and ignore the resonant circuit. Near resonance, the impedance of the resonant circuit drops, providing a path to ground.

Sample problem 13.

Using the values of the circuit shown in figure 24-23, determine: center frequency, inductive reactance, circuit Q, and bandwidth.

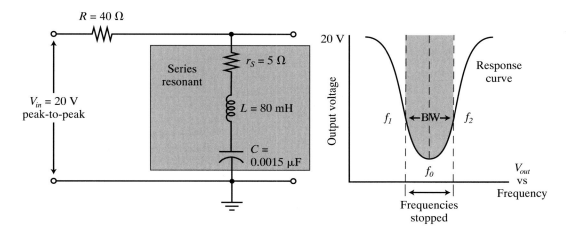

Figure 24-23. Series resonant band stop filter and its response curve. (Sample problem 13)

1. Center frequency.

   Formula: $f_0 = \dfrac{1}{2\pi\sqrt{LC}}$

   Substitution: $f_0 = \dfrac{1}{2 \times \pi \times \sqrt{80 \text{ mH} \times 0.0015 \text{ μF}}}$

   Answer: $f_0 = 14.5$ kHz

2. Inductive reactance.

   Formula: $X_L = 2\pi f L$

   Substitution: $X_L = 2 \times \pi \times 14.5 \text{ kHz} \times 80 \text{ mH}$

   Answer: $X_L = 7.29$ kΩ

3. Circuit Q.

   Formula: $Q = \dfrac{X_L}{r_s + R}$

Substitution: $Q = \dfrac{7.29 \text{ k}\Omega}{5 \; \Omega + 40 \; \Omega}$

Answer: $Q = 162$

4. Bandwidth.

Formula: $BW = \dfrac{f_0}{Q}$

Substitution: $BW = \dfrac{14.5 \text{ kHz}}{162}$

Answer: $BW = 89.5 \text{ Hz}$

Parallel Resonant Band Stop Filter

The parallel resonant circuit also stops frequencies near resonance. By placing this type of circuit in series with the filter's output, off-resonant frequencies pass through, while those near resonance are stopped.

Sample problem 14. _____

Using the circuit in figure 24-24, determine: center frequency, inductive reactance, circuit Q, and bandwidth.

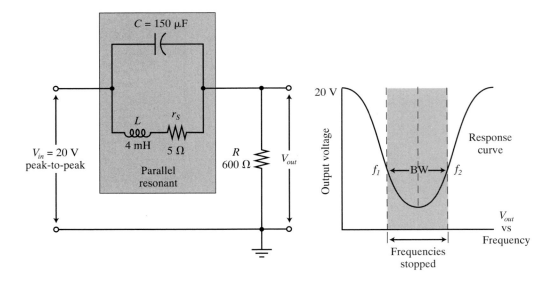

Figure 24-24. Parallel resonant band stop filter and its response curve. (Sample problem 14)

1. Center frequency.

    Formula: $f_0 = \dfrac{1}{2\pi\sqrt{LC}}$

    Substitution: $f_0 = \dfrac{1}{2 \times \pi \times \sqrt{4 \text{ mH} \times 1.5 \text{ }\mu\text{F}}}$

    Answer: $f_0 = 2055$ Hz

2. Inductive reactance.

    Formula: $X_L = 2\pi f L$

    Substitution: $X_L = 2 \times \pi \times 2055 \text{ Hz} \times 4 \text{ mH}$

    Answer: $X_L = 51.6 \text{ }\Omega$

3. Circuit Q.

    Formula: $Q = \dfrac{X_L}{r_s}$

    Note, only r_s is considered in a parallel circuit.

    Substitution: $Q = \dfrac{51.6 \text{ }\Omega}{5 \text{ }\Omega}$

    Answer: $Q = 10.3$

4. Bandwidth.

    Formula: $\text{BW} = \dfrac{f_0}{Q}$

    Substitution: $\text{BW} = \dfrac{2055 \text{ Hz}}{10.3}$

    Answer: $\text{BW} = 200$ Hz

SUMMARY

- When a dc voltage is present with an ac signal, the ac signal will center itself at the dc voltage level.
- A capacitor stops dc while passing the ac signal.
- A transformer can also be used to stop dc while passing ac.
- A capacitor can provide a different path for the ac to follow without changing the dc.

- An inductor passes low frequencies while providing greater reactance to higher frequencies.
- A capacitor passes high frequencies while providing greater reactance to lower frequencies.
- A low pass filter can have an inductor in series or a capacitor in parallel. The filter may contain both.
- On a frequency response curve, the point equal to 70.7 percent of maximum is also known as the cutoff frequency, –3 dB point, or half-power point.
- A resonant circuit can be used to reject or pass a band of frequencies.

KEY WORDS AND TERMS GLOSSARY

band pass filter: A filter that blocks all but one band of frequencies from passing through.

band stop filter: A filter that blocks one specific band of frequencies from passing through.

blocking capacitor: A capacitor placed in series to pass an ac signal while stopping a dc voltage. Also called a coupling capacitor.

bypass capacitor: A parallel connected capacitor to provide a path for the ac signal around a resistor.

center frequency: The resonant frequency (f_0) for filters with bell curve characteristics.

coupling capacitor: A capacitor placed in series to pass an ac signal while stopping a dc voltage. Also called a blocking capacitor.

cutoff frequency (f_c): The frequency where the response curve for a filter circuit falls to 70.7 percent.

decibel (dB): A unit of measure related to the logarithmic response of the human ear. It is a useful measure for the frequency response of filters for two reasons: response curves are usually plotted with frequency on a logarithmic (factor of 10) scale, and it is a method of comparing the input to the output.

–3 dB point: The point on the response curve of a resonant circuit equal to 70.7 percent of the maximum voltage.

pi-type filter: A type of filter that uses two capacitors. See figure 24-15.

saturated: More current is applied to a transformer primary than can be converted to magnetic energy. The excessive current is changed into heat and creates losses.

T-type filter: A type of filter that uses two inductors. See figure 24-15.

KEY FORMULAS

Formula 24.A

$$dB = 20 \log \frac{V_{out}}{V_{in}}$$

Formula 24.B

$$dB = 10 \log \frac{P_{out}}{P_{in}}$$

TEST YOUR KNOWLEDGE

Do not write in this text. Please use a separate sheet of paper.
1. List three uses for filter circuits.
2. Describe how a capacitor is used to remove a dc voltage from an ac signal.
3. Describe how a transformer is used to remove a dc voltage from an ac signal.
4. Describe how a capacitor is used to separate an ac signal for a dc voltage.
5. Which component (inductor or capacitor) offers the least opposition to low frequencies?
6. Which component (inductor or capacitor) offers the least opposition to high frequencies?
7. An inductor and capacitor are connected in parallel as a filter circuit.
    a.   Which path is taken by the dc and low frequencies?
    b.   Which path is taken by the high frequencies?
8. Draw three simplified schematic diagrams of low pass filters.
9. An audio signal of 1 kHz is transmitted with a 50 kHz radio frequency carrier. The audio signal is removed using the low pass filter shown in figure 24-25. Calculate X_L and X_C at both frequencies to show the effectiveness of the filter.

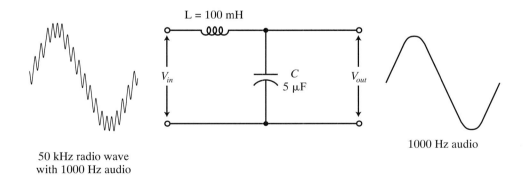

Figure 24-25. Problem #9.

10. Describe how to determine the value to be used for the reactive component when a low pass filter is used as a reactive voltage divider.

11. Determine the cutoff frequency of the low pass filter in figure 24-26. Also determine the output voltage.

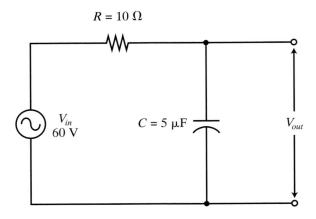

$R = 10\ \Omega$

V_{in}
60 V

$C = 5\ \mu F$

V_{out}

Figure 24-26. Problem #11.

12. A 50 Hz hum is mixed with an audio signal of 20 kHz in an amplifier. The hum is removed using the high pass filter shown in figure 24-27. Calculate X_L and X_C at both frequencies to show the effectiveness of the filter.

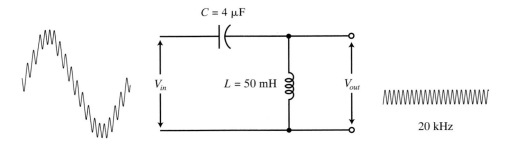

$C = 4\ \mu F$

V_{in} $L = 50$ mH V_{out}

50 Hz mixed with 20 kHz

20 kHz

Figure 24-27. Problem #12.

13. Describe how to determine the value to be used for the reactive component when a high pass filter is used as a reactive voltage divider.

14. Determine the cutoff frequency of the high pass filter in figure 24-28. Also determine the output voltage.

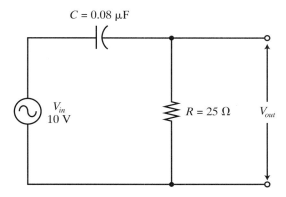

$C = 0.08\ \mu\text{F}$

V_{in}
10 V

$R = 25\ \Omega$ V_{out}

Figure 24-28. Problem #14

15. Describe how a series resonant circuit is used as a band pass filter.
16. Describe how a parallel resonant circuit is used as a band pass filter.
17. Describe how a series resonant circuit is used as a band stop filter.
18. Describe how a parallel resonant circuit is used as a band stop filter.

**Interference on telephone lines introduces errors into
computer data transferred by phone. Filtering circuits
allow the computer data to be transported error free.
(Command Communications, Inc.)**

Chapter 25
Single- and
Polyphase Systems

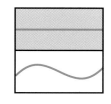

Upon completion of this chapter, you will be able to:
- Describe how a single-phase transformer can produce a three-wire single-phase output.
- Describe the typical three-wire single-phase electrical distribution system used in residential applications.
- Describe a typical two-phase electrical system.
- Calculate total current in a two-phase system.
- Describe the general principles of a three-phase system.
- Draw the connections for a four-wire wye connected three-phase system.
- Calculate the voltage at the output of a wye connected system.
- Draw the connections for a three-wire delta connected three-phase system.
- Calculate the voltage at the output of a delta connected system.
- Draw the generator to output combinations of wye and delta connections.

In **polyphase systems** multiple sine waves are produced. Each of these phases are separated by a given number of electrical degrees. Advantages of polyphase systems include the feeding of several loads with a choice of two voltages: either from one phase to neutral, or across two phases. This chapter discusses three different types of systems: single-phase, two-phase, and three-phase.

25.1 SINGLE-PHASE SYSTEMS

A **single-phase system** produces one sine wave. All of the prior chapters in this book have used single phase ac. Single-phase is the type of electricity normally used in residential applications.

A generator that produces single-phase electricity uses a single coil, shown in figure 25-1. This figure shows only one pair of magnetic field poles. Additional pole pairs can be added to this generator. You may want to refer to Chapter 17 for further details on single-phase generators.

All electrical systems require a minimum of two wires. The single-phase system shown in figure 25-1 is a typical two-wire system. Note that in some electrical circuits,

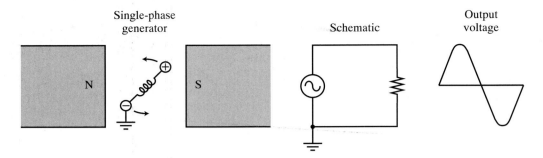

Figure 25-1. Single-phase generator producing a single sine wave output voltage.

such as an automobile, one of the wires is the chassis of the vehicle. In 120 volt residential house wiring, there appears to be three wires for many electrical fixtures. However, only two of the wires are used for current carrying. One wire is a ground.

Three-Wire Single-Phase Systems

The single-phase system can be used to supply a transformer that splits the voltage into two single-phase voltages. Each of the voltages is 180° apart. To achieve this, a center-tapped transformer is used. See figure 25-2. The turns ratio of the transformer allows it to produce 240 volts from one side to the other. Notice that the center-tap is connected to ground, holding it at zero volts. This system is a *three-wire single-phase systems.*

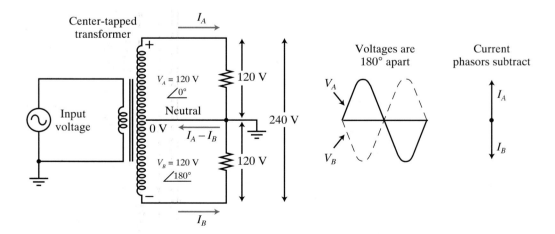

Figure 25-2. Three-wire single-phase systems are typical in residential applications.

Each side of the transformer produces 120 volts in reference to ground. If the incoming sine wave is in the positive half cycle, the top of the transformer has a polarity of positive, while the bottom of the transformer is negative. The center-tap remains at zero volts, which results in the opposite sides of the transformer having an opposite polarity.

The drawing of the two sine waves shows their relationship. The phasor diagram of the currents also shows their 180° separation. Each 120 volt load has a current running from the outside of the transformer to center, which is neutral. With the two currents being of opposite polarities, they subtract in the neutral wire. The neutral wire, therefore, does not need to be large enough to carry the current of both lines. The neutral wire only has to be large enough to carry the maximum current of one line.

Residential Applications of Single-Phase

In residential applications, the electricity supplied from the power company to the circuit breaker panel is a three-wire single-phase system. The ground wire is not counted as one of the wires when identifying a two- or three-wire system. The wire colors used in figures 25-3 to 25-5 refer to the standard color code. See Chapter 2 for more details of house wiring.

There are three basic ways of bringing power to the load circuits.

1.   Individual 120 volt two-wire circuits supply the majority of lights and duplex outlets, shown in figure 25-3.

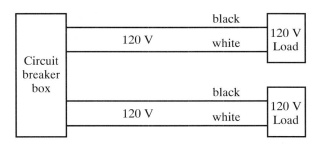

Figure 25-3. Two-wire, 120 volt circuits feed most residential loads.

2.   Three-wire cables supply two 120 volt circuits. This technique might be used to bring the power from the circuit breaker box in a basement to a junction box in the attic. See figure 25-4. Two circuits are then split from the junction box to supply individual loads. Using this three-wire cable to supply two circuits saves copper in comparison to making a long run with two, two-wire cables.

3.   Three-wire cables are also used to supply 240 volts to certain appliances, as shown in figure 25-5. Electric clothes dryers and electric stoves are two examples of appliances that may need the additional power supplied by a 240 volt line. In these cases, the appliances combine the two 120 volt lines for 240 volts.

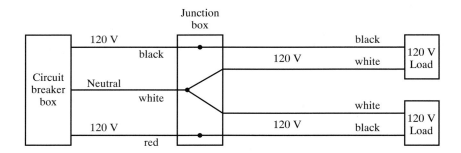

Figure 25-4. Three-wire circuits save wire.

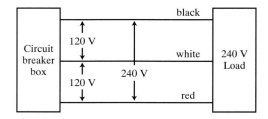

Figure 25-5. Three-wire cables supply 240 volts to appliances that need a great deal of power.

25.2 TWO-PHASE SYSTEMS

When large amounts of current are required from a generator, the single-phase system is not efficient enough. By using two identical coils, a generator is able to produce twice the power without a significant increase in losses. This is the **two-phase system.**

The basic two-phase generator, shown in figure 25-6, has the two coils spaced 90 electrical degrees apart. This produces two independent sine waves with a phase sepa-

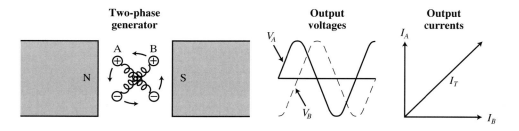

Figure 25-6. Basic two-phase generator. A two-phase system produces two sine waves that are 90° apart.

ration of 90°. The currents from each phase are added using vectors. This results in a total current that is *less* than the sum of the two independent currents.

Types of Two-Phase Systems

The two-phase system can be connected in either of two configurations, four-wire or three-wire. See figure 25-7.

In the four-wire system, the two generator coils are independent. Each coil feeds its own load. With the three-wire system, the coils are connected with a shared neutral.

Calculating Total Current in Two-Phase Systems

The same circuit calculations apply to a polyphase system as applied to a single-phase system. Addition must be performed using phasors.

Sample problem 1. _____

Determine the total current for each of the two circuits shown in figure 25-7.

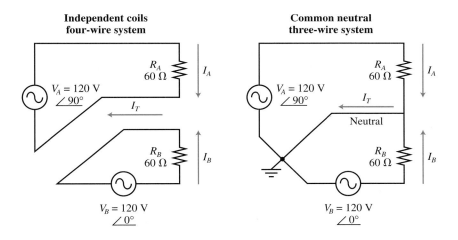

Figure 25-7. Schematic diagrams of two-phase systems. (Sample problem 1)

1. Four-wire system.

    Formula: $I = \dfrac{V}{R}$

    a. Phase A:

    Substitution: $I_A = \dfrac{120\ \text{V}\ \angle 90°}{60\ \Omega}$

    Answer: $I_A = 2\ \text{A}\ \angle 90°$

b. Phase B:

Substitution: $I_B = \dfrac{120 \text{ V } \angle 0°}{60 \text{ }\Omega}$

Answer: $I_B = 2 \text{ A } \angle 0°$

c. Total current:

Formula: $I_T = I_A + I_B$

Substitution: $I_T = 2 \text{ A } \angle 90° + 2 \text{ A } \angle 0°$

Intermediate step: $I_T = \sqrt{2^2 + 2^2}$

Answer: $I_T = 2.828 \text{ A } \angle 45°$

2. Three-wire system (shared neutral).

Formula: $I = \dfrac{V}{R}$

a. Phase A:

Substitution: $I_A = \dfrac{120 \text{ V } \angle 90°}{60 \text{ }\Omega}$

Answer: $I_A = 2 \text{ A } \angle 90°$

b. Phase B:

Substitution: $I_B = \dfrac{120 \text{ V } \angle 0°}{60 \text{ }\Omega}$

Answer: $I_B = 2 \text{ A } \angle 0°$

c. Total current:

Formula: $I_T = I_A + I_B$

Substitution: $I_T = 2 \text{ A } \angle 90° + 2 \text{ A } \angle 0°$

Intermediate step: $I_T = \sqrt{2^2 + 2^2}$

Answer: $I_T = 2.828 \text{ A } \angle 45°$

Notice, as seen by these calculations, that the total current is the same with either system. In a two-phase system, if the loads are balanced, as in this example, the total current is equal to 1.414 or $\sqrt{2}$ times one phase current.

25.3 THREE-PHASE SYSTEMS

The universal means of electrical power production and distribution is the **three-phase system.** The basic three-phase generator has three coils, spaced 120° apart, as shown in figure 25-8. Three sine waves are produced. Each has a phase difference of 120°.

When examining the two methods of connecting a generator, wye and delta, you will see that the most distinct advantage of three-phase systems is a reduction in the need for two wires per phase. With three-phases systems, power companies save on the cost of long distance power lines.

The two basic three-phase generators are the **four-wire wye connection,** shown in figure 25-9 and the **three-wire delta connection,** shown in figure 25-10. The loads shown in these two figures can represent a variety of applications.

The relative polarity of the generator coils is critical for proper operation. Note the positions of the + and − signs, indicating polarity.

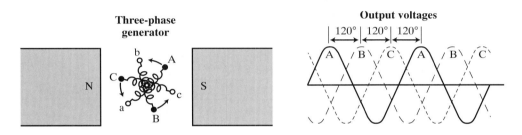

Figure 25-8. Simple three-phase generator. A three-phase generator produces three sine waves 120° apart.

Four-Wire Wye Connected Three-Phase System

In a single-phase system, each line to ground produces 120 volts. A load connected across both lines receives double the voltage, 240 volts. In the **wye (Y) connected three-phase system,** a load connected across two lines does not receive double the voltage. Instead, the load receives $\sqrt{3}$ times the phase voltage. Each line to ground is called the **phase voltage.** Figure 25-9 shows how the load is connected across two coils. The voltages do not simply add because of the phase shifts.

Current in the four-wire wye connected generator adds using vectors. If the loads are equal, each line has an equal current with a phase difference of 120 degrees. The cancellation of vectors results in the current in the neutral wire equaling zero. In a balanced system, it is possible to remove the neutral wire without affecting the performance of the generator. However, it is best if the wye system is operated with four wires. This prevents problems if the system becomes unbalanced. The current in each line is equal to the current in each phase.

The voltage in the wye connected three-phase system can be expressed as:

Formula 25.A

$$V_L = \sqrt{3}\, V_\theta$$

V_L is the voltage across the load, measured in volts.

V_θ is the phase voltage, measured in volts.

Sample problem 2. _____

Calculate the voltage from one line to another of the wye connected generator shown in figure 25-9 if each phase produces a voltage of 120 volts.

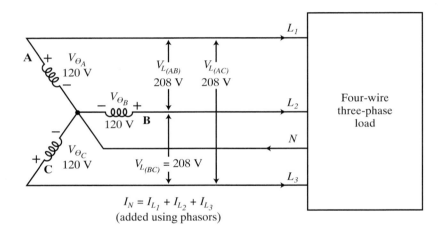

$$I_N = I_{L_1} + I_{L_2} + I_{L_3}$$
(added using phasors)

Figure 25-9. A four-wire three-phase system uses a common neutral in a wye connection. (Sample problem 2)

Formula: $V_L = \sqrt{3}\, V_\theta$

Substitution: $V_L = \sqrt{3} \times 120$ V

Answer: $V_L = 208$ V

Three-Wire Delta Connected Three-Phase System

The **delta (Δ) connected three-phase system** uses only three wires. It has no neutral. The voltage from one line to another is equal to the voltage of one phase. Figure 25-10 shows the phase winding is in parallel with two lines. The current in each line is equal to the $\sqrt{3}$ times the phase current. The phase currents are separated by 120°.

The current in the delta connected three-phase system can be expressed as:

Formula 25.B

$$I_L = \sqrt{3}\, I_\theta$$

I_L is the current across the load, measured in amperes.

I_θ is the phase current, measured in amperes.

Sample problem 3. _____

The delta circuit in figure 25-10 is supplying a balanced load with a current in each phase winding of 15 amps. What is the value of line current?

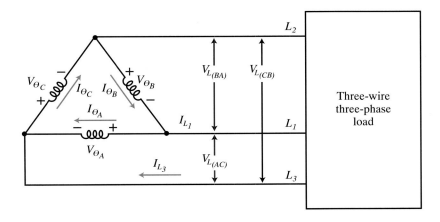

Figure 25-10. The delta connected three-phase system does not use a neutral wire. (Sample problem 3)

Formula: $I_L = \sqrt{3}\, I_\theta$

Substitution: $I_L = \sqrt{3} \times 15\ \text{A}$

Answer: $I_L = 26\ \text{A}$

25.4 THREE-PHASE SOURCE AND LOAD CONFIGURATIONS

There are four possible combinations of source and load connections with the wye and delta systems. These combinations are: Y to Y, Y to Δ, Δ to Y, and Δ to Δ.

There are three possible load configurations:
1.   A three-phase balanced load, such as a three-phase electric motor, where each of the load impedances represents the coils of the motor.
2.   Three independent single-phase loads. All of the loads have the same value, resulting in a balanced load.
3.   Three independent single-phase loads. All of the loads have different values, resulting in an unbalanced load and different line currents.

Y to Y

A Y connected source can be connected to any of the three load configurations. If the load impedances are equal, the currents in each phase and line are equal. The neutral line can be removed allowing the system to become a three-wire. Figure 25-11 shows a Y source connected to a Y load. This figure does not specify load values, therefore the four-wire system is used. The load voltage is equal to $\sqrt{3}$ times the phase voltage.

A very important feature of this Y source connection is the availability of two different voltages. One voltage comes from one phase to neutral. A second voltage can be taken across two phases.

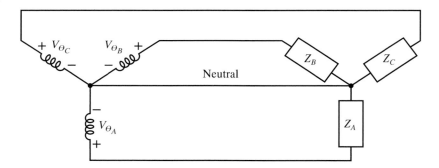

Figure 25-11. Wye source connected to a wye load.

Y to Δ

When the wye source is connected to a delta load, figure 25-12, the neutral wire is not used. This type of configuration does not give the option of using two different voltages.

Δ to Y

A delta source feeding a wye load, figure 25-13, applies each phase voltage across two loads. Each load has the voltage from two phases applied to it. The result of the 120° phase shift is a load voltage equal to the phase voltage divided by $\sqrt{3}$.

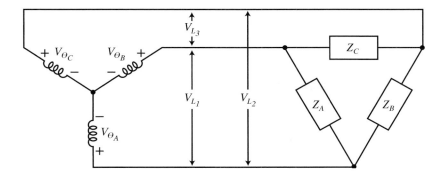

Figure 25-12. Wye source connected to a delta load.

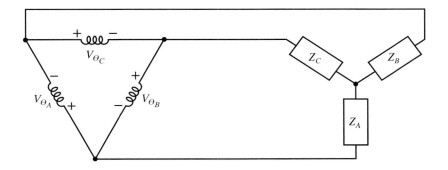

Figure 25-13. Delta source connected to a wye load.

Δ to Δ

With a delta source and delta load configuration, figure 25-14, each load is connected in parallel with a phase voltage. Therefore, the load voltage equals the phase voltage. Each line current is equal to $\sqrt{3}$ times the load current.

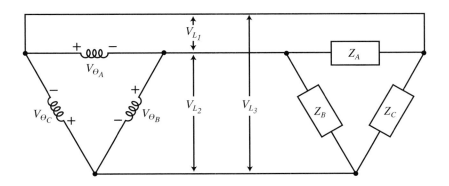

Figure 25-14. Delta source connected to a delta load.

SUMMARY

- Polyphase systems produce multiple sine waves each separated by a fixed phase shift.
- Residential electricity is typically single-phase.
- A single-phase transformer can split the output into two sine waves, each 180° apart.
- A typical two-phase system produces two sine waves with a phase shift of 90°.
- Three-phase systems are the most common form of large scale electrical distribution.

KEY WORDS AND TERMS GLOSSARY

delta (Δ) connected three-phase system: One type of three-phase wiring system that uses only three wires. There is no neutral wire.

four-wire wye connection: A three-phase generator that uses four wires and can produce 3 times the phase voltage.

phase voltage: The voltage between any one line and ground.

polyphase system: An electrical system in which multiple sine waves are produced, with each phase separated by a given number of electrical degrees.

single-phase system: Power system in which only one sine wave is produced.

three-phase system: The universal power distribution system. Three coils, usually 120 degrees apart, produce three sine waves.

three-wire delta connection: A three-phase generator that uses three wires and can produce 3 times the phase current.

two-phase system: Power system with two coils that can produce twice the power of a single-phase system without significantly increasing losses.

wye (Y) connected three-phase system: One type of three-phase wiring system that can use either three or four wires.

KEY FORMULAS

Formula 25.A

$$V_L = \sqrt{3}V_\theta$$

Formula 25.B

$$I_L = \sqrt{3}I_\theta$$

TEST YOUR KNOWLEDGE

Do not write in this text. Please use a separate sheet of paper.

1. Describe how a single-phase transformer can produce a three-wire single-phase output.
2. Describe the typical three-wire single-phase electrical distribution system used in residential applications.
3. Describe a typical two-phase electrical system.
4. Calculate total current in a two-phase system with a load of 40 ohms on each phase. Both phases have an output voltage of 80 volts. Phase A has a phase shift of 90° and phase B has a phase shift of 0°.
5. Describe the general principles of a three-phase system.
6. Draw the connections for a four-wire wye connected three-phase system.
7. Calculate the voltage at the output of a wye connected system if each phase produces 100 volts.
8. Draw the connections for a three-wire delta connected three-phase system.
9. Calculate the current at the output of a delta connected system supplying a balanced load with 20 amps in each phase.
10. Draw the generator to output combinations of wye and delta connections.

Meters, like the one shown here, can measure the current in power lines without the danger of having to connect the meter into the circuit. These meters use the magnetic field around the wire to find circuit values. (Amprobe Instrument)

Chapter 26
Diodes and Power Supplies

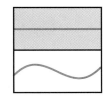

Upon completion of this chapter, you will be able to:
- Define technical terms related to semiconductors.
- Identify the characteristics of P-type and N-type semiconductor materials.
- Describe the current in a semiconductor.
- Use a series of diagrams to demonstrate the operation of a junction diode.
- Describe typical applications for diodes.
- Identify how diodes are rated.
- Test a diode with an ohmmeter.
- Describe the operation of various types of diode circuits.
- Describe the operation of rectifier circuits.
- Calculate the dc output voltage of a rectifier circuit.
- Describe the operation of a simple capacitive power supply filter.
- Calculate the output voltage of a power supply with a capacitive filter.
- Calculate output voltage ripple.
- Recognize the schematic diagrams of voltage multipliers.
- Describe the function and operation of a zener diode.
- Calculate the circuit values of a zener regulator.
- Identify other zener diode circuits.

All electronic equipment is powered by some sort of power supply. The power can come from small batteries or from a cord plugged into a wall socket. Either way, most electronic circuitry needs a dc voltage to operate. A battery supply is adequate as it is. Equipment plugged into a wall socket needs the voltage changed from ac to dc. **Rectification** is the process of changing ac to dc. Rectification and the semiconductor devices that do this job are explored in this chapter.

26.1 SEMICONDUCTORS

Semiconductors are materials with an electrical conductivity that falls between conductors and insulators. They are useful because they are nonlinear devices. A *nonlinear* device is a device that has an unequal amount of resistance over a range of applied voltages.

Semiconductor diodes are used as rectifiers, voltage limiters, and voltage regulators. Although not discussed in this book, the vacuum tube was the predecessor of the semiconductor. The vacuum tube is larger and more fragile than semiconductor diodes. Vacuum tubes also need to be heated, so they are much less efficient than semiconductor diodes.

The semiconductor has been able to revolutionize the electronics industry. Components are available in extremely small packages. In addition, semiconductors are very efficient. They use very little electricity.

Valence Electrons

An atom is made up of a nucleus surrounded by electrons orbiting in shells. The outermost orbit is called the **valence shell.** Electrons in this shell are called **valence electrons.** Valence electrons, figure 26-1, are the ones of primary interest in electronics.

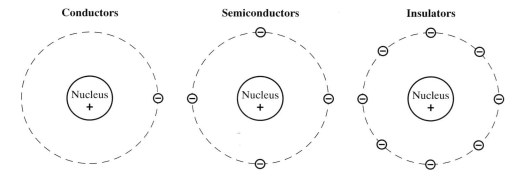

Figure 26-1. Electrons in the valence band of different types of materials.

Electrons are held in orbit around the nucleus by a force similar to the forces holding the planets in orbit around the sun. The valence shells of the atoms in figure 26-1 have a maximum of eight electrons. The electrons in a full shell have a much stronger bond than a single electron occupying a shell.

Figure 26-1 shows the valence bands for a conductor, semiconductor, and insulator. Conductors have only one valence electron, which easily moves out of orbit. This makes it readily available for current flow.

Semiconductors have four electrons, making them fairly stable. This produces some resistance to current. The most commonly used materials for semiconductors are germanium and silicon.

An insulator has a full valence band. It is extremely difficult to break an electron free from an insulator. They offer the maximum possible resistance to electrical current.

Doping

Doping is the process of adding an impurity to semiconductor material. This impurity changes the semiconductor to either a P-type or an N-type. A doped semiconductor is classified as **extrinsic.** A pure semiconductor is classified as **intrinsic.**

P-Type Semiconductor

A **P-type semiconductor** is doped with an impurity that has three valence electrons. Gallium (Ga), figure 26-2, is often used. The **trivalent impurity,** an impurity with three electrons in the valence shell, forms covalent bonds with the semiconductor. With **covalent bonds** the valence electrons of two atoms are shared. When doping a semiconductor with a trivalent impurity holes are produced.

Holes occur in atoms that are missing an electron in their valence bands. They carry a positive charge. Holes are the majority current carrier in P-type semiconductors. A **majority carrier** is used as the primary source of current within a semiconductor.

P-type semiconductors also have a small number of free electrons unable to make a covalent bond. These electrons are **minority carriers.**

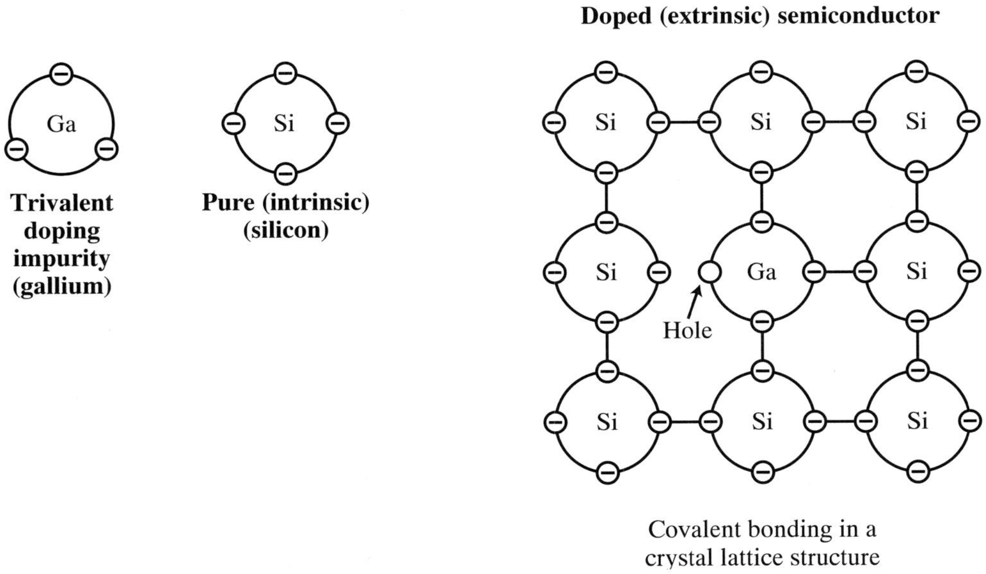

Figure 26-2. P-type semiconductor material has a shortage of electrons, called holes.

N-Type Semiconductor

An **N-type semiconductor** is doped with a **pentavalent impurity,** such as Arsenic, which has five electrons in the valence band. When pentavalent impurities form covalent bonds with a semiconductor material, there are extra electrons that are unable to find an orbit. See figure 26-3. The **free electrons** are used as majority current carriers in the N-type semiconductor. The minority carriers in an N-type semiconductor are holes.

Extrinsic (doped) semiconductor

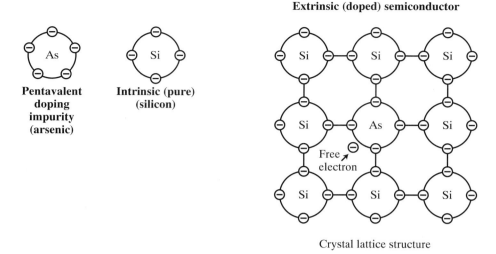

Crystal lattice structure

Figure 26-3. N-type semiconductor material has unshared, free electrons.

Current in Semiconductors

Current flowing through a semiconductor is regulated by the conditions in the circuit it is connected to, called the external circuit. A semiconductor offers resistance to a current by causing a voltage drop. However, the voltage drop remains the same regardless of the amount of current flowing. Compare this to a resistor, where the voltage drop increases and decreases with an increasing and decreasing current.

The current running through the semiconductor is made up of majority current carriers. Certain conditions in diodes and transistors allow some small amounts of minority current to flow. In figure 26-4, a P-type semiconductor is connected to a battery. The series resistor is used as a current limiter.

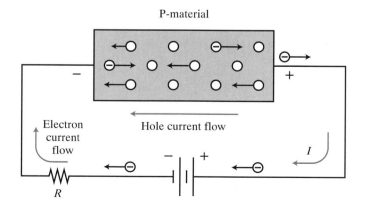

Figure 26-4. Holes are the majority current carriers through the P-type material. Electrons are minority current carriers.

Current in the external circuit is made up of electrons flowing from the battery. As the electrons enter the P-material, they fill the vacant holes. Keep in mind, a complete circuit requires that an equal number of electrons return to the battery. Therefore, for every electron filling a hole on the negative side, an equal number of electrons leave from the positive side of the P-semiconductor.

The polarity of the battery also has an attraction/repulsion effect on the holes and electrons in the semiconductor. The holes, carrying a positive charge, are attracted to the negative terminal. The electrons, having a negative charge, are drawn towards the positive terminal. As a result, hole flow through the P-semiconductor is from positive to negative.

The N-type semiconductor material, shown in figure 26-5, has a majority of electrons. When connected to the battery, the electrons flow from the negative terminal to the semiconductor. Electrons are forced in one side and out the other. The negative and positive polarities cause an attraction/repulsion promoting current flow.

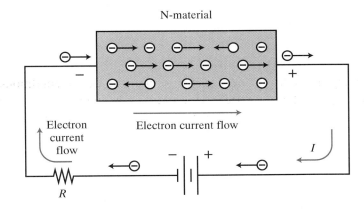

Figure 26-5. Electrons are the majority current carriers through N-type semiconductor material. Holes are minority carriers.

26.2 PN JUNCTION DIODES

The joining of a P-type semiconductor with an N-type semiconductor forms a **diode.** A diode is a commonly used electronic component. The schematic symbol of a diode has the arrow pointing at the N-material. The N is the **cathode** and the P is the **anode.** The arrow in semiconductor symbols points in a direction opposite to the flow of electrons.

A diode can be operated in three possible states: no voltage applied, forward bias, and reverse bias.

No Voltage Applied

Figure 26-6 shows the PN junction diode connected to a circuit with the switch open. No voltage is applied. The point at which the P-type and N-type materials meet is the **junction.** At this junction, the holes from the P-material and the electrons from

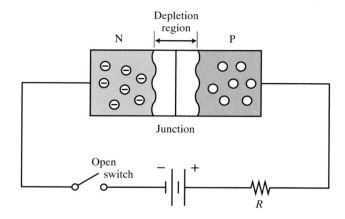

Figure 26-6. With no bias applied to a PN junction diode, the diode remains in a static state.

the N-material cross. The holes and electrons combine to eliminate the current carriers. The elimination of current carriers around the junction produces a **depletion region.** In the depletion region, free current carriers cannot cross. With no voltage applied, a diode remains in this static state.

Forward Biased Diode

A **forward biased** diode allows majority current to flow. The term **bias,** in reference to semiconductors, can be defined as the voltage applied. To forward bias a diode, a negative voltage is applied to the N-material and a positive voltage is applied to the P-material. See figure 26-7.

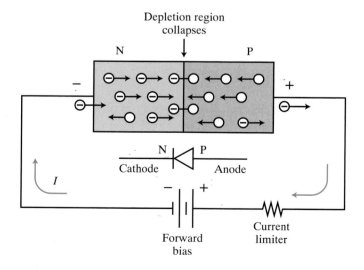

Figure 26-7. Current flows in a forward biased diode like a closed switch.

A negative voltage applied to the N-material repels the electrons, pushing them towards the junction. A positive voltage on the P-material repels the holes, pushing them towards the junction. With current carriers pushed into the junction, the depletion region collapses. This allows the current carriers, holes and electrons, to easily cross the junction and flow to the opposite terminal.

Collapse of the depletion region results in a voltage drop across the junction. This voltage drop is called the **barrier potential.** Once the applied voltage is increased enough to overcome the potential of the depletion region, the diode voltage drop will remain almost constant even if a large voltage is applied to the circuit. In forward bias, current flows in the circuit. All the voltage is dropped across the load, except for the small barrier potential.

The characteristic curves for germanium and silicon, the most widely used semiconductor materials, are shown in figure 26-8. Forward current is plotted vertically. Forward voltage is plotted horizontally. As the voltage across the diode is increased from zero, only a slight amount of current flows. The sharp bend in the curve occurs at a value equal to the barrier potential. For germanium (Ge) the barrier potential is between 0.2 V and 0.3 V. For silicon (Si), the barrier potential is between 0.5 V and 0.7 V.

When the voltage is increased beyond the barrier potential, there is a sudden increase in the current. As the voltage is increased, there is a large change in current with only a very slight change in voltage drop across the diode. The voltage in excess of 0.2 V (Ge) or 0.6 V (Si) is dropped across the other resistances of the circuit. A forward biased diode acts like a closed switch.

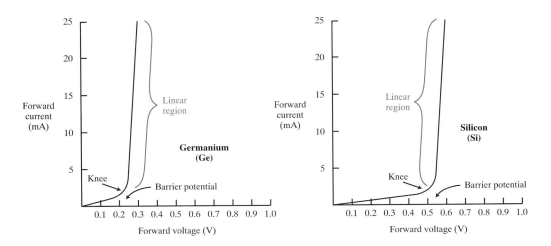

Figure 26-8. Forward characteristic curves for germanium and silicon diodes.

Reverse Biased Diode

A **reverse biased** diode acts like an open switch. In reverse bias, positive is connected to the N-material and negative to the P-material. Majority carriers are attracted to the voltage source. This expands the depletion region. The depletion region then acts like an insulator, allowing almost no current to flow. Figure 26-9 is a block diagram showing the reverse voltage applied to a PN junction diode.

Minority current carriers, electrons in the P-material and holes in the N-material, are drawn across the junction. However, minority current is extremely small. The current is small enough to be ignored except in special applications. Examine the characteristic curve in figure 26-10. A large reverse voltage allows only a slight current.

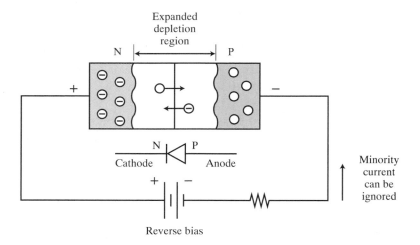

Figure 26-9. With a reverse bias applied to a PN junction, diode current is so small it is considered zero. The diode acts like an open switch.

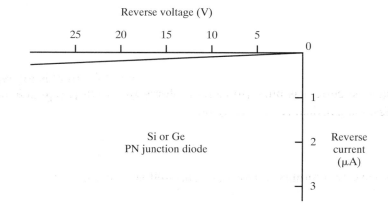

Figure 26-10. Reverse characteristic curve for a PN junction diode shows a very small minority current.

26.3 TYPICAL DIODES

There are two typical categories of applications. Diodes can act as a one-way switch, allowing current in only one direction. This is generally called rectification. Diodes also can be used to produce a voltage reference. Diodes produce a constant voltage drop of either 0.2 V (Ge) or 0.6 V (Si). In this application, the diode can clip off portions of a sine wave.

Diodes, as components, have two leads. One is the anode and the other is the cathode, figure 26-11. The leads of the diode can not be interchanged. Therefore, there is always some form of marking on a diode to indicate the correct polarity. Most diodes have a band on one end. The band end is the cathode, with the arrow of the schematic symbol pointing at the band.

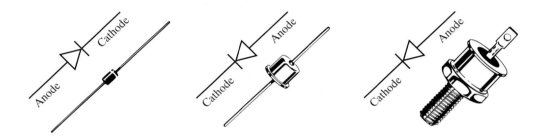

Figure 26-11. Diodes come in many different packages. A larger package can handle more current.

Characteristic Curves of Typical Diodes

Figure 26-12 combines the forward and reverse characteristic curves of the PN junction diode. This example is for a silicon diode. Forward current is majority current carriers. Reverse current is minority carriers. Forward current is typically many thousands of times larger than reverse current.

Ratings of Diodes

Diodes have two ratings, forward current and reverse voltage. Reverse voltage is also called PIV for peak inverse voltage. Both of these rating are the maximum allowable values for the diode. When considering what is the proper rating of a diode for your circuit, consider the maximums for that particular circuit. Many designers add a safety factor to their maximum calculations, such as an extra 20 percent.

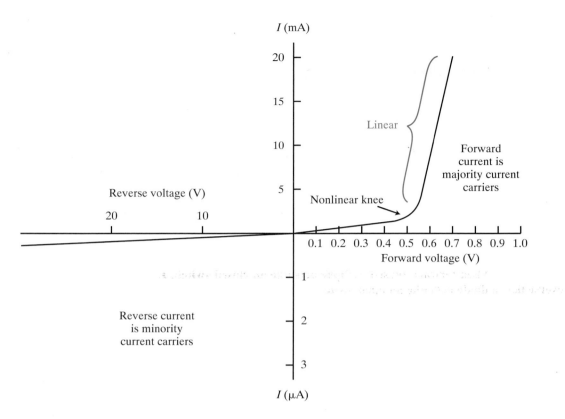

Figure 26-12. Typical volt-ampere characteristic curve of a PN junction diode.

Diode as a One-Way Switch

With a dc power source connected, a diode either conducts well or it does not conduct at all. Using a voltmeter in the circuit demonstrates the operating characteristics of a diode. Figure 26-13 compares the differences between a diode that is forward or reverse biased.

The forward biased diode drops 0.6 volts (Si) and the remaining supply voltage is dropped across the load. This circuit should be treated like a dc circuit when calculating the current and voltage drops. Notice that 0.6 V is the most commonly used value for a forward diode drop. Silicon is the most commonly used diode.

The reverse biased diode acts like an open switch. It has an infinite resistance. All of the supply voltage is dropped across the infinitely high resistance of the diode. There is no current, and no voltage is dropped across the load.

Testing a Diode with an Ohmmeter

When using an ohmmeter to test semiconductors, the polarity of the ohmmeter is very important. The + and – marked on the meter refer to the polarity of the voltmeter. The batteries in an ohmmeter are frequently connected with their negative to the +

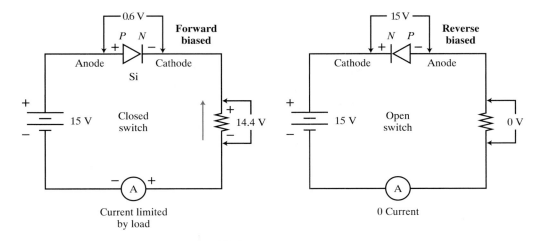

Figure 26-13. When forward biased, a diode acts like an closed switch. In reverse bias, a diode acts like an open switch.

meter jack and their positive to the − meter jack. The ohmmeter polarity can be checked by placing the meter on ohms and connecting a voltmeter to measure the ohmmeter's voltage. Observe the polarity of the voltmeter to determine the polarity of the ohmmeter.

An ohmmeter is quite useful in testing a diode. By measuring the forward and reverse resistance, the ohmmeter identifies the anode and cathode. Also, the meter will indicate if the diode is normal or defective.

Use figure 26-14 as an example. Notice how the leads of the ohmmeter are interchanged, switching the polarity. The ohmmeter's internal battery applies a voltage to the diode sufficient to forward or reverse bias the diode.

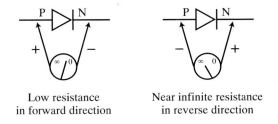

Low resistance in forward direction   Near infinite resistance in reverse direction

Figure 26-14. Testing a diode with an ohmmeter.

When the circuit is forward biased, the ohmmeter measures a fairly low resistance, but not zero. The barrier potential voltage appears as a resistance of a few hundred ohms on the meter. The actual amount of resistance is not important, just as long as there is some. If the diode does not have any resistance in the forward direction, it is a short circuit and considered useless. If a near infinite resistance is measured, the diode is defective (or you mistakenly have the diode in reverse bias.)

Switching the leads of the ohmmeter applies a reverse bias to the diode. The meter should now read infinity. If the actual resistance measured is considerably less than infinity, the diode is defective. Again, if a high resistance is measured in both directions, forward and reverse, the diode is an open circuit and useless.

26.4 DIODE CIRCUITS

Later in this chapter the diode is used as a rectifier to change ac into dc voltages. This section examines four applications of the diode.
- The diode as a one-way switch in dc steering circuits.
- The diode as a one-way switch in ac steering circuits.
- The diode as a voltage reference in clipping circuits.
- The diode as a voltage reference in clamping circuits by combining the diode with a capacitor.

DC Voltage Diode Steering Circuit

The diode is useful as a one-way switch for selecting which load is used depending on the polarity of the applied dc voltage. It can also be used in series with power supply wires to prevent accidental connection to the wrong polarity.

Figure 26-15 shows a steering circuit and its response to a change in the battery's polarity. With positive connected to the top of the circuit and negative to the bottom, diode D_1 is forward biased, allowing load #1 to receive voltage and current. In this case load #2 is prevented from operating. When the battery polarity is switched, with negative at the top, load #1 is turned off and load #2 is turned on.

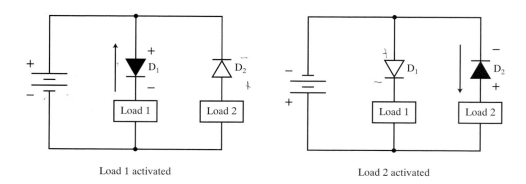

Load 1 activated   Load 2 activated

Figure 26-15. Polarity of the battery determines which diode switch will activate a load.

AC Voltage Diode Steering Circuit

Figure 26-16 is a steering circuit with an ac voltage applied. In this application, the diodes are used in combination with switches to select lamp A, lamp B, or both.

When both switches are open, both lights are off and D_2 and D_3 are effectively removed from the circuit. D_1 and D_4 are pointing in opposite directions. This prevents them from both being forward biased at the same time. The diodes alternately reverse bias as the ac sine wave changes from positive to negative. Current cannot flow unless both diodes are forward biased.

When switch S_1 is closed and S_2 is open, lamp B is on and lamp A off. Closing S_1 places D_2 in the circuit. D_2 faces the same direction as D_4. Both are forward biased during one half of the ac cycle. The current runs through lamp B and diodes D_2 and D_4. Lamp A is blocked from operation.

With S_2 closed and S_1 open, lamp A is on and lamp B off. Diodes D_3 and D_1 both face the same direction, allowing current to flow through lamp A during one half of the cycle.

Closing both switches turns both lights on, each for opposite half cycles of the input sine wave. As the polarity of the ac source switches between positive and negative, the diode pairs turn on and off. Note that the lights appear to be connected in series when they are both on. If this were the case, when only one light is on the brilliance will be twice as much as when both lights are on. However, the diodes cause the lights to take turns being on. Each lamp is on for one half of the ac cycle.

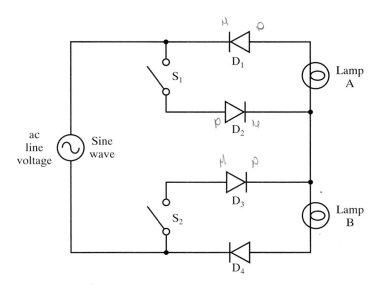

Figure 26-16. This diode steering circuit allows the selection of either or both lamps.

Voltage Reference in a Clipper Circuit

When forward biased, a diode has a constant voltage of 0.6 volts for silicon and 0.2 volts for germanium. This constant voltage drop can be used as a voltage reference when it is placed in parallel with a load. The load voltage cannot exceed the diode voltage. This **clipper circuit** clips off part of a sine wave signal making it look more like a square wave.

In figure 26-17 an ac voltage is applied to a circuit where it is desired to have a square wave of 0.6 volts peak. During the positive half cycle, D_1 is turned on resulting in a voltage drop of 0.6 volts. The remainder is dropped across series resistor R_1. The positive portion of the sine wave is clipped, shown with dotted lines. During the negative half cycle D_2 turns on, clipping the negative portion to a peak of 0.6 volts.

Diodes can be added in series, as shown in figure 26-18 to increase the reference voltage to a desired amount. This figure uses two series diodes for each half cycle. This produces peak values of 1.2 volts.

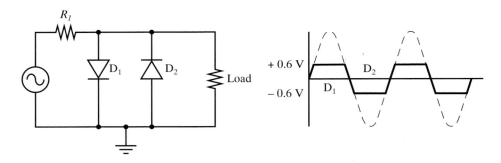

Figure 26-17. This voltage clipper circuit limits the voltage to the load to the diode threshold of 0.6 volts.

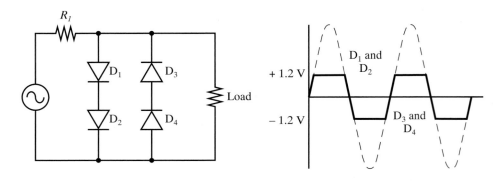

Figure 26-18. Diodes added in series increase the voltage across the load in increments of 0.6 volts.

Later in this chapter, zener diodes, which have a higher reference voltage, are used in place of rectifier diodes. The advantage of a using zener diode is that a wide selection of reference voltages are offered.

Diode Clamper Circuits

A **clamper circuit** changes the centerline of an ac signal to some dc voltage level. Figures 26-19 and 26-20 are examples of clamper circuits. The diode can be connected to either raise the centerline to a positive level or lower the centerline to a negative level.

Figure 26-19 is a positive clamper. During the negative half cycle the diode is forward biased, allowing it to charge the capacitor with the polarity as shown. During the positive half cycle the capacitor voltage is added to the ac signal. If a long time constant is used for the capacitor-load combination, the capacitor will act like a battery. An equivalent circuit with the capacitor and diode replaced by a battery is shown for comparison. This battery circuit is not an exact substitution. It does not account for the 0.6 volts dropped across the diode.

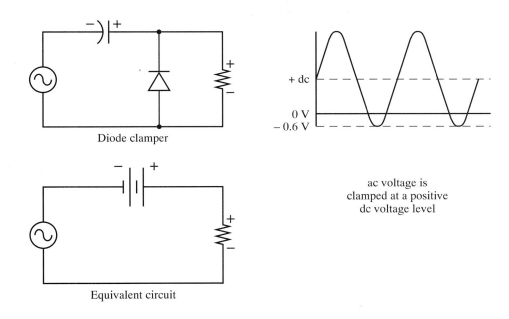

Figure 26-19. Positive diode clamper circuits lift an ac input above ground.

Figure 26-20 is a negative clamper. The positive half cycle charges the capacitor with the polarity shown. The capacitor discharges during the negative half cycle, adding to the ac voltage. This results in dropping the center line to a negative value.

With both of these clampers, the dc voltage is limited by the 0.6 volt drop across the diode. The output sine wave still has a portion equal to the diode drop across the zero volt line.

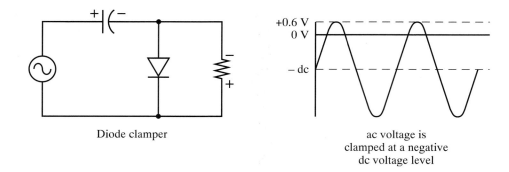

Diode clamper

ac voltage is
clamped at a negative
dc voltage level

Figure 26-20. Negative diode clamper circuits lower ac inputs below ground.

26.5 HALF-WAVE RECTIFIER

The function of a rectifier circuit is to change an ac sine wave into a dc voltage, a signal with all of the voltage having the same polarity. A **half-wave rectifier** blocks half of the ac cycle, while passing the other half.

Figure 26-21 shows a diode connected in a circuit so as to allow the positive half cycle to pass. Observe the polarity across the diode, with positive on the anode and negative on the cathode. The diode drops 0.6 volts across it. The remainder is dropped across the load. Note the polarity of the load. Positive is at the top and negative is at the bottom. The waveform diagram shows the positive half cycle as a pulsating dc.

Figure 26-22 shows the same circuit as figure 26-21. However, the negative half cycle is shown here. The negative half cycle places a negative on the anode of the diode.

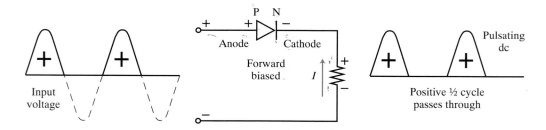

Input
voltage

Forward
biased

Pulsating
dc

Positive ½ cycle
passes through

Figure 26-21. Half-wave rectifier passes half of the sine wave input voltage.

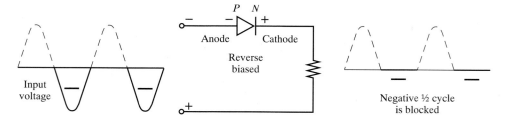

Input
voltage

Reverse
biased

Negative ½ cycle
is blocked

Figure 26-22. Half-wave rectifier blocks half of the sine wave input voltage.

This means that the diode is reverse biased during the negative half cycle. The waveform diagram shows that the negative half cycle is blocked.

Direction of the Diode Determines the Output Polarity

By changing the direction of the diode in the circuit, the output voltage can be either positive or negative. Figure 26-23 shows a positive half-wave rectifier. The waveform diagram compares the input sine wave to the output pulsating dc.

Figure 26-24 shows a negative half-wave rectifier. By turning the diode to face the opposite direction, the output is negative. Notice that in the waveform diagram the positive half cycle is blocked while the negative is passed.

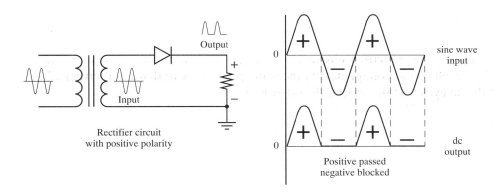

Figure 26-23. Diode polarity determines which half of the wave passes through the half-wave rectifier. This circuit is set to pass the positive half of the input voltage.

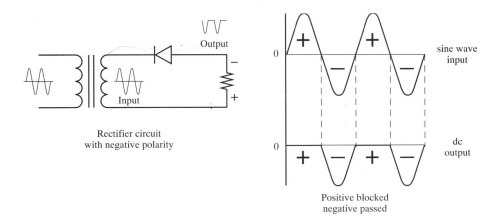

Figure 26-24. This circuit is set to pass the negative half of the input voltage.

Combination Positive and Negative Rectifier Circuit

Connecting two rectifier circuits to the same input sine wave produces two separate outputs. Figure 26-25 shows a dual half-wave rectifier circuit. One output produces a positive output dc voltage. The other produces a negative dc voltage. The two loads are also independent.

Later in this chapter, full-wave rectifier circuits are discussed. The full-wave circuit combines two half-wave rectifiers in a different way than shown here. The full-wave circuit has a *common* load and both of its half-wave circuits produce the *same* polarity of output voltage.

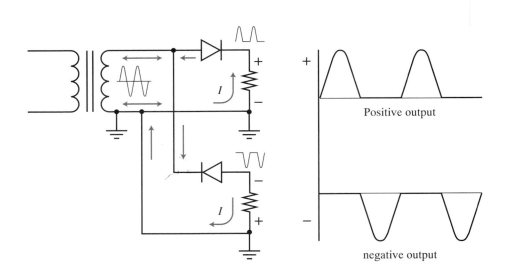

Figure 26-25. The dual half-wave rectifier produces both positive and negative dc voltage outputs.

Practical Half-Wave Rectifier Circuit

The half-wave circuit has many applications. In AM radio circuits, it is used as an audio detector. The half-wave rectifier circuit is most often used to convert the 120 volt ac to a dc voltage needed to operate electronic equipment. The half-wave circuit can be connected directly to the ac line or through a transformer.

The relationships of rms values, average values, and peak values of sine waves are all used when calculating the performance of power supplies. Review Chapter 15, if necessary, for the methods of converting from one value to another. The 120 volts available at a wall socket is expressed in rms. Also, an ac voltmeter reads in rms. An oscilloscope reads in peak or peak-to-peak. The peak value is the best to use with rectifier circuits. The diode passes an entire peak. Calculations for the dc output are from the average and are based on peak values.

Figure 26-26 diagrams the process of going from the ac line voltage to a load that needs a dc voltage. First, the 120 volts ac input is applied to a 10:1 step-down transformer. The secondary of the transformer produces 12 volts rms (17 V peak). Next, the reduced ac is the input to a diode. There is 0.6 volts dropped across the diode, leaving the load to receive 16.4 volts peak pulsating dc.

Calculating DC Voltage Output of a Half-Wave Rectifier

A dc voltmeter connected across a load measures the average of the pulsating dc voltage coming out of a half-way rectifier. To calculate this dc voltage, convert the peak value to the average value and divide by two. You divide by two because only one half of the sine wave is used.

Formula 26.A

$$V_{dc} = \frac{\text{peak}}{2} \times 0.636 = \text{peak} \times 0.318$$

V_{dc} is the average voltage coming from a half-wave rectifier, measured in volts.

peak is the peak input voltage, measured in volts.

Sample problem 1.

Calculate the voltage that will be measured by a dc voltmeter across the load in figure 26-26.

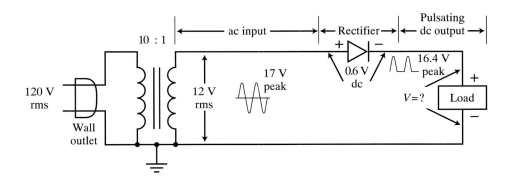

Figure 26-26. Circuit for a practical half-wave rectifier. (Sample problem 1)

Formula: $V_{dc} = \dfrac{\text{peak}}{2} \times 0.636$

Substitution: $V_{dc} = \dfrac{16.4 \text{ V}}{2} \times 0.636$

Answer: $V_{dc} = 5.2 \text{ V}$

26.6 FULL-WAVE RECTIFIER

The **full-wave rectifier** uses two diodes to pass both halves of a sine wave. One of the halves is flipped over so that both halves have the same polarity. The output is a dc voltage with twice the value of a half-wave rectifier. Most full-wave rectifier circuits use two diodes and require a center-tapped transformer. The center-tap of the transformer is a zero volt reference. One diode is connected to each side of the transformer, as shown in figure 26-27. Each diode operates as a half-wave rectifier. Each diode passes one half of the sine wave. Their outputs are connected together and applied to the load.

The transformer has a 180° phase shift at opposite ends of the secondary. The diodes, connected as shown in the figure, each require a positive voltage from the sine wave to be forward biased. Since the polarities of the top and bottom of the transformer are opposite, when one side is in the positive half cycle, the other is in the negative half cycle.

As a result of the 180° phase shift, the diodes take turns operating. Each diode passes a positive half cycle. This results in two positive half cycles combined at a common point. The pulsating dc output has twice as many pulses as the signal emerging from a half wave circuit.

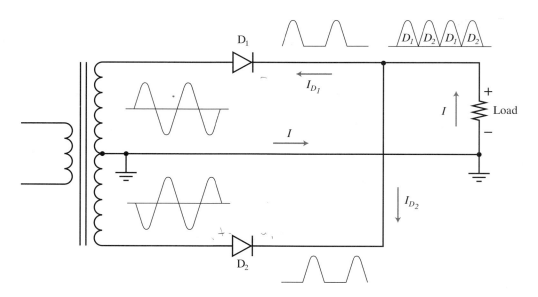

Figure 26-27. A full-wave rectifier circuit combines two half-wave circuits for a continuous pulsating output.

A Practical Full-Wave Rectifier

A typical application of a full-wave circuit will have it connected to a 120 volt line. The circuit shown in figure 26-28 uses a step-down transformer. The output voltage is 20 volts rms (28.3 V peak) between the center-tap and either side.

With 0.6 volts dropped across the diode, the peak value is 27.7 volts. The two peaks combine to form a waveform with a positive peak for every half cycle of the sine wave. The peak voltage remains the value of one peak.

Calculating DC Voltage in a Full-Wave Rectifier

A dc voltmeter connected across the output of a full-wave rectifier measures the average value. The full-wave rectifier output has as many peaks as the input sine wave, so the average value is calculated as it would be for a sine wave. The dc voltage at the load is twice the value of a half-wave circuit.

Formula 26.B

$$V_{dc} = \text{peak} \times 0.636$$

V_{dc} is the average voltage coming from a full-wave rectifier, measured in volts.

peak is the peak input voltage, measured in volts.

Sample problem 2.

Determine the dc voltage measured at the load of the full-wave rectifier circuit in figure 26-28.

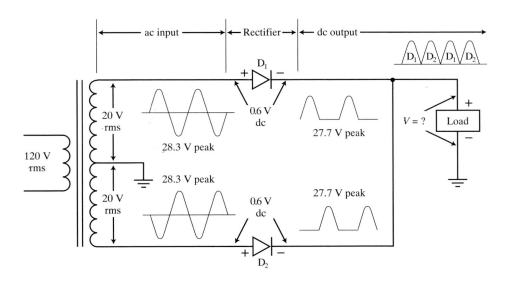

Figure 26-28. Circuit for a practical full-wave rectifier. (Sample problem 2)

Formula: $V_{dc} = \text{peak} \times 0.636$

Substitution: $V_{dc} = 27.7 \text{ V} \times 0.636$

Answer: $V_{dc} = 17.6 \text{ V}$

26.7 BRIDGE RECTIFIER

The **bridge rectifier** circuit has a full-wave rectifier output waveform. The bridge rectifier uses four diodes and does *not* require a center-tapped transformer. The rectifier can be connected directly to the ac line in the same manner as the half-wave circuit. Eliminating the cost of the transformer can be a significant savings.

Refer to the schematic shown in figure 26-29. When the input sine wave is in its positive half cycle, the top of the circuit (labeled ac_1) has a positive polarity. The bottom of the circuit (ac_2) has a negative polarity. Diodes D_1 and D_3 are turned on (forward biased) during the positive half cycle. Diodes D_2 and D_4 are turned off (reverse biased) at this time.

When the ac sine wave is in the negative half cycle, the ac_2 input has a positive polarity and ac_1 is negative. Diodes D_2 and D_4 are turned on and D_1 and D_3 are turned off.

The + and – polarities labeled on the bridge rectifier indicate the polarity of the dc output voltage. *Regardless of which half cycle the input sine wave is in, the dc voltage maintains the same polarity.*

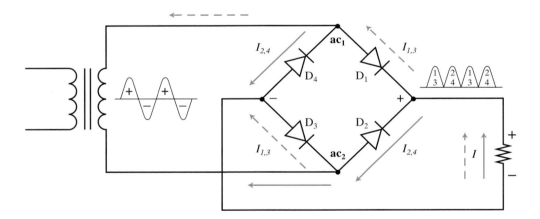

Figure 26-29. Bridge rectifier circuit produces a full-wave output using four diodes. Two diodes conduct each half cycle.

Bridge Rectifier As a Single-Unit Component

Many electronic circuits that require no modification for a wide range of applications are manufactured as a single-unit package. The bridge rectifier is such a circuit.

Figure 26-30 shows the bridge rectifier as a single-unit component. It is available in different package styles. The wattage rating has the greatest influence on the package size and shape. The wattage rating is necessary because of the voltage dropped across it. The current to the load produces heat, which is destructive to electronic components. Semiconductors are often very sensitive.

Figure 26-30. Bridge rectifiers are available as single-unit components.

The bridge rectifier has four leads, each having a specific label. Two leads are labeled as the ac input and two are labeled as the +/– dc output. The leads are not interchangeable.

Practical Bridge Rectifier

The 120 volt line can be connected directly to a bridge rectifier, figure 26-31. Approximately 1.2 volts is dropped across the rectifier circuit. The dc output voltage is calculated using the same formula as the full-wave circuit.

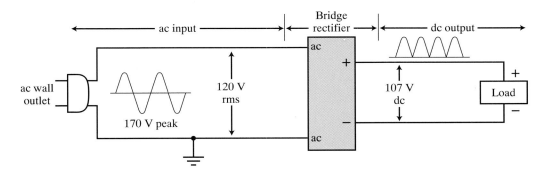

Figure 26-31. Circuit for a practical bridge rectifier.

Extreme caution must be exercised when using test equipment, especially an oscilloscope, to measure the bridge rectifier. The oscilloscope lead is grounded. It must not be connected in the bridge rectifier if there already exists an ac ground. For example, figure 26-32 shows a transformer where the primary and secondary share a common ground. The ground is also connected to one ac input. When the oscilloscope is connected at the output, its ground places another ground at the negative lead. With this connection D_4 has one side connected to ac and the other to ground without any load resistance to limit the current. This will destroy the component. A problem with grounding can also occur when connecting directly to the ac line.

The use of a center-tapped transformer, as shown in figure 26-33, may also include a grounding problem. Some circuits connect the center-tap to ground. If the bridge is also connected to the center-tap, which may be the case, the ground causes problems with test equipment.

One safe method of connecting a bridge rectifier circuit is to use an isolation transformer, figure 26-34. This type of transformer can have any turns ratio, including 1:1.

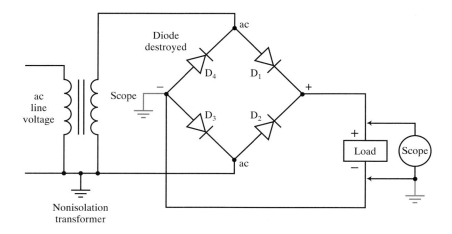

Figure 26-32. Danger. Oscilloscope ground will short D₄ when using a nonisolation transformer and an ac ground already exists.

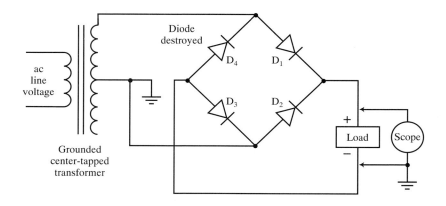

Figure 26-33. Danger. In some circuits the center tap is connected to ground.

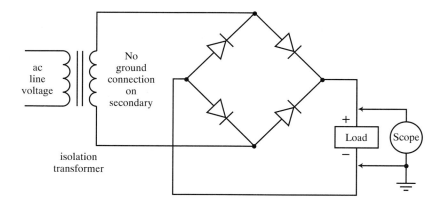

Figure 26-34. An isolation transformer makes it safe to connect an oscilloscope.

The secondary does not share a common connection with the primary. When the oscilloscope ground is connected in the circuit, it will not short the diodes.

Using an Ohmmeter to Test a Bridge Rectifier

The ohmmeter is an effective tool for testing to see if a bridge rectifier is defective. Using figure 26-35 as a guide, measure each diode separately. Next, measure the component as a complete unit. Remember, the ohmmeter must not be connected in a circuit with voltage present.

To test each diode independently, part A, connect one side of the ohmmeter from one ac input to one dc output. If the polarity of the ohmmeter forward biases the diode, there is a low resistance reading. Switch the ohmmeter leads to change polarity. The diode should now be reverse biased, indicated by infinite resistance.

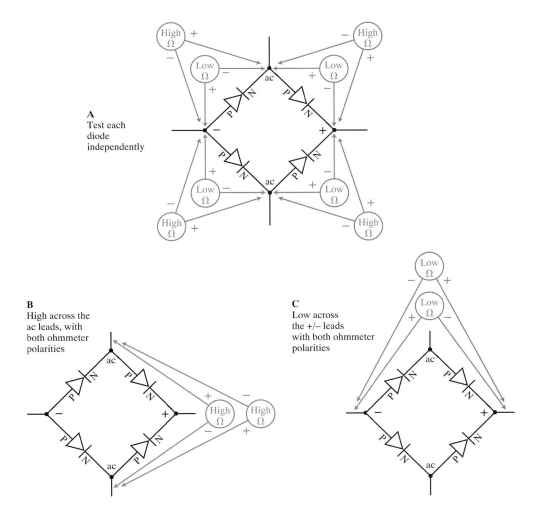

Figure 26-35. Testing a bridge rectifier with an ohmmeter.

Check each of the four diodes, checking both forward and reverse resistance. A faulty diode is indicated by either low resistance in both directions or high resistance in both directions.

Then, using part B of figure 26-35 as a guide, measure the resistance from one ac input to the other. Switch the meter leads to check in both directions. The resistance should be infinity in both directions.

Use part C as a guide to check between the + and − leads of the bridge. The resistance should be low in both directions. Note that this resistance is the measurement of two diodes.

26.8 POWER SUPPLY FILTERS

The purpose of a power supply filter is to smooth the pulsating dc produced by a rectifier. An ideal dc voltage is perfectly flat, a perfectly constant voltage. A good battery is an example of a perfect dc voltage.

Filter circuits were discussed in Chapter 24. A power supply needs a low pass filter to remove the ac fluctuations while passing a constant dc voltage. For simplicity, this chapter uses only single capacitive filters.

Block Diagram of a Power Supply

Figure 26-36 shows a block diagram of an unfiltered power supply. It has an ac sine wave input, a rectifier circuit, and a load that uses the dc voltage. Figure 26-37 adds a capacitor to the diagram. The capacitor is placed in parallel with the output of the rectifier circuit, which is also in parallel with the load.

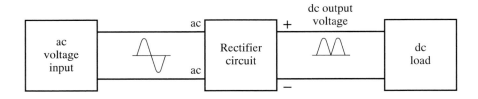

Figure 26-36. Block diagram of an unfiltered power supply. The output is a pulsating dc.

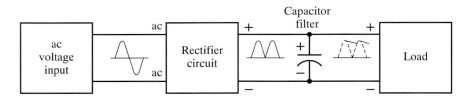

Figure 26-37. Block diagram of a power supply with a simple filter. The capacitor smooths the fluctuations in the pulsating dc.

Capacitor Filter Smooths the Pulsating DC

The rectifier circuit charges the capacitor filter to the peak value of the pulsating dc voltage, figure 26-38. Then, as the voltage pulse rounds the peak, the capacitor begins to discharge. This discharge fills in the voltage being supplied to the load.

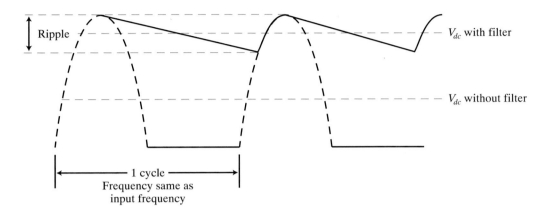

Figure 26-38. Filtering a half-wave rectifier output smoothes the dc output voltage.

The amount of the capacitor filter discharge is dependent on the RC time constant between the capacitor and the load resistor. The capacitor voltage remains higher (producing a smoother dc signal) with a larger value capacitor or with a lower load current.

A full-wave rectifier, including the bridge rectifier, has peaks occurring twice as often as the half-wave rectifier. Therefore, as shown in figure 26-39, the capacitor does not have to supply voltage for as long a period of time. The end result is a smoother dc voltage.

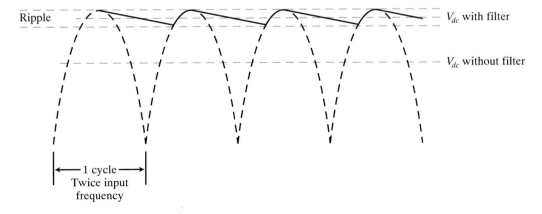

Figure 26-39. Full-wave rectifier output is smoothed by filtering. V_{dc} is increased.

DC Voltage Calculations

The dc voltage in a filtered power supply is very difficult to calculate accurately because it changes with the load conditions. If the load current is very small, the capacitor discharges only slightly. This causes the output voltage to be essentially equal to the peak value. As the load current increases, a greater demand is placed on the capacitor and the output voltage drops somewhat.

As an approximation, the no-load dc voltage is equal to the peak value. This applies to a filtered power supply, either half-wave or full-wave. A loaded circuit has a dc voltage between the peak value and the dc value without a filter. These formulas can be written as follows:

Formula 26.C

$$V_{dc} = \text{peak voltage}$$

V_{dc} is the average voltage coming from a filtered power supply under no-load conditions, measured in volts.

Formula 26.D

$$V_{dc} = \text{a value between peak and peak} \times 0.636$$

V_{dc} is the average voltage coming from a filtered power supply (full-wave) with a load, measured in volts.

Formula 26.E

$$V_{dc} = \text{a value between peak and peak} \times 0.318$$

V_{dc} is the average voltage coming from a filtered power supply (half-wave) with a load, measured in volts.

Ripple Calculations

Ripple is the ac content of the dc voltage produced by a filtered power supply. Ripple has an ac voltage and a frequency content. A half-wave rectifier produces a ripple frequency equal to the input frequency. If the input is 60 Hz, the half-wave output has a ripple of 60 Hz. The full-wave rectifier circuit produces two peaks per cycle. Therefore, the ripple frequency is twice the input frequency. In the 60 Hz example, the ripple frequency would be 120 Hz. Refer to figures 26-38 and 26-39.

The ripple value of a power supply is calculated in percentage. The percent value of ripple increases as the load current increases because the dc voltage becomes less smooth as the filter is called upon to supply more current. The ripple value is close to zero with no load and increases up to five percent typically. The lower the percent ripple, the smoother the dc voltage. The percent ripple is calculated by dividing the output peak-to-peak ac ripple by the output dc voltage.

Formula 26.F

$$\% = \frac{ac}{dc} \times 100\%$$

ac is the peak-to-peak value of the ripple, measured in volts.

dc is the average voltage of the waveform, measured in volts.

Sample problem 3. _____

The output of a 24 V dc power supply has a ripple content of 100 mV peak-to-peak at full load. Determine the percent ripple.

Formula: $\% = \dfrac{ac}{dc} \times 100\%$

Substitution: $\% = \dfrac{100 \text{ mV}}{24 \text{ V}} \times 100\%$

Answer: $\% = 0.4\%$

26.9 VOLTAGE MULTIPLIERS

Transformers, studied in Chapter 16, are one way of stepping up the voltage from a 120 V ac line. However, transformers are fairly expensive, especially if a dc voltage is needed. Diodes and capacitors can be used to step up voltage instead of a transformer. At the same time, they can change the ac voltage to dc.

This section examines three multiplication circuits: the doubler, the tripler, and the quadrupler. Voltage multipliers do have one disadvantage to transformers. They have high unloaded voltages with a very sharp decrease as the load current increases.

Half-Wave Voltage Doubler

The schematic of the half-wave voltage doubler is shown in figure 26-40. The negative half cycle charges C_1 through the forward biased D_1. In the negative half

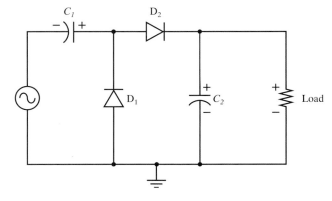

Figure 26-40. Circuit for a half-wave voltage doubler.

cycle, the load is blocked from receiving current by D_2. C_2 is also blocked during the negative half cycle.

During the positive half cycle, D_1 is turned off and D_2 turned on. The charge across C_1 adds to the ac line voltage. This results in a peak value of twice the input peak voltage. C_2 is a filter capacitor, charging to the double-sized peak. C_2 fills in the voltage across the load during the negative half cycle.

Full-Wave Voltage Doubler

Refer to the schematic shown in figure 26-41. During the positive half cycle, D_1 allows C_1 to charge to the peak value of 170 volts, assuming a line voltage of 120 volts rms. The negative half cycle charges C_2, through D_2, to the peak of 170 volts. The capacitor voltages add to supply the load. Each capacitor is recharged during the appropriate half cycle.

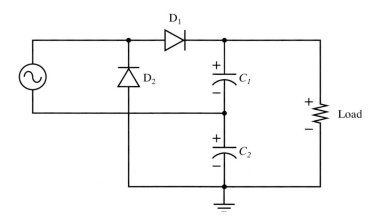

Figure 26-41. Circuit for a full-wave voltage doubler.

Voltage Tripler

The voltage tripler, figure 26-42, produces an output voltage equal to three times the peak value of the line voltage. This schematic also includes a low value surge resistor to protect the diodes from a high initial current. The uncharged capacitors appear as an open circuit when the power is first applied. The **surge resistor** drops voltage in response to the initial high current. Then, as the current levels off with the charging of the capacitors, there is almost no voltage dropped due to the small resistance value. Most voltage multipliers include a surge resistor.

In figure 26-42, C_1 is charged through D_1, during the positive half cycle, to the peak value. The next positive half cycle recharges C_1, whose voltage will be added to the line voltage through D_1 and D_2 to charge C_2 to double the peak. The next positive half cycle

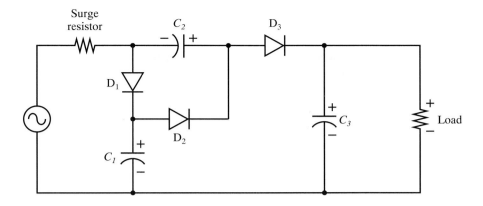

Figure 26-42. Circuit for a voltage tripler.

will recharge C_1 and C_2, whose voltage will be added to the line voltage through D_1, D_2, and D_3 to charge C_3 to three times the peak value.

After the initial charging, each positive half cycle recharges each of the three capacitors. The load is supplied with a half-wave filtered dc voltage that is approximately equal to three times the peak value of the line voltage.

Voltage Quadrupler

The circuit in figure 26-43 is a voltage quadrupler. It produces a voltage equal to four times the input peak. During the positive half cycle, the polarity of the ac input is + on top and – on bottom. This polarity allows C_3 to charge through D_3 and C_4. The negative half cycle has an ac input polarity of – on top and + on bottom. During this time C_1 charges through D_1.

The next positive half cycle recharges C_3, which will add its voltage to C_1. D_2 and D_4 turn on, charging C_2 and C_4. The load will have a voltage of four times the input.

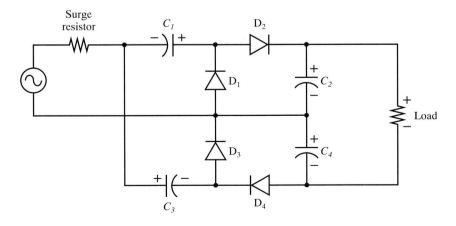

Figure 26-43. Circuit for a voltage quadrupler.

26.10 ZENER DIODES

A **zener diode** is a special type of diode that is designed to operate in the reverse direction. When a reverse bias is applied to a zener diode, it maintains a fairly constant voltage drop across it for a wide range of currents.

One application for the zener is regulating the voltage to the load. In a dc power supply with no regulation, the load voltage drops as the load current is increased. This drop can be significant enough to cause problems in the operation of the circuit. Zener diodes supply the necessary regulation to prevent problems.

Characteristic Curve of a Zener Diode

The characteristic curves of a rectifier diode were shown in figure 26-11. The characteristic curve of the zener diode is shown in figure 26-44. These curves are similar in the forward direction, but they are very different in reverse bias.

In the forward direction, the rectifier diode and zener diode have essentially the same characteristics. To start conduction the zener diode needs 0.6 V (0.2 V for germanium). Once conduction starts, the diode maintains approximately the same voltage drop while the load current increases.

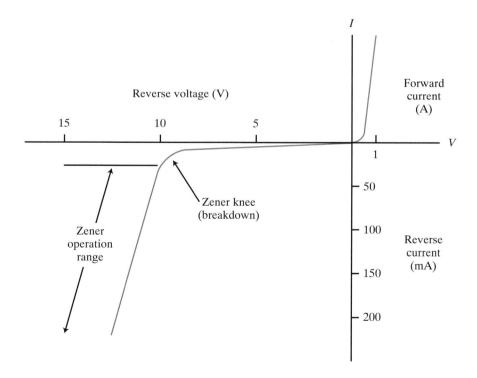

Figure 26-44. Characteristic curve of a 10 volt zener diode. In the operating range the diode holds a fairly constant voltage with a changing current.

In the reverse direction, a rectifier diode has a current of essentially zero. There is only a very small minority current. A zener diode, however, has a point where it will breakdown. Breakdown in reverse is normal for a zener diode. **Breakdown** occurs when the reverse voltage is increased enough to allow a large amount of current. The current is a minority current, but large enough to supply the circuit's needs.

The **zener knee** is the name for the section of the curve where there is a sudden change in the current. Beyond the zener knee is the **zener operating range.** On this portion of the curve, the voltage increases only slightly as the current is increased. The ideal condition for a regulator would be to have a perfectly constant voltage, regardless of the amount of current. The zener is not a perfect device, so there is a slight change.

Zener Diode As a Voltage Regulator

As the load current demands change, the changing current drain on the dc power supply causes the voltage to change. When a zener diode is placed in parallel with a load, it maintains a constant voltage across the load.

The zener voltage regulator is made up of a series resistor, for dropping the excess voltage, and the zener diode in parallel. See figure 26-45.

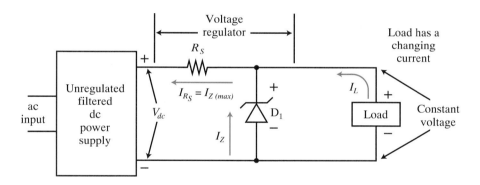

Figure 26-45. A zener diode used in a voltage regulator maintains a constant voltage across the load with a changing load current.

Calculating a Zener Regulator

There are two types of calculations needed for zener regulators, current and series resistance. The voltage requirement is determined by the load. The load also determines the starting point for current calculations. Once the voltage to the load and the current requirements are known, the value of R_S can be determined.

The zener voltage is in parallel with the load, therefore its voltage is equal to the load voltage. For example, if the load is a six volt portable tape player, the zener voltage should be six volts.

A tape player has different current demands. The current needed when the tape player is in fast forward or rewind is different than the current required when in play or the current required when the player is off. The zener circuit is designed to maintain a constant current through R_S. The zener current and the load current add together to equal the current through R_S. If the load current goes down, zener current goes up and visa-versa. The current through R_S remains constant. The zener diode must be able to handle the full current when the load is turned off and it must have a slight reserve when the load is at maximum current.

It is a good design practice to assign a zener current value of the maximum load current plus 10 percent. This value will be the full zener current when the load is off. When the load is at maximum, the zener will have a reserve of 10 percent to maintain it in the operating region.

Zener diodes are rated for voltage and wattage. The zener wattage is determined using the power formulas with the voltage and maximum current. When purchasing a component, it is always best to provide a safety factor of 20 percent or more above the calculated wattage value.

Series resistor, R_S, has a voltage drop across it equal to the difference between the supply voltage and the zener voltage. Knowing the value of voltage and current, the value of resistance can be calculated. The wattage rating is also needed. This is found using the power formulas. As with the zener diode, when purchasing a resistor, include a safety factor of 20 percent.

The dc power supply connected to a zener voltage regulator must have a high enough voltage to supply the zener voltage and the voltage drop across the series resistor. The supply voltage, therefore, must be higher than the zener voltage. Keep in mind also, too high of a voltage from the supply will result in an excessively large voltage drop across the series resistor. The resistor may need too high of a wattage rating to be practical. If the supply voltage is too high, rather than drop the excess across the resistor, a voltage divider using zener diodes can be used. Connect the diodes in the form of a voltage divider by placing two or more in series. The load would then be connected across the proper voltage.

Sample problem 4.
_____

A certain portable electronic video game is connected to the household wall socket through a nine volt adaptor. Its maximum current rating is 750 mA. Design a zener regulator circuit to enable connection to the cigarette lighter in a car. Note, a car voltage system is 13.8 volts (not 12 volts).

1. Zener voltage.

   Formula: $V_Z = V_L$

   Answer: $V_Z = 9$ V

2. Maximum zener current.

   Formula: $I_{Z(max)} = I_L + 10\%$

Substitution: $I_{Z(max)} = 750 \text{ mA} + (0.1 \times 750 \text{ mA})$

Answer: $I_{Z(max)} = 825 \text{ mA}$

3. Zener power rating.

Formula: $P_Z = I_{Z(max)} \times V_Z$

Substitution: $P_Z = 825 \text{ mA} \times 9 \text{ V}$

Answer: $P_Z = 7.4 \text{ W}$

4. Current through the series resistor.

Formula: $I_{R_S} = I_{Z(max)}$

Answer: $I_{R_S} = 825 \text{ mA}$

5. Voltage across the series resistor.

Formula: $V_{R_S} = V_{dc} - V_Z$

Substitution: $V_{R_S} = 13.8 \text{ V} - 9 \text{ V}$

Answer: $V_{R_S} = 4.8 \text{ V}$

6. Ohmic value of the series resistor.

Formula: $R_S = \dfrac{V_{R_S}}{I_{R_S}}$

Substitution: $R_S = \dfrac{4.8 \text{ V}}{825 \text{ mA}}$

Answer: $R_S = 5.8 \ \Omega$

7. Power dissipated in the series resistor.

Formula: $P_{R_S} = V_{R_S} \times I_{R_S}$

Substitution: $P_{R_S} = 4.8 \text{ V} \times 825 \text{ mA}$

Answer: $P_{R_S} = 3.96 \text{ W}$

Zener Diodes As a Voltage Divider

Zener diodes can be connected in series, as shown in figure 26-46. Any number can be connected in this manner, allowing a choice of voltages for the load connection. The voltage ratings of the zener do not need to be equal. However, the current must not exceed the lowest current rating of any of the diodes.

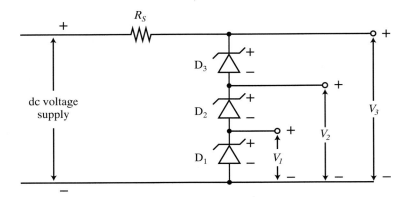

Figure 26-46. These zener diodes are connected to create a voltage divider. Any number of diodes can be connected in series.

Using a voltage divider also helps if the dc input voltage is too high to be dropped entirely across the series resistor. A smaller portion of the voltage can go to R_S, while the remainder is distributed across the zener diodes.

Zener Diode As a Clipper

Rectifier diodes were shown in figure 26-17 in a clipper circuit. Using rectifier diodes, the voltage is clipped to 0.6 V. The zener diode, with its constant voltage characteristics, can also be used as a clipper.

Figure 26-47 is the schematic of a voltage clipper using zener diodes. The rectifier diodes are used in this circuit to stop the zeners from conducting in the forward direction. If they were not used, the zener would act like a rectifier diode and clip at 0.6 V, in the forward direction.

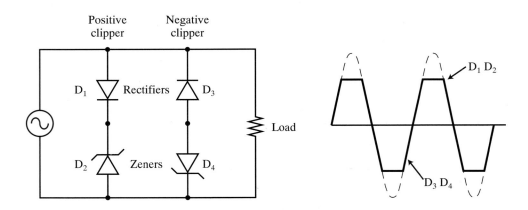

Figure 26-47. Schematic of zener diodes used as voltage clippers.

The amount of the input sine wave to be clipped is equal to the zener voltage plus 0.6 V (for the rectifier diode in series.) Zeners are available in a wide range of voltage ratings.

SUMMARY

- Semiconductors are used in many applications, including diodes.
- Valence electrons are those electrons in the outermost orbit of an atom. They are the particles used for current flow.
- P-type semiconductors have a majority of holes and a minority of electrons.
- N-type semiconductors have a majority of electrons and a minority of holes.
- Current flows through a semiconductor through the use of the majority current carriers.
- Current flows in only one direction through a junction diode.
- Forward bias, to allow current to flow in a diode, has a positive on the P-material and negative on the N-material.
- A diode is checked with an ohmmeter by connecting the leads first in one direction, and then the other. High resistance should be measured in reverse bias and low resistance in forward bias.
- A diode performs like a one-way switch, which can be used in many dc and ac applications.
- A diode conducts only one-half of a sine wave. A single diode is classified as a half-wave rectifier. The output is a dc voltage.
- Two diodes can pass both halves of a sine wave, with one of the halves flipped over so that both halves have the same polarity. The output is a dc voltage with twice the value of a half-wave rectifier.
- A bridge rectifier circuit is formed with four diodes. Its major advantage over the two-diode, full-wave circuit is it does not need a transformer.
- A power supply filter is used to smooth the output voltage, producing a better quality dc voltage.
- Combinations of diodes and capacitors are used to produce voltage multipliers.
- Zener diodes are used as simple voltage regulators.

KEY WORDS AND TERMS GLOSSARY

anode: The positive element of diode and other electronic devices. The anode of a diode is the N-type semiconductor material. Electron current flows towards the anode.

barrier potential: The voltage drop across junction of a P- and an N-type material.

bias: The voltage levels applied to two points in a circuit for the purposes of controlling the behavior of the circuit.

breakdown: In semiconductor diodes, it is the point at which there is a sudden increase in current in the reverse direction.

bridge rectifier: One type of full-wave rectifier. It uses four diodes for rectification instead of a center-tapped transformer.

cathode: The electron-emitting element of a diode and other devices. The N-type material of a semiconductor diode is the cathode.

clamper circuit: Places the centerline of an ac signal to a dc voltage level.

clipper circuit: Clips off part of a sine wave, making it look more like a square wave.

covalent bond: The sharing of valence electrons between two atoms to fill the valence shell.

depletion region: An area near the junction of a semiconductor where the free current carriers cannot cross.

diode: A semiconductor device that allows current to flow in only one direction.

doping: The process of adding an impurity to semiconductor material to change the semiconductor to either a P-type or an N-type.

extrinsic: A semiconductor material containing impurities.

forward biased: The voltage applied to a diode to allow a majority current.

free electron: An electron that moves easily between atoms to provide current.

full-wave rectifier: A rectifier that passes both halves of an ac cycle, inverting one half of the wave to make the entire wave the same polarity.

half-wave rectifier: A rectifier that passes one half of an ac cycle while blocking the other half of the cycle.

hole: An atom missing an electron in its valence band. It carries a positive charge. A hole is the majority current carrier in a P-type semiconductor.

intrinsic: A pure semiconductor material.

junction: In reference to semiconductors, it is the surface where the P- and N-type materials are connected.

majority carrier: The current carrier used as the primary source of current within a semiconductor when forward biased.

minority carrier: The current carrier in a semiconductor of a polarity opposite that of the material in which they are present. Minority current flows in the reverse direction of the majority current.

N-type semiconductor: A semiconductor material doped with an impurity having five valence electrons, such as arsenic (As). The pentavalent impurity forms covalent bonds with the semiconductor leaving free electrons.

pentavalent impurity: A semiconductor doping impurity with five electrons in the valence band. When it forms covalent bonds with the semiconductor material, there are extra electrons available to be used as current carriers.

P-type semiconductor: A semiconductor material doped with an impurity having three valence electrons, such as gallium (Ga). The trivalent impurity forms covalent bonds with the semiconductor producing holes.

rectification: The process of changing ac to dc by changing an ac waveform to have only voltages of the same polarity.

reverse biased: The voltage applied to a diode to allow only a minority current.

ripple: The ac content of the dc output voltage of a power supply.

semiconductor: Electrical materials with conductivity that falls between conductors and insulators.

surge resistor: A small resistor used to drop a voltage in response to a high initial current.

trivalent impurity: A impurity having three electrons in the valence band that is used for doping a semiconductor material to become P-type.

valence electron: Electron found in the valence shell. The electrons used in electrical current.

valence shell: The outer-most orbit of an atom, containing the electrons used in electrical current.

zener diode: A type of diode that is designed to operate in the reverse direction.

zener knee: When voltage is increased in the reverse direction through a zener diode, there is zero current up to the *zener knee,* which occurs at the breakdown point. The zener knee is the name for the section of the curve where there is a sudden change in the current. From this point on, there is only a slight change in voltage for wide changes in current.

zener operating range: The range beyond the zener knee. In this range there is only a slight increase in voltage in response to a large increase in the current.

KEY FORMULAS

Formula 26.A

$$V_{dc} = \frac{\text{peak}}{2} \times 0.636 = \text{peak} \times 0.318$$

Formula 26.B

$$V_{dc} = \text{peak} \times 0.636$$

Formula 26.C

$$V_{dc} = \text{peak voltage}$$

Formula 26.D

$$V_{dc} = \text{a value between peak and peak} \times 0.636$$

Formula 26.E

$$V_{dc} = \text{a value between peak and peak} \times 0.318$$

Formula 26.F

$$\% = \frac{\text{ac}}{\text{dc}} \times 100\%$$

TEST YOUR KNOWLEDGE

Do not write in this text. Please use a separate sheet of paper.

1. List three applications for semiconductors.
2. What type of doping impurity is used to make a P-type semiconductor?
3. What type of doping impurity is used to make an N-type semiconductor?
4. What are the majority current carriers in a P-type material?
5. What are the majority current carriers in an N-type material?
6. What are the minority current carriers in a P-type material?
7. What are the minority current carriers in an N-type material?
8. Use a series of diagrams to demonstrate the operation of a junction diode.
9. No voltage is applied to a diode. Describe the depletion region.
10. A forward voltage is applied to a diode. Describe the depletion region.
11. A reverse voltage is applied to a diode. Describe the depletion region.
12. When a forward voltage is applied to a diode, what is the typical voltage measured across it?
13. When a reverse voltage is applied to a diode, what is the typical voltage measured across it?
14. Describe the current when a forward biased voltage is applied to a diode.
15. Describe the current when a reverse biased voltage is applied to a diode.
16. List some applications for diodes.
17. What are the two most common ratings for a diode?
18. "A diode is a one-way switch." What is meant by this statement?
19. When testing a diode with an ohmmeter, what is the expected reading when the leads are connected in forward bias?
20. When testing a diode with an ohmmeter, what is the expected reading when the leads are connected in reverse bias?
21. How can two diodes protect a circuit from the wrong polarity dc voltage?
22. Draw the schematic symbol of a diode. Use + and – symbols to show the correct polarity for forward bias.
23. What does a diode clipper circuit do to the shape of a sine wave?
24. What effect does a diode clamper circuit have on a sine wave?
25. Draw a basic half-wave rectifier circuit. Also draw the output waveform when a sine wave is applied.
26. Describe the operation of a half-wave rectifier circuit.
27. Calculate the dc output voltage of a half-wave rectifier circuit with 24 volts rms applied.
28. Draw a basic full-wave rectifier circuit. Also draw the output waveform when a sine wave is applied.
29. Describe the operation of a full-wave rectifier circuit.
30. Calculate the dc output voltage of a full-wave rectifier circuit with 24 volts rms applied.
31. Draw a basic bridge rectifier circuit. Also draw the output waveform when a sine wave is applied.
32. Describe the operation of a bridge rectifier circuit.

33. Calculate the dc output voltage of a bridge rectifier circuit with 30 volts rms applied.
34. Draw the output of an unfiltered half-wave rectifier. On this drawing, show the effect of a simple capacitive power supply filter.
35. Calculate the no-load output voltage of a power supply with a capacitive filter on a half-wave rectifier with an input of 48 volts peak.
36. Calculate percent ripple of a 36 volt power supply with 200 mV peak-to-peak ac ripple content.
37. Draw the schematic diagrams of each of the following voltage multipliers:
    a.   Half-wave.
    b.   Full-wave.
    c.   Tripler.
    d.   Quadrupler.
38. Draw the characteristic curve of a zener diode. Use the curve to describe the function and operation of a zener diode.
39. Calculate the circuit values of a zener regulator for a 12 volt output at 250 mA. The input voltage is 20 volts dc.
40. Draw the schematic diagrams of these zener diode circuits:
    a.   Voltage divider.
    b.   Clipper.

Power supplies for cordless phones reside in the base units. The power supplies convert the ac voltage from a wall outlet into a dc voltage, which is used to charge the batteries in the handset.

Chapter 27
Bipolar Junction Transistors

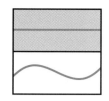

Upon completion of this chapter, you will be able to:
- Define technical terms associated with bipolar junction transistors.
- Test a transistor with an ohmmeter.
- Describe the current relationships in a transistor.
- Use the beta formula to calculate current.
- Describe the voltage relationships in a transistor.
- Define the range of transistor operation.
- Calculate and plot the dc load line on a transistor characteristic curve.
- Calculate the Q point and dc operating parameters for different biasing arrangements.
- Describe how an ac signal passes through a common emitter amplifier.
- Describe the differences between common emitter, common base, and common collector amplifiers.

The **bipolar junction transistor,** usually just called a **transistor,** is a single component capable of a wide range of operating characteristics. Transistors are used as electronic switches for digital circuits, amplifiers for audio and radio frequency circuits, and to regulate current and voltage.

This chapter examines general transistor theory and simplified applications. How the transistor is structured, general transistor testing, and the operation of basic amplifiers are explored.

27.1 TYPICAL TRANSISTORS

The transistor has many different sizes, shapes, and applications. The examination of a typical transistor, however, shows how they are all similar. They are structured the same and have the same range of operating characteristics.

Different transistor models and ratings allow transistors to find applications in both low current and high current circuits. Transistors are also used in high voltage applications as well as very low voltage applications. Transistor power ratings range from a few milliwatts to many watts.

Transistor Structure

A transistor is a three-layered device. It is similar to two diodes manufactured back-to-back, figure 27-1. Depending on how the diodes are facing, the transistor can be either an NPN or a PNP.

A transistor has three elements: the emitter, the base, and the collector. On the schematic symbol for a transistor, the emitter lead has an arrow. The arrow points toward the N-type material. The arrow either points at the N-type base (PNP transistor) or the N-type emitter (NPN transistor).

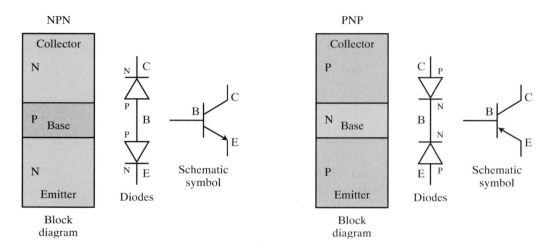

Figure 27-1. A bipolar junction transistor is made by placing two diodes together.

In the structure of the transistor, the **base** is the layer sandwiched between the **emitter** and the **collector.** A transistor has three legs (leads) and is available in many package styles, as shown in figure 27-2. Some packages have a screw type mounting. The mounting is the collector. Other package types have different arrangements for the three elements. It is necessary to check in a catalog to determine the correct locations of B, E, and C for each transistor you use.

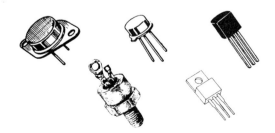

Figure 27-2. Typical transistor packaging.

Testing a Transistor with an Ohmmeter

The two diodes that makeup the transistor are tested using a similar procedure to that used to test a rectifier diode. The diodes are tested by first checking the resistance with one ohmmeter polarity, then checking with the other polarity.

The polarity of the ohmmeter and the amount of resistance measured in the transistor depends on whether the transistor is NPN or PNP. The ohmmeter has a low reading when the negative lead is connected to the N-material and positive lead to the P-material. A transistor is tested with an ohmmeter in three steps:

1. Clip one lead to the base of the transistor. Measure the resistance between the collector and the base, then between the emitter and the base.
2. Reverse the polarity of the ohmmeter leads. Again, clip one lead to the base. Measure from the collector to the base, and then measure from the emitter to the base.
3. Attach a resistor of approximately one kilohm between the base and collector. Now measure the resistance between the collector and emitter.

With either type of transistor, there should be a high resistance between the emitter and collector, except when the resistor is attached. The resistor allows the transistor to turn on, if the polarity is correct, giving a low resistance reading.

The NPN transistor, shown in figure 27-3, has a high resistance with the negative lead on the base. With the positive lead connected to the base, there is a low resistance reading to both the collector and emitter. Observe the polarity of the collector-emitter test with the resistor attached. The positive lead must be on the collector for this test to work correctly.

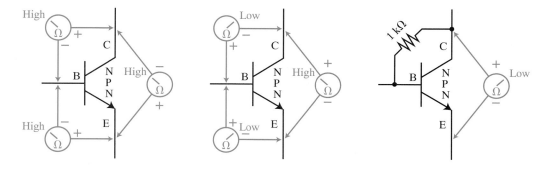

Figure 27-3. Testing an NPN transistor with an ohmmeter.

The PNP transistor, shown in figure 27-4, has the same resistance readings as the NPN, except the polarity of the meter is opposite. With either type of transistor, the base is the key element.

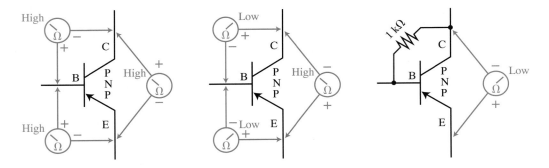

Figure 27-4. Testing a PNP transistor with an ohmmeter.

27.2 TRANSISTOR CHARACTERISTICS

A transistor's basic function is to control the amount of current flowing in the collector circuit. The control of collector current is performed by regulating the transistor's effective resistance between emitter and collector. *The controlling element is the base.*

Current Relationships

A transistor must have a dc voltage source, regardless of the application. The dc voltage, V_{CC}, is connected between the collector and emitter with a biasing branch to the base. See figure 27-5.

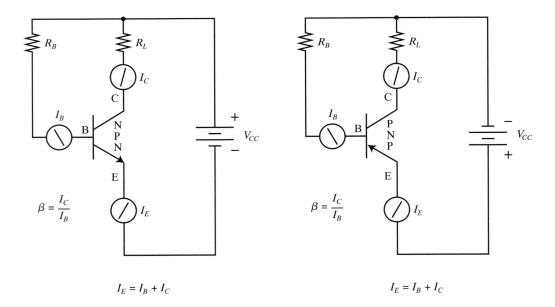

Figure 27-5. Current relationships in a transistor. A transistor circuit must have a dc voltage source.

Current through a semiconductor device flows in a direction determined by the type of material. In an NPN transistor, the majority current is electrons, supplied by the negative side of V_{CC}. In a PNP transistor, the majority current is holes, supplied by the positive side of V_{CC}.

The emitter is the input for dc current. Emitter current splits, like a parallel circuit, to the base and the collector. The ratio of base current to collector current is determined by the **beta** (β) of the transistor. Collector current is *beta times larger* than the base current. If beta is 100, I_C will be 100 times greater than I_B.

The current relationships in a transistor can be represented by two formulas. The formula for emitter current is the same as the formula for total current in a parallel circuit. Branch currents add together to equal the total current. The formula for beta is the ratio of collector to base currents.

Formula 27.A

$$I_E = I_B + I_C$$

I_E is the emitter current, measured in amperes.

I_B is the base current, measured in amperes.

I_C is the collector current, measured in amperes.

Formula 27.B

$$\beta = \frac{I_C}{I_B} \text{ or } I_C = \beta \times I_B$$

β is the ratio of the collector to the base current, a number with no units.

I_C is the collector current, measured in amperes.

I_B is the base current, measured in amperes.

Voltage Relationships

The voltages of a transistor are labeled according to which elements the voltmeter leads are connected to. The voltages shown in figure 27-6 are the three basic points to be considered, V_{CB}, V_{BE}, and V_{CE}.

The voltage source for a transistor circuit is labeled V_{CC}. This stands for **voltage collector-to-common.** The collector is connected to the highest voltage in the circuit, with the emitter connected closest to zero. In an NPN transistor, V_{CC} has a positive, with negative as common. A PNP transistor uses a negative V_{CC}, with positive as common.

Voltage collector-to-base, V_{CB}, has a higher reading on the collector than the base, considering polarities. In the NPN transistor, the collector is more positive than the

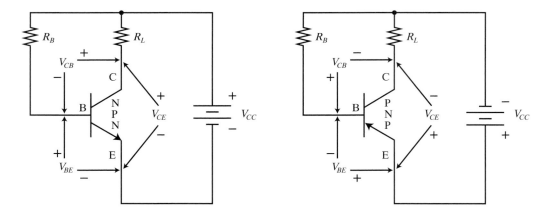

Figure 27-6. Voltage relationships in a transistor.

base. In the PNP transistor, the collector is more negative than the base. The collector should have the voltage closest to V_{CC}. V_{CB} is a voltage of a polarity such that it will reverse bias the base-to-collector diode. Current from the base to the collector is actually minority current.

Voltage base-to-emitter, V_{BE}, is a voltage equal to that found across a forward biased diode. If the transistor is silicon, V_{BE} is approximately 0.6 V. If the transistor is germanium, V_{BE} is approximately 0.2 V. Observe the polarity of V_{BE}. It depends on whether the transistor is NPN or PNP.

Voltage collector-to-emitter, V_{CE}, is the voltage across the transistor. As will be demonstrated, this is the voltage that shows the transistor in operation. V_{CE} can be very close to zero, the full applied voltage, or any value in between. The voltage reading depends on the operating state of the transistor.

Characteristic Curves

The characteristics curves in figure 27-7 show that a transistor is controlled by base current. Although base current is much smaller than collector current, I_B varies the transistor's resistance. This controls the flow of I_C. Also, if the supply voltage is varied while the base current is held constant, the collector current still remains nearly constant.

The transistor characteristic curves are produced by setting a value of base current. This can be done using a base voltage that is separate from the collector voltage. With the constant base current, such as 150 µA, the collector current is measured. While monitoring I_C the supply voltage, V_{CC}, is increased and the voltage across the transistor is measured. Notice in the characteristic curves, the flat lines representing I_B show a slight increase in collector current as V_{CE} increases.

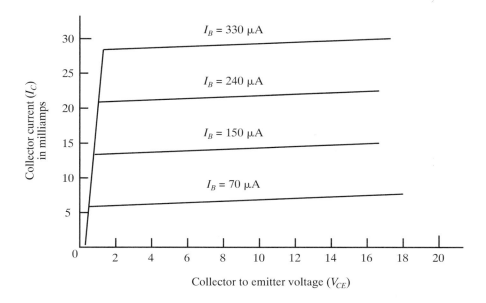

Figure 27-7. Typical transistor characteristic curves.

27.3 DC OPERATING CONDITIONS

The actual operating conditions of a transistor are determined by the values of resistors in the circuit. If enough base current is applied, a transistor reaches a point where it offers no resistance to the circuit. The collector current is at a maximum. It is limited only by the other circuit resistances and the transistor is in **saturation.** The opposite extreme is when the transistor offers so much resistance that it stops the current. This point is called **cutoff.**

DC Load Line

The dc load line is drawn across the transistor's characteristic curves. The end points of the **load line** are the operating extremes. A sample dc load line is drawn in figure 27-8. The lines representing base current have been left out.

The value of I_C at saturation is one end of the load line. The opposite end of the line is the point where the transistor has turned off. Its voltage drop is equal to the supply voltage. The **Q point** is the actual operating point of the transistor circuit. A transistor's **active region** is the operating range between saturation and cutoff.

Defining the Range of Transistor Operation

The range of transistor operation is predetermined by the values of the resistors in the circuit. Based on the circuit values, the transistor has an upper limit, with maximum collector current, at the saturation point. The lower limit, with no collector current, is

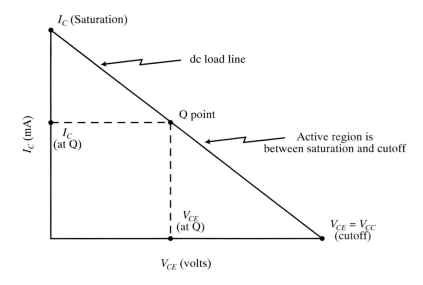

Figure 27-8. Transistor dc load line.

cutoff. The values of collector current between these two limit points is the active region. Refer to figure 27-9.

At the point of saturation, the transistor acts like a closed switch. The transistor offers virtually no resistance to the circuit. All of the circuit voltage is dropped across the resistors in the collector and emitter circuits. Voltage across the transistor, V_{CE}, is approximately zero. In actual practice, it is about 0.2 V. Collector current is maximum, limited by the circuit resistance and not the transistor.

At cutoff the transistor turns off and has an effect on the circuit of an open switch. There is no current through the transistor and the applied voltage is measured at V_{CE}. If the transistor is to be used as an electronic switch, its operating range would be set so that it is either on or off, with nothing between.

The active region of a transistor is used in amplifier applications and in some circuits as a current or voltage regulator. A transistor varies its resistance according to the amount of base current flowing. If the base current is changing at a sine wave rate, such as with an audio amplifier, the transistor will vary its resistance at the same rate. The changing resistance changes the collector current, producing a changing output voltage.

27.4 TRANSISTOR APPROXIMATIONS: BASE BIAS

Figures 27-10 through 27-14 show the steps used to calculate the current, voltage, and load line values of a base bias circuit. A base bias circuit is used to measure transistor characteristic curves. The base bias circuit is also used as a transistor switch.

The base bias circuit is *beta-dependent.* This means that the value of beta has an effect on collector current. Transistors with the same part number can have differ-

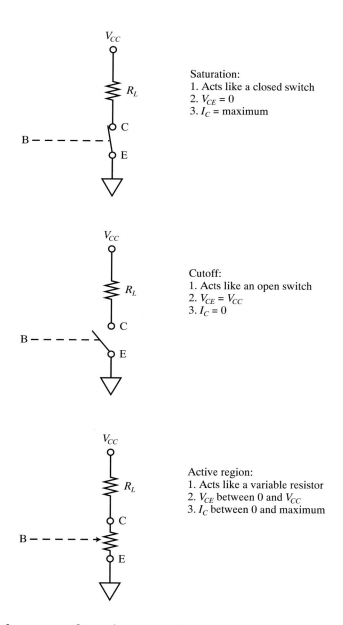

Saturation:
1. Acts like a closed switch
2. $V_{CE} = 0$
3. I_C = maximum

Cutoff:
1. Acts like an open switch
2. $V_{CE} = V_{CC}$
3. $I_C = 0$

Active region:
1. Acts like a variable resistor
2. V_{CE} between 0 and V_{CC}
3. I_C between 0 and maximum

Figure 27-9. There are three areas of transistor operation.

ences in their betas. The differences can be enough to cause a change in operating characteristics. In addition, the beta of a transistor in a circuit changes with the circuit temperature. Because of these beta problems, the base bias circuit is seldom used as an amplifier.

Figure 27-10 is the base bias circuit. The value for beta and resistor values are given. The voltmeters drawn and the current arrows show the values to be calculated. In the following problems, Ohm's law and the beta formula are needed.

Sample problem 1. _____

Solve the circuit shown in figure 27-10.

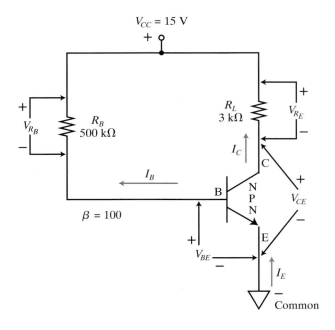

V_{CC} = 15 V

Figure 27-10. Base bias, a beta-dependent circuit. (Sample problem 1)

1. Calculate and plot the load line. Refer to figure 27-11.

   a. Collector current at saturation, $I_{C(sat)}$.

      Formula: $I_{C(sat)} = \dfrac{V_{CC}}{R_L}$

      Substitution: $I_{C(sat)} = \dfrac{15 \text{ V}}{3 \text{ k}\Omega}$

      Answer: $I_{C(sat)} = 5 \text{ mA}$

   b. Transistor voltage, V_{CE} at cutoff.

      $$V_{CE} = V_{CC} = 15 \text{ V (at cutoff)}$$

   c. The load line.

      Locate $I_{C(sat)}$ and V_{CE} at cutoff. Draw a straight line connecting the two limits.

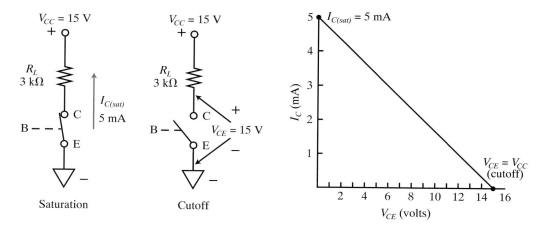

Figure 27-11. Calculating the end points of the dc load line.

2. Calculate the Q point.

   The Q point is the actual operating point for I_C and V_{CE}. The base current determines these values. To determine collector current at the Q point, it is necessary to know either the value of base current or the values of base resistance and beta. The formula for beta states that collector current is beta times larger than base current. Therefore, the collector resistance must be equal to the base resistance divided by beta. The collector resistance includes the resistance of the transistor. Refer to figure 27-12.

   a. Collector resistance, including the transistor.

   Formula: $R_C = \dfrac{R_B}{\beta}$

   Substitution: $R_C = \dfrac{500 \text{ k}\Omega}{100}$

   Answer: $R_C = 5 \text{ k}\Omega$

   b. Collector current at the Q point.

   Formula: $I_{C(Q)} = \dfrac{V_{CC}}{R_C}$

   Substitution: $I_{C(Q)} = \dfrac{15 \text{ V}}{5 \text{ k}\Omega}$

   Answer: $I_{C(Q)} = 3 \text{ mA}$

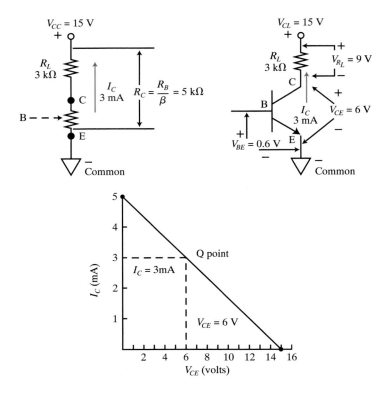

Figure 27-12. Calculating the Q point.

3. Calculate collector circuit voltage drops.

The collector current produces a voltage drop across the load resistor, R_L, and across the transistor, V_{CE}. Use Ohm's law to find the voltage across the resistor. Subtract the voltage across R_L from the supply voltage to find the voltage across the transistor. Refer to figure 27-12.

a. Voltage across the load resistor.

    Formula: $V_{R_L} = I_{C(Q)} \times R_L$

    Substitution: $V_{R_L} = 3\ \text{mA} \times 3\ \text{k}\Omega$

    Answer: $V_{R_L} = 9\ \text{V}$

b. Voltage across the transistor.

    Formula: $V_{CE} = V_{CC} - V_{R_L}$

    Substitution: $V_{CE} = 15\ \text{V} - 9\ \text{V}$

    Answer: $V_{CE} = 6\ \text{V}$

c. Adding the Q point to the load line.

   Locate the value of the collector current at the Q point. Locate the value of V_{CE} at the Q point.

4. Perform base circuit calculations.

   The base circuit includes the base biasing resistor and the forward biased base-to-emitter diode. The calculations for collector current are approximations. They do not take into account the base-to-emitter diode. When calculating the base circuit, the slight error is again ignored. Refer to figure 27-13.

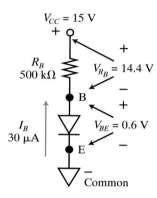

Figure 27-13. Base circuit calculations.

a. Base current at the Q point is found using beta.

   Formula: $\qquad I_B = \dfrac{I_C}{\beta}$

   Substitution: $I_B = \dfrac{3 \text{ mA}}{100}$

   Answer: $\qquad I_B = 30 \text{ μA}$

b. Base-to-emitter forward biased diode voltage drop.

   No calculation is necessary. Here, the transistor is assumed to be silicon. $V_{BE} = 0.6$ V.

c. Voltage across the base resistor.

   Formula: $\qquad V_{R_B} = V_{CC} - V_{BE}$

   Substitution: $V_{R_B} = 15 \text{ V} - 0.6 \text{ V}$

   Answer: $\qquad V_{R_B} = 14.4 \text{ V}$

5. Label the Circuit.

The results of the circuit calculations are shown in figure 27-14. All of the circuit voltages and currents shown are for the Q point.

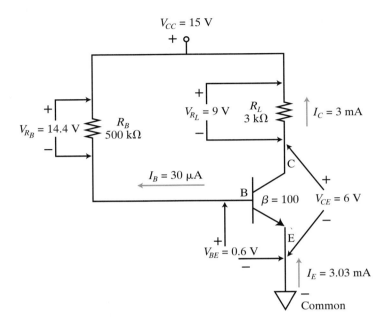

Figure 27-14. Combining the results of all circuit calculations.

27.5 TRANSISTOR APPROXIMATIONS: BASE BIAS WITH EMITTER FEEDBACK

Figures 27-15 and 27-16 show the steps in solving a base bias with an emitter resistor. The emitter resistor develops a voltage drop as a result of the collector current. This voltage drop helps to stabilize the transistor, which can fluctuate because of heat changing the value of beta. This circuit is used in some applications as an amplifier.

An examination of figure 27-15 shows this circuit requires the same calculations as the base bias circuit of figure 27-10, with one exception. The exception is the voltage drop across the emitter. In this problem a PNP transistor is used. The calculations are the same as those for an NPN, except the polarities are opposite.

Sample problem 2. _____

Solve the circuit in figure 27-15. The results of the calculations are shown in figure 27-16.

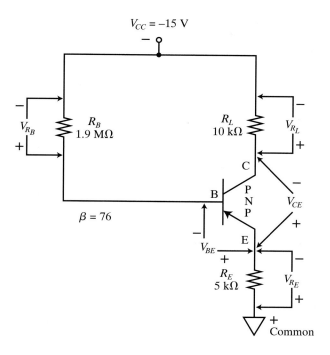

$V_{CC} = -15$ V

V_{R_B} R_B
                  1.9 MΩ

R_L V_{R_L}
10 kΩ

C

B   P
           N V_{CE}
β = 76   P

            E
V_{BE}

R_E V_{R_E}
5 kΩ

Common

Figure 27-15. Base bias, with emitter feedback, a beta-dependent circuit. (Sample problem 2)

1. Find transistor voltage at cutoff.

    $$V_{CE} = V_{CC} = 15 \text{ V (at cutoff)}$$

2. Calculate the collector current at saturation.

    Both the load resistor and emitter resistor are considered.

    Formula: $I_{C(sat)} = \dfrac{V_{CC}}{R_L + R_E}$

    Substitution: $I_{C(sat)} = \dfrac{15 \text{ V}}{10 \text{ k}\Omega + 5 \text{ k}\Omega}$

    Answer: $I_{C(sat)} = 1$ mA

3. Calculate the collector current at the Q point.

   a. Collector resistance, including the transistor.

      Formula: $R_C = \dfrac{R_B}{\beta}$

      Substitution: $R_C = \dfrac{1.9 \text{ M}\Omega}{76}$

      Answer: $R_C = 25 \text{ k}\Omega$

   b. Collector current at Q point.

      Formula: $I_{C(Q)} = \dfrac{V_{CC}}{R_C + R_E}$

      Substitution: $I_{C(Q)} = \dfrac{15 \text{ V}}{25 \text{ k}\Omega + 5 \text{ k}\Omega}$

      Answer: $I_{C(Q)} = 0.5 \text{ mA}$

4. Voltage drops at the Q point.

   a. Voltage across the load resistor.

      Formula: $V_{R_L} = I_{C(Q)} \times R_L$

      Substitution: $V_{R_L} = 0.5 \text{ mA} \times 10 \text{ k}\Omega$

      Answer: $V_{R_L} = 5 \text{ V}$

   b. Voltage across the emitter resistor.

      Formula: $V_{R_E} = I_{C(Q)} \times R_E$

      Substitution: $V_{R_E} = 0.5 \text{ mA} \times 5 \text{ k}\Omega$

      Answer: $V_{R_E} = 2.5 \text{ V}$

   c. Voltage across the transistor.

      Formula: $V_{CE} = V_{CC} - (V_{R_L} + V_{R_E})$

      Substitution: $V_{CE} = 15 \text{ V} - (5 \text{ V} + 2.5 \text{ V})$

      Answer: $V_{CE} = 7.5 \text{ V}$

5. Plot the load line.

   Locate the end points and place the Q point on the load line at the proper location. See figure 27-16.

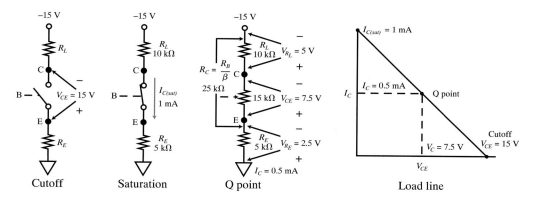

Figure 27-16. Steps in circuit calculations.

27.6 TRANSISTOR APPROXIMATIONS: VOLTAGE DIVIDER BIASING

Voltage divider biasing is a common circuit configuration for audio amplifier applications. This circuit is stable over a range of changing supply voltages and collector currents. Also, the most important characteristic of this setup is that it is independent of beta. The collector current at the Q point is determined by a voltage divider in the base circuit, as shown in figure 27-17. The voltage divider establishes a base voltage, which is reflected to the collector current through the base-to-emitter diode.

Sample problem 3.

Solve the circuit in figure 27-17.

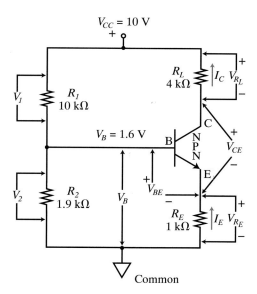

Figure 27-17. Voltage divider biasing. Beta independent. (Sample problem 3)

1. Base voltage is equal to the voltage across R_2 in the voltage divider. Use the voltage divider formula.

    Formula: $\quad V_B = V_{CC} \times \dfrac{R_2}{R_1 + R_2}$

    Substitution: $V_B = 10 \text{ V} \times \dfrac{1.9 \text{ k}\Omega}{10 \text{ k}\Omega + 1.9 \text{ k}\Omega}$

    Answer: $\quad V_B = 1.6 \text{ V}$

2. Draw the load line. See figure 27-18.

    a. Transistor voltage at cutoff.

    $$V_{CE} = V_{CC} = 10 \text{ V (at cutoff)}$$

    b. Collector current at saturation.

    Formula: $\quad I_C = \dfrac{V_{CC}}{R_L + R_E}$

    Substitution: $I_C = \dfrac{10 \text{ V}}{4 \text{ k}\Omega + 1 \text{ k}\Omega}$

    Answer: $\quad I_C = 2 \text{ mA}$

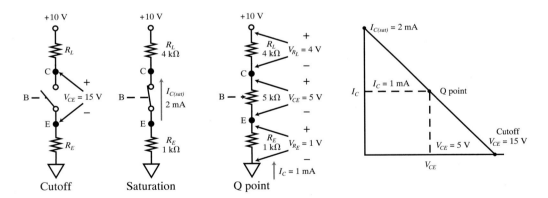

Figure 27-18. Collector portion of a voltage divider circuit.

3. Collector current at the Q point is found using two steps. First, subtract the drop across the base-to-emitter diode from the base resistor voltage drop to find the voltage across the emitter resistor. Second, use the voltage across the emitter resistor to determine the collector current.

a. Voltage across the emitter resistor. (Base-to-emitter diode drop is assumed to be 0.6 V.)

Formula: $V_{R_E} = V_B - V_{BE}$

Substitution: $V_{R_E} = 1.6 \text{ V} - 0.6 \text{ V}$

Answer: $V_{R_E} = 1.0 \text{ V}$

b. Collector current.

The collector current is much larger than base current. In this circuit approximation, collector current is equal to emitter current.

Formula: $I_{C(Q)} = I_E = \dfrac{V_{R_E}}{R_E}$

Substitution: $I_{C(Q)} = \dfrac{1 \text{ V}}{1 \text{ k}\Omega}$

Answer: $I_{C(Q)} = 1 \text{ mA}$

4. Voltage drops at the Q point.

a. Voltage across the emitter resistor.

Answer was found in step 3a.

$$V_{R_E} = 1.0 \text{ V}$$

b. Voltage across the load resistor.

Use Ohm's law.

Formula: $V_{R_L} = I_{C(Q)} \times R_L$

Substitution: $V_{R_L} = 1 \text{ mA} \times 4 \text{ k}\Omega$

Answer: $V_{R_L} = 4 \text{ V}$

c. Voltage across the transistor.

Found by subtracting.

Formula: $V_{CE} = V_{CC} - (V_{R_L} + V_{R_E})$

Substitution: $V_{CE} = 10 \text{ V} - (4 \text{ V} + 1 \text{ V})$

Answer: $V_{CE} = 5 \text{ V}$

27.7 PASSING AN AC SIGNAL THROUGH A COMMON EMITTER AMPLIFIER

A transistor can be used as a switch, or it can be used in its active region. If a transistor is used as a switch, it is controlled by a dc signal that turns it on (into saturation) or off (into cutoff). A sine wave applied to a transistor switch is changed into a square wave. The transistor clips the peaks of the wave.

A transistor in the active region finds applications in both dc and ac circuits. With a dc voltage, the transistor can be used as an electronically controlled variable resistor. Placed in series with a load, it can regulate the current and voltage. With an ac signal, the transistor amplifies the signal to a larger value. There are many applications.

AC Signal Changes the Base Voltage

In the common emitter amplifier, the signal is applied to the base through a coupling capacitor, figure 27-19. (Note that the values in this circuit are from sample problem 3.) The amplitude of the input voltage should not be so large to cause the transistor to go into saturation or cutoff. Output from the common emitter circuit is taken from the collector to ground. The output has a 180° phase shift as compared to the input signal.

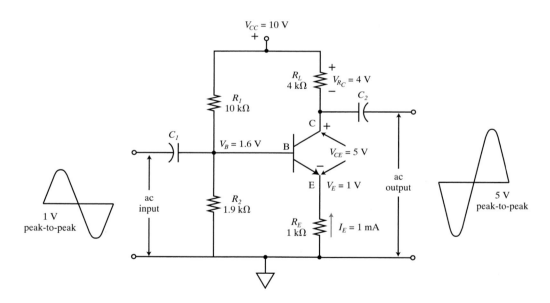

Figure 27-19. AC signal amplified by a common emitter circuit.

In the sample circuit, the base voltage at the Q point is 1.6 volts. The **coupling capacitor** prevents the ac signal from containing a dc voltage that comes from another part of the circuit. When the ac passes the capacitor, it goes to a dc level equal to the base voltage at the Q point. See figure 27-20. The base voltage varies to follow the one volt peak-to-peak input sine wave. The changing base voltage is referred to as ΔV_B.

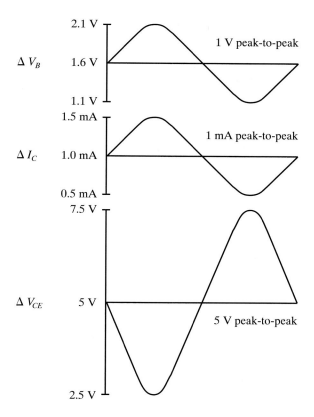

Figure 27-20. Examine an ac signal passing through the common emitter amplifier.

Collector Current Follows Base Voltage

Refer to the calculations shown in sample problem 3. The voltage across the emitter resistor is found by subtracting the base-to-emitter diode drop from the base voltage. Collector current is then found using Ohm's law with the voltage across the emitter resistor.

The collector current changes by the same amount as the changing base voltage. Figure 27-20 shows ΔI_C to have a peak-to-peak value of one mA with ΔV_B of one volt peak-to-peak and an emitter resistor of one kΩ.

Amplified Output Voltage with 180° Phase Shift

As the base voltage increases through the positive half cycle, the collector current also increases in a positive half cycle. The negative half cycle of the base voltage decreases the collector current by the same sine wave shape.

An increase in collector current results in a greater voltage drop across the emitter and load resistors. The increased voltage drops across the resistors decreases the voltage across the transistor. A decreasing collector current decreases the voltage drops across the emitter and load resistors and places a greater voltage drop across the transistor. The result is the voltage across the transistor, ΔV_{CE}, has a 180° phase shift, figure 27-20, from the input signal. Also notice that the input signal has been amplified by five.

Collector-to-Emitter Resistance

Figure 27-21 examines the transistor being used as a variable resistor. An increasing base current causes a decrease in transistor resistance to allow the increase in collector current. In the same manner, a decrease in the base current causes an increase in collector-to-emitter resistance, resulting in a decreasing collector current.

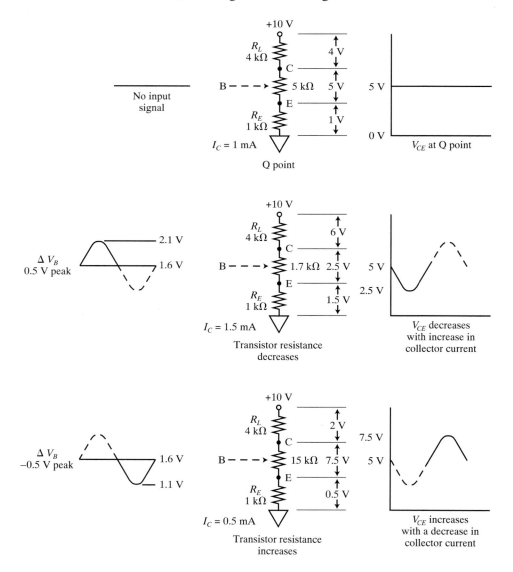

Figure 27-21. Resistance of the transistor changes in response to base current.

The collector-to-emitter resistance of the transistor in figure 27-21 is determined to be five kΩ at the Q point. This is based on the values of collector current and the collector-to-emitter voltage drop. The resistance will vary opposite the base voltage.

The base voltage increases in the positive half cycle to a peak value of 2.1 volts. The collector current increases to 1.5 mA. The voltage drops across the load and emitter resistors increase by the same proportion. The transistor's collector-to-emitter resistance drops to a peak value of 1.7 kΩ.

When the base voltage decreases in the negative half cycle to a peak value of 1.1 volts, the transistor's collector-to-emitter resistance rises to 15 kΩ. The rising and falling of the collector-to-emitter resistance and collector current produce a sine wave output across the collector.

27.8 APPLYING THE AC SIGNAL TO THE LOAD LINE

A load line is used to show the dc biasing conditions of a transistor. A load line shows the maximum collector current, at saturation, based on the circuit resistance. The load line also shows the maximum voltage drop across the transistor. This is the value of the supply voltage. The active region of the transistor is found between the two maximum limit points.

The load line can also be used to show the placement of the Q point and the effects of an ac signal applied to the circuit. The location of the Q point determines how the circuit responds to the incoming ac signal.

Q Point in the Center of the Load Line

The voltage divider biasing circuit used to produce the load line in figure 27-22 has the Q point located in the center. The input sine wave is ΔI_C with the center line at the value of I_C at the Q point.

Dashed lines, projected left from the load line, are used to show how the ac signal slides the operation up and down the load line. The positive half cycle of the input drives the transistor closer to saturation. The negative half cycle drives the transistor closer to cutoff.

Dashed lines, projected down from the load line, show the shape of the ac output voltage. As the transistor is driven closer to saturation, the output voltage, ΔV_{CE}, goes closer to zero volts. As the transistor is driven closer to cutoff, the output voltage goes closer to the maximum of V_{CC}.

Misplaced Q Point

A resistor of the wrong size can change the location of the Q point. If the Q point is located too close to saturation, a much smaller input signal will drive the transistor to the maximum collector current. As shown in figure 27-23, when the input sine wave is applied, the output voltage is clipped. In the same manner, if the Q point is placed too close to cutoff, the output voltage will again be clipped. Clipping causes distortion. This distortion is especially noticeable with audio amplifiers.

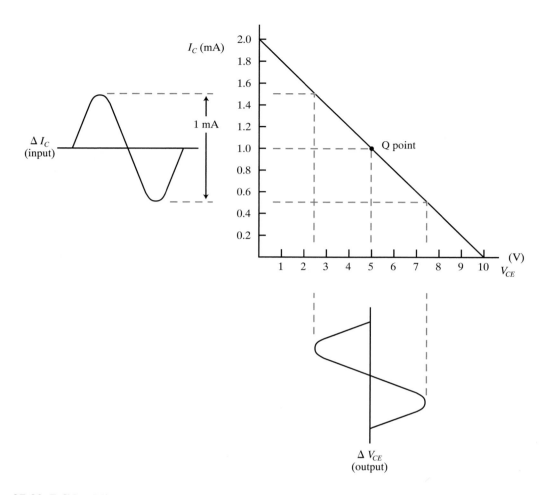

Figure 27-22. DC load line with input and output waveforms.

Some applications require the Q point to be located at a point other than the center. Two examples are the transistor switch and the RF amplifier. A transistor switch needs the transistor to be off (cutoff) until a voltage is applied to the base. In the RF amplifier, where distortion is not important, the Q point is placed well below the cutoff point. This placement allows a much larger amplification.

27.9 TRANSISTOR CIRCUIT CONFIGURATIONS

The ac conditions of a transistor circuit depend on where the input sine wave is connected and where the output is taken from. There are three types of circuit configurations, figure 27-24.

- Common emitter.
- Common collector.
- Common base.

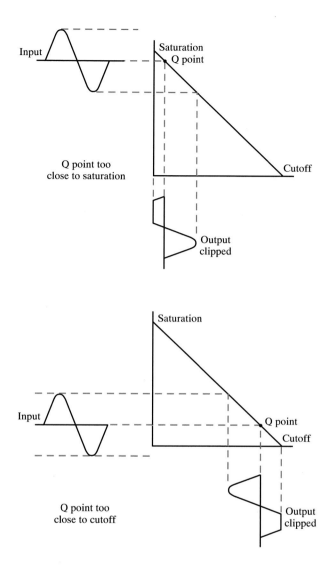

Figure 27-23. Misplacement of the Q point results in clipping of the output waveform.

The types are identified by which transistor element is common to both the input and output signals. The common element is not connected directly to either ac signal.

Common Emitter

The circuits discussed to this point are the type classified as common emitter circuits. The input voltage is on the base and the output is taken from the collector. Characteristics of the **common emitter circuit** are a 180° phase shift and good voltage amplification. The common emitter is used for audio amplifiers where distortion must be kept to a minimum.

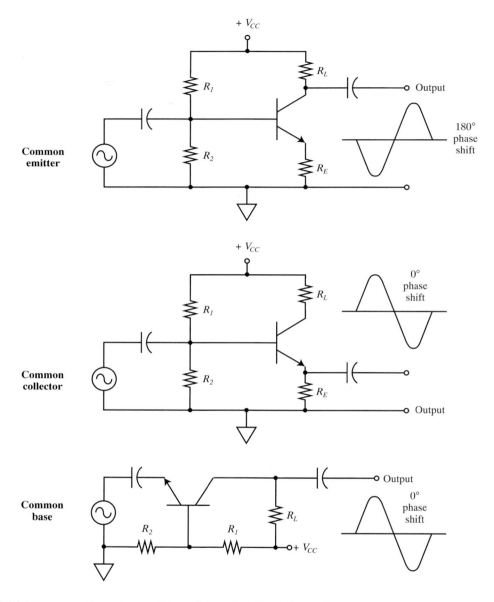

Figure 27-24. There are three types of transistor circuit configuration.

Common Collector

The **common collector circuit** has the input sine wave on the base and the output across the emitter resistor. The ac output voltage is slightly *less* than the ac input voltage. This is a voltage gain of less than one. There is no phase shift between the input and output sine waves.

The common collector circuit is used as a buffer circuit. In certain applications, maximum transfer of power cannot be obtained when connecting directly to a common emitter amplifier. The common collector provides isolation.

Common Base

The **common base circuit** receives its input on the emitter, and the output is connected to the collector. There is no phase shift between the input and output signals. Voltage gain is good. Its most useful characteristic is a high output impedance.

The common base circuit is used as an RF amplifier. It is also used in dc circuits for current and voltage regulation.

SUMMARY

- The transistor is a three-layered device similar to two diodes back-to-back.
- The three elements of a transistor are the emitter, base, and collector.
- Transistors are either NPN or PNP.
- An ohmmeter is used to test a transistor using a procedure similar to testing diodes.
- The dc power supply is connected with its polarity such that the emitter supplies majority current carriers to the base and collector, positive to a P-emitter and negative to an N-emitter.
- Beta is the ratio of collector current to base current.
- The dc load line is drawn across the transistor's characteristic curves. It has one end at collector current saturation. The other end is at the cutoff voltage, equal to V_{CC}.
- The maximum operation range of a transistor is from saturation to cutoff. The active region is between the two limits.
- An ac signal passes through a transistor by changing the base voltage and current. This changes the emitter voltage and current, which changes the collector voltage and current. The changing voltages and currents follow the input ac signal.
- Amplification in a common emitter amplifier occurs in the collector and has a 180° phase shift.

KEY WORDS AND TERMS GLOSSARY

active region: The range of operation of an active component between cutoff and saturation. In amplifier applications, it is the range where the device varies its resistance in relation to an ac input signal.

base: The element of a bipolar junction transistor sandwiched between the emitter and the collector. Base current controls collector current.

beta(β): The current gain of a bipolar junction transistor. Collector current is beta times larger than the base current.

bipolar junction transistor: A transistor composed of two back-to-back PN junctions.

collector: One of the three elements of a bipolar junction transistor. Collector current is controlled by base current.

common base circuit: Transistor circuit in which the input is applied between the emitter and the base and the output is taken between the collector an the base.

common collector circuit: Transistor circuit in which the input is applied between the base and the collector and the output is taken between the emitter and collector.

common emitter circuit: Transistor circuit in which the input is applied between the base and the emitter and the output is taken between the collector and the emitter.

coupling capacitor: A capacitor placed in series to pass an ac signal while stopping a dc voltage. Also called a blocking capacitor.

cutoff: The point where a transistor is off. The effect on the circuit is that of an open switch, with no current.

emitter: One of the three elements of a bipolar junction transistor. The emitter provides current to both the base and collector.

load line: Line drawn across a transistor's characteristic curves. The end points of the load line are the transistor's operating extremes.

Q point: The actual operating point of the transistor circuit.

saturation (transistor): The point at which the transistor acts like a closed switch, offering virtually no resistance to the circuit. All of the circuit voltage is dropped across the resistors in the collector and emitter circuits. Voltage across the transistor, V_{CE}, is approximately zero. In actual practice, it is about 0.2 V. Collector current is maximum, limited by the circuit resistance and not the transistor.

transistor: Semiconductor device capable of a wide range of operating characteristics.

KEY FORMULAS

Formula 27.A

$$I_E = I_B + I_C$$

Formula 27.B

$$\beta = \frac{I_C}{I_B} \text{ or } I_C = \beta \times I_B$$

TEST YOUR KNOWLEDGE

Do not write in this text. Please use a separate sheet of paper.
1. The arrow on the schematic for a transistor points toward what type of material?
2. Give a procedure to test a transistor with an ohmmeter.
3. Which element is the supplier of majority current carriers?
4. In a NPN transistor, what is the majority current?
5. In a PNP transistor, what is the majority current?
6. Which element has the largest current?

7. Which element has the smallest current?
8. What is the relationship of collector current to base current?
9. What is the relationship of emitter current to collector and base currents?
10. What is the formula for emitter current in terms of base and collector currents?
11. What is the beta formula?
12. Calculate emitter current with a base current of 80 µA and collector current of 10 mA.
13. Calculate the beta of the transistor in question 12.
14. What is the approximate voltage V_{BE} for a silicon transistor? For a germanium transistor?
15. What is the range of voltage for V_{CE}?
16. What is the polarity for V_{CC} with an NPN transistor? With a PNP transistor?
17. What are the limits to the range of transistor operation?
18. Use the circuit shown in figure 27-25 to complete the following:
    a. Calculate the dc load line: I_C at saturation and V_{CE} at cutoff.
    b. Plot the load line on graph paper.
    c. Calculate collector current at the Q point.
    d. Calculate the voltage drop across the collector resistor.
    e. Calculate V_{CE} at the Q point.
    f. Show the Q point on the load line.
    g. Calculate base current at the Q point.

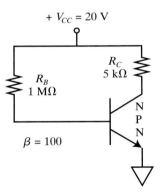

$+ V_{CC} = 20$ V

R_C
5 kΩ

R_B
1 MΩ

N
P
N

$\beta = 100$

Figure 27-25. Problem #18.

19. Use the circuit shown in figure 27-26 to complete the following:
    a. Calculate the dc load line: I_C at saturation and V_{CE} at cutoff.
    b. Plot the load line on graph paper.
    c. Calculate collector current at the Q point.
    d. Calculate the voltage drop across the collector resistor.
    e. Calculate the voltage drop across emitter resistor.
    f. Calculate V_{CE} at the Q point.
    g. Show the Q point on the load line.

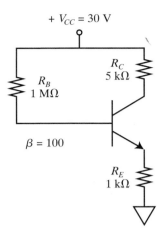

Figure 27-26. Problem #19.

20. Use the circuit shown in figure 27-27 to complete the following:
    a.   Calculate the dc load line: I_C at saturation and V_{CE} at cutoff.
    b.   Plot the load line on graph paper.
    c.   Calculate the base voltage at the Q point.
    d.   Calculate collector current at the Q point.
    e.   Calculate the voltage drop across the collector resistor.
    f.   Calculate the voltage drop across the emitter resistor.
    g.   Calculate V_{CE} at the Q point.
    h.   Show the Q point on the load line.

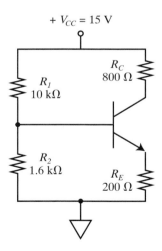

Figure 27-27. Problem #20.

21. Describe how an ac signal passes through a common emitter amplifier.

Chapter 28
Other Semiconductor Devices

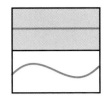

Upon completion of this chapter, you will be able to:
- Define technical terms related to semiconductor devices.
- Describe the operation of field-effect transistors.
- Describe the operation of basic operational amplifiers (op-amps) and calculate output gain.
- Describe the operation of the 555 timer as an integrated circuit and calculate the output waveform frequency and duty cycle.
- Describe the operation of the unijunction transistor (UJT).
- Describe the operation of the silicon controlled rectifier (SCR).
- State typical applications for the DIAC, TRIAC, LED, and light-sensitive transistor.

Semiconductor devices of all types have a very important part in the operation of electronic equipment. Diodes and transistors, Chapters 26 and 27, are the basic semiconductors. This chapter introduces other semiconductor devices and some simple applications.

There are many changes being made in the electronics industry. Most of the changes are found in the more advanced types of equipment, such as the computer industry. The basic devices discussed in here are the backbone of electronics. The only changes are in their applications.

28.1 FIELD-EFFECT TRANSISTORS

The **field-effect transistor (FET)** is a type of transistor that is used in many of the same applications as the bipolar junction transistor. However, with some applications, the two devices are not interchangeable.

There are two general categories of FET, junction field-effect transistors (JFET) and insulated-gate field-effect transistor (IGFET). One common type of IGFET is the metal oxide semiconductor field-effect transistor (MOSFET). All FETs operate using the same basic principle. An electric field controls current through the device by opposing or enhancing the flow of majority carriers.

Block Diagram of a JFET

The **junction field-effect transistor (JFET)** is made of a continuous block of semi-conductor material with small pieces of the opposite type (N-type or P-type) on its sides. It has three elements: the source, the drain, and the gate. When comparing to the bipolar transistor: the **source** compares to the emitter, the **drain** to the collector, and the **gate** to the base.

JFETs is available in two types, *N-channel* and *P-channel.* The N-channel JFET is shown in figure 28-1. The P-channel is shown in figure 28-2. Both types operate using the electric field concept to regulate the flow of majority carriers.

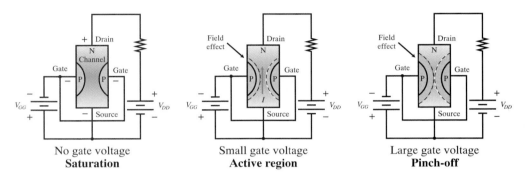

Figure 28-1. Block diagram of an N-channel JFET showing the FET's three areas of operation.

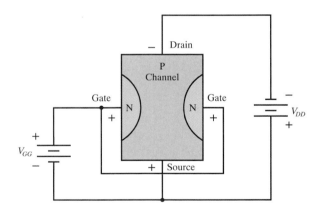

Figure 28-2. Block diagram of a P-channel JFET.

The **channel** is continuous from the source to the drain. The source receives a supply voltage that has the same polarity as the type of channel. An N-channel receives a negative voltage on the source. A P-channel receives a positive voltage on the source. The other side of the supply voltage is connected to the drain.

The FET biasing is accomplished with the gate by producing an electric field radiating into the channel. The field does not require gate current. Therefore, the FET is considered more efficient than the bipolar transistor. The bipolar transistor biases the base with a small amount of base current, as well as base voltage. There is a small amount of power wasted in this base biasing.

Operating a JFET

The biasing voltage on the gate has an opposite polarity from the type of semiconductor material. The biasing voltage controls the amount of electric field imposed on the channel. In most applications, biasing resistors are used rather than a separate biasing battery.

Referring to figure 28-1, the FET has three ranges of operation: saturation, active, and pinch-off, similar to the bipolar transistor's saturation, active, cutoff.

When there is no gate voltage, current flows through the channel with no opposition. The FET is saturated. As the gate voltage increases, majority carriers from the gate are forced into the channel. The majority carriers from the gate are opposite from the majority carriers of the channel. This produces an area which opposes current in the channel. The FET operates in the active region with the gate voltage controlling channel current.

If enough gate voltage is applied, the field becomes strong enough to stop the current in the channel. This condition is called **pinch-off.**

Amplifier Configurations

Using the same format as the bipolar transistor, the FET is connected in three different configurations:

• Common source.

• Common drain.

• Common gate.

The common source amplifier, shown in figure 28-3 produces a 180° phase shift. This circuit has similar characteristics to the common emitter amplifier using a bipolar

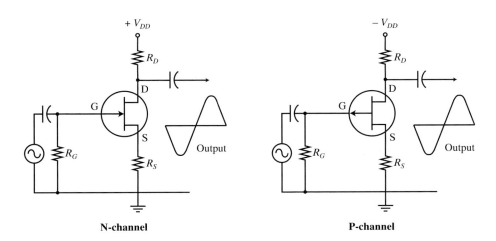

Figure 28-3. JFET connected as a common source amplifier is similar to the common emitter circuit.

transistor. The biasing resistor, R_G, is connected between the gate and ground. This resistor provides the necessary negative voltage on the P-type gate.

Also, note the schematic symbol for the JFET. In keeping with the standard for semiconductors, the arrow points at the N-type material. An N-channel JFET has the arrow pointing at the section between the drain and source. A P-channel JFET has the arrow pointing away from the channel, toward the gate.

Figure 28-4 shows the schematic diagrams for the common drain and common gate configurations. These circuits have a 0° phase shift. Their counterparts using bipolar transistors are common collector and common base, respectively.

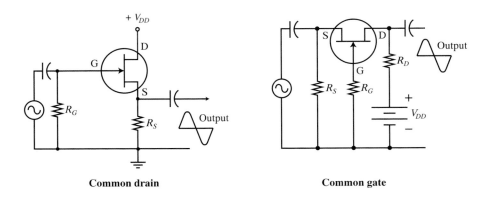

Common drain   **Common gate**

Figure 28-4. Common drain and common gate configurations.

Metal Oxide Semiconductor Field-effect Transistor (MOSFET)

The **metal oxide semiconductor field-effect transistor (MOSFET)** has its gate insulated from the channel by a very thin layer of silicon dioxide. There are four terminals on a MOSFET: source, drain, gate, and substrate, figure 28-5.

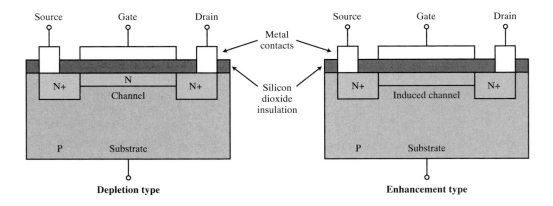

Depletion type   **Enhancement type**

Figure 28-5. There are four terminals on a metal oxide semiconductor field-effect transistor (MOSFET).

The source and drain of the MOSFET function the same as the JFET. Their terminals are made of metal and attached to the channel material. The source and drain are very heavily doped, as indicated by the + symbol in the block drawing. The gate is a metal plate rather than semiconductor material. The **substrate** is a massive block of semiconductor material that is made of the opposite type from the channel.

MOSFET Operating Modes

The MOSFET is designed to be operated in either of two modes, depletion or enhancement. In the **depletion mode,** a channel of the same material type as the source and drain is used. The gate voltage has the same polarity as the channel. As the gate voltage is increased, the current carriers are forced out of the channel region. The channel becomes *depleted* of current carriers. The substrate has the same polarity as its type, positive for P and negative for N. The substrate is used as a place for the displaced current carriers.

An **enhancement mode** device does not have a continuous channel. Instead, the gate has the same polarity as the substrate. As the gate voltage is increased, minority carriers from the substrate form an induced channel allowing current to flow. The substrate is used as a supply for current carriers.

The schematic symbols for MOSFET devices are shown in figure 28-6. The depletion mode type uses a solid line for the channel. The enhancement mode device uses a broken line for the channel.

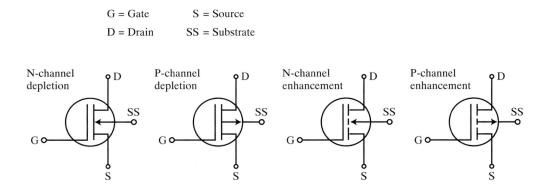

G = Gate   S = Source

D = Drain   SS = Substrate

Figure 28-6. Schematic symbols for four different varieties of MOSFET.

28.2 OPERATIONAL AMPLIFIER (OP-AMP)

The **operational amplifier,** or **op-amp** is a high gain amplifier circuit containing many transistors enclosed in one integrated circuit package. The package is usually rectangular with eight legs, four legs on each side. However, other packages are available. An op-amp is designed to have gains as high as 100,000. Gains that large are not used in practical applications, however, due to other problems.

Schematic Symbol

The schematic symbol for the op-amp is shown in figure 28-7. A triangle is a common symbol used for amplifiers. This figure shows $+V$ and $-V$. These represent sources for a positive voltage and a negative voltage. The negative voltage is not the same as ground (0 volts). There must be two separate voltage sources. The $+V$ has its negative connected to ground. The $-V$ has its positive connected to ground.

The schematic symbol also shows two inputs, inverting and noninverting. An ac signal connected to the **inverting input** will have an output with 180° phase shift. An ac signal connected to the **noninverting input** has a 0° phase shift at the output.

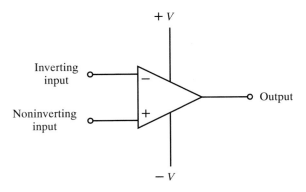

Figure 28-7. Schematic symbol for an operational amplifier.

Op-Amp Circuits

The inverting op-amp circuit is shown in figure 28-8. This circuit has several advantages. One advantage is that the connection made at the minus input is virtual ground. **Virtual ground** is a point in the circuit equal to zero volts. The advantage of having a

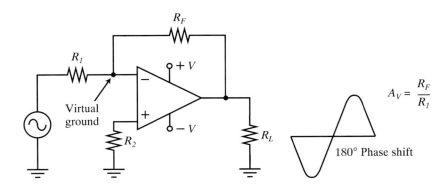

Figure 28-8. This op-amp uses negative feedback with inverting input.

virtual ground point is the ease at setting an input impedance. The ac signal is connected to R_1, making this resistor the load for the ac signal. This resistor can be set to the best value for maximum transfer of power.

The noninverting op-amp amplifier, shown in figure 28-9 has a 0° phase shift between the input and output signals. The ac signal in this circuit is not connected to a virtual ground point. The plus input has nearly infinite impedance. The impedance seen by the ac source is determined by the value of R_2. With this resistor being in parallel, rather than series, its value can be quite high.

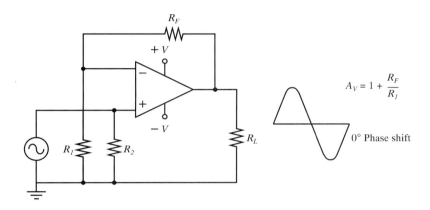

Figure 28-9. Op-amp using the noninverting input.

Op-Amp Gain

Typical specifications for op-amps give a very high open-loop gain. An open-loop circuit has no feedback. This gain, however, is not equal over a range of input frequencies. In fact, the specified open-loop gain may be valid for only dc voltages. As the frequency is increased, the gain drops off rapidly.

The closed loop gain, which uses a negative feedback resistor, R_F, sets the gain to a value that produces an equal gain for all frequencies required. An op-amp provides charts showing the frequency response for various gains. A gain of 10 to 20 is typical for most applications. Gain can be calculated using one of these formulas.

Formula 28.A (Inverting input)

$$A_V = \frac{R_F}{R_1}$$

A_V is the amplifier gain, a number with no units.

R_F is the feedback resistance, measured in ohms.

R_1 is the resistance of resistor R_1, measured in ohms.

Formula 28.B (Noninverting input)

$$A_V = 1 + \frac{R_F}{R_1}$$

A_V is the amplifier gain, a number with no units.

R_F is the feedback resistance, measured in ohms.

R_1 is the resistance of resistor R_1, measured in ohms.

Sample problem 1.

An inverting op-amp circuit has a 10 kΩ resistor for R_F and a 1 kΩ resistor for R_1. Determine the gain.

Formula: $A_V = \dfrac{R_F}{R_1}$

Substitution: $A_V = \dfrac{10 \text{ k}\Omega}{1 \text{ k}\Omega}$

Answer: $A_V = 10$ (Note that gain has no units.)

Sample problem 2.

An op-amp circuit with the ac input connected to the noninverting input has a feedback resistor of 100 kΩ and R_1 of 5 kΩ. What is the circuit gain?

Formula: $A_V = 1 + \dfrac{R_F}{R_1}$

Substitution: $A_V = 1 + \dfrac{100 \text{ k}\Omega}{5 \text{ k}\Omega}$

Answer: $A_V = 21$

28.3 555 TIMER

The 555 timer is an integrated circuit that contains a complete circuit within one package, figure 28-10. It is used in many different applications. The primary function of the **555 timer** is to produce a rectangular wave at a specified frequency. The 555 timer can be used to trigger other circuits. If its output is fed to a speaker, it makes a chirping sound.

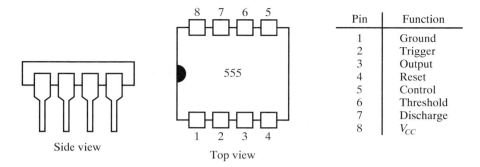

Figure 28-10. The 555 timer is an integrated circuit.

555 Control Components

The circuit shown in figure 28-11 connects the 555 as a free-running multivibrator. This produces a continuous rectangular or square wave. Three components eternal to the timer control the shape and frequency of the output signal: R_A, R_B, and C_1. C_2 is used as a decoupling capacitor, which is a noise filter. It has no effect on the output waveform.

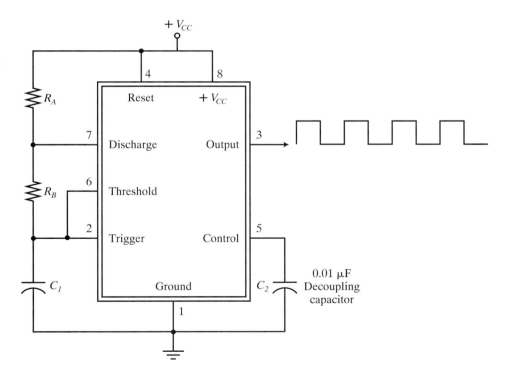

Figure 28-11. 555 timer connected as a free-running multivibrator. It produces a continuous rectangular wave.

The timing of the output waveform is determined by the RC time constant circuit. The RC time constant circuit is formed by R_A, R_B, and C_1. The charge path is through both resistors. The discharge path is through only R_B. When the capacitor charges to approximately one time constant, (2/3 of V_{CC}) the output switches to low. The capacitor discharges through only R_B (to approximately 1/3 of V_{CC}), and the output switches on. This cycle repeats continuously.

Frequency of the 555

The frequency is the rate at which the output switches on and off. It is controlled by the combination of the resistors and capacitor.

Formula 28.C (Refer to figure 28-11)

$$f = \frac{1.44}{(R_A + 2R_B) \times C_1}$$

f is the frequency of the 555 timer, measured in hertz.

R_A, R_B, and C_1 are individual circuit values.

Sample problem 3. _____

Calculate the output frequency of a 555 timer circuit with the following component values: R_A = 3.3 kΩ, R_B = 5.6 kΩ, C_1 = 0.01 μF

Formula: $f = \dfrac{1.44}{(R_A + 2R_B) \times C_1}$

Substitution: $f = \dfrac{1.44}{(3.3 \text{ kΩ} + (2 \times 5.6 \text{ kΩ})) \times 0.01 \text{ μF}}$

Answer: f = 9931 Hz

Sample problem 4. _____

Using the same resistors as those of sample problem 3, find the value of capacitance required to produce an output frequency of five kilohertz.

Formula: $f = \dfrac{1.44}{(R_A + 2R_B) \times C_1}$

Rearranging: $C_1 = \dfrac{1.44}{(R_A + 2R_B) \times f}$

Substitution: $C_1 = \dfrac{1.44}{(3.3 \text{ kΩ} + (2 \times 5.6 \text{ kΩ})) \times 5 \text{ kHz}}$

Answer: C_1 = 0.019 μF

Duty Cycle of the 555 Oscillator

The **duty cycle** is the relative width of the on and off times of the output waveform. Using the circuit shown in figure 28-11, the duty cycle will always be 50 percent or greater. This means there will always be an on time equal to or longer than the off time. A 50 percent duty cycle, on and off are times equal, is achieved by selecting a value of R_B much greater than R_A. A value approximately 10 times greater should be sufficient. The value of the capacitor does not affect the duty cycle.

Formula 28.D

$$\text{Duty cycle} = \frac{R_A + R_B}{R_A + 2R_B} \times 100\%$$

R_A and R_B are individual circuit values.

Sample problem 5.

Calculate the duty cycle of the circuit in sample problem 3.

Formula: $\text{Duty cycle} = \dfrac{R_A + R_B}{R_A + 2R_B} \times 100\%$

Substitution: $\text{Duty cycle} = \dfrac{(3.3\ \text{k}\Omega + 5.6\ \text{k}\Omega)}{(3.3\ \text{k}\Omega + (2 \times 5.6\ \text{k}\Omega))} \times 100\%$

Answer: $\text{Duty cycle} = 61.4\%$

28.4 UNIJUNCTION TRANSISTOR

The unijunction transistor is not used as an amplifier, as its name would suggest. Instead, the **unijunction transistor (UJT)** is used as part of a timing circuit. The UJT makes use of the characteristics of an RC time constant circuit. The UJT allows the capacitor to charge and discharge at a set frequency.

UJT Construction

The UJT is formed by attaching a P-type emitter to a block of N-material, as shown in figure 28-12. The N-material has a lead at each end. The leads are labeled base 1 (B_1) and base 2 (B_2). The emitter is located closer to base 2, approximately 2/3 of the distance from base 1 to base 2. The schematic symbol, also shown in figure 28-12, appears almost the same as the JFET symbol. The only difference is that the JFET symbol has its arrow drawn straight and the UJT arrow is slanted down.

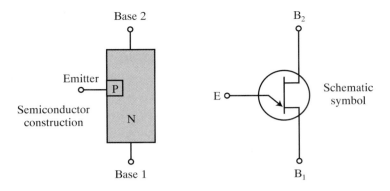

Figure 28-12. In the unijuction transistor (UJT), the emitter is located closer to base 2 than to base 1.

UJT Relaxation Oscillator

The UJT is a normally-closed switch. The operating contacts of the switch are between the emitter and base 1. Referring to figure 28-13, with V_{BB} applied, capacitor C_1 charges through resistor R_1. When the charge across the capacitor equals approximately 2/3 of the voltage at B_2 the UJT turns on and the switch between E and B_1 closes. This is approximately one time constant of charge time.

When the UJT turns on, there is approximately 0.6 V drop from E to B_1. The PN junction is acting as a forward biased diode. The UJT remains turned on, discharging the capacitor through R_3. When the capacitor voltage drops below 0.6 V the UJT turns off and the cycle repeats.

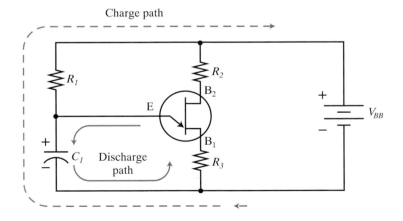

Figure 28-13. Charge and discharge paths of a unijunction transistor relaxation oscillator.

UJT Output Waveforms

The UJT relaxation oscillator has two outputs available, as shown in figure 28-14. Output A is the voltage across the capacitor. Notice how this waveform has the shape of the capacitive time constant, as seen in Chapter 14. Output B is the shape of the capacitor discharging. Output A is a sawtooth waveform. Output B produces timing spikes.

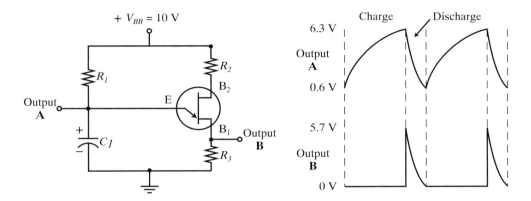

Figure 28-14. Two outputs are available from the UJT relaxation oscillator. The outputs have the shape of an RC time constant circuit.

28.5 SILICON CONTROLLED RECTIFIER

The **silicon controlled rectifier (SCR)** is a rectifier diode with a gate to turn it on at a specified time. The SCR can be used in both dc and ac applications. The SCR conducts in only one direction, in the same manner as a rectifier diode. However, even though it is forward biased, it will not conduct until the gate triggers it on. Once it is turned on, the gate loses control of the diode. The SCR is turned off by removing the forward bias voltage to the anode and cathode.

SCR in a DC Circuit

The SCR can be used to act like a relay circuit, where a control signal operates the relay switch applying power to a load. Figure 28-15 shows the SCR in a dc circuit with two different power supplies.

Voltage is applied to the load circuit by closing S_2. No load current can flow, however, until the SCR is triggered. The polarity of the load voltage is proper to forward bias the rectifier diode.

When S_1 is closed, the positive voltage is applied to the gate triggering the SCR, allowing load current to flow. Gate resistor R_G provides a current to the gate.

When the push button switch S_1 is opened, holding current continues to flow through R_G and the forward biased diode. The diode is held on as long as there is forward bias applied to the anode and cathode. To stop the current, S_2 is opened.

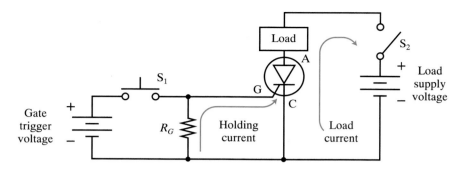

Figure 28-15. A silicon controlled rectifier is switched on by the positive gate trigger voltage.

One application for this SCR circuit is a burglar alarm. The load would be the alarm bell, powered by a rechargeable battery. The momentary gate trigger switches would be the magnetic switches on doors and windows, powered by a battery. The opening of a window switch would trigger the alarm. It would stay on until power is removed, even if the window switch was closed again.

SCR in an AC Circuit

The SCR can be used as a half-wave rectifier with a power control on the gate, figure 28-16. The gate can delay the rectifier from turning on at any point from 0° to 90°. By delaying the turn-on point, load power is controlled by allowing load current for a smaller portion of the cycle.

When the ac signal switches to the negative half cycle, the rectifier is reverse biased turning it off. The gate automatically regains control during the positive half cycle.

Diode D_1 is used to protect the gate from a reverse bias. R_1 protects the gate from too much current. Variable resistor R_2 adjusts the turn-on time and gate holding current.

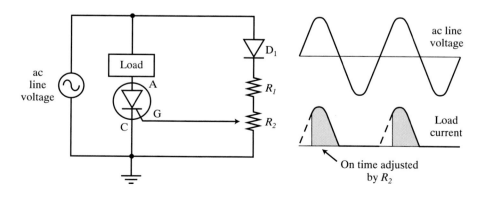

Figure 28-16. The SCR can be used as a half-wave rectifier. Its on time is controlled by the gate.

28.6 TRIACS AND DIACS

The TRIAC and DIAC are two devices used to control ac power in a manner similar to an SCR, except as full wave devices. The schematic symbols are shown in figure 28-17. The **TRIAC** operates like two SCRs in opposite directions with a common gate. The **DIAC** operates like two opposite facing zener diodes (refer to Chapter 26).

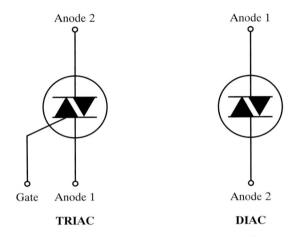

Figure 28-17. Schematic symbols of two common thyristor power control devices.

As shown in figure 28-18, the ac line voltage is controlled on both halves of the cycle. The load power is reduced by having less on-time, while still receiving the same line voltage. This type of circuit is used for motor speed control.

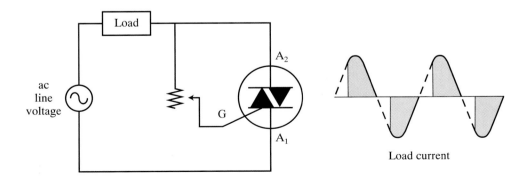

Figure 28-18. The TRIAC can be used to control load current in an ac circuit.

The DIAC is a voltage reference device. It can be purchased in a wide range of voltage ratings. It is an off switch until the reference voltage is reached, then it acts like an on switch. The DIAC is often used in conjunction with a TRIAC to provide a more sophisticated gate control, figure 28-19. This type of circuit is used in lighting control.

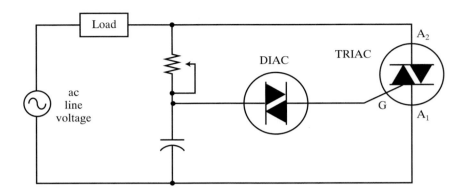

Figure 28-19. DIAC used to trigger a TRIAC.

28.7 LIGHT-EMITTING DIODE

The **light-emitting diode (LED)** is a semiconductor device that is used in place of small indicator lamps. It has the same operating characteristics as a rectifier diode, except that when it is forward biased it emits light. A simple circuit is shown in figure 28-20.

The LED has many advantages over a typical lamp. It only drops 0.6 V and has practically no power consumption. It is very small and can be used in such applications as phones and computers.

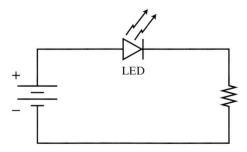

Figure 28-20. A light-emitting diode gives off light when it is forward biased.

Figure 28-21 uses a transistor switch to control the operation of the LED. The LED in this circuit is used to indicate when some operation has taken place. For example, it can indicate that power to a machine has been turned on.

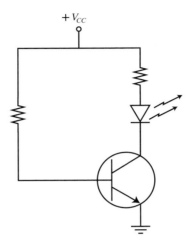

Figure 28-21. LED operated by a transistor switch.

Seven-Segment LED Display

Figure 28-22 shows the layout of a seven-segment LED display. This is available in a single package in many different sizes and colors. As shown, combinations of the segments determine what digit is formed.

Each segment is a separate LED and is controlled by its own circuit. The control circuit may be separate transistors for each segment, or it may be a single-unit decimal decoder. A *decimal decoder* receives information from another circuit telling it which decimal number to display.

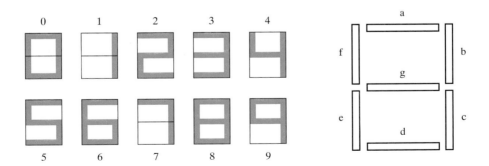

Figure 28-22. Seven-segment LED displays come in packages. These displays are used to form digits.

Seven-segment LED displays are commonly used in such applications as digital clocks. The liquid crystal display, found in digital watches, operates on the same basic principal as the LED display, except it does not need as much current.

28.8 LIGHT-SENSITIVE TRANSISTOR

Certain semiconductor devices are sensitive to light. One example is the **light-sensitive transistor,** or **phototransistor.** Light strikes the base element. The amount of base current is related to the amount of light striking the surface.

Some applications use an LED mounted close to a phototransistor. The LED operates the transistor, which in turn operates another circuit. This application is used in the end-of-tape indicators in VCRs. At the end and beginning of the VCR tape is a section of clear plastic leader. An LED shines through the clear portion but not through the normal tape. This is how the machine knows when it has arrived at the end or beginning of tape and allows the VCR to stop before the tape breaks.

Light-sensitive transistors are also used to control circuits in response to the presence or absence of light. Figure 28-23 shows two possible circuits of this type. In the darkness operated circuit, Q_1 is on when light is present, which holds Q_2 off. With Q_2 off, the relay is deactivated. When the light is removed, Q_1 turns off and Q_2 turns on, activating the relay. In the light operated circuit, Q_2 is turned on when light strikes Q_1.

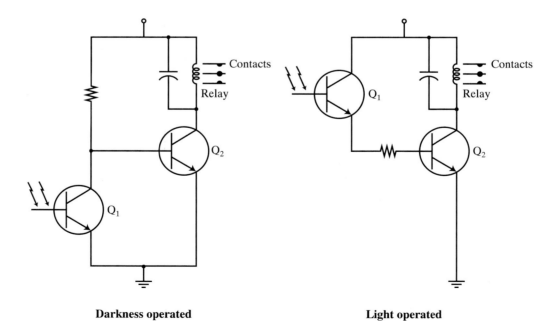

Darkness operated   **Light operated**

Figure 28-23. A light-sensitive transistor is used to operate a transistor switch.

There is an advantage to using a second transistor operated by the light-sensitive transistor, instead of using a light-sensitive transistor on its own. Many light-sensitive devices do not have high current capacity. These devices are best used to operate a power transistor, which operates the load.

SUMMARY

- The FET is close to the bipolar junction transistor in its applications. The advantage of the FET is higher efficiency.
- The FET has a drain, source, and gate. It is available with either an N-channel or P-channel.
- The range of operation of the FET is from saturation to pinch-off with the active region in between.
- A MOSFET is an FET with an insulated gate.
- MOSFETs can be operated in either the depletion mode or enhancement mode.
- An op-amp is an integrated circuit amplifier designed to have very high efficiency and high gain.
- Op-amp gain is controlled by feedback resistors.
- The 555 timer is an integrated circuit used to produce a rectangular or square wave.
- The UJT is a semiconductor switch used in timing circuits and to produce a sawtooth waveform.
- The SCR is a rectifier diode with a semiconductor switch controlling when it turns on and off.
- TRIAC and DIAC are devices similar to the SCR except they are used to control ac circuits.
- The light-emitting diode produces light when current flows through it.
- Light-sensitive devices produce current when struck by light.

KEY WORDS AND TERMS GLOSSARY

channel: Area where electrons flow between the source and drain.

depletion mode: One of the operating modes of a MOSFET. The channel is made of the same type of material as the source and drain.

DIAC: A semiconductor device that operates like two opposite facing zener diodes.

drain: One lead on a FET. Current is controlled by the gate. Similar to the collector on a BJT.

duty cycle: The amount of time a circuit is on, as compared to the time off.

enhancement mode: One of the operating modes of a MOSFET. There is no continuous channel and the gate has the same polarity as the substrate.

field-effect transistor: A class of transistor that includes the JFET, MOSFET, and IGFET.

555 timer: Device that produces a rectangular wave at a specified frequency.

gate: One lead on a FET. Controls current through the drain. Similar to the base on a BJT.

junction field-effect transistor (JFET): A type of transistor made up of a continuous piece of semiconductor material with small pieces of the opposite type on either side. Has three elements: source, gate, and drain.

light-emitting diode (LED): A diode that emits light when forward biased.

light-sensitive transistor: A transistor that has a base current that can be controlled with light. Also called a phototransistor.

metal oxide semiconductor field-effect transistor (MOSFET): A type of transistor with four terminals: source, gate, drain, and substrate.

operational amplifier (op-amp): A high gain amplifier circuit containing many transistors in one integrated circuit package.

phototransistor: A transistor that has a base current that can be controlled with light. Also called a light-sensitive transistor.

pinch-off: Condition in a FET when a high enough gate voltage stops current in the channel.

silicon controlled rectifier (SCR): A rectifier diode with a gate to turn it on at a specified time.

source: One lead on a FET. Provides current to the drain and gate. Similar to the emitter on a BJT.

substrate: Block of semiconducting material that makes up a MOSFET.

TRIAC: A semiconductor device that operates like two SCRs in opposite directions that have a common gate.

unijunction transistor (UJT): One type of transistor that is used in timing circuits.

virtual ground: A point in an op-amp circuit equal to zero volts.

KEY FORMULAS

Formula 28.A

$$A_V = \frac{R_F}{R_1}$$

Formula 28.B

$$A_V = 1 + \frac{R_F}{R_1}$$

Formula 28.C

$$f = \frac{1.44}{(R_A + 2R_B) \times C_1}$$

Formula 28.D

$$\text{Duty cycle} = \frac{R_1 + R_2}{R_1 + 2R_2} \times 100\%$$

TEST YOUR KNOWLEDGE

Do not write in this text. Please use a separate sheet of paper.
1. What are the two general categories into which FETs can be divided?
2. Draw a block diagram of an N-channel JFET. Label the source, drain, gate, and channel. Show the polarities of the supply and biasing voltages. Describe the operation of the channel.
3. Draw a block diagram of a P-channel JFET. Label the source, drain, gate, and channel. Show the polarities of the supply and biasing voltages. Describe the operation of the gate.
4. When the JFET is saturated, is the gate voltage large, small, or medium? What is the relative amount of drain current?
5. When the JFET is in pinch-off, is the gate voltage large, small, or medium? What is the relative amount of drain current?
6. When the JFET is in the active region, is the gate voltage large, small, or medium? What is the relative amount of drain current?
7. What type of material is the gate made of in a MOSFET?
8. What type of material is the substrate made of in a MOSFET?
9. What is the function of the substrate in a depletion-mode MOSFET?
10. What is the function of the substrate in an enhancement-mode MOSFET?
11. When voltage is applied to the gate of a depletion-mode MOSFET, what is the effect on the channel and its relative amount of current?
12. When voltage is applied to the gate of an enhancement-mode MOSFET, what is the effect on the channel and its relative amount of current?
13. Draw the schematic symbol of a depletion-mode MOSFET. Label the source, drain, gate, substrate, and channel.
14. Draw the schematic symbol of an enhancement-mode MOSFET. Label the source, drain, gate, substrate, and channel.
15. Draw the schematic diagram of a basic op-amp circuit with an inverting input. Label the resistors as: R_F, R_1, R_2, R_L. Label the virtual ground point.
16. With the circuit of Problem #15, calculate the output gain if $R_F = 25$ kΩ and $R_1 = 5$ kΩ.
17. Draw the schematic diagram of a basic op-amp circuit with a noninverting input. Label the resistors as: R_F, R_1, R_2, R_L.
18. With the circuit of Problem #17, calculate the output gain if $R_F = 25$ kΩ and $R_1 = 5$ kΩ.

19. What is the purpose of negative feedback in an op-amp circuit? Which resistors are used for negative feedback?

20. Draw the schematic diagram of a 555 timer connected as an astable multivibrator. Label R_A, R_B, C_1, C_2. Also label the pin numbers and their functions.

21. What is the expected output of an astable multivibrator?

22. Using the circuit of Problem #20, determine the output frequency with the following component values: R_A = 4.7 kΩ, R_B = 10 kΩ, C_1 = 0.22 μF.

23. Using the resistance values given in Problem #22, what should the value of C_1 be changed to for an output frequency of 10 kHz?

24. Calculate the duty cycle of the 555 timer with resistance values given in Problem #22.

25. What is the primary application for a unijunction transistor?

26. Draw the schematic diagram of a UJT relaxation oscillator. Label R_1, R_2, R_3, C_1, E, B_1 and B_2. Draw arrows to indicate the charge and discharge paths.

27. Draw the expected output waveforms of the circuit in Problem #26. Explain the shape of each waveform.

28. Draw the schematic diagram of an SCR used as a relay in a dc circuit. Label the gate, cathode, and anode. Draw an arrow to indicate the path for load current.

29. What is the function of the gate in the SCR circuit of Problem #28? What is the purpose of the holding current?

30. Draw the expected output waveform when an SCR is used in an ac circuit. What is the purpose of having a portion of the output sine wave missing? What controls how much of the waveform will be missing?

31. What is typical application of a TRIAC? What device has a function similar to the TRIAC (in this chapter)?

32. What is typical application of a DIAC? What device has a function similar to the DIAC (see Chapter 26)?

33. What is an LED? What is a typical application for an LED?

34. What is a light-sensitive transistor? What is a typical application of a light-sensitive transistor?

Chapter 29
Introduction to
Digital Electronics

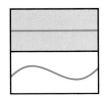

Upon completion of this chapter, you will be able to:
- Contrast digital circuits with analog circuits.
- Define the binary numbering system for use in digital circuits.
- Convert decimal numbers to binary and binary numbers to decimal.
- Analyze basic binary logic gates such as AND, OR, NOT, NAND, and NOR.

All of the circuitry discussed to this point, with the exception of transistor switches, has been analog circuitry. **Analog** circuits are circuits that have a continuous range of voltages, from zero to the maximum applied voltage. **Digital** circuits operate using only discrete steps. Often, digital circuits have only two voltage states, on and off. These digital circuits produce an output of either zero volts or the applied voltage.

Digital circuits are used primarily with computers and decision-making circuits. A **decision-making circuit** can choose between a yes and no answer. The on/off voltages states make it very easy for a computer to use decision-making circuits. The on/off can be used to represent yes/no answers.

The digital concept has also been applied to other areas of electronics technology. Digital clocks and watches use digits to indicate the time, as opposed to the analog-type clocks with sweep hands. An example of a decision-making circuit can be seen in an electronic combination lock.

29.1 BASIC NUMBERING SYSTEMS

We, as human beings, use the customary decimal system. The decimal system is based on 10 digits, 0 through 9. These digits can be placed in an infinite number of combinations to produce any desired result. The digital computer does not use the decimal system directly. It is translated to a numbering system containing only two states, on and off. The **binary** system, base 2, has only the digits zero and one. The **base** of a system states how many digits are available. The decimal system is base 10.

Place Values

The position a digit holds in a number determines its value. Each place, starting from the decimal point, is found by raising the base to an increasingly larger power. Notice in the place value charts, figures 29-1 and 29-2, that the exponents start with zero. Any number raised to the zero power equals one. Any number raised to the one power equals the number itself.

In figures 29-1 and 29-2, the rows labeled "Value" show the decimal equivalent value for each digit's place. To determine the decimal equivalent value for each digit in a number, multiply that digit by the digit's place value. When working with numbers in various bases, a small subscript following each number represents the base the number is written in. The number 11_{10} represents 11 in the base 10 system. The number 11_2 represents 11 in the base 2 system, which is equal to 3 in the base 10 system. These subscripts can be important.

Relating to the Decimal System

The place value chart for the decimal system is partially shown in figure 29-1. The first place is the one's column, with the base raised to the zero power. Any number raised to the zero power equals one. The exponents increase one digit with each place value. The base is raised to that power. A number raised to a power states how many times the number is multiplied by itself. For example: $10^3 = 10 \times 10 \times 10$.

Figure 29-1. Place value chart for the decimal system.

Place	10^5	10^4	10^3	10^2	10^1	10^0
Value	100,000	10,000	1,000	100	10	1

Sample problem 1.

Find the value of each digit in the decimal number $720{,}316_{10}$. Then add individual values to find the total value of the complete number.

$$720{,}316_{10} =$$

$$
\begin{aligned}
7 \times 100{,}000 &= 700{,}000 \\
2 \times 10{,}000 &= 20{,}000 \\
0 \times 1{,}000 &= 0 \\
3 \times 100 &= 300 \\
1 \times 10 &= 10 \\
6 \times 1 &= 6 \\
\hline
&720{,}316_{10}
\end{aligned}
$$

Binary System

As mentioned earlier, the binary system uses only the digits zero and one. Every number in the binary number system is made up of a combination of zeros and ones. A partial place value chart for the binary system is shown in figure 29-2. As with the decimal system, the exponent indicates how many times the base is multiplied by itself. For example: $2^4 = 2 \times 2 \times 2 \times 2 = 16$. Each place value column is named after its decimal value. For example: 2^0 is the one's column, 2^1 is the two's column, and 2^4 is the sixteen's column.

Figure 29-2. Place value chart for the binary system.

Place	2^5	2^4	2^3	2^2	2^1	2^0
Value	32	16	8	4	2	1

Sample problem 2.

Find the decimal equivalent value of each digit in the binary number: 110011_2. Also, add to find the decimal equivalent value of the complete number.

$$110011_2 =$$

$$1 \times 32 = 32$$
$$1 \times 16 = 16$$
$$0 \times 8 = 0$$
$$0 \times 4 = 0$$
$$1 \times 2 = 2$$
$$1 \times 1 = 1$$
$$\overline{ 51_{10}}$$

Counting in Binary

When counting, the place value of digits is very important. Zeros are used as place holders, to keep each digit in its proper place value column. Compare the counting in decimal to the equivalent binary numbers in the chart shown in figure 29-3.

Figure 29-3. Decimal numbers are shown to the left with their binary equivalents to the right.

Decimal	Binary	Decimal	Binary
0	0	16	10000
1	1	17	10001
2	10	18	10010
3	11	19	10011
4	100	20	10100
5	101	21	10101
6	110	22	10110
7	111	23	10111
8	1000	24	11000
9	1001	25	11001
10	1010	26	11010
11	1011	27	11011
12	1100	28	11100
13	1101	29	11101
14	1110	30	11110
15	1111	31	11111

When counting in binary, each place value column uses the digits 0 and 1. In decimal we count 0 through 9, then the next column is used to make 10. Then, the cycle is repeated, 0 through 9 again. In binary, 0 through 9 is: 0, 1, 10, 11, 100, 101, 110, 111, 1000.

29.2 CONVERTING FROM ONE SYSTEM TO ANOTHER

The binary place value chart can be used to convert a binary number to a decimal number or a decimal number to a binary number. When converting from binary to decimal, the individual binary digits, called **bits,** are written in their proper place in the place value chart. The decimal equivalents for each place are added together. To convert from decimal to binary, each place value is subtracted.

Decimal to Binary

To convert from decimal to binary, a series of steps can be used. These steps are as follows:

1. Make a binary place value chart.
2. Determine the largest place value that can be subtracted from the decimal number you are converting. If the decimal number is 70, the largest place value that can be subtracted is 64 (2^6).
3. Subtract that largest place value and record a "1" in the proper location on the place value chart. The result of the subtraction is used with the next place value. If 70 was the decimal value, six would be the result.
4. Examine the next largest place value on the chart. This is the value immediately to the right of the place value just used. If this place value is too large to subtract from the result, a "0" is written in for that place value. No subtraction is performed. If the place value is smaller than the result, the place value is subtracted and a "1" is written in the place value chart.
5. Step 4 is repeated until you reach the final spot on the place value chart (2^0).

Sample problem 3. _____

Convert the decimal number 350_{10} to binary. Write the digits for the binary number in the proper location on the place value chart. Refer to figure 29-4.

Binary	Subtraction
2^8 = 1	350 – 256 = 94
2^7 = 0	94 is smaller than 128: no subtraction
2^6 = 1	94 – 64 = 30
2^5 = 0	30 is smaller than 32: no subtraction
2^4 = 1	30 – 16 = 14
2^3 = 1	14 – 8 = 6
2^2 = 1	6 – 4 = 2
2^1 = 1	2 – 2 = 0
2^0 = 0	

Answer: 350_{10}= 101011110_2

Figure 29-4. Solution to sample problem 3.

Binary Place	2^8	2^7	2^6	2^5	2^4	2^3	2^2	2^1	2^0
Decimal Value	256	128	64	32	16	8	4	2	1
Binary Number	1	0	1	0	1	1	1	1	0

Binary to Decimal

Another series of steps converts a binary number to decimal number. These steps are as follows:

1. Write each digit of the binary number in its proper location in the binary place value chart.
2. Write the decimal equivalent value for each "1" in the place value chart.
3. Add up all of the decimal equivalents.

Sample problem 4. _____

Convert the binary number 10111001_2 to a decimal number. Write the binary bits in their proper location on the place value chart. Refer to figure 29-5.

Binary Place	Decimal Value
2^7	128
2^5	32
2^4	16
2^3	8
2^0	1
	185_{10}

Answer: 10111001_2= 185_{10}

Figure 29-5. Solution to sample problem 4.

Binary Place	2^8	2^7	2^6	2^5	2^4	2^3	2^2	2^1	2^0
Decimal Value	256	128	64	32	16	8	4	2	1
Binary Number	0	1	0	1	1	1	0	0	1

29.3 BASIC BINARY LOGIC GATES

Binary logic gates are used in digital circuits for the purpose of decision making. The gates are the foundation of all digital circuits. They are used in many applications with circuits ranging from very simple to quite complex.

Five basic gates are discussed in this chapter: AND, OR, NOT, NAND, and NOR. Each of the gates is shown with a simple circuit to represent its performance, a schematic symbol, a truth table, and Boolean expression. Also included with each logic gate is a simplified written expression of how the gate works. Read these expressions carefully.

AND Gate

A circuit using switches to represent the **AND** gate is shown in figure 29-6. The switches are in series, requiring both switch A *and* switch B to be closed at the same time to light the lamp.

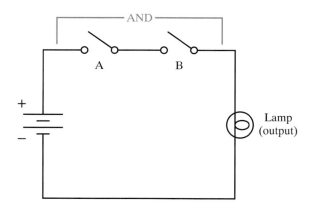

Figure 29-6. AND gates requires both switches to close to complete the circuit.

Figure 29-7 is the schematic symbol of the AND gate. It is round on the output side and flat on the input side. The inputs are labeled with an A and B to represent the switches. The output is labeled with the Boolean expression: A • B. A **Boolean** expression is used to write the combination needed to solve a digital logic circuit.

Figure 29-7. Schematic symbol for the AND gate along with its Boolean expression.

Figure 29-8 shows the truth table for the AND gate. The inputs for truth tables are set up in the same manner. All possible input combinations are listed on the left side. The result of each combination is listed on the right. All of the inputs must equal 1 to get a 1 on the output in an AND gate.

To determine the maximum number of input combinations, raise two to a power equal to the number of inputs. This sample is a two-input gate. The maximum possible number of combinations is $2^2 = 4$.

Inputs		Output
A	B	A • B
0	0	0
0	1	0
1	0	0
1	1	1

Written expression
———
Both A and B must be present to produce an output.

Boolean expression
———
A • B

Figure 29-8. Truth table for an AND gate with its written and Boolean expressions.

OR Gate

The switch circuit for the **OR** gate is shown in figure 29-9. With the switches in parallel, the lamp turns on when either switch A *or* switch B is closed.

The schematic symbol for the OR gate is shown in figure 29-10. Note the shape of the symbol is more pointed on the output side and curved on the input side. The

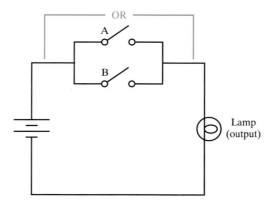

Figure 29-9. OR gates operate when either of two switches are closed or when both switches are closed.

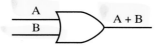

Figure 29-10. Schematic symbol for the OR gate along with its Boolean expression.

inputs represent the names of the switches. The output is the Boolean expression for an OR gate: A + B. Caution, a common mistake is to confuse the + symbol of the OR expression with an addition sign in arithmetic. As a result, the + sign is incorrectly read as AND.

Figure 29-11 shows the truth table with the written expression and the Boolean expression. A 1 on either input or both inputs gives a 1 at the output.

Inputs		Output
A	B	A + B
0	0	0
0	1	1
1	0	1
1	1	1

Written
expression
―――――――
Either A or B
or both
will produce
an output.

Boolean
expression
―――――――
A + B

Figure 29-11. Truth table for an OR gate with its written and Boolean expressions.

NOT Gate

The **NOT** gate inverts the input or output of another gate. A NOT gate is also called an **inverter.** The common emitter transistor switch circuit, figure 29-12, represents the action of the NOT gate. The base of the transistor is connected to a positive voltage through switch A. The output is on the collector.

If switch A is open, representing a 0, there is no base current. The transistor is turned off, which puts a high voltage, representing a 1, on the output.

When switch A is closed, representing a 1, positive voltage is applied to the base. The base resistor is low enough to allow the transistor to turn on and drive into saturation. In saturation, there is no voltage dropped across the transistor, giving the output zero voltage.

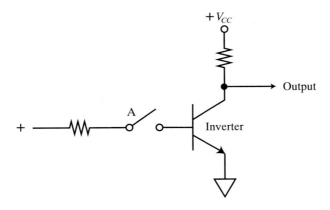

Figure 29-12. The NOT gate produces an output opposite the input.

The schematic symbol for the NOT gate is shown in figure 29-13. The triangle is also used to represent an amplifier. However, a bubble on the output indicates a 180° phase shift from the input. A bubble is also used in other digital circuit gates to represent NOT. In addition, notice the bar written over the letter on the output. This is the Boolean expression for NOT.

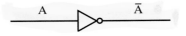

Figure 29-13. Schematic symbol for the NOT gate along with its Boolean expression. The bubble indicates the inverting action.

The truth table, written expression, and Boolean expressions are shown in figure 29-14. With only one input the possible number of combinations is $2^1 = 2$.

Written expression	Input		Output	Boolean expression
	Switch	A	$\overline{A}$	
The output is NOT equal to the input.	Open	0	1	$\overline{A}$
	Closed	1	0	

Figure 29-14. Truth table for a NOT gate with its written and Boolean expressions.

NAND Gate

The **NAND** gate is a combination of AND and NOT, as shown by the circuit in figure 29-15. Switches A and B are in series. This forms an AND gate. The transistor inverts the input.

When either switch A or switch B is open, the base has no current and the transistor is turned off. The output voltage is equal to V_{CC}, which is a logic 1.

When both A and B are closed, the base of the transistor has a positive voltage. The transistor turns on, and with a small base resistor the transistor is driven into saturation. The output drops to ground, giving it a logic 0.

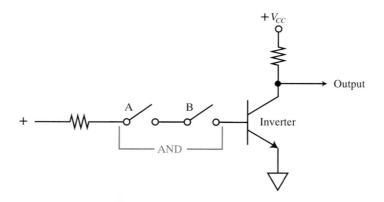

Figure 29-15. The NAND gate is an inverted AND.

The schematic symbol for a NAND gate is shown in figure 29-16. The symbol has the same shape as the AND gate, with a bubble added to the output. The bubble represents a NOT gate. The Boolean expression has a bar, indicating the inversion.

The truth table, shown in figure 29-17, has outputs exactly the opposite the outputs of an AND gate. Both inputs must be 1 to get a 0 output.

Figure 29-16. Schematic symbol for the NAND gate along with its Boolean expression.

	Inputs		Output
	A	B	$\overline{A \cdot B}$
	0	0	1
	0	1	1
	1	0	1
	1	1	0

Written expression

$\overline{\text{With both A and B}}$ present the output is low.

Boolean expression

$\overline{A \cdot B}$

Figure 29-17. Truth table for a NAND gate with its written and Boolean expressions.

NOR Gate

The **NOR** gate is an inverted OR, as shown in the circuit of figure 29-18. The input switches are in parallel. Either switch will apply a positive to the base of the transistor. The transistor inverts the input.

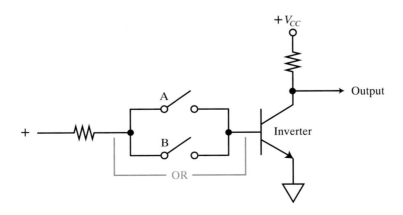

Figure 29-18. The NOR gate is an inverted OR.

The schematic symbol, figure 29-19, has the same shape as the OR gate with a bubble added. The bubble indicating an inversion. The Boolean expression has a bar over the OR statement.

The truth table for the NOT gate is shown in figure 29-20. It has the outputs exactly opposite those of the OR gate. A 1 on any of the inputs gives a 0 output.

Figure 29-19. Schematic symbol for the NOR gate along with its Boolean expression.

	Inputs		Output
A	B	$\overline{A + B}$
0	0	1
0	1	0
1	0	0
1	1	0

Written expression
———————
With either A or B the output is low.

Boolean expression
———————
$\overline{A + B}$

Figure 29-20. Truth table for a NOR gate with its written and Boolean expressions

SUMMARY

- Most digital circuits operate with two states, on or off.
- The binary numbering system is used to represent the possible combinations of outputs for a digital circuit.
- A binary zero represents a digital logic off.
- A binary one represents a digital logic on.

- Decimal numbers are converted to a binary number by finding the highest binary place value contained in the decimal number. Binary place values which can be subtracted from the decimal number receive a binary one. Place values which cannot be subtracted receive a binary zero.
- Binary numbers are converted to decimal by adding each of the place values containing a binary one.
- The five basic logic gates discussed in this chapter have two inputs, except the NOT gate. Two inputs have a total of four possible combinations. To determine the number of combinations the binary base of two is raised to a power equal to the number of inputs. For the gates in this chapter: $2^2 = 4$.

KEY WORDS AND TERMS GLOSSARY

analog: Circuits that are variable over a continuous range. Analog circuits have a range of voltages, from zero to the maximum applied voltage. Analog-type meters have a needle that moves along a scale.

AND: Logic that requires all inputs to be ones for an output of one.

base: States how many numbers are available in a number system.

binary: A numbering system that uses only the digits 0 and 1.

bits: The individual digits of a binary number.

Boolean expression: An algebraic expression used to describe the output of a digital logic gate.

decision-making circuit: A circuit that can choose between a yes and no answer. Typically found in digital applications.

digital: A circuit that contains a discrete number (often two) of possible voltage states. Instruments with a digital display have numbers that change with the reading.

inverter: Inverts the input signal. Also called a NOT gate.

NAND: Combination of AND and NOT. Logic that requires all inputs to be zero for and output of one.

NOR: Combination of OR and NOT. Logic that produces a zero if one or more inputs are one.

NOT: Logic that inverts the input signal. Also called an inverter.

OR: Logic that requires at least one input to be a one for an output of one.

TEST YOUR KNOWLEDGE

Do not write in this text. Please use a separate sheet of paper.

1. What is the difference between digitals and analog circuits?
2. Write a decimal place value chart up to 10^6.
3. Write each digit in the following decimal numbers multiplied by its place value. (Refer to sample problem 1.)
   a. 4,302,916.
   b. 57,008.

    c. 256.

    d. 7029.

    e. 2,682,530.

4. Write the binary place value chart up to 2^8.

5. Number your paper from 1 to 32 (decimal). Match each number with its binary equivalent.

6. Write each digit in the following binary numbers multiplied by its place value. (Refer to sample problem 2.) Also find the decimal equivalent of each.

    a. 101110.

    b. 11101.

    c. 1100110.

    d. 00111001.

    e. 10101011.

7. Convert the following decimal numbers to binary.

    a. 29.

    b. 256.

    c. 150.

    d. 100.

    e. 35.

8. Convert the following binary numbers to decimal.

    a. 101.

    b. 110011.

    c. 101100.

    d. 1110111.

    e. 10101010.

9. For each of the following binary logic gates, write the schematic symbol and an expression in words how the gate operates.

    a. AND.

    b. OR.

    c. NOT.

    d. NAND.

    e. NOR.

10. For each of the following binary logic gates, write the Boolean expression and truth table.

    a. AND.

    b. OR.

    c. NOT.

    d. NAND.

    e. NOR.

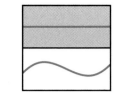

Appendix

APPENDIX A

Standard Color Code

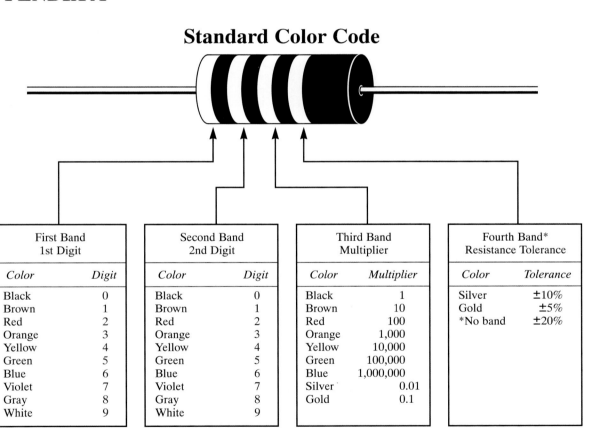

First Band 1st Digit		Second Band 2nd Digit		Third Band Multiplier		Fourth Band* Resistance Tolerance	
Color	*Digit*	*Color*	*Digit*	*Color*	*Multiplier*	*Color*	*Tolerance*
Black	0	Black	0	Black	1	Silver	±10%
Brown	1	Brown	1	Brown	10	Gold	±5%
Red	2	Red	2	Red	100	*No band	±20%
Orange	3	Orange	3	Orange	1,000		
Yellow	4	Yellow	4	Yellow	10,000		
Green	5	Green	5	Green	100,000		
Blue	6	Blue	6	Blue	1,000,000		
Violet	7	Violet	7	Silver	0.01		
Gray	8	Gray	8	Gold	0.1		
White	9	White	9				

APPENDIX B

Schematic Symbols

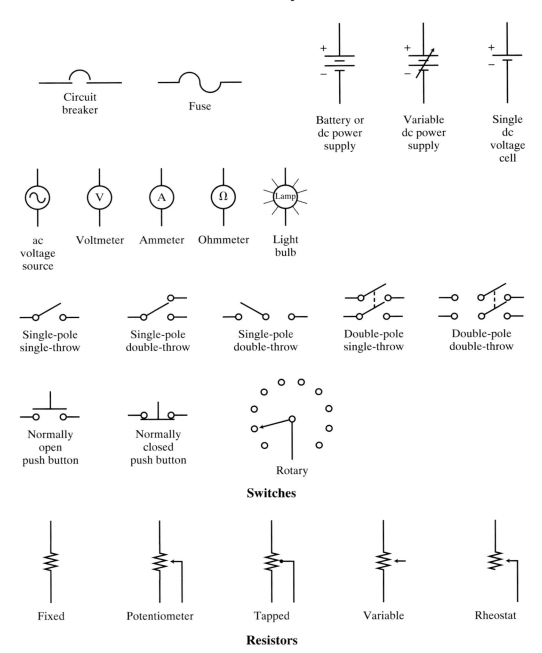

Circuit breaker

Fuse

Battery or dc power supply

Variable dc power supply

Single dc voltage cell

ac voltage source

Voltmeter

Ammeter

Ohmmeter

Light bulb

Single-pole single-throw

Single-pole double-throw

Single-pole double-throw

Double-pole single-throw

Double-pole double-throw

Normally open push button

Normally closed push button

Rotary

Switches

Fixed

Potentiometer

Tapped

Variable

Rheostat

Resistors

Schematic Symbols

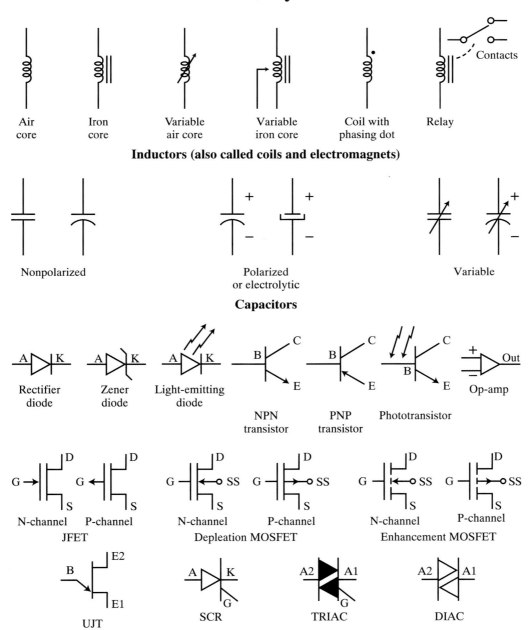

Air core   Iron core   Variable air core   Variable iron core   Coil with phasing dot   Relay   Contacts

Inductors (also called coils and electromagnets)

Nonpolarized   Polarized or electrolytic   Variable

Capacitors

Rectifier diode   Zener diode   Light-emitting diode   NPN transistor   PNP transistor   Phototransistor   Op-amp

N-channel   P-channel   N-channel   P-channel   N-channel   P-channel
JFET   **Depleation MOSFET**   **Enhancement MOSFET**

UJT   SCR   TRIAC   DIAC

Semiconductors

Schematic Symbols

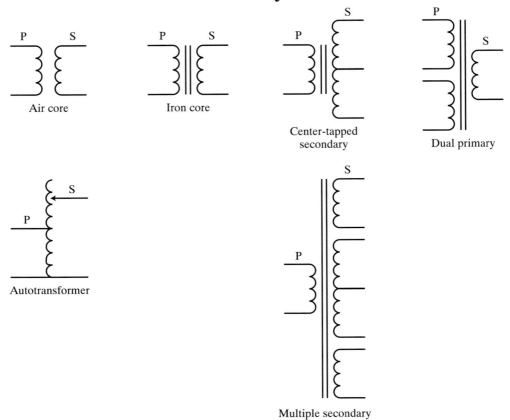

Air core

Iron core

Center-tapped
secondary

Dual primary

Autotransformer

Multiple secondary

Transformers

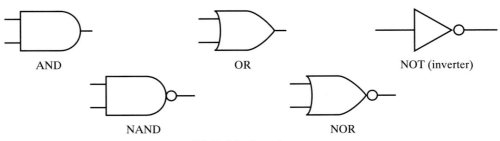

AND

OR

NOT (inverter)

NAND

NOR

Digital logic gates

APPENDIX C

U.S. Customary — Metric Conversion Factors and Prefixes

Prefix	Symbol	Multiplication Factor	
exa	E	10^{18} =	1,000,000,000,000,000,000
peta	P	10^{15} =	1,000,000,000,000,000
tera	T	10^{12} =	1,000,000,000,000
giga	G	10^{9} =	1,000,000,000
mega	M	10^{6} =	1,000,000
kilo	k	10^{3} =	1,000
hecto	h	10^{2} =	100
deca	da	10^{1} =	10
(unit)		10^{0} =	1
deci	d	10^{-1} =	0.1
centi	c	10^{-2} =	0.01
milli	m	10^{-3} =	0.001
micro	μ	10^{-6} =	0.000001
nano	n	10^{-9} =	0.000000001
pico	p	10^{-12} =	0.000000000001
femto	f	10^{-15} =	0.000000000000001
atto	a	10^{-18} =	0.000000000000000001

APPENDIX D

CONVERSION TABLE
U. S. CONVENTIONAL TO METRIC

WHEN YOU KNOW	MULTIPLY BY:		TO FIND
	VERY ACCURATE	APPROXIMATE	

Length

WHEN YOU KNOW	VERY ACCURATE	APPROXIMATE	TO FIND
inches	*25.4		millimeters
inches	*2.54		centimeters
feet	*0.3048		meters
feet	*30.48		centimeters
yards	*0.9144	0.9	meters
miles	*1.609344	1.6	kilometers

Weight

WHEN YOU KNOW	VERY ACCURATE	APPROXIMATE	TO FIND
grains	15.43236	15.4	grams
ounces	*28.349523125	28.0	grams
ounces	*0.028349523125	0.028	kilograms
pounds	*0.45359237	0.45	kilograms
short ton	*0.90718474	0.9	tonnes

Volume

WHEN YOU KNOW	VERY ACCURATE	APPROXIMATE	TO FIND
teaspoons	*4.97512	5.0	milliliters
tablespoons	*14.92537	15.0	milliliters
fluid ounces	29.57353	30.0	milliliters
cups	*0.236588240	0.24	liters
pints	*0.473176473	0.47	liters
quarts	*0.946352946	0.95	liters
gallons	*3.785411784	3.8	liters
cubic inches	*0.016387064	0.02	liters
cubic feet	*0.028316846592	0.03	cubic meters
cubic yards	*0.764554857984	0.76	cubic meters

Area

WHEN YOU KNOW	VERY ACCURATE	APPROXIMATE	TO FIND
square inches	*6.4516	6.5	square centimeters
square feet	*0.09290304	0.09	square meters
square yards	*0.83612736	0.8	square meters
square miles	*2.589989	2.6	square kilometers
acres	*0.40468564224	0.4	hectares

Temperature

WHEN YOU KNOW	MULTIPLY BY	TO FIND
fahrenheit	* 5/9 (after subtracting 32)	Celsius

* = Exact

CONVERSION TABLE
METRIC TO U. S. CONVENTIONAL

WHEN YOU KNOW	MULTIPLY BY:		TO FIND
	VERY ACCURATE	*APPROXIMATE*	
Length			
millimeters	0.0393701	0.04	inches
centimeters	0.3937008	0.4	inches
meters	3.280840	3.3	feet
meters	1.093613	1.1	yards
kilometers	0.621371	0.6	miles
Weight			
grains	0.00228571	0.0023	ounces
grams	0.03527396	0.035	ounces
kilograms	2.204623	2.2	pounds
tonnes	1.1023113	1.1	short tons
Volume			
milliliters	0.20001	0.2	teaspoons
milliliters	0.06667	0.067	tablespoons
milliliters	0.03381402	0.03	fluid ounces
liters	61.02374	61.024	cubic inches
liters	2.113376	2.1	pints
liters	1.056688	1.06	quarts
liters	0.26417205	0.26	gallons
liters	0.03531467	0.035	cubic feet
cubic meters	61023.74	61023.7	cubic inches
cubic meters	35.31467	35.0	cubic feet
cubic meters	1.3079506	1.3	cubic yards
cubic meters	264.17205	264.0	gallons
Area			
square centimeters	0.1550003	0.16	square inches
square centimeters	0.00107639	0.001	square feet
square meters	10.76391	10.8	square feet
square meters	1.195990	1.2	square yards
square kilometers	0.3861019	0.4	square miles
hectares	2.471054	2.5	acres
Temperature			
Celsius	*9/5 (then add 32)		Fahrenheit

* = Exact

APPENDIX E

Units of Measure

Unit Name	Symbol	Unit of Measure for...
ampere (amp)	A	current
circular mil	cmil	cross-sectional area
coulomb	C	charge
farad	F	capacitance
henry	h	inductance
hertz	Hz	frequency
kilowatt-hour	kWh	power consumption
mho	℧	conductance/admittance
mil	mil	small distances
ohm	Ω	resistance/reactance/impedance
second	s	period
siemen	S	conductance/admittance
volt	V	voltage
watt	W	power

Magnetic Units of Measure

Term	Symbol	cgs system	SI mks system
area	A	square centimeter (cm^2)	square meter (m^2)
flux	Q	maxwell (Mx) or lines	weber (Wb)
flux density	B	gauss (G) or Mx/cm^2	tesla (T) or Wb/m^2
magnetic field intensity	H	oersted (Oe) or Gb/cm	ampere-turn per meter (At/m)
magnetic force	f	dyne and unit pole	
magnetomotive force	mmf	gilbert (Gb)	ampere-turn (At) or (NI)
permeability	μ	gauss per oersted (g/Oe)	henry per meter (H/m)
reluctance	$\mathcal{R}$	gilbert per maxwell (Gb/Mx)	ampere-turn per weber (At/Wb)

APPENDIX F

The Greek Alphabet

A	α	alpha
B	β	beta
Γ	γ	gamma
Δ	δ	delta
E	ϵ	epsilon
Z	ζ	zeta
H	η	eta
Θ	θ	theta
I	ι	iota
K	κ	kappa
Λ	λ	lambda
M	μ	mu
N	ν	nu
Ξ	ξ	xi
O	o	omicron
Π	π	pi
P	ρ	rho
Σ	σ	sigma
T	τ	tau
Y	υ	upsilon
Φ	ϕ	phi
X	χ	chi
Ψ	ψ	psi
Ω	ω	omega

APPENDIX G

Natural Trigonometric Fuctions

Angle	Sine	Cosine	Tangent	Angle	Sine	Cosine	Tangent
1°	.0175	.9998	.0175	46°	.7193	.6947	1.0355
2°	.0349	.9994	.0349	47°	.7314	.6820	1.0724
3°	.0523	.9986	.0524	48°	.7431	.6691	1.1106
4°	.0698	.9976	.0699	49°	.7547	.6561	1.1504
5°	.0872	.9962	.0875	50°	.7660	.6428	1.1918
6°	.1045	.9945	.1051	51°	.7771	.6293	1.2349
7°	.1219	.9925	.1228	52°	.7880	.6157	1.2799
8°	.1392	.9903	.1405	53°	.7986	.6018	1.3270
9°	.1564	.9877	.1584	54°	.8090	.5878	1.3764
10°	.1736	.9848	.1763	55°	.8192	.5736	1.4281
11°	.1908	.9816	.1944	56°	.8290	.5592	1.4826
12°	.2079	.9781	.2126	57°	.8387	.5446	1.5399
13°	.2250	.9744	.2309	58°	.8480	.5299	1.6003
14°	.2419	.9703	.2493	59°	.8572	.5150	1.6643
15°	.2588	.9659	.2679	60°	.8660	.5000	1.7321
16°	.2756	.9613	.2867	61°	.8746	.4848	1.8040
17°	.2924	.9563	.3057	62°	.8829	.4695	1.8807
18°	.3090	.9511	.3249	63°	.8910	.4540	1.9626
19°	.3256	.9455	.3443	64°	.8988	.4384	2.0503
20°	.3420	.9397	.3640	65°	.9063	.4226	2.1445
21°	.3584	.9336	.3839	66°	.9135	.4067	2.2460
22°	.3746	.9272	.4040	67°	.9205	.3907	2.3559
23°	.3907	.9205	.4245	68°	.9272	.3746	2.4751
24°	.4067	.9135	.4452	69°	.9336	.3584	2.6051
25°	.4226	.9063	.4663	70°	.9397	.3420	2.7475
26°	.4384	.8988	.4877	71°	.9455	.3256	2.9042
27°	.4540	.8910	.5095	72°	.9511	.3090	3.0777
28°	.4695	.8829	.5317	73°	.9563	.2924	3.2709
29°	.4848	.8746	.5543	74°	.9613	.2756	3.4874
30°	.5000	.8660	.5774	75°	.9659	.2588	3.7321
31°	.5150	.8572	.6009	76°	.9703	.2419	4.0108
32°	.5299	.8480	.6249	77°	.9744	.2250	4.3315
33°	.5446	.8387	.6494	78°	.9781	.2079	4.7046
34°	.5592	.8290	.6745	79°	.9816	.1908	5.1446
35°	.5736	.8192	.7002	80°	.9848	.1736	5.6713
36°	.5878	.8090	.7265	81°	.9877	.1564	6.3138
37°	.6018	.7986	.7536	82°	.9903	.1392	7.1154
38°	.6157	.7880	.7813	83°	.9925	.1219	8.1443
39°	.6293	.7771	.8098	84°	.9945	.1045	9.5144
40°	.6428	.7660	.8391	85°	.9962	.0872	11.4301
41°	.6561	.7547	.8693	86°	.9976	.0698	14.3006
42°	.6691	.7431	.9004	87°	.9986	.0523	19.0811
43°	.6820	.7314	.9325	88°	.9994	.0349	28.6363
44°	.6947	.7193	.9657	89°	.9998	.0175	57.2900
45°	.7071	.7071	1.0000	90°	1.0000	.0000	∞

APPENDIX H

Gate Symbols and Truth Tables

GATE SYMBOLS	INPUTS		OUTPUT
	A	B	X
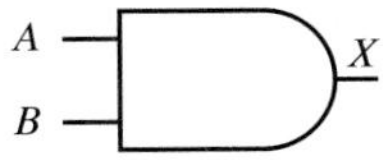 **AND**	0	0	0
	0	1	0
	1	0	0
	1	1	1
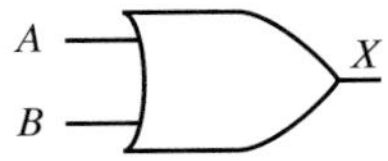 **OR**	0	0	0
	0	1	1
	1	0	1
	1	1	1
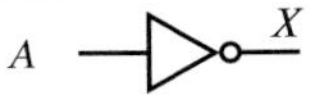 **INVERTER**	1		0
	0		1
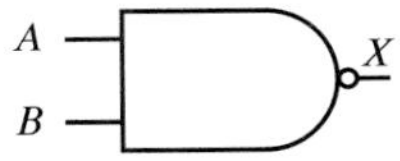 **NAND**	0	0	1
	0	1	1
	1	0	1
	1	1	0
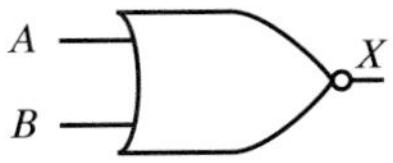 **NOR**	0	0	1
	0	1	0
	1	0	0
	1	1	0
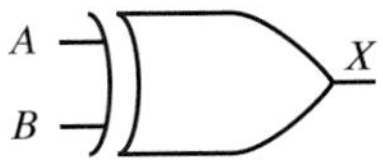 **EXCLUSIVE OR**	0	0	0
	0	1	1
	1	0	1
	1	1	0
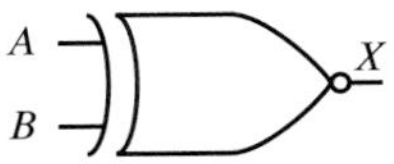 **EXCLUSIVE NOR**	0	0	1
	0	1	0
	1	0	0
	1	1	1

BOOLEAN ALGEBRA

= stands for equal

A is A

$\overline{A}$ is *not* A

A + B stands for A or B, *not* A plus B

A•B is the same as AB and stands for A *and* B

Glossary

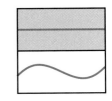

A

absorbed power: Power that is absorbed by the load.

ac: The abbreviation for alternating current.

active component: An electronic component that requires power to operate. Examples: transistors, diodes, integrated circuits.

active region: The range of operation of an active component between cutoff and saturation. In amplifier applications, it is the range where the device varies its resistance in relation to an ac input signal.

ac voltage: A voltage that changes polarity.

admittance (Y): The ease with which alternating current flows in a circuit. It is the reciprocal of impedance. The unit of measure is siemens (S) or mhos ($\mho$).

alternating current (ac): Current in an electric circuit that periodically changes direction due to the voltage changing polarity. Examples of ac voltage sources include: household electricity, all forms of radio signals, magnetic tape recordings, sound converted to an electronic signal.

American Wire Gauge (AWG): The standard for wire sizes. Wire is commonly available from approximately one-half inch diameter (AWG 0000) to three one-thousandths inch diameter (AWG 40).

ampere (A): Unit measure of current.

ampere hour (Ah): The unit of measure for current capacity of a battery. The ampere-hour rates the life of the battery for an amount of current for a period of time. For example, if a battery has a rating of 200 mAh (milli-ampere-hour), 200 mA can be drawn for one hour. A smaller current drain results in a longer time. For example, 20 mA of load current can be drained for 10 hours.

ampere-turn (At or NI): The unit of measure in the SI mks system for magnetomotive force.

ampere-turns per meter (At/m): The unit of measure in the SI mks system for magnetic field intensity.

amplitude: The strength of a signal, measured in either voltage, current, or power. When viewed on an oscilloscope, it is the height of the waveform and is measured as either peak or peak-to-peak.

analog: Circuits that are variable over a continuous range. Analog circuits have a range of voltages, from zero to the maximum applied voltage. Analog-type meters have a needle that moves along a scale.

AND: Logic that requires all inputs to be ones for an output of one.

anode: The positive element of diode and other electronic devices. The anode of a diode is the N-type semiconductor material. Electron current flows towards the anode.

apparent power (P_s): The total power in an ac circuit containing both resistance and reactance. The hypotenuse of the power triangle represents the apparent power. Apparent power is measured in Volt-Amperes (VA).

arc: A discharge of electricity through the air, approximately 25,000 volts per inch.

armature: The moving coil of wire in a generator spinning past the magnetic fields.

armature reaction: Magnetic resistance to the armature created when the armature passes through the magnetic field.

atom: The smallest portion of an element that still maintains the properties of the element. An atom is made up of electrons, protons, and neutrons.

atomic number: The number of protons in a single atom of an element.

autotransformer: A transformer that has only one coil used for both the primary and the secondary.

average value: In a sine wave, the average is calculated by multiplying 0.636 times the peak. A half-wave rectified sine wave has a dc voltage equal to one half the average value of the sine wave. A full-wave rectifier produces a dc voltage equal to the average.

B

B: The abbreviation for flux density.

back emf: Counterelectromotive force.

band pass filter: A filter that blocks all but one band of frequencies from passing through.

band stop filter: A filter that blocks one specific band of frequencies from passing through.

bandwidth: The range of frequencies close to the resonant frequency. Bandwidth is measured at the points on the response curve equal to 70.7 percent of maximum current, which is also equal to the -3 dB points.

barrier potential: The voltage drop across junction of a P- and an N-type material.

base (number system): States how many numbers are available in a number system.

base (transistor): The element of a bipolar junction transistor sandwiched between the emitter and the collector. Base current controls collector current.

bell curve: The impedance and current curves used to describe characteristics of a resonant circuit. The curves are given the name because of their bell shape.

beta: The current gain of a bipolar junction transistor. Collector current is beta times larger than the base current.

bias: The voltage levels applied to two points in a circuit for the purposes of controlling the behavior of the circuit.

bimetallic strip: Two pieces of metal, made of different materials, joined at one end. It is also called a thermocouple. When the junction is heated, a voltage is produced at the other end.

binary: A numbering system that uses only the digits 0 and 1.

bipolar junction transistor: A transistor composed of two back-to-back PN junctions.

bits: The individual digits of a binary number.

blocking capacitor: A capacitor placed in series to pass an ac signal while stopping a dc voltage. Also called a coupling capacitor.

Boolean expression: An algebraic expression used to describe the output of a digital logic gate.

branch current: Current in the independent branches of a parallel circuit.

breakdown: In semiconductor diodes, it is the point at which there is a sudden increase in current in the reverse direction.

breaker: A breaker, also called a circuit breaker, is used to protect a circuit from excessive current.

bridge rectifier: One type of full-wave rectifier. It uses four diodes for rectification instead of a center-tapped transformer.

brush: Spring-mounted conductive material for making contact with the slip rings in ac generators or commutators of dc generators. The brushes connect the external load circuit to the armature.

buzzer/bell: An electromagnetic device that uses a relay to vibrate a striker arm to ring a bell.

bypass capacitor: A parallel connected capacitor to provide a path for the ac signal around a resistor.

C

C: Letter symbol for capacitance.

capacitance (C): The ability of a device to store an electric charge. Through the storing and discharging of a charge, capacitance opposes a change in voltage. The unit of measure is farad (F). Common units of measure are μF and pF.

capacitive reactance (X_C): The resistance a capacitor offers to an air conditioning signal. The unit of measure for capacitive reactance is ohms.

capacitor-start induction motor: A split-phase motor with a capacitor in series with the starting winding.

cathode: The electron-emitting element of a diode and other devices. The N-type material of a semiconductor diode is the cathode.

center frequency: The resonant frequency (f_0) for filters with bell curve characteristics.

centimeter-gram-second (cgs): A system of measurement.

centrifugal force: The force of nature which tries to keep an object moving in a straight line. When the object is revolving in orbit, centrifugal force exerts energy away from the center of the orbit.

centrifugal switch: A switch operated by the rotating motion of a motor. When the rotor reaches a certain speed, the switch contacts open.

channel: Area where electrons flow between the source and drain.

charging resistance: A voltage applied to a battery in the reverse direction forcing electrons through the electrolyte.

choke: An inductor placed in series as a filter to pass dc voltages while stopping ac signals.

circuit breaker: An electromagnetic overcurrent protection device.

circuit *Q*: The ratio of inductive reactance to the circuit resistance.

circuit theorem: Mathematical tools that make it easier to solve complicated circuits.

circumference: The distance around a circle.

clamper circuit: Places the centerline of an ac signal to a dc voltage level.

clipper circuit: Clips off part of a sine wave, making it look more like a square wave.

cmil: The abbreviation for circular mil, a measurement for the area of very small circles, such as the cross section of a wire.

cold solder joint: The name used to describe a bad solder connection. It can be recognized by its dull appearance and it may have tiny cracks.

collector: One of the three elements of a bipolar junction transistor. Collector current is controlled by base current.

common base circuit: Transistor circuit in which the input is applied between the emitter and the base and the output is taken between the collector and the base.

common collector circuit: Transistor circuit in which the input is applied between the base and the collector and the output is taken between the emitter and collector.

common emitter circuit: Transistor circuit in which the input is applied between the base and the emitter and the output is taken between the collector and the emitter.

commutator: A mechanical device used with generators to produce pulsating dc from a sine wave, similar to rectification. Commutators are also used with motors to change the applied sine wave to dc.

complex number: A number used to represent the quantities of an electronic circuit containing both resistance and reactance.

compound generator: A generator with both a series field winding and a shunt field winding.

conductance (*G*): The ease with which a conductor allows a current to flow. It is the reciprocal of resistance. The unit of measure is the siemen (S) or mho ($\mho$).

conductor: A material which allows the easy flow of electricity.

coulomb (C): Unit measure of electrical charge.

counterelectromotive force (cemf): The property of an inductor to oppose any change in the instantaneous building of the magnetic field. It also opposes a change in the current.

coupling capacitor: A capacitor placed in series to pass an ac signal while stopping a dc voltage. Also called a blocking capacitor.

covalent bond: The sharing of valence electrons between two atoms to fill the valence shell.

cps: Abbreviation for cycles per second, an obsolete unit of measure for frequency.

cross-sectional area: The surface of an object that would be exposed by slicing the object.

current (I): The flow of electricity through a conductor. The unit of measure is the ampere (A).

current rating: States maximum current allowed through a device.

cutoff: The point where a transistor is off. The effect on the circuit is that of an open switch, with no current.

cutoff frequency (f_c): The frequency where the response curve for a filter circuit falls to 70.7 percent.

cycle: On an ac waveform, from one point to the point where the waveform next repeats itself.

cycles per second (cps): An obsolete unit of measure for frequency.

D

damping: The reduction of energy. In resonant circuits, resistance is added to increase the bandwidth.

dB: Abbreviation for decibel.

dc: Abbreviation for direct current.

dc voltage: A voltage that has a continuous polarity.

dead circuit: A circuit containing no voltage.

decibel (dB): A unit of measure related to the logarithmic response of the human ear. It is a useful measure for the frequency response of filters for two reasons: response curves are usually plotted with frequency on a logarithmic (factor of 10) scale, and it is a method of comparing the input to the output.

decision-making circuit: A circuit that can choose between a yes and no answer. Typically found in digital applications.

delta (Δ) connected three-phase system: One type of three-phase wiring system that uses only three wires. There is no neutral wire.

delta network: A three-terminal network that is connected in series in a closed loop. Also called a pi network.

depletion mode: One of the operating modes of a MOSFET. The channel is made of the same type of material as the source and drain.

depletion region: An area near the junction of a semiconductor where the free current carriers cannot cross.

DIAC: A semiconductor device that operates like two opposite facing zener diodes.

diamagnetic material: Classified as nonmagnetic material. However, it is actually very slightly repelled by a magnetic field. Examples are: copper, lead, gold, antimony, bismuth, mercury.

dielectric: An electrical insulator between the plates of a capacitor.

dielectric constant: The rating for materials used as dialectrics. The number reflects how many times better the material is than a vacuum.

dielectric strength: States the maximum voltage that can be applied to a capacitor before breakdown occurs.

differentiator: An electronic circuit whose output waveform has two portions to consider, the positive charge portion and the negative discharge portion. The charge portion is the difference between the input voltage and the voltage dropped across other circuit components. The discharge portion goes negative as result of the energy being released in the circuit. The average of the differentiator output is zero.

digital: A circuit that contains a discrete number (often two) of possible voltage states. Instruments with a digital display have numbers that change with the reading.

diode: A semiconductor device that allows current to flow in only one dirrection.

direct current (dc): Current that is always in the same direction. The voltage and current can fluctuate up and down, but the polarity of the voltage remains the same.

direct relationship: When one quantity changes, it causes a change in another quantity in the same direction. An increase causes an increase and a decrease causes a decrease.

doping: The process of adding an impurity to semiconductor material to change the semiconductor to either a P-type or an N-type.

double insulated tool: Tools on which the outside surfaces are plastic coated. They require only a two-prong plug.

double-pole: Switches with two common positions.

double-throw: Switch makes contact in either of two positions.

drain: One lead on a FET. Current is controlled by the gate. Similar to the collector on a BJT.

duty cycle: The amount of time a circuit is on, as compared to the time off.

dynamic resistance: The ac resistance of a circuit.

dyne: The unit of measure in the cgs system for magnetic force.

E

eddy current: Electrical current flowing within the core of an electromagnet resulting from voltage induced in the iron core. Eddy currents oppose the current producing the magnetic field resulting in a loss of power.

efficiency: The ratio of power output available to the load to the power input from the source.

electrical shock: The physical sensation of the nerves and muscles reacting to electricity passing through the body.

electrocution: The term used when the exposure to electric shock results in death.

electrolyte: A substance which produces ions when it conducts electricity.

electrolytic capacitor: A type of capacitor that is polarized.

electromagnet: Magnet produced with the aid of electricity.

electromotive force (emf): Electrical pressure applied to a circuit. It is another name for voltage and has the same unit of measure, the volt.

electron: A particle of an atom having a negative charge. Electrons orbit the nucleus in a manner similar to that of planets orbiting the sun.

electron drift: The result of the electrons going from one atom to the next, in a random manner, toward the positive side of the voltage source.

electron shell: The positions the electrons hold in orbit around the nucleus. Electrons fill the shells closest to the nucleus first, then they fill the outer shells.

electrostatic field: The attraction between negative and positive voltages.

element: The smallest particle of a substance containing atoms with the same atomic number.

emf: Abbreviation for electromotive force.

emitter: One of the three elements of a bipolar junction transistor. The emitter provides current to both the base and collector.

energizing current: The slight amount of current needed to develop the magnetic field in a coil.

engineering notation: A system in mathematics which uses multiplier names and powers of 10 to move the decimal point and label quantities.

enhancement mode: One of the operating modes of a MOSFET. There is no continuous channel and the gate has the same polarity as the substrate.

excited: The current producing a magnetic field. There are five basic methods of producing the magnetic fields in a generator: permanent magnets, independent dc voltage, self-excited shunt connections, self-excited series connections, combination of shunt and series connections.

exponential notation: A system in mathematics which writes the power of 10 as E followed by two digits. This format is used by computers and calculators.

extrinsic: A semiconductor material containing impurities.

F

farad (F): Unit measure of capacitance.

ferrites: Materials that are strongly attracted to a magnetic field but will not conduct electricity. These are chemical compounds, made with a magnetic material combined with a ceramic material.

ferromagnetic: Materials that are strongly attracted to a magnetic field and are also good conductors of electricity. Examples include: iron, nickel, cobalt.

field: The part of a generator containing the magnetic field. They may be made of permanent magnets or electromagnets.

field-effect transistor: A class of transistor that includes the JFET, MOSFET, and IGFET.

field intensity (H): The measure of the strength of the magnetic field.

field strength: The strength of a magnetic field.

filter: A circuit that offers varying degrees of opposition to different frequencies.

555 timer: Device that produces a rectangular wave at a specified frequency.

fixed resistor: Resistor with an ohmic value determined during manufacture. The values available range from close to one ohm to several megohms. The ohmic value of a resistor has nothing to do with its physical size. The size of the resistor is determined by its wattage rating.

flux density (*B*): The intensity of a magnetic field.

forward biased: The voltage applied to a diode to allow a majority current.

four-wire wye connection: A three-phase generator that uses four wires and can produce 3 times the phase voltage.

free electron: An electron that moves easily between atoms to provide current.

frequency: The number of cycles in one second. The unit of measure is hertz (Hz) or a multiple such as kilohertz (kHz), and megahertz (MHz). Cycles per second, cps, is an older unit of measure, generally considered obsolete.

frequency response: The ability to respond to a range of frequencies.

frequency response rating: A rating that indicates the best frequency applications for a transformer.

full-wave rectifier: A rectifier that passes both halves of an ac cycle, inverting one half of the wave to make the entire wave the same polarity.

fuse: Overcurrent protection device that melts open to protect a circuit.

G

galvanometer: A voltmeter with the zero in the center. If there is a difference in voltages between its terminals, the needle swings toward the higher voltage.

gate: One lead on a FET. Controls current through the drain. Similar to the base on a BJT.

gauss (G): Unit of measure in the cgs system for flux density.

gauss per oersted (G/Oe): Units for measuring permeability in the cgs system.

generator: A device that uses mechanical energy to produce electrical energy.

gilbert: Unit of measure in the cgs system for magnetomotive force.

ground fault interrupter (GFI): A circuit breaker that protects against electric shock by sensing an unbalanced condition between the two current carrying wires. The circuit is interrupted with very small amounts of current difference.

ground wire: The bare wire in house wiring. Its main functions are to prevent fire and to protect against electrical shock. If the hot wire comes loose, it would contact the bare wire, which would trip the circuit breaker stopping all current.

H

***H*:** The abbreviation for magnetizing force.

half-power point: The point on the response curve of a resonant circuit equal to 70.7 percent of the maximum voltage. The half-power point is also equal to the -3 dB point, and it is the point used to measure the bandwidth.

half-wave rectifier: A rectifier that passes one half of an ac cycle while blocking the other half of the cycle.

harmonic frequency: A multiple of a sine wave's fundamental frequency. The second harmonic is twice the fundamental fequency and the third harmonic is three times the fundamental frequency.

head: Device the reads the magnetic pulses stored on tapes or disks. Also called a magnetic pickup.

heat sink: Protects sensitive components by dissipating excess heat.

henry per meter: Units for measuring permeability in the SI mks system.

hertz (Hz): The unit measure of frequency.

high pass filter: A filter that offers little opposition to high frequency signals and greater opposition to low frequency signals.

hole: An atom missing an electron in its valence band. It carries a positive charge. A hole is the majority current carrier in a P-type semiconductor.

horsepower (hp): The force required to move 33,000 pounds one foot in one minute, or 550 pounds one foot in one second. One horsepower also equals 746 watts.

hot wire: The voltage and current carrying wire. It is the hot wire that causes all electrical shocks. In house wiring, the standard color code for hot is black, and it is attached to the brass-colored screw on the electrical device.

hysteresis: The amount of magnetization or flux density (*B*) that lags the magnetizing force (*H*) because of molecular friction.

hysteresis loop: Represents the amount of energy needed to create a magnetic field in the core of a transformer.

I

I: Letter symbol for current.

imaginary number: The j-term of a complex number, used to represent the reactive component.

impedance: The total ac resistance of a circuit containing both resistance and reactance. Unit of measure is ohms. Impedance has a phase angle equal to the phase angle of the circuit voltage.

independently excited generator: A generator in which the voltage to the field windings comes from a separate source.

inductance: The property of a circuit to oppose a change in current due to a counterelectromotive force. It is the result of converting electrical energy to magnetic energy or magnetic energy to electrical energy.

induction motor: Motor having a magnetically revolving stator field. Induction motors do not require stator windings.

inductive kick: The high voltage produced when the discharge path for an inductor is an open circuit.

inductive reactance (X_L)**:** The ac resistance of an inductor. Unit of measure is ohms.

instantaneous value: Any point on a sine wave at an instant in time. The sine wave is made up of an infinite series of points, each having an instantaneous value.

insulation: Protection placed on wiring.

insulator: Materials which offer an infinite resistance, allowing no current to flow. An example is the plastic insulation on a wire.

integrator: An application of a time constant circuit. When an all-positive square wave is applied to the input, the output has the shape of the universal time constant curve. The steepness of charge/discharge depends on the value of the time constant.

internal resistance: Resistance within a voltage source.

intrinsic: A pure semiconductor material.

inverse relationship: When one quantity increases, the related quantity decreases. Also, when one quantity decreases, the related quantity increases.

inverter: Inverts the input signal. Also called a NOT gate.

ion: An atom or molecule (group of atoms) which is electrically charged. Positive ions are lacking electrons. Negative ions have excess electrons.

IR drop: A name for the voltage drop in loop equations.

I²R losses: Losses due to the copper wire in a transformer.

isolation transformer: A transformer with the function of isolating a portion of a circuit.

J

j-operator: A mathematical tool used to represent reactances as complex numbers.

j-term: The imaginary part of a complex number.

junction: In reference to semiconductors, it is the surface where the P- and N-type materials are connected.

junction field-effect transistor (JFET): A type of transistor made up of a continuous piece of semiconductor material with small pieces of the opposite type on either side. Has three elements: source, gate, and drain.

K

kilowatt-hour (kWH): The equivalent of using 1000 watts in a one hour time period.

L

lagging phase shift: A phase shift resulting from the waveform appearing after the reference.

LC product: The combination of inductance and capacitance values in a resonant circuit.

leading edge: The portion of a wave switching from zero to a maximum.

leading phase shift: A phase shift resulting from the waveform appearing prior to the reference.

light-emitting diode (LED): A diode that emits light when forward biased.

light-sensitive transistor: A transistor that has a base current that can be controlled with light. Also called a phototransistor.

like phasors: Phasors in circuits containing more than one of the same type of component. They are plotted end-to-end, with addition used to find the resultant.

live circuit: A circuit with voltage present, even if it is not connected to a load.

load line: Line drawn across a transistor's characteristic curves. The end points of the load line are the transistor's operating extremes.

lodestone: A material with magnetic properties in its natural state.

long time constant circuit: A circuit that has its full charge/discharge time greater than the time of one pulse width.

low pass filter: A filter that offers little opposition to low frequency signals and greater opposition to high frequency signals.

M

magnet: A substance which produces a magnetic field.

magnet wire: The type of wire used to make magnetic coils.

magnetic domain: Tiny magnetic particles making up a magnet. Each of these tiny magnets has a north and south pole.

magnetic field: The area influenced by a magnet.

magnetic field intensity: The measure of the magnetic field inside of the magnet.

magnetic flux (ϕ): The lines of force, as a group, making up the magnetic field.

magnetic lines of force: The invisible lines making up the magnetic field. Magnetic lines of force flow from north to south. This is synonymous to electron current flowing from negative to positive.

magnetic pickup: Device that reads the magnetic pulses stored on tapes or disks. Also called a head.

magnetic shield: A material that conducts a magnetic field around an area that should not be exposed to magnetic flux.

magnetomotive force (mmf): The strength of the source of magnetism. The unit of measure is the gilbert (Gb). Magnetomotive force is synonymous to the voltage of a battery.

mainline buss-bar: The wires in electrical equipment connecting all of the positive connections or all of the negative connections.

majority carrier: The current carrier used as the primary source of current within a semiconductor when forward biased.

market survey: An investigation to find out if there is any significant interest in a product.

maxwell (Mx): Unit of measure in the cgs system for magnetic flux.

mechanical connection: Connections made to components by twisting or attaching firmly before soldering.

medium time constant circuit: A circuit that has its full charge/discharge time slightly smaller than or equal to the time of one pulse width.

metal oxide semiconductor field-effect transistor (MOSFET): A type of transistor with four terminals: source, gate, drain, and substrate.

mho ($\mho$): Unit measure of conductivity.

microprocessor: An integrated circuit that is the central processing unit in a computer.

mil: Unit of measure equal to 0.001 inches.

minority carrier: The current carrier in a semiconductor of a polarity opposite that of the material in which they are present. Minority current flows in the reverse direction of the majority current.

mmf: Abbreviation for magnetomotive force.

motor: A device that uses mechanical force to produce electrical energy.

motor action: The attraction and repulsion of like and unlike magnetic fields.

motor load: A mechanical force that is applied to the output shaft, which is connected to the rotor.

multimeter: A meter for electrical measurements, usually capable of four different types of measurements: ohms, milliamps, dc volts, ac volts.

multiplier resistor: A resistor connected internally in a voltmeter. It is connected in series with the meter movement to allow selection of different ranges.

mutual inductance: The effect of a magnetic field from one inductor crossing the turns of a different inductor.

N

NAND: Combination of AND and NOT. Logic that requires all inputs to be zero for an output of one.

negative temperature coefficient: In certain materials, the resistance decreases with an increase in temperature. As current flows, heat is produced, which increases the temperature and further decreases the resistance. Examples are semiconductor materials, such as silicon and germanium.

−3 dB point: The point on the response curve of a resonant circuit equal to 70.7 percent of the maximum voltage.

net reactance: The effective reactance in a circuit containing both X_L and X_C.

neutral wire: In house wiring, it is a current carrying wire with a voltage potential equal to ground. The standard color code for the neutral wire is white. It is wired to the silver-colored screw on the electrical device.

neutron: A particle in an atom with neutral charge and a mass approximately equal to that of the proton.

nominal value: The ideal value.

nonlinear device: A semiconductor whose operating characteristics have an unequal amount of resistance over a range of voltages.

nonlinear scale: The spacing between each line is not the same across the scale. An example is the Ohms scale on a multimeter.

NOR: Combination of OR and NOT. Logic that produces a zero if one or more inputs are one.

normally closed: A switch that requires operation to open.

normally open: A switch that requires operation to close.

Norton equivalent circuit: Circuit where all but the load is turned into an equivalent current source and parallel resistor.

Norton equivalent current: Current that would be measured by removing the load and replacing it with a short.

Norton equivalent resistance: Resistance that would be measured across the load terminals if the load resistor was removed and the voltage source was replaced with a short.

NOT: Logic that inverts the input signal. Also called an inverter.

nucleus: The combination of the protons and neutrons of an atom. The large mass of the nucleus makes it very difficult to move in comparison to the lightweight mobility of the electrons.

O

oersted (Oe): Unit of measure in the cgs system for magnetic field intensity.

ohm (Ω): Unit measure of resistance.

ohms adjust potentiometer: Adjusts a meter to read zero when no resistance is being measured.

open circuit: A point in a circuit where the current path is supposed to be that has an infinite amount of resistance. An open circuit will not allow a current.

operational amplifier (op-amp): A high gain amplifier circuit containing many transistors in one integrated circuit package.

opposite angle(ϕ): When comparing the shift of the inductor voltage with the applied voltage, V_L will lead V_a by an amount equal to 90 - θ. This is referred to as the *opposite angle* because in the triangle, it is at the opposite side from θ.

opposite phasors: Phasors in circuits containing both capacitors and inductors. They are plotted end-to-end, with subtraction used to find the resultant.

optoelectric: Electronic components that use light to operate.

OR: Logic that requires at least one input to be a one for an output of one.

ordered pair: Points located on a four-quadrant graph. The points are located with paired numbers, called ordered pairs, which contain a real term written first and an imaginary term written second.

oscilloscope: Device that draws a picture of the voltage in a circuit.

overcurrent protection device: Device that prevents too much current from flowing in a circuit. Fuses and circuit breakers are examples of overcurrent protection devices.

oxidation: The production of a high resistance surface on a material caused by exposure to air. This can lead to circuit failure and poor soldering surfaces. Rust on iron is an example of oxidation.

P

parallel circuit: Circuit characterized by: the same voltage throughout the circuit, current splitting into branches, total circuit current equaling the sum of the branches, the total resistance being smaller than the smaller branch resistor, the total power equaling the sum of the branch powers.

parallel resonant circuit: A circuit that has an inductor in parallel with a capacitor.

paramagnetic: Materials that are nonmagnetic. Examples are: aluminum, platinum, oxygen, copper sulfate. These materials actually display a very slight magnetic attraction.

peak: A measurement of an ac waveform from zero to either the maximum positive value or the maximum negative value.

peak-to-peak: A measurement of an ac waveform from the maximum positive value to the maximum negative value.

pentavalent: A semiconductor doping impurity with five electrons in the valence band. When it forms covalent bonds with the semiconductor material, there are extra electrons available to be used as current carriers.

period: The length of time required for an ac waveform to complete one cycle. Unit of measure is seconds or part of a second, such as a millisecond or microsecond.

permeability (μ): Measure of ease with which magnetic lines can flow through a material. It is the reciprocal of reluctivity. It is actually the ratio of lines of force passing through the material as compared to the lines of force passing through the air. Generally, permeability is used as means of comparing the quality of a magnetic material. A perfectly nonmagnetic material has a permeability of one. Typical values for iron and steel are 100 to 9000.

permeance: The ability of a material to carry magnetic lines of force. It is the reciprocal of reluctance and corresponds to the electrical term conductance.

phase angle (θ): The time difference between the applied voltage and the resistor voltage. Also called phase shift.

phase shift (θ): The difference in time between two waveforms. Phase shift is measured in electrical degrees ranging from 0° to 180°. Also called phase angle.

phase shifter: A circuit that shifts the phase of an electrical signal.

phase voltage: The voltage between any one line and ground.

phasing dots: Dots on a schematic symbol used to indicate the direction in which the coil is wound.

phasor: A vector used to show the magnitude (size) and direction of an electrical quantity.

photosensitive: Materials responding when light strikes by either producing a voltage or changing resistance.

phototransistor: A transistor that has a base current that can be controlled with light. Also called a light-sensitive transistor.

piezoelectric effect: Voltage produced in certain types of crystals when a mechanical pressure is applied.

pinch-off: Condition in a FET when a high enough gate voltage stops current in the channel.

pi network: A three-terminal network that is connected in series in a closed loop. Also called a delta network.

pi-type filter: A type of filter that uses two capacitors. See figure 24-15.

plate: In reference to capacitors, it is a conductive surface with the functions of collecting electrons on the negative side of the capacitor and give up electrons on the positive side.

plunger: Moveable portion of a solenoid.

polar form: A number expressed as a length and angle. It is used to describe the magnitude and direction of a phasor.

polarity: The positive and negative relationships of a voltage.

polarized plug: A two-prong plug with one of the blades larger than the other. The larger blade is the neutral wire and the smaller is the hot wire.

poles: The origins of the north and south of a magnet. The north and south poles are synonymous with the positive and negative of a battery.

polyphase system: An electrical system in which multiple sine waves are produced, with each phase separated by a given number of electrical degrees.

potential difference: The voltage measured between two points in a circuit. When current flows through a resistor, or any component that uses power, there is a difference in potential (voltage) from one side to the other. Unit of measure of potential difference is the volt.

potentiometer: A variable resistor with three terminals.

power (P): Electrical work performed. Unit of measure is the watt.

power factor (PF): The ratio of real power to apparent power. Power factor can be found using either of two formulas: the ratio of the powers or the cosine of the phase angle. Power factor is a ratio so it has no units of measure. The value of power factor is always between zero and one. A PF of zero is purely reactive, and a PF of one is purely resistive.

primary cell: A single-cell of a battery which cannot be recharged. The chemical making up the electrolyte is consumed when changed into ions.

primary windings: The input of a transformer. The current through the primary windings develops a magnetic field. The amount of current in the primary is determined by a combination of primary resistance and secondary current.

proton: The part of an atom that has a positive charge. There is an equal number of protons as electrons for each atom. In terms of physical size, it is 1845 times larger than an electron.

prototype: A working model of a product built for a customer's approval to ensure its design is satisfactory. The prototype is tested under the conditions in which it will be used, allowing the engineers to work out any problems.

P-type semiconductor: A semiconductor material doped with an impurity having three valence electrons, such as Gallium (Ga). The trivalent impurity forms covalent bonds with the semiconductor producing holes.

pulse width: The length of time that a square wave is either on or off.

pure capacitive reactance: A theoretical condition of a capacitor in which the reactance contains no resistance.

pure inductive reactance: A theoretical condition of a inductor in which the reactance contains no resistance.

Pythagorean theorem: A mathematical formula used to calculate the length of the hypotenuse of a right triangle, such as formed by the phasors of an ac circuit.

Q

*Q***:** The ratio of inductive reactance to the circuit resistance.

Q point: The actual operating point of the transistor circuit.

quality factor (*Q*): The ratio of inductive reactance to the circuit resistance.

R

radian: A term from geometry referring to the distance from the center to the circumference of the circle. The circumference is the distance around the circle and can be calculated by the formula: $c = 2 \times \pi \times r$. Therefore, there are 2π radians forming the radius of a circle. A radian can be defined as the distance along the circumference of the circle equal to the radius of the circle. One radian equals 57.3 degrees.

radius: Distance from the center to the circumference of a circle.

ramp waveform: A type of repetitive waveform. See figure 15-6. Also called a sawtooth wave.

range: The maximum value that can be read on an analog meter scale.

reactance: The ac resistance of an inductor or capacitor. The unit of measure is the ohm. Letter symbol is X_L for inductive reactance and X_C for capacitive reactance.

reactive power: Power dissipated by pure reactance.

real power: The power dissipated in a pure resistance. Usually the term is applied to ac circuits, but it also applies to dc circuits. Unit of measure is the watt.

rectangular coordinates: The values to locate a point on a graph divided into four quadrants by the real axis and imaginary axis.

rectification: The process of changing ac to dc by changing an ac waveform to have only voltages of the same polarity.

reflected power: Power that is reflected back to the power source.

relay: A electromagnetically operated switch.

reluctance: The opposition to the flow of magnetic flux. It corresponds to the term resistance in electricity. Reluctance for a magnetic material is not a constant but varies with flux density. The reluctance of a nonmagnetic material is a constant.

reluctivity: The specific reluctance or the reluctance per cubic centimeter. Reluctivity corresponds to the electrical characteristic resistivity.

research and development (R & D): Involves the design, building, and testing of new products.

residual magnetism: Magnetism remaining in a temporary magnet after the magnetizing force has been removed.

resistance (R): The opposition to the flow of electrical current. Unit of measure is ohm. It does not slow down the current, which flows at the speed of light. Instead, resistance restricts the *volume* of current flowing in a circuit. Larger resistance results in less current and smaller resistance results in more current.

resistance ratio: One method of calculating the primary resistance.

resonance: A condition that occurs in an ac circuit when capacitive reactance is equal to inductive reactance. The reactances cancel, leaving the circuit to display purely resistive characteristics.

resonant frequency (f_r): Frequency at which resonance occurs in a given circuit.

resistivity: The resistance for a one-foot length of a material, with a cross-sectional area of one mil.

resistor: Electrical component used to oppose the flow of electricity.

retentivity: The ability of a material to retain magnetism after the magnetizing field has been removed. Permanent magnets have a high retentivity. Temporary magnets have a low retentivity.

reverse biased: The voltage applied to a diode to allow only a minority current.

rheostat: A variable resistor with only two terminals.

ripple: The ac content of the dc output voltage of a power supply.

RLC circuit: Circuits that contain at least one resistor, one inductor, and one capacitor.

rms value: Stands for *root mean square*. The effective value of a sine wave. When a sine wave is applied to a resistive circuit, it produces the same amount of heat as a dc voltage equal to the rms. AC voltmeters read the rms value of a sine wave.

robot: A machine that can be programmed to do work to make life simpler for humans.

rotor: The part of an electric motor that spins. It is the same as the armature in a generator. In a motor, the rotor drives the gears connected to the output shaft. The rotor is the driving force of the mechanical energy produced. The rotor may be a series of laminated permanent magnets.

rotor slip: The difference between the speed of the rotating magnetic field of the stator and the running speed of the motor.

rounding: Changing a number to its approximate equivalent to make the number easier to read and to manipulate. Zeros are used as place holders to maintain the value of the significant figures.

S

safety: A life skill in which a person protects himself and others from possible harm. In a classroom and laboratory setting, it is the responsibility of the teacher to demonstrate safe techniques. It is the responsibility of the student to make every effort to learn and use safe operating procedures at all times.

saturation: 1. The point at which the transistor acts like a closed switch, offering virtually no resistance to the circuit. All of the circuit voltage is dropped across the resistors in the collector and emitter circuits. Voltage across the transistor, V_{CE}, is approximately zero. In actual practice, it is about 0.2 V. Collector current is maximum, limited by the circuit resistance and not the transistor. 2. More current is applied to a transformer primary than can be converted to magnetic energy. The excessive current is changed into heat and creates losses.

sawtooth wave: A type of repetitive waveform. See figure 15-6. Also called a ramp waveform.

scale: In reference to a meter, it is the set of numbers belonging to the selected range.

schematic diagram: A picture that shows how circuit components are electrically connected.

scientific notation: A mathematic process of moving the decimal point and multiplying by 10 raised to a power to maintain an equivalent value.

secondary cell: A single-cell battery which can be easily recharged.

secondary windings: The output coil of the transformer. The magnetic field developed in the primary induces voltage in the secondary, through magnetic induction.

Seebeck effect: Effect by which a current is created when the joint between two dissimilar metals is heated.

self-excited series generator: A generator that has the field winding connected in series with the armature and the load.

self-excited shunt generator: A generator that has the field winding connected in parallel with the armature winding.

self inductance: The property of a conductor to induce voltage within itself. This is a result of the magnetic field developed from current flow crossing the conductor to reverse the energy conversion. The amount of self inductance determines how much counter electromotive force will be developed. Self inductance is usually called simply inductance.

semiconductor: Electrical materials with conductivity that falls between conductors and insulators.

sensitivity: A term used to describe the bandwidth of a resonant circuit. Selectivity is usually described as the strength of the frequency, measured in volts, or microvolts. This measurement is used to determine the ability of the resonant circuit to select a specified band of frequencies, no more and no less.

series circuit: An electrical circuit with only one path for the current to flow. Voltage will drop across each resistor. Total resistance is the sum of the individual resistors.

series resonant circuit: A circuit that has an inductor in series with a capacitor.

shading ring: A small portion of the stator field coil has a notch cut into it in a shaded pole motor. Around the notch is a copper ring, called a shading ring. Current is induced in the shading ring.

shielded transformer: A transformer with a magnetic shield on the outside of the windings. The shield prevents the magnetic flux from interfering with nearby circuits.

shock: In reference to electrical shock, is the physical sensation of the nerves and muscles reacting to electricity passing through the body.

short circuit: A defect in a circuit that causes the current to by-pass a portion or the load, or the entire circuit. If the entire load is shorted, there will be no limit to the amount of current flow. The fuse will blow.

short time constant circuit: A circuit that has its full charge/discharge time much smaller than the time of one pulse width.

shunt: 1. A connection made in parallel. 2. In an ammeter, the resistors for changing the range are shunt resistors, connected in parallel. 3. In a shunt-type generator, the field winding is connected in parallel with the armature winding.

siemen (S): Unit measure of conductance.

significant figures: All non-zero digits and zeros used as place holders between two non-zero numbers. Significant figures determine the accuracy of the number.

silicon controlled rectifier (SCR): A rectifier diode with a gate to turn it on at a specified time.

sine wave: A periodic alternating waveform. It is the shape of the voltage used in residential electricity.

single-phase system: Power system in which only one sine wave is produced.

single-pole: A switch with only one common connection.

single-throw: A switch with only one direction to make contact.

slip rings: The armature coil wires are attached directly to a pair of slip rings. The slip rings spin with the armature.

solenoid: An electromagnet with a moveable core.

source: One lead on a FET. Provides current to the drain and gate. Similar to the emitter on a BJT.

speed rating: The speed at the output shaft of the motor.

split-phase induction motor: A motor with two types of stator coils. A starting coil placed 90 electrical degrees apart from the running coil. The running coil has low resistance with high inductance. The starting coil has high resistance with low inductance. The different characteristics of the coil induces a rotating field in the stator, starting the motor.

split ring commutator: Rotates with the coil as the coil produces the sine wave to pass the generated electricity to the load.

standard international meter-kilogram-second (SI mks): A system of measurement.

static electricity: A build up of electrons or a shortage of electrons creating a difference in potential until a discharge path can be provided.

stator: The stationary magnetic fields of a motor. In almost all motors, the stator is an electromagnetic. It serves much the same function as the fields windings in a generator.

steady state resistance: The dc resistance of a circuit.

step-down transformer: A transformer with a smaller number of turns in the secondary than the primary. In a step-down transformer, secondary voltage is lower than primary voltage.

step-up transformer: A transformer with a larger number of turns in the secondary than the primary. In a step-up transformer, the secondary voltage is higher than primary voltage.

substrate: Block of semiconducting material that makes up a MOSFET.

super conductor: A conductor with zero resistance. Recent experiments in technology have lead to promising developments with super-conductive materials.

surge: A sudden burst of high voltage.

surge resistor: A small resistor used to drop a voltage in response to a high initial current.

symmetrical square wave: A square wave that has the on pulse with equal to the off pulse width.

symmetrical waveform: A waveform with equal values of positive and negative voltages. With a symmetrical waveform, the peak-to-peak value is equal to twice the peak value.

synchronous motors: The rotor matches its speed to the line frequency.

synchronous speed: The speed at which the rotor speed equals the speed of the rotating magnetic field.

T

tank circuit: A parallel resonant circuit.

technician: A skilled worker trained in basic electronics with advanced training in a specialized area.

tee network: A three-terminal network where the components have one end connected in common and the other ends are connected to the power source and load. Also called a wye network.

temperature coefficient: A mathematical factor used to predict how a material will respond to changes in temperature.

tesla (T): Unit of measure in the cgs system for flux density.

thermal runaway: A condition found in semiconductors. As current flows, heat is produced, which causes an increase in current due to the lower resistance. The increase in current produces more heat, less resistance, and more current.

thermistor: A resistor that varies its resistance with changes in temperature.

thermocouple: Two pieces of metal, made of different materials, joined at one end. It is also called a bi-metalic strip. When the junction is heated, a voltage will be produced at the other end.

thermostat: A heat-activated switch.

Thevenin equivalent circuit: Circuit where all but the load is turned into an equivalent voltage source and series resistor.

Thevenin equivalent impedance: Impedance that would be measured at the load terminals if the load was removed and the voltage source was removed and replaced with a short.

Thevenin equivalent resistance: Resistance that would be measured at the load terminals if the load resistor was removed and the voltage source was removed and replaced with a short.

Thevenin equivalent voltage: Found by removing the load resistor and measuring the voltage at the load terminals.

three-phase system: The universal power distribution system. Three coils, usually 120 degrees apart, produce three sine waves.

three-wire delta connection: A three-phase generator that uses three wires and can produce 3 times the phase current.

time constant: The length of time for a resistor-capacitor circuit or resistor-inductor circuit to reach 63.2% of full charge or discharge. Full charge is reached in five time constants.

time difference: The difference between the starting points of two sine waves.

tolerance: The acceptable accuracy of the assigned value.

toriod: The best shape for an inductor. Shaped like a doughnut.

torque: 1. Rotational force. 2. Torque in a motor is related to the force produced on the conductors when rotating through the magnetic field. 3. In terms of a rating, it is a measure of the force produced at the output shaft. 4. Torque is measured in force × distance: foot-pounds (ft-lbs) or ounce-inches (oz-in).

trace: The line on the oscilloscope produced by the incoming voltage signal.

trailing edge: The portion of the wave switching from a maximum to zero.

transformer: An electrical device which uses magnetism to link one coil of wire to another. Magnetic coupling of coils can result in an increase in voltage, decrease in voltage or equal voltages from the input to the output.

transistor: Semiconductor device capable of a wide range of operating characteristics.

TRIAC: A semiconductor device that operates like two SCRs in opposite directions that have a common gate.

triangular wave: A type of repetitive waveform. See figure 15-7.

trivalent: An impurity having three electrons in the valence band, used for doping a semiconductor material to become P-type.

true power: The power dissipated in a pure resistance. Also called real power. Unit of measure is the watt.

T-type filter: A type of filter that uses two inductors. See figure 24-15.

two-phase system: Power system with two coils that can produce twice the power of a single-phase system without significantly increasing losses.

U

unijunction transistor (UJT): One type of transistor that is used in timing circuits.

unit pole: Unit of measure in the cgs system for magnetic force.

unity power factor: A power factor of one. This provides maximum power efficiency to a load.

V

valence electron: Electron found in the valence shell. The electrons used in electrical current.

valence shell: The outer-most orbit of an atom, containing the electrons used in electrical current.

Van de Graaff generator: An electrostatic machine that produces extremely high voltages.

VAR: Unit of measure for power in the pure reactance portion of an ac circuit. VAR stands for Volt-Ampere-Reactive.

variable resistor: A resistor which can have its resistance adjusted. It has a contact which can be moved to change its resistance in relation to the total resistance of the device.

virtual ground: A point in an op-amp circuit equal to zero volts.

volt (V): Unit measure of voltage.

voltage (V): 1. The driving force that causes electricity to flow through a conductor. Without voltage, there can be no work performed by the electricity. Unit of measure is volt (V). 2. Potential energy of the electric source. Also called electromotive force (EMF).

voltage divider: Reduces the voltage from a power supply as needed by a load.

voltage drop: The difference in potential from one side of a resistor to the other. The amount of voltage drop is directly related to the amount of current and size of resistor. The voltage drop will have a polarity (+/–) equivalent to the direction of current flow.

voltage rating: The maximum voltage a device can safely handle.

volt-ampere (VA): Unit of measure of apparent power.

volt-ampere-reactive (VAR): Unit of measure for power in the pure reactance portion of an ac circuit.

W

watt (W): Unit measure of power.

wattage rating: The maximum power a device can safely handle.

wattmeter: Meter that records how many watts of power are used.

waveform: The shape of the voltage over a period of time. Waveforms are usually thought of as the picture that would be obtained if the voltage were to be viewed with an oscilloscope.

weber (Wb): Unit of measure in the SI mks system for magnetic flux.

Wheatstone bridge: A circuit that places the load between two parallel resistor branches.

wye (Y) connected three-phase system: One type of three-phase wiring system that can use either three or four wires.

wye network: A three-terminal network where the components have one end connected in common and the other ends are connected to the power source and load. Also called a tee network.

Z

zener diode: A type of diode that is designed to operate in the reverse direction.

zener knee: When voltage is increased in the reverse direction through a zener diode, there will be zero current flow up to the zener knee, which occurs at the breakdown point. The zener knee is the name for the section of the curve where there is a sudden change in current flow. From this point on, there is only a slight change in voltage for wide changes in current.

zener operating range: The range beyond the zener knee. In this range there is only a slight increase in voltage in response to a large increase in the current.

zero volt reference point: The reference point on a voltage divider.

Index

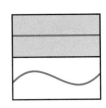